The Passion of GEORGE SARTON

A Modern Marriage and Its Discipline

The Passion of GEORGE SARTON

A Modern Marriage and Its Discipline

Lewis Pyenson

AMERICAN PHILOSOPHICAL SOCIETY
Philadelphia 2007

Memoirs
of the
American Philosophical Society
Held at Philadelphia
For Promoting Useful Knowledge
Volume 260

ISBN-13: 978-0-87169-260-3
US ISSN: 0065-9738

Design and Composition
Book Design Studio

Library of Congress Cataloging-in-Publication Data

Pyenson, Lewis.
The passion of George Sarton : a modern marriage and its discipline / Lewis Pyenson.
p. cm.
ISBN 978-0-87169-260-3
1. Sarton, George, 1884–1956—Biography. 2. Historians of science—United States—Biography. 3. Science—Historiography. I. Title.

Q143.S228.P94 2007
509—dc22
[B]

2007031940

S, in Arcadia cultrix

Contents

List of Illustrations

Fig. 52. Ex-libris, designed by Mabel Sarton for George Sarton, as recorded in George Sarton's diary, 31 August 1920. We may read the prominent nipple profile as an assertion of the artist's independence. By the 1970s, the History of Science Society preferred a sexless hieroglyph. Rosemary Regan and Richard H. Schallenberg, ed., *History of Science Society Directory of Members and Guide to Graduate Study* (New York, 1977), the cover illustration.

Biographers insist too much on success which is accidental, and too little on intrinsic virtue or greatness. "Success" is sometimes a decent woman but in most cases she is nothing but a whore.

George Sarton, diary entry for 9 June 1953.

Preface

George Sarton is the central figure in the crystallization of the discipline of history of science in America. The introduction to a recent collection on the history of the city of Ghent, where Sarton was born and educated, cites Sarton and Nobel-laureate Maurice Maeterlinck in the same breath as internationally known *gantois* in the 20th century.[1] Testimonials notwithstanding, however, Sarton is the subject of no significant éloge. A generation ago, L. Pearce Williams criticized Sarton for historical naïveté and Mirko D. Grmek challenged Sarton's sense that the history of science, before the modern period, had to transcend the boundaries of scientific disciplines.[2] A widely-cited historian of science, Thomas Kuhn, offhandedly branded Sarton's synthetical approach as an embarrassment; "somewhat naive and surprisingly unhistorical," is the judgment of Sarton's writings by another prominent historian of science, Helge Kragh.[3] The authority of mathematics and astronomy in the ancient and medieval world, Noel M. Swerdlow, concludes: "What Sarton provides is mainly names of authors and books interspersed with pieties and platitudes."[4] Still another study, by Arnold

1. Johan Decavele, ed., *Gand: Apologie d'une ville rebelle* (Antwerp, 1989), p. 11.
2. C. Doris Hellman, "George Sarton," *International Encyclopedia of the Social Sciences* (New York, 1968), *s.v.* Sarton, for Williams and Grmek.
3. Thomas Kuhn, "The Relations between History and the History of Science," in Kuhn, *The Essential Tension* (Chicago, 1977), pp. 127–61, on p. 148: "Historians of science owe the late George Sarton an immense debt for his role in establishing their profession, but the image of their specialty which he propagated continues to do much damage even though it has long since been rejected." Helge Kragh, *Introduction to the Historiography of Science* (Cambridge, 1987), p. 18.
4. Noel M. Swerdlow, "Montucla's Legacy: The History of the Exact Sciences," *Journal of the History of Ideas, 54* (1993), 299–328, on p. 327.

Thackray and Sarton's student Robert K. Merton, ignores the unity of the world that dominates his work by emphasizing the "paradoxes" of his life.[5] The paradoxes extend to May Sarton, whose career as a writer borrows from that of her father, even though she is clear about rejecting George in favor of her artist mother Mabel.[6]

The Passion of George Sarton originates in a desire to correct a number of misapprehensions (first among them my own) and recover the origins of a scholarly field in remarkable personalities.[7] The book draws on the memoirs of Mabel and George's daughter, May Sarton, one of America's premier poetic personalities;[8] on George's correspondence, more than 40,000 pages of letters and personal diaries, across more than 50 years; and on more than 4,000 pages of intimate correspondence between the spouses. In what follows I contend that the discipline of history of science is best understood from the point of view of the marriage that sustained and, at a number of points, overwhelmed it.

Introducing a short collection of Mabel's letters to her, May Sarton appeals to Shakespeare's Sonnet cxvi with the thought that her parents' union "was in its particular way 'a marriage of true minds' and they shared everything that really matters, and enjoyed each other mightily."[9] The following pages elaborate the sentiment of Shakespeare's sonnet, that love is the guiding star in life's storm. Love is not time's fool. It remains when everything else has gone, and it stands fast in the face

5. Arnold Thackray and Robert K. Merton, "On Discipline Building: The Paradoxes of George Sarton," *Isis, 63* (1972), 473–95.

6. "Lewis Pyenson, "Elegant Sartons: Platonic Science, Platonic Letters," in *Elegance: Beauty and Truth*, ed. Lewis Pyenson (Lafayette, La., 2001), pp. 5–13, for a correction.

7. For an example of misapprehensions, the eminent historian of science Stanley Jaki attributed George Sarton's neglect of Pierre Duhem's historical writings to "the free-mason Sarton's deep-rooted anticlericalism coupled with his dogmatic socialism born out of a youthful cavorting with Marxism." Jaki quoted in Kragh, *Introduction*, p. 111. Largely as a social gesture, George followed his father as a Freemason, but the present pages provide evidence of his religiosity and persistent attraction to people of faith, his social-democratic approach to socialism (where Karl Marx makes virtually no appearance), and his resentment that Duhem declined to become a founding patron of *Isis*, his periodical. *Cf* George Sarton, "Acta atque agenda," *Archives internationales d'histoire des sciences, 4* (1951), 323–56, pp. 334–7 for a generous assessment of Duhem. Henry Guerlac, an authority on science in France, offered that George Sarton "was genuinely opposed to the history of technology, as somehow, he believed, unworthy and unimportant for the growth of science." Guerlac, "A Backward View," in his *Essays and Papers in the History of Modern Science* (Baltimore, 1977), pp. 54–65, on p. 62. The following pages amply demonstrate George's respect for technology.

8. Katharine Strelsky, "Bibliography of the Publications of George Sarton," *Isis, 48* (1957), 336–46, followed by a thematic index, is a basic resource that includes May Sarton's early commentaries.

9. Eleanor Mabel Sarton, *Letters to May*, ed. May Sarton (Orono, 1986), p. xi.

of oblivion. In the concluding, dissonantly rhymed couplet, the Bard speaks about his craft as a writer. If his affirmation of love is proved wrong, then there is no sense to any line of his, and, worse, no meaning to humanity's deepest longing. The marriage of George and Mabel Sarton is a meditation upon the meaning of love. They thought deeply about it, their intensity of feeling and questioning expressed with unusual candor. They did not live to read Albert Camus, but in their lives they approach Camus's godless sainthood more closely than either one would have been comfortable acknowledging. Their love may help guide a reader who, in modest or obscure circumstances, maintains the view that a life devoted to pursuing beauty, knowledge, and truth is indeed worth living. Love, the thing that we can locate but upon which no value can be placed, animates it all.

The Passion of George Sarton devotes attention to the vicissitudes of a marriage. If the daily changes of mind of an experimental scientist merit detailed reporting, if it is significant to know how a theoretician moves through all the equations in a seminal paper, if relevance is ascribed to the twists and turns of obtaining funds for constructing a laboratory, if the meticulous travel accounts of a geographical explorer can occupy the center of a historical narrative, then studying a scholar's married life should be an unproblematic undertaking.[10] What could be more important than the person with whom a scholar spends more time than any other? This book shall have achieved its goal if a reader concludes that the marriage of George and Mabel Sarton is as rich and complex as a stack of laboratory notebooks or a theory about microscopic matter.

The following pages recount the formative years of the Sartons' marriage, which were also formative years for the discipline of history of science in America. The book concludes with the Sartons at the height of their creative energy: Mabel is established as a prize-winning designer in America and Europe; George has finished the manuscript of the first volume of his *Introduction to the History of Science*, where he provides the most complete illustration of his approach to his discipline.[11] The Sartons were never physically robust, and over the next decade, infirmities

10. For examples of the former genres written by major voices: Frederic Lawrence Holmes, *Claude Bernard and Animal Chemistry* (Cambridge, Mass., 1974); Arthur I. Miller, *Albert Einstein's Special Theory of Relativity: Emergence (1905) and Early Interpretation (1905–1911)* (Reading, Mass., 1981); John L. Heilbron and Robert W. Seidel, *Lawrence and His Laboratory: A History of the Lawrence Berkeley Laboratory, volume 1* (Berkeley, 1989); D. Graham Burnett, *Masters of All They Surveyed: Exploration, Geography, and a British El Dorado* (Chicago, 2000).

11. NYPL, Madeleine van Thorenburg to Mabel Sarton, 18 December 1925, reporting that George's history of science is finished.

increasingly compromise their activity.[12] I contend that public features of the discipline—notably in its writings, lectures, and organization—derive from anxieties, tensions, and affirmations in the marriage. In the history of science, it is not uncommon to publish the story of a life in parts, each of which may exert special attraction on a reader.[13] The years recounted in the present volume are precisely those that would have interested George Sarton, for he writes in his diary on First August 1954: "I love to read biographies of great men but it is only the beginning that interests me, the years of struggle and more often than not, of misery. When success comes in I am ready to leave." In a subsequent volume, *Deo volente*, I intend to consider the last part of the Sartons' life in the context of ecumenical thought during the middle of the 20th century.[14]

Every period of history, no less than every life, is a time of transition.[15] So it is with the first half of the 20th century, the Sartons' time, and also with our own time. In their reflections on matters in culture and world affairs, their belief in social justice, their attempt to endow a unique contribution to humanity, the Sartons recount euphoria along with profound disappointment. From this rich and varied body of writing we may take inspiration. I suspect that over the past decades we have focused on avarice, stupidity, and revelation to the exclusion of reason, clarity, and elegance. By the quality of its writing, the present time does compare favorably with all past ages. The Sartons wrote and wrote. Their lives give courage to people who, excluded from powerful councils and absent from glittering assemblies, believe that future generations will see our own time as sensible, productive, and enlightened.

Because recovering the past is a central theme in the Sartons' marriage, at the outset some words about history are appropriate. From their young adulthood, Mabel and George Sarton took special pleasure in museums and exhibits. Museums today contain objects sacred to the

12. NYPL, Mabel Sarton to Marthe Patyn, 15 December 1935, in French, where Mabel observes that George is sometimes tired. "He follows the advice of the doctor carefully since the *accroc* he had 3 years ago." YMS, George Sarton to May Sarton, 3 July 1938, for Mabel's second heart attack; NYPL, Mabel to Evelyn Van Loon, 28 April [1940]: "I had a nasty heart-attack a few days ago (in town—I was alone & no-one knows how ill I was; probably not dangerous but I certainly feel for a moment or two that I am dying:—the doctor assures me however I can have many more before a final one!)."

13. For example, two highly-acclaimed biographies: Martin J. Klein, *Paul Ehrenfest, volume 1: The Making of a Theoretical Physicist* (Amsterdam, 1970), volume 2 has not yet appeared; Janet Browne, *Charles Darwin: Voyaging* (New York, 1995), followed by *Charles Darwin: The Power of Place* (Princeton, 2002).

14. A beginning has appeared: Lewis Pyenson, "Western Historians of Science and Oriental Science in the Age of Imperialism,"*Historia Scientiarum, 15* (2005), 97–124.

15. George Sarton, *Introduction to the History of Science, volume 3: Science and Learning in the Fourteenth Century* (Baltimore, 1947), p. 11: "Each age is a middle age."

muses, but they also continue a tradition dating from antiquity by fostering new knowledge about old things. Although objects on display are framed in their context, a successful exhibit allows viewers to form their own conclusions. There is a dialectic between sentiment and understanding. The Greeks believed that sentiment has priority, for they portrayed Mnemosyne as mother of the muses, among whom is Clio. Now we think in reverse fashion that historians give meaning to memory. In exercising their responsibility, historians assemble documentation, conflate various accounts, and prepare inventories of the human world. Recounting the past serves the future, and along the way it provides entertainment and instruction. Who knows the past commands the moment, or in the notion of classicist Ulrich von Wilamowitz-Moellendorff, clarifies the hazy outlines of monuments present everywhere.

The power of history in the present accounts for a persistent urge to fabricate it. No amount of facts and figures could have substituted for Winston Churchill's reciting "Barbara Frietchie" to charm Franklin D. Roosevelt as they drove toward Shangri-La (known today as Camp David) in 1943 :

Up from the meadows rich with corn,
Clear in the September morn,
The clustered spires of Frederick stand,
Green-walled by the hills of Maryland.

Maybe Stonewall Jackson (who apparently did not pass Frietchie's house) was indeed cowed by nonagenarian Frietchie in 1862, and, if so, John Greenleaf Whittier caught the moment; maybe there was a denizen of Frederick who, as Roosevelt and Churchill drove by, could recall Jackson and Frietchie. Churchill, a Nobel-laureate historian, used myth to make history.[16]

Capitulation to myth, in the view of Mircea Eliade, is required to provide meaning to life and to avoid the terror of history, which he imagined as one thing after another, in sequence but without reason. "The historical event in itself, whatever its importance, is held in popular memory and inflames poetical imagination only to the extent that it is strongly connected with a mythical model."[17] Eliade held that a belief in God is sufficient to avoid hopelessness. But that belief cannot deal with the existence of evil—the problem that severed reason from religion in the Enlightenment. As one of their responsibilities, historians have

16. Eleanor Roosevelt, "Churchill at the White House," *Atlantic Monthly, 215* (1965), 77–80, on p. 78.

17. Mircea Eliade, *Le Mythe de l'éternal retour* (Paris, 1969), p. 58; the terror of history on p. 168. "Consciousness of a real and meaningful world is intimately linked with the discovery of the sacred." Eliade, *Ordeal by Labyrinth: Conversations with Claude-Henri Roquet*, trans. Derek Coltman (Chicago, 1982), p. 153

traditionally recounted good and evil. Graduate seminars in history used to discourage mythologizing the past, although times are changing.[18]

Eliade himself found hope not in invisible powers but in expressing thoughts for other people through scholarship. He wrote to George Sarton upon the death of Mabel Sarton: "I learned with pain about the death of your wife. I experienced the same tragedy, losing my first wife: For a long time I was able to survive only in and for work."[19] Eliade meant to say that love encapsulates hope. Time dissolves as the object of love appears then, now, and forever. History is formed from the common experience of the lovers. It is made, or invented, and we mark it by events like weddings, funerals, and birthdays. Love is everywhere in everyday life. For the majority of the world, love gives life significance, meaning, and purpose. We ask for dispassion in works of history, but passion moves the writing. George Sarton notes: "To write without passion or prejudice does not mean to write without feeling."[20]

There is no meter-stick for measuring time; the best that can be done is to "determine" it—establish it, make it definite, abstract it from sentiment.[21] Scientists have a great deal to say about the matter, but to the historian is granted the privilege of identifying or defining temporal consciousness.[22] From this comes the problem of characterizing the time in which we live. Shall we call it the Postmodern Age, following philosopher Jean-François Lyotard, or the Age of Mouldwarp, following novelist and critic Peter Ackroyd? Perhaps the Golden Age is still due to materialize, as molecular biologist Gunther Stent imagined more than 30 years ago.[23]

---

18. A. S. Byatt, *On Histories and Stories: Collected Essays* (Cambridge, Mass., 2000), p. 10, for recent interweaving of fact and fiction.

19. HGS, Mircea Eliade to George Sarton, 24 October 1951.

20. George Sarton, *Introduction to the History of Science, volume 3*, p. 10.

21. According to the formulation of Suzanne Débarbat, President of the Bureau of Longitudes in Paris, in a presentation at the Institut de France during the summer of 2004.

22. Diverse specialists claim parts of time, sometimes relating it to space: Fritz Gelhar, "Raum und Zeit," in *Philosophie und Naturwissenschaften: Wörterbuch zu den philosophischen Fragen der Naturwissenschaften*, ed. Herbert Hörz, Rolf Löther, and Siegfried Wollgast (Berlin, 1978), pp. 783–90; P. C. W. Davies, "Time," in *Encyclopedia of Physics*, ed. Rita G. Lerner and George L. Trigg (New York, 1991), pp. 1284–5; Ernst Pöppel and Marc Wittmann, "Time in the Mind," in *The MIT Encyclopedia of the Cognitive Sciences*, ed. Robert A. Wilson and Frank C. Keil (Cambridge, Mass., 1999), pp. 841–3; Tian Yu Cao, "Space and Time," in *The Oxford Companion to the History of Modern Science*, ed. John L. Heilbron (Oxford, 2003), pp. 765–7.

23. Jean-François Lyotard, *The Postmodern Condition: A Report on Knowledge*, trans. Geoff Bennington and Brian Massuimi (Minneapolis, 1984); Peter Ackroyd, *The Plato Papers: A Prophecy* (New York, 1999); Gunther Stent, *The Coming of the Golden Age: A View of the End of Progress* (New York, 1969).

For a historian, abstracting time from sentiment can itself become a consuming passion, as the past beyond personal experience continually makes itself felt—a book or manuscript, record or artifact, image or event always there, just below the surface.

Here is ordnance for parents to direct against a prospective in-law: A dreamer who can't earn a living. George Sarton suffered from attacks of this kind—before, during, and after his marriage; in her own way, Mabel Sarton was also impractical. But as happens, children grow up, and under these circumstances, thoughts turn to the snows of yesteryear. For this, the grown-up dreamer has a clear advantage. The historian improves with age, seeing connections, avoiding pitfalls, commanding imagery. Among the greatest historical writings of the 20th century are books written by older men in time of war—Marc Bloch's *Apologie pour l'Histoire* and Erich Auerbach's *Mimesis*. Arguably the most significant historical work of the past one hundred years, the multi-volume *Science and Civilisation in China*, was begun at mid-century also in war-time by middle-aged Joseph Needham.[24]

It is no accident that these historians wore their morality on their sleeve. As shall be seen in the following pages, war encourages historians to free the past for a better future. Needham, for example, called past knowledge a watershed of many rivers, all eventually flowing together into a great sea.[25] In war or in peace, however, morality emerges immanently from historical deliberation. Both heroes and villains know it. In condemning the Old Regime, the Jacobins recounted nearly a millennium of royal oppression; in defending their actions, the accused at Nuremberg contended that history would vindicate them.

George Sarton writes in 1921: "What is life itself if not a continual compromise between the past and the present?"[26] We live on the tip of time's arrow, anticipating an arc we only dimly project. The conditions just behind the turbulence of point, shaft, and feather we commonly call the present. Ironically, we know where we are by observing the surroundings of our rushing passage. Passing by once, we return in our imagination, where the snows drift higher, the bedclothes are softer, and the afternoons have a soft breeze and languor that envelop us in security. Memory

---

24. Mansel Davies, "Joseph Needham (1900–95)," *British Journal for the History of Science, 30* (1997), 95–100, where Needham is identified as "one of the greatest scholars in this or any country, of this or any century," whose *Science and Civilisation in China* is "perhaps the greatest work of scholarship achieved by one individual since Aristotle."

25. Needham's river metaphor is explicitly challenged by anthropologist Clifford Geertz, who sees in history "no general story to be told, no synoptic picture to be had." Geertz also challenges Bloch's and Lucien Febvre's notion of recovering long-term historical trends. Geertz, *After the Fact: Two Countries, Four Decades, One Anthropologist* (Cambridge, Mass., 1995), p. 2.

26. George Sarton, "Science and Style," *Scribner's Magazine, 69* (1921), 755–9, on p. 756.

can also be unsettling. Mabel writes to George on 8 August 1907, about her father's death: "Memory has a way of out-doing, & escaping all one's reasoning, a way of holding up to you, not one big fact but little vivid, *unimportant* details, ways of looking, tones of voice, atoms of every-day life of last year, of life that *was* last year, & that is why they hurt so." We remember the past, and that memory, distilled, is the source of human qualities like courage and honor.

Historians are the tribe responsible for sorting out memories. In exercising their responsibility, they assemble documentation, conflate various accounts, and prepare inventories of the human world. They twist the tiger's tail, asking questions that have not been posed before, but they avoid extracting unnatural confessions. Historians are closer to Benjamin Franklin than to Tómas de Torquemada, more Diego Velázquez than Peter Paul Rubens. The following account is cautioned by the elegance and clarity of Franklin's electrical research and by the generality radiating from Velázquez's portraits. It rejects precious concatenation in favor of direct narrative. Historians have many responsibilities and a number of privileges, but none of these involve a mystification of the past. George Sarton shared Albert Einstein's view that mystery lies at the origin of art and science, but George writes: "Obscure ideas are good means of enslaving and exploiting credulous people; in the long run they are not compatible with freedom and justice." He quotes Cicero: "Who does not know that the first law of history is not to say anything false nor to hide anything true?"[27]

George Sarton imagined great thinkers as cabbages, rising to their destiny regardless of environment.[28] Yet he would have been the first to agree that to grow, cabbages require water and soil and sunlight, and that particular circumstances allow some cabbages to obtain large proportions, while others a few rows away may languish or expire. To look closely at a life is to know that each of us is born into a world not of our own choosing. In the vocabulary of the 17th century, we glide between *virtù* and *fortuna*, or between will and fate. In George Sarton's terms, there is a counterpoint of cabbages and fields.[29] Character may be present at an early age, but the circumstances of life change. George Sarton, an admirer of homespun wisdom from the State of Georgia, might have been amused

---

27. George Sarton, *Introduction to the History of Science, volume 3*, pp. 9, 18, Einstein on p. v.

28. George Sarton, *A History of Science: Ancient Science through the Golden Age of Greece* (Cambridge, Mass., 1952), p. xiii.

29. Alistair C. Crombie, *Styles of Scientific Thinking in the European Tradition* (London, 1994), the introduction to volume 1; Lewis Pyenson and Susan Sheets-Pyenson, *Servants of Nature: A History of Scientific Institutions, Enterprises and Sensibilities* (New York, 1999), p. 22.

by the Georgian work-ethic, where a time-consuming task is reduced to how the cow ate the cabbages—one cabbage at a time. George dispatched cabbages by the truckload. And as with Robert Louis Stevenson's cow, known to generations of writers through William Strunk and E. B. White's *Elements of Style*, fortitude brought him a form of apotheosis.

Works of history are like George Sarton's cabbages. The present book follows upon a career in the French cultures of North America—Quebec and Louisiana. Among institutions that supported my research at various times are Université de Montréal, McGill University, and Concordia University, all in Montreal, and the University of Louisiana at Lafayette, also known as Université des Acadiens. The director and the staff at the University Library of Ghent have been generous with their assistance. For their help and encouragement I should like to thank Richard S. Dunn, Liliane Beaulieu, Russell McCormmach, Thomas F. Glick, Chikara Sasaki, Susan Sherman, Paulette Blanchet, May Sarton, Sylvia van Peteghem, Jean-François Gauvin, Nathalie Martimbeau, Marie-Brunette Spire, Charles C. Gillispie, Marc de Mey, Fernand Vandamme, Michel Thiery, and A. L. J. van de Walle. Christophe Verbruggen has shared the fruit of his dissertation at the University of Ghent on the intellectual climate of Ghent at the beginning of the 20th century. My work could not have proceeded without the friendly assistance of archivists at Houghton Library of the Harvard College Library and at the Berg Collection of the New York Public Library. The book has been written on the banks of a bayou, at moments snatched between raising children and teaching undergraduates, and completed with the help of a Franklin Research Grant awarded by the American Philosophical Society. The writing has helped me appreciate a phrase in Thomas Carlyle's *Past and Present*: "All Works, each in their degree, are a making of Madness sane."

Fernewood, Lafayette, Louisiana
December 2005

On Conventions and Sources

The marriage of Mabel and George Sarton is well documented, notably in their correspondence. In the early years they wrote to each other almost entirely in French, George's mother tongue, and a language in which Mabel was fluent if not always completely at ease. Beginning with the couple's exile from Belgium in 1914, it is for the most part in English, Mabel's native tongue, with occasional letters in French; Mabel herself sometimes chooses French when she deals with French situations, or when she tries to express a subtle emotional point. I have generally translated everything into English, although sometimes I retain turns of phrase from the French text. Mabel, for example, on occasion refers to "delicious" weather and to George's "forces" (we might understand power or energy). I follow the Sartons in referring to Mabel Sarton and Céline Dangotte's feminist club in Ghent as the *Flinke*, rather than to its Dutch plural, *De Flinken*.

The letters between Mabel and George are largely in pen and ink. George learned about the advantages of a typewriter from his epistolary conscience Emile Masson in 1905 (when Masson tried one out at the home of Alexander Carlyle in Scotland) and in 1911, and he experimented with a typewriter beginning in 1913, but he never really accepted the machine.[1] George writes to Mabel on 26 June 1913 that he is considering purchasing a typewriter, but

> I will only decide to buy one after having assured myself by several trials of the possibility of writing quicker with it than without it—which seems to me very doubtful. I don't have the time now to use a machine, nor the money. I don't think that I would ever use a machine myself. I dream of having a machine available, and a typist, but when? Already now I would have material to give her for several hours per day.

---

1. HGS, Emile Masson to George Sarton, 20 November 1905, 18 November 1906, and 27 December [1911]. From the first: "I need it [a manuscript lent to George] for the machine which I have begun to try out during the vacation and which I will probably use soon." From the last, referring to his wife: "We are now, Elsie & I, learning to use a typewriter."

Mabel serves as George's secretary—a traditional rôle for the wife of an academic—and later, at Harvard University, George often dictates letters to a typist-secretary. Both George and Mabel are most comfortable expressing their thoughts with pen or pencil (Mabel, the artist, finds pencil convenient) for their entire life. It is remarkable that the correspondence is as orderly as it is. Ideas are sometimes elaborated in the writing, and the writer will have second thoughts about what has just been set down. George writes from Florence, Italy, to Mabel in a suburb of Brussels on 9 November 1919: "I have your beloved little letter written with pencil on All Saints & All Souls. I only received it yesterday. A part of it was cut out." Paper was expensive, leading to marginal postscripts and occasionally cross-writing. Time was also precious, arguing against recopying, and when Mabel and George have second thoughts, they may strike out a sentence or paragraph. A writer who is confident with a correspondent will not hesitate, under the circumstances, to set down short impressions and incomplete notions. The letters make much use of the em-dash, variously for emphasis and to introduce a new topic, and also the ampersand; there are abbreviations in French, for example, *qq* for *some;* Mabel often distinguishes small groups of sentences within paragraphs by leaving a long blank space; in English she seldom uses apostrophes—difficult marks to introduce quickly with a nibbed ink pen—and she passes over them in French, as well. In the following pages, these notations have sometimes been altered.

Spelling, for George and Mabel Sarton, is irregular. The infelicity may derive from the nature of bilingual Belgium early in the 20th century. In Flanders at the time, streets and places carry both French and Flemish (that is to say, Dutch) names, and these sometimes have no standard form in either language.[2] An example is Sint-Martens-Latem, or Laethem St-Martin, also known as Lathem St-Martin, a suburb of Ghent where distinguished artists and poets lived. Family names and given names also had several forms, and not merely concerning the æ-a variant in French and the ij–y variant in Flemish. A member of the artists' colony at Laethem St-Martin, Mabel's design teacher for several years in Zurich, was Julius de Praetere, also known as Jules de Pratere; in Switzerland, the signature on his paintings is Prater, and Mabel initially knew him as Jules de Praat. In their correspondence, George and Mabel indicate George's early flame as Mélanie Loreyn; she married the Polish socialist Victor Alter as Lorein. Political meaning attached to the

2. The distinction of Flemish as a language separate from Dutch carried and still carries political connotation. English writers *circa* 1900 did refer to the Flemish language, as did Francophone Belgian writers. While in Belgium and even after residing in the United States, Mabel and George Sarton generally write about Flemish, although George occasionally refers to the Dutch language in Belgium (on occasion George refers to both the Flemish language and the Dutch language). I retain the Sartons' usage in the present text.

choice of a name. Although multilingual Europe was less than careful about spelling proper names, much depended on whether one referred to the socialist De Man with the given name of Henri, Hendrik, or Henry.[3] George Sarton received a French given name at birth, Georges, but from his teens he preferred the English George; in stepping away from the Walloon élite (he stemmed from both Walloon and Flemish stock), he avoided the traditional Flemish Joeris or Jorge, but he did on occasion receive and sign letters "Jooris Sarton."[4] In their correspondence, George and Mabel misspell the name of acquaintances and even friends; these slips have generally been corrected silently.

The Sartons change in the course of their life—change is indeed fundamental to their common credo, itself rich and complex—while remaining integral personalities. One instance is their attachment to vegetarianism, which dissolved early in their marriage. Another instance, which made it possible for George to find a position in philosophy at Harvard in 1916, is the way their belief in pacifism broke following the German invasion of Belgium in 1914.[5] In middle age George was far from a loner, although during the Second World War, after the events considered in the following pages, he retreated into his own world of scholarship.[6]

George Sarton emphasizes in the introduction to the final volume of his *Introduction to the History of Science*: "The views and style of a man—a growing and living man, not a fossil—change repeatedly throughout his life; a great many examples of unimpeachable authority prove it, every man's own experience confirms it. The argument of

3. Examples from wider Europe in the following pages include the Swiss polymath Auguste (August) Forel and the Swedish historian of mathematics Gustaf (Gustav) Eneström.

4. Roger Calcoen, repeating Paul von Oye, suggests that young George shed the ultimate "s" because for George, "Georges" seemed to be a plural proper noun, and he was a single person; Calcoen also for Jooris. Roger Calcoen, "George Sarton," *Handelingen der Maatschappij voor Geschiedenis en Oudheidkunde te Gent, 34* (1980), 181–91; Van Oye, p. 19. Léopoldine Dangotte, Mabel's adoptive mother in Ghent, writes to George as Joores and as Jooris de Saye. HGS, Léopoldine Dangotte to George Sarton, 18 November 1911.

5. The Philosophy Department at Harvard proposed Bertrand Russell as Josiah Royce's replacement in 1916, but the president, A. Lawrence Lowell, rejected him as an unsuitable pacificist. Louis Menand, *The Metaphysical Club* (New York, 2001), pp. 417–18.

6. Students close to George during the 1940s report on his isolation at Harvard University. One student from this period, Henry Guerlac (who took a doctorate at Harvard University and knew George there), commented that George's "contacts...with Harvard scientists seem to have been virtually non-existent." Guerlac, "Sartoniana and Forward," *Lychnos*, 1986, pp. 1–27, on p. 9. George's doctoral student I. Bernard Cohen recalled: "Sarton was a lonely man. He had no colleagues at Harvard, very few close intellectual friends." I. Bernard Cohen, "Introduction: The Impact of the Merton Thesis," in *Puritanism and the Rise of Modern Science: The Merton Thesis*, ed. Cohen (New Brunswick, N.J., 1990), pp. 1–111, on p. 26. In 1940 George was 56 years old, and he had been urged by his physician to moderate his labor.

consistency is a very weak one in a world full of inconsistencies."[7] I have tried to present evidence for the evolution of the Sartons' common striving in an unobtrusive yet unambiguous way. This accounts for the unusual decision to provide, in the text, dates for many items of correspondence and many diary entries. At the same time, in an age when identification of names and institutions is readily made by way of the Internet, reference to secondary works about personalities relevant to the Sartons has been kept to a minimum. A reader interested in learning more about Francisco (Pancho) Villa, Feodor Ivanovich Chaliapin, or Paul von Hindenburg, all appearing in the following pages, will have no trouble finding reliable sources.

The art and design of Mabel Sarton are frequently discussed in *The Passion of George Sarton.* For this reason, it is appropriate to include a portfolio of her work as an appendix. The paintings, sketches, designs, embroidery, and furniture illustrated there, as indeed all the illustrations in the present book, come from the private collection of Susan Sherman in Riverdale, Bronx, New York. Ms Sherman, a distinguished editor and teacher, has been generous with her time and knowledge. Readers are in her debt for preserving much of the Sartons' heritage.

7. George Sarton, *Introduction to the History of Science, volume 3: Science and Learning in the Fourteenth Century* (Baltimore, 1947), p. 27.

Abbreviations in the Text and Notes

Agenda—George Sarton's pocket agendas (accessioned as appointment books): MS Am 2117, *60M–178. Houghton Library of the Harvard College Library, Cambridge, MA.

ALS—Lucy May Stanton Papers (Ms 392, Box 1, folders 7 and 8), Hargrett Rare Book and Manuscript Library, University of Georgia, Athens, GA.

BFG—Fielding Garrison Papers, History of Medicine Division, National Library of Medicine, Bethesda, MD.

CAJ—Horace Meyer Kallen Papers, Jacob Rader Marcus Center, American Jewish Archives, Cincinnati, OH.

CIW—Archives of the Carnegie Institution of Washington, Washington, DC.

Diary—George Sarton's diaries: MS Am 2117, *60M–177. Houghton Library of the Harvard College Library, Cambridge, MA.

HGS—George Sarton Papers: bMS Am 1803.1 (letters by George Sarton), bMS Am 1803.2 (letters to George Sarton). Houghton Library of Harvard College Library, Cambridge, MA.

LM I—University of Ghent, *Liber memorialis: Notices biographiques*, 2 vols (Ghent, 1913).

LM II—Theo Luykx, ed., *Liber memorialis 1913–1960*, 4 vols (Ghent, 1960).

MO—Sir William Osler Papers, Osler Library for the History of Medicine, Faculty of Medicine, McGill University, Montreal, Quebec, Canada.

NYPL—Papers of May Sarton, Series XV Family Papers 1846–1993, Boxes 171–182, The Berg Collection of English and American Literature, The New York Public Library, Astor, Lenox and Tilden Foundations, New York. Material cited in the text without an indication of location is from the separately accessioned correspondence between George and Mabel Sarton in the Berg Collection.

Phoenix—May Sarton, *I Knew a Phoenix: Sketches for an Autobiography* (New York, 1959).

"Scholar's Wife."—George Sarton, "Adventures of a Scholar's Wife." Notes for a biography of Mabel Elwes Sarton, in YMS and NYPL.

Van Oye—Paul van Oye, *George Sarton: De mens en zijn wert uit brieven an vrienden en kennissen* (Brussels, 1965) [*Verhandelingen van de Koninklijke Vlaamse Academie voor Wetenschappen, Letteren en Schone Kunsten van België, Klasse der Wetenschappen, 27*, no. 82].

YMS—George Sarton Papers, consulted at the residence of May Sarton, York, ME. Since 1997 the material has been dispersed, much of it to NYPL.

Introduction

Love and devotion animate the past, and not only where the beloved is a passive object of desire. There are writers and artists: Eloise's Abelard, de Châtelet's Voltaire, Stephen's Leonard, Kahlo's Diego, de Beauvoir's Jean-Paul, and Plath's Ted. We would live in a gray and horizonless present without the evocative creations inspired by eros, the Belgian philosopher Suzanne Lilar emphasized nearly half a century ago. In her work of 1963, *Le Couple*, she champions the "cause of *unreasonable* love," going against the principal affliction of the 20th century, the demythologization that "attacks all values." She speaks for "a whole and responsible love" now possible for the first time in history. In Lilar's view, love is self-conscious and reflective: "A great love is above all an awareness of a deep-rooted *nostalgia*, which it passionately tries to satisfy. This, whatever may be thought to the contrary, presupposes a strong sexuality, a motive power capable of assuring communication, able to join as well as to transport." Lilar begins her enquiry with the marriages of Peter Paul Rubens and his different wives, Isabella Brant and Hélène Fourment, following the inspired discussion of Eugène Fromentin.[1] The theme, the painter, and the art critic all figure in the following pages.

To describe those infected with love today, we would need a list where the possessed and the possessing are interchangeable, for we see the partners in a marriage as intellectual and financial equals. The equality has been achieved in the world of capitalism under the intense pressure of a persistent egalitarian model stretching from medieval times through the English, French, Russian, and Chinese revolutions and with elaboration by millenarian communities elsewhere. The achievement has not been without strain. Surveys report that into the early years of the 21st century, most adults in America are married, but they suggest that a substantial portion of this group are not happily married. About half of all marriages seem to result in divorce; a large number of the remaining unions suffer estrangement; most married couples, it seems, are unfaithful to each other.[2]

1. Suzanne Lilar, *Aspects of Love in Western Society*, trans. Jonathan Griffin (London, 1965), "Introduction" and p. 23.

2. These are some of the issues discussed by Phyllis Rose in her study, *Parallel Lives: Five Victorian Marriages* (New York, 1983). The marriages are those of the Carlyles, the Ruskins, the Mills, the Dickenses, and George Eliot and George Henry Lewes.

Even if marriage can be distinguished from love, alliances of woman and man do take distinct forms over time. When did the modern pattern, with its new sensibility, emerge? Demographers have mapped the precipitous decline in European fertility as a way to trace modernization across the 19th and 20th centuries; the decline, they conclude, is based on birth control—women's empowerment over their own body.[3] Freed from perpetual child-rearing, ordinary women could become industrial workers or, more optimistically, imitate the aristocracy by studying nature and writing poetry. Whether women's political rights, no less than the spread of birth-control techniques, trickled down from the highest levels of society or percolated up from the lowest, at the end of the 19th century women entered institutions of higher learning in substantial numbers, and they often married—for love—fellow students. Formed in women's secondary schools by talented teachers unable to obtain a university position, young women *circa* 1900 learned to wield the instrument of reason. However posterity views their radiance, the intellectual firmament early in the 20th century displays a large number of binary stars: the Einsteins (Mileva and Albert), Curies (Marie and Paul), Ehrenfests (Tatiana and Paul), Clays (Tettje and Jacob), Webers (Marianne and Max), Webbs (Beatrice and Sidney), and Woolfs (Virginia and Leonard).[4] The unions were often complex and unstable. Without seeking to generalize, the following pages document one of these unions in which each partner pioneered a new field of inquiry in art and scholarship.

The subjects are Mabel Elwes and George Sarton. The domains are decorative arts and history of science. The decorative-arts movement intended to allow sensitive design to enter the life of laborers and farmers, whether in fabrics, wall-hangings, domestic utensils, or furniture; for better or worse, the interior of our homes today owes much to these

3. Edward Shorter, "Female Emancipation, Birth Control, and Fertility in European History," *American Historical Review, 78* (1973), 605–40; Ansley J. Coale and Roy Treadway, "A Summary of the Changing Distribution of Overall Fertility, Marital Fertility, and the Proportion Married in the Provinces of Europe," in *The Decline of Fertility in Europe: The Revised Proceedings of a Conference on the Princeton European Fertility Project*, ed. Ansley J. Coale and Susan Cotts Watkins (Princeton, 1986), pp. 31–181.

4. The best treatment of women in science around 1900 is José Manuel Sánchez Ron, *El Poder de la ciencia* (Seville/Madrid, 1992), pp. 171–92, which presents comparative statistics and data about the European and United States experience. The marriage of scientists and physicians is considered in Helena M. Pycior, Nancy G. Slack, and Pnina G. Abir-Am, eds, *Creative Couples in the Sciences* (New Brunswick, N.J., 1996). The book focuses on English-speaking people in the North Atlantic World during the late 19th century and early 20th century, omitting seminal figures like Emil Bose and Margrete Heiberg de Bose in Germany and Argentina, Etienne Vassy and Arlette Vassy-Tournaire in Morocco and France, and Jacob Clay and Tettje Clay-Jolles at Bandung and Bernardus Jan van der Plaats and Agathe van der Plaats-Keyzer at Batavia, the latter couples on Java. Generally, Marilyn Ogilvie and Joy Harvey, ed., *The Biographical Dictionary of Women in Science* (New York, 2000), with 2500 entries.

disciples of William Morris. History of science, an established if fractious department of higher learning, emerged to celebrate industrial progress (the very notion of progress depends on a recapitulation of the past) and also to organize and analyze the contents of enormous museums and more particularly libraries that had come into being over the generations before the First World War; we still hope for new gadgets and medical cures, and, for better or worse, few scholars today can completely avoid the shadow cast by the Harvard College Library or the Louvre. Among the general reading public, Mabel and George Sarton are known as the parents of the feminist poet May Sarton, who has written eloquently and evocatively about them and their times.[5]

"Manifestly no Macaulay," is the assessment of George Sarton by several biographers, who identify his legacy "in the creation of tools, standards, and critical self-awareness in a discipline."[6] Mabel Sarton's designs and creations can be found in great collections, notably the Smithsonian Institution and the New York Public Library, but her name is associated with no particular motif or structure. The following pages contend that the Sartons are creators centrally situated in major cultural currents. Equally significant is the emotional side of their marriage, which is documented in detail. Even if their intellectual accomplishments had been dimmer, their affection testifies to the emergence of a new pattern of family life, the effects of which are felt for much of the 20th century.

The story of a minor academic discipline might seem restricted in appeal [7] The discipline of the history of science in America, however, is

---

5. May Sarton, *The Bridge of Years* (New York, 1946); *I Knew a Phoenix* (New York, 1959); *A World of Light* (New York, 1976). George and Mabel Sarton figure in May's fiction as well as in May's non-fiction. See Margot Peters, *May Sarton: A Biography* (New York, 1997).

6. Arnold Thackray and Robert K. Merton, "On Discipline Building: The Paradoxes of George Sarton," *Isis, 63* (1972), 473–95, p. 490. By this assessment, the authors mean that George was not a great prose stylist. The phrase is ironic, given that they emphasize George's positivistic, progressivist view of history of science, lending support, inaccurately I believe, to identifying George as a "Whig" historian. In a manifesto about history of science, George cites Thomas Babington Macaulay to the effect that judging the past requires freedom from the prejudices of the present: George Sarton, "The New Humanism," *Isis, 6* (1924), 9–34, on p. 30.

7. Among the brightest introductions are Noel M. Swerdlow, "Montucla's Legacy: The History of the Exact Sciences," *Journal of the History of Ideas, 54* (1993), 299–328; José María López Piñero, "Las Etapas iniciales de la historiografía de la ciencia: Invitación a recuperar su internacionalidad y su integración," *Arbor, 142*, nos 448/559/560 (1992), 21–67. Earlier appraisals include Helge Kragh, *An Introduction to the Historiography of Science* (Cambridge, 1987); Arnold Thackray, "History of Science," in *A Guide to the Culture of Science, Technology, and Medicine*, ed. Paul T. Durbin (New York, 1980), pp. 3–69; I. Bernard Cohen, "History of Science as an Academic Discipline," in *Scientific Change*, ed. A. C. Crombie (London, 1961), pp. 769–80; and José Babini, *Ciencia, historia e historia de la ciencia* (Buenos Aires, 1967).

indissociable from the life of its animator, George Sarton. For more than half a century, much of its activity derived from his large vision of humanity, as well as from his anxieties and personality. His early programmatic statements as well as his popular publications—that is to say, a majority of his writings before the late 1920s—are designed to attract funding for his enterprise and to secure a better position for himself and his family.[8] From the time of his youth, George claimed to privilege his studies over his family, but his family is embedded in his work. Mabel Sarton, in particular, is the agent of George's success. Her kind and sympathetic social sense acted as a counterweight to George's impulsive and abrupt manner. She was able to establish strong connections with the Protestant Establishment who supported them both; she helped secure George's lifetime post with the Carnegie Institution. The birth of George's discipline required both husband and wife.

A portion of the present book derives from correspondence between George Sarton and Mabel Sarton. The young married couple held their early exchanges so dear that, before they left Belgium for the Netherlands as refugees, they buried their correspondence in the garden. Friends subsequently deposited the letters in a vault at the Banque de Flandres in Ghent.[9] In 1920 George brought the letters to Cambridge without opening them. "To-day I opened one of the packages, hers—but could not bear to analyze the contents," George writes in his agenda on 5 May 1950, following Mabel's death. "I lack the courage to destroy them as they perhaps deserve to be destroyed," he continues, and he prefers to leave them to the scrutiny of "indifferent and objective hands."

The first part of the correspondence is almost entirely in French, until in 1910 George begins to write in English. George records, in his manuscript biography of Mabel, that early in their relationship "we always used *my* language, which she knew, not *hers*, which I did not." He continues that in 1909 he took a Berlitz course in English, but the instructor advised him: "You will never be able to speak English properly." (His diary indicates English lessons from September 1907 into January 1908.) George writes to Mabel in 1910 (perhaps on 16 May) to thank Mabel for her beautiful book bindings: "Never fear to speak to me in English, for I adore to hear it spoken. The few words that you have spoken to me have given me pleasure." Yet George indicates in his biography of Mabel that he could not read English books aloud, in the early years of his marriage: "I could not read well enough nor could I understand what was read to me."

8. At the end of his life, George listed 27 of his popular articles for a collection, *Crochets and Pleasures*, to be edited by May Sarton. YMS.

9. George gave power of attorney to his biologist friend Paul Pelseneer, who opened George's box in the bank vault in 1919. HGS, Paul Pelseneer to George Sarton, 8 May 1919.

George's American correspondent David Eugene Smith compliments him on his English prose in 1912—"It is almost without a flaw"—although Mabel is likely responsible for the result.[10] As late as 26 February 1925, Mabel wants to make sure that George understands her recent letters, where she identifies in him "a frightening vagueness concerning, not only future obligations but *existing ones*." Mabel "thought you were worried very much & *afraid to tell me the whole truth*." She wonders if the problem is George's style: "In spite of wonderful ease you do still use English expression queerly." At the end of his life, George confides in his diary that he has perfect facility only in French and English.

Polyglot Europe intrudes on the Sartons' life together. As George learns English, Mabel learns Flemish, perhaps largely through dealing with servants and tradespeople in Wondelgem, a suburb of Ghent, although she is never comfortable writing it.[11] On 9 December 1910 Mabel writes in French to George that she has had to correct the colors on a poster in Antwerp, "the two workers with whom I worked did not know a word of French, but due to the good intentions of both parties, everything is fine!" On 11 January 1911 George writes to Mabel that a young socialist designer knows only Flemish, and so she would be unable to speak with him. Volunteering himself late in December 1914 to Guy Eden as an interpreter for the British army, George indicates that he speaks "French, German, Dutch, Flemish, and English," in addition to "the Walloon dialects of Eastern Belgium."[12] Writing on 15 January 1915 to offer himself as a lecturer to the president of Columbia University, George emphasizes that French is "the only language I feel completely master of," although in an appended vita he lists as his languages Greek, Latin, Flemish, Dutch, German, English, French, and Italian.[13] In November 1914, George, writing from London in English prose constructed by Mabel, presents himself to Wickliffe Rose, director-general of the Rockefeller Foundation's General Health Commission, offering his services and claiming perfect knowledge of both Belgian languages. But on 28 October 1919 George in Florence forwards a letter in Flemish to Mabel in Rhode St-Genèse, a suburb of Brussels. In French: "I should like to ask you to take care of it because you know Flemish much better than I." On 9 November 1919, Mabel informs George that she is taking care of responding "but it is extremely

10. HGS, David Eugene Smith to George Sarton, 21 December 1912.

11. People who use French habitually in Ghent are a distinct minority, rising from about 5% of the population early in the 19th century to about 10% on the eve of the First World War. Herman Balthazar, "Genèse de la métropole industrielle," in *Gand: Apologie d'une ville rebelle*, ed. Johan Decavele (Antwerp, 1989), pp. 155–178, on p. 168.

12. YMS.

13. NYPL, George Sarton to Nicholas Murray Butler, 15 January 1915.

difficult for me to *write* Flemish in spite of my facility in speaking!!" George is less kind in his manuscript biography of Mabel: "She never learned or bothered to speak Spanish or Flemish (Dutch) correctly." He recalls listening at Wondelgem to Mabel speaking with their gardener, who knew only Flemish: "It was very amusing to hear Mabel explaining everything to him in her own Anglo-Dutch mixture which he listened with some incredulity and astonishment."

Mabel and George negotiated around the Babel of language. They admired writers of diverse origin formed in the 19th century—Carlyle, Ruskin, Maeterlinck, Romain Rolland, Fromentin, Tolstoy—and they read omnivorously, but their own writing is for the most part the uninflected, workaday prose favored among practical people early in the 20th century.[14] As the publications of Albert Einstein suggest, style among younger academic writers declined dramatically at this time. One is tempted to associate the decline with the great democratization of universities, which saw enrollment growing in leaps and bounds; increasing teaching burdens on the shoulders of younger scholars; and a desperate contest for visibility in the periodical literature, required for landing a university chair. The plain-speaking that echoes in the Sartons' correspondence with each other radiates earnestness and authenticity; it is when they try for allusion that the words fall flat. George and more often Mabel are confused or upset by figures of speech sent by their correspondent. Each knows that something is not right when the other dwells too long on abstract matters.

They were no strangers to conventions of their time. Race is a word appearing often in the following pages, and attention shall be devoted to it. Mabel, consistently attracted to the harmonies of pre-Raphaelite portraits, places emphasis on outward appearances. She writes to George about Leo Kallen, physician brother of George's philosopher friend Horace Kallen: "I don't like that brother—I have a real repugnance for him but cant explain why. His neck is too thick & the back of his head too *flat* & his eyes so small & he is fat & German in the wrong way, all of which are superficial things but I feel are outward & visible signs of a lack of inward & spiritual grace."[15] George, in his English publications, refers to mankind and men of letters, and in French to l'Homme (which I translate as humankind). George is conventional in viewing women as objects of desire, and early

14. Diary, 4 April 1921. George and Mabel draw up a list of books for their Ideal Library, not being able to afford to buy them. Included are Sir Henry Maine's *Ancient Law*, Louis Hémon's French-Canadian classic *Marie Chapdelaine*, and works by Gustaf af Geijerstram, Giovanni Papini, Selma Lagerlöf, Lytton Strachey, Serge Aksakoff, Ferenc Molnar, Edmund Gosse, Evangeline Wilbour Blashfield, Godfrey Rathbone Benson (Lord Charnwood), Herbert G. Ponting, Jérome and Jean Tharaud, Felix Timmermans, and George Borrow.

15. NYPL, Mabel Sarton to George Sarton, 23 January 1933.

bugbears, in popular publications, are "snobbish women" who have a bit of university-level education, but the following pages provide considerable evidence of his ability to engage women as intellectual equals.[16] As with gender, so with sex. Commentators interested in sexing prose, as one might sex a chick or kitten, can find grist here for their mill. I prefer to think that a more credible picture of George's and Mabel's sexuality, as well as their views on the gendering of knowledge and art, emerges from their intimate notes and correspondence. In his early publications in English where women snobs appear, George wrote for money, and he certainly knew what would sell. There is not the slightest suggestion that George ever resented Mabel's artistic success as a designer or, later, his daughter May's success as a woman of letters. All evidence supports the contrary assertion.

Mabel and George were devoted to the epistolary art. Mabel's mother observes to Mabel's friend Céline Dangotte around 1901: "I do not like writing letters. (Where does Mabel get her love for it from I wonder?!!)"[17] They would plan their day or week to be able to write to each other at length. On occasion, they would write twice or even three times in a day, the letter carrier becoming almost a private courier. They wrote frequently to each other when they were separated by long distances, and also, before they were married, when they both lived in

---

16. CIW, George Sarton to Robert S. Woodward, 25 August 1918, for the "loathsome counterfeit" of admiration for heroic figures "by *snobbish* & crazy women." Men are also objects of George's derision: "No amount of knowledge will give generosity and intelligence to a man who has none. On the contrary, it will set his pettiness in greater contrast, in the same way that wealth makes mediocrity more conspicuous." George Sarton, "Secret History," *Scribner's Magazine, 67* (1919), 187–92, on p. 189. George also declaims against male snobs, "the vanity and the snobbism of so many men, who are aroused by the incessant desire to be spoken about." George Sarton, "L'Organisation scientifique du journalisme," *La Vie internationale,* 5 (1913), 391–428, on p. 412. And later: "But even this wretch of a snob is not as depersonalized as he seems....Once his snobbery has been pierced through, he is as unable to hide his real self as any professional actor." George Sarton, "Transparency: A Way of Looking at People," *Scribner's Magazine,* 77 (1925), 308–313, on p. 311. Sarton's appreciation of women in America bears a resemblance to Albert Einstein's. In 1921 Einstein spoke in an interview about marital life in America, which he had recently visited: American husbands "are toy dogs for their wives, who spend the money in the most excessive fashion and who shroud themselves in a veil of extravagance. They will do anything that's in vogue and in fashion, and, as it happens, have thrown themselves among the throngs of the 'Einstein-craze.'" Einstein later contended that he was misquoted. "An Interview with Professor Albert Einstein," in *The Collected Papers of Albert Einstein*, volume 7, ed. Michel Janssen, Robert Schulmann, József Illy, Christoph Lehner, and Diana Kormos Buchwald (Princeton, 2002), p. 624; *cf* pp. 443–4.

17. NYPL, Eleanor Elwes to Mabel Elwes, ca 1901. Mabel's father wrote regularly to England when in his youth he worked as an engineer in India. No correspondence survives from the Elwes parents to their children when they lived apart from their daughter and son in connection with the overseas employment of Mabel's father.

Ghent. (Their residences in the core of the medieval part of Ghent were a 10-minute walk apart.) In the period before the First World War, mail traveled efficiently across Western Europe. A letter mailed by George in Ghent could arrive later the same day with Mabel in a small Swiss town. Urban delivery could take a matter of only hours. The system allowed George and Mabel to elaborate thoughts with precision and conduct a dialogue when ideas were still fresh in mind. Mabel's and George's lodgings in Ghent and in America had a telephone. But in preference to speaking, each liked to set down intimate thoughts by hand. When they were apart, letters continued a conversation of theirs, whether during the day or weeks previously. On occasion, letters over the course of many days were posted together. George sometimes cried for letters from Mabel, if only a few lines on a postcard; Mabel felt a mixture of fear and excitement when she opened a letter of George's. A long letter from Mabel to George on 26 February 1925 begins: "I must come & talk to you. Letters are dreadful things are they not? & yet so precious! (Even when they upset me of course I would rather be upset than not have any letter that week!)"

A reader will find their exchanges unusually frank and forthright. On 14 July 1923 in Ipswich, Mabel is happy to receive George's and May's short letters, even though, in French, "they told me little of importance." In their correspondence, they generally bare their deepest concerns. This feature is especially interesting because the language of their early letters is French, by then a highly inflected literary vehicle. The straightforward prose results in part because they set down emotions without the advantage of sharing a mother tongue, although it is also an expression of their character. They both sought, above all, essential qualities in life. For example, on 10 May 1915 Mabel writes to George. Mabel is in England, convalescing from measles; George is in Washington, perilously beginning his academic career in the United States. Mabel returns to the matter of her illegitimate birth (her parents married a short time afterwards). She wonders if George spoke about it to a colleague, with whom he worked for a number of months in the War Office in London:

> Tell me, did you perhaps tell him the circumstances of my birth? All that wretched thing, it has been haunting me. Can you understand how grieved I am that these facts, against which *I can do nothing*, can be an instrument against *you*? I did not imagine they could be of such importance, even break your career. That is a terrible thought. I loathe having *something to hide*, something one is *afraid* should be known! I feel almost that if I were *alone* I would go to the other extreme & sift my friends & acquaintances by telling them the truth as quietly as possible.

She needs to obtain travel documents from the Belgian Embassy, and when she goes there she will do her best to keep the matter secret. "One thing I have decided: When she is big enough, May shall be told. I myself will tell her."

Sigmund Freud casts a long shadow across the first part of the 20th century, and especially so when the intimacy of man and woman is in question. George Sarton observes psychoanalysis in the second number of his periodical, *Isis*, in 1913. In 1920, reviewing Freud's study of Leonardo da Vinci (on whom George focused a great deal of attention), he delivers a stunning rebuke:

> Those who wish to know to what degree of absurdity Freudian psychoanalysis can be carried—and that by the Master himself!—have only to read this book. These complicated and extravagant theories based upon the scantiest and most uncertain facts are a credit to Freud's ingeniousness rather than to his critical spirit. It might be time to make a psychopathological study of Freud himself![18]

Dreams, childhood deprivation, and sexuality all have a place in the following pages; psychoanalysis and psychological theory generally do not. Abundant documentation is provided for readers with a psychologizing turn of mind. Sensational books notwithstanding, history is best crafted from the scruples and traditions of historians.

May Sarton, daughter of Mabel and George, is present through much of the story of their life together, as a commentator and as a participant. She is widely known today as a feminist poet and writer, in part because she is one of the earliest women in the United States to acknowledge in public and to celebrate her own lesbian, sexual orientation. In fact there is very little sex in her writings, which are almost Platonic in their search for purity and essence; May shares this stylistic orientation with Mabel and George.[19] A reader may seek the origin of May's sexuality in the account of her parents. Is it nature or nurture? Both May and her authorized biographer have emphasized her unusual upbringing—lacking a permanent home, ignored by a possibly neurotic mother and a self-confessed, emotionally-arrested father.[20] I suggest in the present book that Mabel and

18. *Isis, 1* (1913), 320: under "Sciences des primitifs" in the analytical bibliography, Sigmund Freud, "Der Wilde und der Neurotiker," *Imago*, nos 1, 3, and 4 (1912); George Sarton, "Sigmund Freud, *Leonardo da Vinci: A Psycho-Sexual Study of an Infantile Reminiscence*," *Isis, 3* (1920), 104.

19. Lewis Pyenson, "Elegant Sartons: Platonic Science, Platonic Letters," in *Elegance: Beauty and Truth*, ed. Lewis Pyenson (Lafayette, La., 2001), pp. 5–13.

20. Peters, *May Sarton*.

George raised May creditably and honorably. Mabel, for example, frets about May's living conditions when her daughter begins an acting career in New York: "I am distressed to find all her things so *messy*, shabby & even dirty. I wonder have I not been severe enough to her about such things? I have always thought that 'example was better than precept' in this as in all things—though, I did train her as a child."[21] Whatever material shortcomings there were in May's youth, she was guided by passionate intellectuals to appreciate the life of the mind—something intellectual parents strive for today. Evidence is presented for Mabel's hormonal imbalance in the years before and after May's birth; to correct the imbalance, she was one of the first women to receive hormonal therapy. Readers may decide for themselves if this circumstance is related to May's constitution.

If psychoanalysis and sexual orientation are not found in the following pages, sex is. Sex occupies a relatively large place in Mabel's and George's early correspondence, as it does in the life of many young people. Documents provide an unusually complete picture of their sexual life, which they observe without prudery—notwithstanding their conventional sensibilities about nakedness.[22] George, a compulsive note-taker, regularly records sexual congress by making a Maltese cross in his agenda. The evidence, which seems conclusive, indicates that at several points in his life, he is unfaithful to Mabel. Should the evidence be passed over, read by the fire and then burned, according to earlier custom? The available records are consistent and substantial, going beyond isolated anecdote.[23]

Concluding his study of historian Thomas Carlyle and Jane Baillie Welsh, Osbert Burdett notes: "Records of any marriage are as rare as they are precious." Burdett emphasizes:

> The story of the Carlyles is an authentic history substantially repeated in a thousand homes. It holds, interests, and at times exasperates us, as living problems do. The love, and the difficulties, the misunderstandings and the unity, the friction on the surface and the devotion at the core, are the mirror of the confusion of daily life.[24]

---

21. NYPL, Mabel Sarton to George Sarton, 25 January 1933.

22. We shall see presently that during their engagement, the Sartons correspond about appearing in the nude. YMS, George Sarton to May Sarton, 18 July 1943: "A few days ago it was so hot I took my tie off, but I apologized to mother and felt a little ashamed of it. What we call 'les bienséances' is more important than comfort and without it life tends to become brutish."

23. Evidence for George's infidelity comes from his agenda. His custom is to indicate sex by a Maltese cross; comments confirm the denotation. The custom goes back at least to Robert Hooke in the 17th century, and George uses the notation as early as 1904. Agenda, 11 and 12 August 1904, "mauvaise journée" marked "+."

24. Osbert Burdett, *The Two Carlyles* (Boston, 1931), pp. 292–3.

But the remarkable, epistolary exchanges of the Carlyles, for whom some 10,000 letters exist, are silent about marital sex, leaving us to infer intimate matters only from the gossip of acquaintances.[25] A recent biographer emphasizes: "No amount of speculation, or reviewing the speculations of others, can take us further into the mystery at the heart of this, as of any other, marriage."[26] The Sartons mirror the Carlyles in a number of ways: their long engagement and turbulent marriage, their sense of ease with the cultures of Western Europe, the self-imposed isolation of the writer-husband and the resentment of the sensitive, invalid wife, and their compulsion to write. Still, the paper trace left by the Woolfs, notably Leonard, is perhaps closer to what we have from the Sartons, especially George. Lord Annan summarizes:

> Leonard Woolf, counting the dates of Virginia's periods because he had detected a correlation between delay in menstruation and her bursts of manic depression, became transformed from the devoted husband into an anal monster whose so-called concern for her health was on a par with his habit of keeping meticulous accounts, logging the number of miles he drove, recording the date he had his hair cut, the number of bushels of apples yielded by each tree in his orchard, the events of every day for fifty years and their exact expenditure and earnings.[27]

George Sarton, 4 years Leonard Woolf's junior, also married a woman with artistic temperament older than he, a woman who also wrestled with depression. Mabel Sarton, who admired the publications of Virginia Woolf, was a creator sustained by a husband who kept detailed records of his moral and material progress.

---

25. Charles Richard Sanders, ed., *The Collected Letters of Thomas and Jane Welsh Carlyle*, vol. 1 (Durham, 1970), p. xxiv for the number. The untrustworthy Frank Harris, for example, reports Jane Carlyle's virginity after 25 years of marriage. John Stewart Collis, *The Carlyles* (New York, 1971), pp. 181–2 for Harris; also Rosemary Ashton, *Thomas and Jane Carlyle: Portrait of a Marriage* (London, 2003), p. 11. In the absence of documentary evidence, the union of sociologist Max Weber and Marianne Weber—both prolific writers—is also suspected to have been unconsummated. Donald G. Macrae, *Max Weber* (New York, 1974), p. 24; *cf* Patricia Madoo Lengermann and Jill Niebrugge-Brantley, *The Women Founders: Sociology and Social Theory, 1830–1930* (Boston, 1998), p. 214 for the contrary assertion.

26. Ashton, *Thomas and Jane Carlyle*, p. 76. Ashton adds: "All we can say is that no positive proof of consummation survives, but then, apart from pregnancies and children, of what could such proof consist?"

27. Noel Annan, *Our Age: English Intellectuals between the World Wars, A Group Portrait* (New York, 1990), p. 86. With the Sartons, George tracks Mabel's period, but it is Mabel who makes the connection between menstruation and her emotional state.

The Sartons were conscious from their young adulthood about the significance of letters for future times. They avidly read biographies and autobiographies as a part of their program for edification; on occasion, biographies motivate their actions. Mabel and George Sarton did not write for posterity, however. They saw their letters as a record, for themselves, of their spiritual and emotional progress. They often reread and commented on past correspondence.

Mabel expresses the notion clearly when she visits the home of the Carlyles in Chelsea. She writes to George on 23 June 1910 that her umbrella was stolen (later she comments that she left it in Chelsea), but she does not mind because she was so impressed by the house. She read some of the love letters between Thomas Carlyle and his wife, Jane Welsh. In French in second-person familiar: "In general it displeases me very much to see such letters published. This time, I think that it was necessary to justify them both. I can't define why just now, but I felt yesterday—intensely—that you resemble Carlyle." George indeed admired Carlyle, by virtue of Carlyle's correspondence with Welsh. He writes in his diary on 31 August 1910 about himself and Mabel: "I can say of us what Carlyle said one day of Jane Welsh: 'Has not a kind Providence created us for one another? Have we not found each other? And might not both of us go round the Planet seeking vainly for a heart we could love so well?'" It is a quotation he sends to Mabel on 18 February 1911, after parting from her in France.

Carlyle was the historian who fired the mind of both the young and the old George Sarton. On 31 August 1905 he notes in his agenda the translation of Carlyle's writings by Emile Masson, who becomes a friend. To Andries MacLeod, son of the Flemish nationalist and biologist Julius MacLeod, George writes on 13 January 1947: "Your life in Vintjärn [Sweden] makes me think of Carlyle's in Craigenputtoch, where the silence was so deep that his wife, Jane, could hear the sheep graze."[28] Thomas Carlyle, whom George finds a kindred spirit, was foreign to many of George's contemporaries, notably Virginia Woolf, who visits the Chelsea house on 24 February 1909 in preparation for reviewing an edition of the love letters for the *Times Literary Supplement*. Woolf writes about the Carlyle marriage: "It taxes our powers to the utmost to understand; the more we see the less we can label, and both praise and blame become strangely irrelevant." Woolf concludes after her visit: "Did one always feel a coldness between them? The only connection the flash of the intellect. I imagine so."[29] A reader will find persistent echoes of the Carlyles in the following pages.

28. Van Oye, pp. 110–11.

29. Virginia Woolf, *Carlyle's House and Other Sketches*, ed. David Bradshaw (London, 2003), pp. 4, 49. Woolf's father Sir Leslie Stephen wrote the entry on Carlyle for the *Encyclopaedia Britannica*.

The action of a life is not chronicled entirely in contemplative correspondence. In times of crisis, the courage or the will to write may fail. Even among people who may seem born to the pen, intense emotions can preclude discursive prose. When people live together, furthermore, they do not generally write to each other about daily occurrences. Supplementing the correspondence are George's diaries and appointment agendas. George kept these accounts for his entire adult life, and he was meticulous to a fault. Mabel also kept a diary. On occasion they showed their diaries to each other. George sometimes had an acquaintance write an address in his agenda. It is striking that George's diaries from 1907–1911 and 1914 are missing, as are his agendas from the early years of his marriage in Belgium. From early 1917 to June 1919 George kept his diary on cards, his "memindex"; he resumes a conventional diary at about the time he returns with his family to Belgium. The cards are unavailable, and Mabel's diary has not been located. George's missing diaries cover turbulent times, but in view of the material he preserved there is no reason to suspect that he destroyed them.

The present book focuses on the emotional side of the Sarton marriage. Mabel and George might have called it a record of their moral progress. Their professional life is part of the marriage. Mabel kept house, painted, and designed clothes and furniture. George published, lectured, and wrote countless professional letters. George referred to professional correspondence as a *besogne*, a task, and he tried to handle it as quickly as possible. In professional matters, he ironically preferred sparkling conversation to letter writing, a sensibility that Mabel observes on 11 January 1933:

> A thought in my mind when I read what you say about your writings versus your lecturings: anyone who hears you speak says to himself "here is a *real* fellow—not a mere peg for ideas" & anything he reads of yours afterwards takes on a new value, a vivid glow & *impact* that it could not have otherwise. You chiefly want to impart truth—facts—but inevitably truth & facts create ideas, reflexions, hopes & hypotheses both in the man who offers the facts & in those who receive them. Here & there are men who like yourself perhaps are not interested in anything *but* the facts but some truths are affected strongly by the way they are offered & your warm, human way of presenting them (which shows also in your writings) awakens & stimulates interest. It is as if those facts were seeds, each one containing germs of larger truths & they often need to be "started" by the warmth of a human personality.

George's friends in Cambridge came from a wide variety of disciplines, as they did throughout his life. George's correspondence relating to his periodical, *Isis*, and his scholarly writings provide information about his personal life, as does Mabel's correspondence with Céline Dangotte, her adoptive Belgian sister, in whose enterprise in *art décoratif* Mabel worked as a designer. Mabel established herself as a leading designer well before George obtained a comparable place in his chosen profession.

The available letters are clear that Mabel and George bare their soul to each other, but they do not reveal everything. In 1928 Mabel writes to the artist Lucy May Stanton, whom she regards as her spiritual American sister, about her partnership with several women in Washington, which was called Belgart:

> I have literally to whip & spur myself on from day to day & even then am dismayed & frightened to see that I cannot make ends meet, of strength—or money. I ought to find some other way of earning (& have tried in vain) because Washington will never pay me I fear more than a "few drops" in the bucket. They can't—they don't make money. Twelve years designing—giving of my best, & here I am where I started, from the financial point of view & I feel too old & too worn out to find a new outlet. I confess I am horribly depressed & have been since June.
>
> *I don't want George* to know too much. It would only upset him & he can't help. I have told him a little because it is not quite fair to him he should not know at all. He is a *dear*, you know—& as blind & full of the conviction that he is the traditional bulwark between me & worries, as any other husband![30]

George is also very close to, possibly intimate with, Stanton.[31]

George is remembered, by those who knew him, as a difficult, distant man. Following a European tradition that has recently come to America, he spent his life scrambling for the leisure and the resources to undertake original scholarship of a synthetic nature. He was exploited by his employers: Harvard kept him as a poorly-paid lecturer for many years, and the Carnegie Institution of Washington, the source of most of his income, decided not to continue its investment in history of science when he retired. His close friends twice brought in a younger historian to sup-

---

30. ALS, Mabel Sarton to Lucy Stanton, 15 December 1928. On 6 April 1930 Mabel calls herself a "sister-friend" in a letter to Stanton; on 20 February 1931 she signs herself "always your loving sister."

31. May Sarton recalls, "Lucy Stanton came into our lives like an angel at a time when perhaps we were all three rather lonely....My father, always sensitive to the charm of women of character, basked in her warmth." May Sarton, "Lucy Stanton, a glimpse of her seen by a little girl," June 1966, ALS.

plant him at Harvard.[32] He persevered. From the time of his youth he tended to be curt in personal relations. In old age, he became curmudgeonly. In her early adulthood, Mabel experienced more than a decade of ill-health and frequent depression, but she won nearly everyone whom she encountered by her warmth and generosity. Mabel was immensely proud of George, and George continually credited Mabel with having licked him into shape, making him less of an *ours mal léché*.

The marriage has been analyzed by May Sarton, who expressed strong feelings about George's single-minded devotion to his scholarly labor and Mabel's self-effacing support of it. (George was amused at May's earliest writings about his "table-talk," appearing in the *New Yorker*, and he corrected them silently.[33]) The following pages show deeper levels in the marriage, which records mutual love and devotion in the face of many kinds of adversity. It is also a record of the perils and rewards of academic life in the first part of the 20th century, when women demanded to be treated as the intellectual equal of men and when academics sought support for their projects from universities. Tensions evident in the marriage later became structural features of family life. The reversals of poverty, war, and perfidy, of hopes unrealized and small courtesies unremembered, did not dim their common credo, established in the years before they married. Love, beauty, self-improvement, and human betterment are consistent themes in their life together.

Judgment is also a theme in the marriage. Mabel and George strove not to judge each other, believing that love transcends critical appraisal; they sought to be generous in dealing with other people. When, on 3 September 1919, Mabel closed up their home in Wondelgem and packed away what remained of their Belgian silver and jewelry, she found time to look through their correspondence:

> I also put all our letters down at the bottom of the trunk. George, I am so happy those letters were not at the mercy of the Germans. I re-read a few of yours of *1908*. They are such dear letters. I wonder do you realize what progress you have made since then on unselfishness & real tenderness

32. "On two occasions a foreign teacher was invited to teach the history of science in H.U. in competition with me—and I was not warned or consulted in either case, except (in the second case) *pro forma, in extremis*." The foreigners were Georgio de Santillana and Willy Hartner. Diary, 5 November 1948. George lived with continual uncertainty about whether the Carnegie Institution would publish his *Introduction to the History of Science*.

33. Diary, 15 January 1954 and 7 February 1954. From the former: "May has published in the latest *New Yorker* a delightful essay derived from my table-talk. It is fairly correct and yet I do not recognize myself, because the period described is too distant in years and vicissitudes. It represents a kind of prehistoric me."

> & generosity not only to me but to every-one? I don't think I realized it myself. Not that you were not already trying hard to be all that, but you were still extremely passionate & intolerant. Now you are much more deeply, truly, tolerant than I am.

Generosity is not frequently ascribed to George in the authorized biography of his daughter, nor in testimonials and remembrances. Yet it would be a mistake, a century later, to look at the Sartons with a disapproving eye. They wrestled honestly with large issues. They succeeded, in the face of daunting obstacles, in raising one of America's most popular women of letters and in establishing a significant academic discipline in America. Their struggles, no less than their achievements, clearly reveal the compromises that are endemic to intellectual and artistic life today.

George and Mabel Sarton's notions of art and literature, science and scholarship, crystallized before the First World War. The Sartons had insiders' awareness of radical breaks with the past during their lifetime. In the following pages, the reader will see quantum physics, relativity, and abstract art. George, in June 1914, reviews the transformations in physics over the preceding decade that have "entirely modified the horizons of science": the revival of atomic theories and the notion of discontinuity.[34] George was interested in the new art and the new science, but he kept a distance from them. For example, on 9 February 1948 George writes in his diary about a "very instructive anecdote" related to him by his mathematician friend at Harvard, Philippe Le Corbeiller. Paul Valéry, whom George admired,

> met Einstein at a party given in Paris in the latter's honor. He asked him whether he carried a little book to note down ideas that occurred to him. As Einstein said "no," he began to explain the great convenience of doing so; he himself always carried such a book in which he had occasion to write many "ideas" every day, which might have been lost otherwise. Einstein replied smilingly, "You know, I have so few ideas."

Instructive also about George, the compulsive note-taker who knew Einstein personally.[35] In 1908 George sees paintings by Pablo Picasso, "a young Spanish painter still unknown who has an extremely original

34. George Sarton, "La Nouvelle Physique," *Isis, 2* (1914), 193–8, focusing on the Solvay Conferences in quantum physics.

35. George Sarton to Mabel Sarton, 28 January 1933, where George reports on talking with Einstein for about half an hour in the Athenaeum at the California Institute of Technology, where they were both staying; Mabel Sarton to George Sarton, 2 February 1933: "So you talked with Einstein: I'm glad he had on his sweater. What a pity he talks such awful English. This will he a stumbling block for the intercourse which Dr [Abraham] Flexner so dreams of in his group of 'advanced study' men." The two men

vision of life, but some of his canvasses are ghastly."[36] Neither George nor Mabel was much impressed with non-representational art, and George expressed his views in publications on art history. In a letter to philosopher Horace Meyer Kallen, George emphasizes: "Our modern artists are rotten with irrelevant ideas. They want to know everything except their own craft, and the result is an awful mess."[37]

In the Sartons' life, war and revolution are omnipresent, but there is relatively little about political elections. One senses that George and Mabel devoted their effort to works of inspiration, and above all to understanding each other. They sought balance and harmony, and they tried to avoid dissonance and iconoclasm. By the 1930s, they each suspect that the world might not be governed by the ideals of their youth: Mabel gives up her career as a designer and, inspired by Virginia Woolf, writes short stories which do not find their way into print;[38] the Carnegie Institution of Washington comes to hold one of its most highly-decorated employees, George, in contempt.[39] Yet the Sartons' ideals reflect the values of the

---

both received an honorary doctorate from Harvard University on 20 June 1935, along with Thomas Mann and George's employer John Campbell Merriam of the Carnegie Institution of Washington—Sarton and Merriam an LLD, Einstein a DSc, and Mann a DLitt. "Einstein and Mann Get Harvard Honor,"*New York Times*, 21 June 1935, p. 8. YMS, Mabel Sarton to May Sarton, 21 June 1935, for a report on the ceremony: "Daddy's position at Harvard is an anomaly in that he is still technically an *instructor* & I imagine the first such in the history of the University to receive an honorary degree." George and Albert Einstein were also early members of the Gesellschaft für positivistische Philosophie, led in 1913 by Joseph Petzoldt. *Zeitschrift für positivistische Philosophie, 1* (1913), membership list tipped into the volume in the Graduate Library, University of Michigan. George and Albert Einstein both contributed to *Romain Rolland, sexagenario ex innumerabilibus amicis paucissimi grates agunt*, eds Maxim Gorki, Georges Duhamel, and Stefan Zweig (Zurich, 1926). At the end of his life, George saluted Einstein as a philosopher. George Sarton, "Henri Berr (1863–1954): La Synthèse de l'histoire et l'histoire de la science," *Centaurus, 4* (1954), 185–97, on p. 192. In 1926 George shares Einstein's plea for tolerance in Palestine on the occasion of the founding of the Hebrew University: *Isis, 8* (1926), 596. George writes in his diary on 2 August 1955: "Einstein was a truly great man, not simply because of his immense discoveries but because of the perfection of his social sense."

36. NYPL, George Sarton to Mabel Sarton, Pentecost 1908.

37. CAJ, George Sarton to Horace Meyer Kallen, 5 October 1930.

38. NYPL, Mabel Sarton, copybook dated Ipswich, July 1932, beginning with a long passage from Virginia Woolf's book *The Waves*, and also one from *A Room of One's Own*; in the copybook, an epigram from Woolf begins Mabel's manuscript "The Orchard," later titled "The Devil in Aunt Mollie."

39. NYPL, Mabel Sarton to Evelyn Van Loon, 1 January 1939 [misdated for 1940], reporting on significant cuts to George's salary. "The sad fact remains that [Vannevar] Bush, the new president of the Carnegie, is absolutely indifferent to George's studies & considers all his work 'irrelevant.' (Julian Huxley on learning this rapped out 'Good wine needs no bush'!)"

first half of the 20th century, just as pessimism, relativity, and materialism dominate the second half. Notwithstanding criticism directed against their life and work, the Sartons' notion of truth, justice, and progress animates the United Nations and the Nobel institutes, assemblies that, despite the odds, continue to radiate authority, respect, and hope. We would not be foolish to follow their lead.

Part I

Parnassus and Pisgah, 1905–1908

CHAPTER 1

What They Brought

Eleanor Mabel Sarton's maternal grandfather, Thomas Cole, a shoemaker and postmaster in Great Oakley, Essex, wore out much shoe-leather delivering the mail. Thomas had 8 children with his wife, Susannah; the seventh, Mabel's mother Eleanor, was born in 1848. She was a talented artisan in needlework, embroidery, and woodcarving, and she was a keen observer of human character. Mabel's father, Richard Gervase Elwes, was born in Poslingford, Suffolk, in 1841. He descended from a 17th-century lieutenant in the Tower of London and an alumnus of St John's College, Cambridge. His grandfather was Lt General John (Timms) Hervey Elwes, who in 1879 placed the family's coats of arms on a stained-glass window in the Elwes Chapel in the Church of St John the Baptist at Stoke College, Stoke-by-Clare, some 70 miles from London in southern Suffolk. Gervase, as he was known, was the eldest child of Timms Hervey Elwes and Louisa Weston. Gervase was educated at Queen Elizabeth's School in Ipswich and then at the Department of Applied Sciences, King's College, London, where he graduated in 1858. At Addiscomb in 1860, he took the examination for the Indian engineering service, and he scored best in it.[1]

In India, Gervase managed various construction jobs. Among his first projects was a bridge at Koksur, north of the Himalayas on the border of Tibet. He wrote to his parents on 26 January 1862 that it would be madness to marry under his conditions in India, "but at the same time I hope to be able to marry some day." He handled a number of projects, rose to serve for a time as secretary for public works in Hyderabad, and retired from the Indian service in 1874 with the rank of Executive Engineer, second grade. Later in life he commented that Rudyard Kipling's story, "The Bridge-Builder," is an accurate portrayal of his profession.[2] After India, he worked on projects in England, Canada, Italy, Malta, Spain, and South Africa. During his employment in the Transvaal in 1897, he invested heavily in mining stocks there.[3]

1. "Scholar's Wife." George gives Mabel's mother as Eleanor. Margot Peters, *May Sarton: A Biography* (New York, 1997), p. 402, gives Elinor. Daniel Rice, "The Elwes Family," *Elsham* ([Derby], 1981), 7–19 for an extensive account of the family.

2. NYPL, Gervais Elwes correspondence and Mabel Elwes to George Sarton, 19 July 1907; Rudyard Kipling, "The Bridge-Builder," in *The Day's Work* (London, 1898).

3. NYPL.

At the time of Mabel's birth in 1878, Gervase and Eleanor were not legally married. The decision was Gervase's. He was half-engaged to a young woman dying of tuberculosis; Mabel's parents postponed marriage until the woman's death, although they lived together and started a family. Discovery of the circumstance, by which she is technically illegitimate, brings Mabel much anguish at the time of her own marriage. When Mabel is born, the Elwes family live in London, at Heath Lodge, Thornton Heath; later they move to a house in Shadowbush at Norbury, within walking distance of Crystal Palace, Upper Norwood. Mabel and her brother Hugh have many pets, and they collect butterflies—their part of London having not yet been completely assimilated into urban life.[4] In George Sarton's view, the siblings share character traits—"extreme generosity and discretion—qualities which became weaknesses in the midst of rascals."[5] They attend concerts and fireworks in the Palace grounds. Gervase is concerned to improve the mind of his children, who beg him to speak to them on lighter topics. About 1885 Gervase and Eleanor leave for Winnipeg, Canada, where Gervase constructs the waterworks, and the children are left with a family, an old woman and her mature daughter, in a Welsh village, Milford. Mabel recalls the time as a dark one, when she does not have enough to eat and when she is sometimes alone at night, sitting at the head of the stairs leading to her room, "paralyzed with night terrors." The experience follows her into adulthood, and late in life Mabel writes a short story about the Welsh time, "The Devil in Aunt Mollie," where the daughter of her host suffers from fits of irrational rage.[6]

Upon the return of Mabel's parents, Hugh goes to Warwick Grammar School and Mabel stays at home. Then, when Gervase is posted abroad again, Mabel stays with the Vicar of Acton in Suffolk, while Hugh is placed with a Doctor Athill in Great Bentley. In 1894, Gervase finds a position in Huelva, Andalusia, Spain, constructing railway bridges to serve the Rio Tinto copper mines in the vicinity. The family travels there together. Fifteen-year-old Mabel spends a year without schooling in Huelva and sees much of the countryside. For a time the family live in a bankrupt hotel, the Colón, constructed for the Columbian quadricentennial.

Mabel's oldest friend is Sophie Jackson, daughter of the physician who brought her and her brother Hugh into the world, a man who admired Gervase Elwes "as the man with about the highest standards of honour of any he had ever known."[7] Mabel writes to Sophie on 20 September 1894 from Spain. Her brother Hugh has learned Spanish and is studying engineering in Clausthal, Germany. Mabel does not like Huelva, a dirty

4. Generally, Donald J. Olsen, *The Growth of Victorian London* (London, 1976).

5. Diary, 23 May 1953.

6. NYPL; May Sarton, "A Wild Green Place," in *Phoenix*.

7. HGS, Sophie B. Jackson Smith to George Sarton, 23 November 1950.

seaport. She traveled to Seville, Granada, and Córdoba. "How enchanted I was with the Alhambra & how I longed to stay for years & paint every nook & corner, every exquisite little view." Mabel is especially taken with Moorish remains. She worries that she shall be behind in school knowledge when she returns to London.[8] In retrospect, Mabel laments her lack of formal education. In fact, by her 20th year she is widely cultured.

Apart from her erratic schooling, Mabel receives instruction at the South Place Ethical Society in Finsbury, under the pacifist minister Moncure Daniel Conway, an American émigré who edited the literary periodical, the *Dial*. Gervase Elwes joins Conway's congregation in 1875 and continues his association at least until 1900, when he publishes an article in the *South Place Magazine* in favor of the British war in South Africa against the Boers, whom Gervase characterizes as venial exploiters of Africans and heartless territorial aggrandizers. The British mine owners, against whom Boer supporters railed, are in reality "hundreds of thousands of small shareholders, whose average income is probably a few hundreds a year." Gervase relates that he, indeed, has shares in a mine along with 6000 or 7000 others (a third of whom are French and Germans); the average holding is £25. Gervase urges readers to rally against the "slavery, tyranny, superstition, corruption, and robbery embodied in the Government of Transvaal."[9] Self-interest lies behind Gervase's principles. Mabel writes to her friend Céline Dangotte on 4 October 1899:

> Listen. You know that there will probably be war in the Transvaal. Perhaps you *don't* know that that may spell "ruin" for many business men in London, those who are interested in S. African mines for instance. Well, altho' Father has not *all* his business there, still he has *some*. So this worries him & when he comes home he just sits & thinks & does not tell us all the little bits of news like he generally does & every-one is quiet & subdued.[10]

Gervase's article brings much favorable attention. Mabel writes to Céline on 7 April 1900 that her father "has received ever so many letters of thanks & congratulations—we feel quite proud!"[11] Gervase Elwes's position on the Boer War goes against the views of pacifist Moncure Daniel Conway, and in 1903, as a result, Gervase votes Conservative, overturning 40 years of loyalty to Radical and Liberal causes.[12]

---

8. NYPL.
9. R. Gervase Elwes, "The Ethics of the South African War," *South Place Magazine*, 5 (1900), 100–108.
10. NYPL.
11. NYPL.
12. R. Gervase Elwes to the Hon. Secretary of the Conservative Association, printed extract in NYPL.

It is likely that Mabel and Hugh hear Conway's sermons during his last years at South Place. Mabel writes on 6 April 1899 to her friend Céline Dangotte:

> I want to tell you that we hardly *ever* go to what you understand by a church. We belong to a society called "South Place Ethical Society." They have a building which is called a chapel but is not at all like one, at Finsbury in London. On Sundays there is a service—no prayers, no regular sermon. The service begins by singing something, not like the hymns in ordinary churches but poems in which the great idea is of One Supreme Being & of universal love & fellowship *on earth*, without expecting you to believe in a conventional *heaven*. After the singing something is read from some book on religion, not only *Protestant* but Roman Catholic, Buddhism or any other where pure & holy thoughts are contained. Then an anthem is sung (always by professionals, they do it for nothing). Then there is the "lecture" or sermon. Every Sunday it is a different man, or nearly so, any-one who has something worth saying may lecture & it is on all sorts of creeds & religions.[13]

The South Place spirit remains important in Mabel's life; it is possible, for example, that she hears the lectures of John M. Robertson in 1891 on, among others, Thomas Carlyle and John Ruskin.[14] She later writes to her mother from Ghent: "I have been looking through the S. P. Magazines I have. I want to have them bound a little later. I find I have not any year complete (not even with those Cé[line] had). The following nos. are missing. If you find you have them twice, any of them, let me know."[15] George Sarton obtains Conway's book, *My Pilgrimage to the Wise Men of the East* (Boston, 1906), soon after he marries Mabel, and he credits the book with awakening his interest in things Oriental.[16]

13. NYPL.

14. John M. Robertson, *Modern Humanists: Sociological Studies of Carlyle, Mill, Emerson, Arnold, Ruskin, and Spencer* (1891; Port Washington, N.Y., 1968).

15. NYPL, Mabel Elwes to Eleanor Elwes, March 1907.

16. "Scholar's Wife." In 1914 George notes a biography of Conway, "the general thinker and noble pilgrim." George Sarton, "John M. Robertson, *The Life Pilgrimage of Moncure Daniel Conway*," *Isis, 2* (1919), 450. Well before then George has looked into things Oriental. In his agenda on 4 May 1904, he notes Charles Holme's article, "Japanese Flower Painting," *The Studio*, 15 April 1904. George is likely the "G.S." from the University of Ghent who authored a poem, "O'Kama d'Osaka," on pp. 378–9 of the *Almanach de l'Université de Gand* for 1904, published by the Société générale (fédération) des étudiants libéraux. It is a Japanese legend about a woman who suffers the death of her lover. In 1934, however, George writes: "I have read the autobiography of Moncure Conway—two large volumes which I bought in Wondelgem 25 years ago. It is well I did not read them before because there is much in them which I would have been unable to appreciate, my knowledge of America for instance being too rudimentary." George Sarton to Mabel Sarton, 29 August 1934.

Upon returning to London from Spain, Mabel attends art classes at Kensington High School, takes the examination in drawing in light and shade on 30 April 1896, and receives first-class honours; in the Drawing Examination at Girls' Public Day School Company in June 1896, Mabel takes honours in Division III.[17] Later in 1896 the Elweses send Mabel to study French at the Institut Charles de Kerchove, in Ghent.[18] Taking its name from the middle 19th-century liberal mayor of Ghent, Charles de Kerchove de Denterghem, it is a private *école primaire supérieure*, or finishing school, seeking to provide girls with a broad and general education. Mabel boards there as an *élève libre*, that is, a student who does not follow the prescribed curriculum for a diploma. The curriculum includes a full range of humanities and sciences, along with domestic arts and fine arts. The school engaged a number of university professors as lecturers, and among Mabel's teachers are Frédéric Swarts, a pioneer in the study of fluorine compounds who would later teach chemistry to George Sarton at the University of Ghent; Ernest Discailles, who teaches French and Romance literatures to George; and philologist Paul Fredericq. Mabel remembers her teacher of French poetry, Adèle Dupuis, and also Joseph de Smet, the historian of art who introduced Lafcadio Hearn to French readers.[19] When Gervase Elwes loses his fortune in South African mining stocks, Mabel continues at the Institut de Kerchove, from 1900 to 1902, as a *surveillante*, or assistant. She is active in founding the school's alumnae association.[20]

Around 1898 Mabel meets Céline Dangotte, a student at the Institut de Kerchove who is 7 years younger than she. The Dangottes own one of the premier retail establishments in Ghent for domestic wares, "Adolphe Dangotte," which expands later into an ambitious undertaking in interior design, "L'Art décoratif," perhaps inspired by the art-nouveau designs made for the home of the Brussels architect Henry van de Velde.[21] The

17. NYPL.

18. Belated birthday greeting from Mabel to her mother, dated Institut de Kerchove, 13 December 1896. Collection of Susan Sherman, Riverview, Bronx, New York.

19. "Scholar's Wife."

20. The Institut de Kerchove around 1900 is recorded in documents contained in Fonds Vliegende Bladen IE28 Ecole in the University Library of Ghent. The documents are a donation from Paul Fredericq. I am grateful to Head Librarian Sylvia van Peteghem for access to this material.

21. NYPL, Mabel Sarton to Cécile Dangotte, 6 April 1901, on a postcard of the firm of Adolphe Dangotte, advertising porcelain, crystal, bronzes, and works of art, along with silver and tableware; the firm had two locations in Ghent and one in Ostende. University Library of Ghent, Fonds Vliegende Bladen F.1.C. 054.40, for a three-color lithograph (Ghent: M. Meins, 1897) advertising the firm's wares and a small engraved illustration of the firm's establishment on Digue de Brabant in Ghent, 1898. Henry van de Velde, *La Maison moderne* (Paris, 1901). Generally, Anne Pingeot and Robert Hoozee, *Paris-Bruxelles, Bruxelles-Paris: Réalisme, impressionisme, symbolisme, art*

girls become close friends. In letters to Céline, Mabel describes her hopes and feelings. Mabel tells about arriving home after crossing the Channel at night:

> London looked so grey & mysterious. The stars were still shining & a few lamps were lit; the sun rose while I was in the train on the way home. It *was* lovely. There was a fog which gradually lifted like a curtain & which was all lighted up with bright pink by the sun. Father met me at New Eltham station & Brownie [Mabel's dog]—Brownie nearly pulled me to pieces he was so excited & he wont leave me a moment. He seems afraid I shall go away again.

Mabel comments to Céline about Alfred Dreyfus:

> What do I say about Dreyfus? Well you know since you wrote he has got his "free pardon"—free pardon for being the bravest most cruelly treated man of our century. It is for France to go down on her knees & beg a *pardon* of *him*. Oh I *loathe* all those traitorous generals & people.[22]

Mabel spends 1899–1900 in London. Her father, she writes to Céline on 20 January 1900, published a well-received article on copper in a London mining periodical. "He has worked very hard for a Copper Company whose mines are in India. If this Company is a success we might get almost rich again." She sends along quotations from John Ruskin. On 29 January 1900 Mabel tells Céline that she has become a student at the Blackheath School of Arts, and that the headmaster will instruct her in painting. Her intent is to prepare for a career as a miniaturist, painting portraits on thin slices of ivory. The friends visit each other. On 8 April 1900 Mabel feels faint and ill, and her mother sends her to see their physician, Dr Jackson. On 9 April, Mabel thinks that the mining projects of her father are illusory: "I am afraid we are all building sand castles." On 13 April, Dr Jackson reports that Mabel is "not in very good health." Mabel returns to Blackheath in the spring of 1900 and sells her first miniature.[23] She goes once more to the Institut de Kerchove and begins providing drawings to Mme Dangotte for the

---

nouveau (Paris/Antwerp, 1997). May Sarton writes about the Dangotte firm in her *roman à clef, The Bridge of Years* (New York, 1946). In the copy presented to her by May Sarton, Céline Limbosch-Dangotte corrected misapprehensions. On p. 20, Limbosch-Dangotte notes that her mother, a successful shopkeeper, was not insulted for keeping a horse-drawn carriage. Michel Thiery, son of George Sarton's friend of the same name, kindly showed me the annotated book. Karel Blondeel, *Vrouwelijke schilders in Gent (1880–1914), een socio-historische studie* (Licentiaat in de Geschiedenis, University of Ghent, 2003), gives the Dangottes as Leopoldine and Adolf.

22. NYPL, Mabel Elwes to Céline Dangotte, 25 December 1898 and 21 September 1899.

23. NYPL, Mabel Elwes to Céline Dangotte, 14 June 1900.

wares sold in her shop.[24] Mabel writes to Céline from Devonport, late in January, 1901, just after the death of Queen Victoria: "When I was sixteen there was no idea that it would ever be necessary so I was kept at home & allowed to travel with my parents, which is in a way a very good education, but *not* a practical aid to earning one's living."

By 3 October 1901 Mabel reports to her Belgian friend: "Father is terribly pre-occupied about business matters just now and talks of nothing else but." Mabel is very short on money at the end of 1901, and she is selling her paintings and miniatures. Her father is "very depressed over business matters." Mabel wants to devote herself to art studies, she writes to Céline on 11 October 1901, but "It is living the two lives—the home & social one taking far the largest share—that seems to destroy all 'suite' in my work—it is done by snatches." On 19 October, Mabel's father has let her resume art classes in anatomy and, she hopes, in nude drawing. On 5 November, the nude drawing classes rivet her attention.[25]

Mabel writes to Céline on 10 November 1901:

> I sometimes can't help wondering *what will become of us* if this war does not stop soon; you see for two years now, the mines from which Father had most of his income have not been working, so we have been living on his savings, on our "capital"—well, but very soon we shall be at end of all resources, & if he has not got some new work by then, what *shall* we do?

Mabel has violent headaches, but she is finding true joy in her drawing from life: "I feel as if I were only just waking up to the fact that it is one of the greatest joys possible." She continues: "The door is open to me, & I feel that all I have to do for the moment is to *work*." At about this time Mabel's mother writes to Céline: "We are much tired and worried over business matters....Altho' we are bitterly disappointed, we take comfort saying things may be better with us next summer." She continues about Mabel: "We are anxious she should be able to earn something, so that if the worst comes she may be the better prepared for it. Mabel is so good about it all, never complaining because she cannot have & do as other girls. Our dearest daughter is a great comfort to us. We look forward to the time when she will make a name for herself."[26]

Mabel writes to her mother on 20 September 1902 that she has sold a drawing for F 35 and hopes to sell another one, and that she is working all day on miniature portraits. Painting miniature portraits, largely

---

24. NYPL, Mabel Elwes to Céline Dangotte, 27 December 1900.

25. NYPL, Mabel Sarton to Céline Dangotte, 29 Dec 1901: "Did I tell you that a whole roll of my 'nude' drawings—3 different ones—disappeared about a week before the end of the term? I am so vexed because they are the best of that sort wh. I have done, & represented several weeks work."

26. NYPL for the letters, and also in the following paragraph.

of children, although including friends and family, is a signal choice. Miniature portrait-painting, where a likeness is for private possession rather than for public display, flourished beginning late in the 16th century. The craft was associated with natural-history illustration, and women in England like Mary Beale, Catharine Da Costa, Anne Foldsone Mee, and Angelica Kauffmann were prominent in it. The genre declined with the availability of daguerreotypes, but it remained, at least in principle, one way to earn a living. Belgium was recognized as a leading place for miniatures, notably the school of Lewis M. J. S. Moreels in Brussels.[27]

Mabel returns to London in 1903, her father having found employment, and attends courses at the Blackheath School of Arts, where she receives the Sketch Club prize. Céline joins the Elwes family as a paying guest to complete her education in English. Mabel imagines that her career will advance more quickly in Belgium. In 1904 she comes to live in the Dangotte household as an adopted daughter. Beginning in 1905, Mabel attends courses at the Académie des beaux-arts in Ghent. The new director of the academy, Jean Delvin, is in the process of revolutionizing art instruction. He favors bringing art to the middle classes and the working classes, he admits women to day classes, and he begins instruction in *art décoratif*, in the history of ornamentation, and in the technology of materials. He himself teaches painting, and in 1905–06 his pupils include the young Flemish artists Constant Permeke, Albert Servaes, and Frits van den Berghe. At the academy, Mabel meets the sculptor Madeleine van Thorenburg, daughter of a wealthy wine and spirits merchant, and the painter Cécile Cauterman, *née* Boonan.[28] The three friends are among the first women admitted to the academy, and especially to Delvin's classes of drawing from live models.[29] Mabel follows Delvin's highly esteemed lectures in Flemish on history of art, attended also by the Symbolist writer Karel van de Woestijne.[30]

---

27. Baron H. Kervyn de Lettenhove, ed., *Exposition de la miniature, Bruxelles, mars-juillet 1912, catalogue général* (Brussels, 1912); George Charles Williamson, ed. Charles Holme, *Portrait Miniatures* (London, 1910), an introduction published by the journal, *Studio*; Max von Boehn, *Miniatures and Silhouettes*, trans. E. K. Walker (London, 1928).

28. May Sarton, "Mabel Elwes Sarton," in May Sarton, *A World of Light: Portraits and Celebrations* (New York, 1976), pp. 49–68, on pp. 56–7; Simone Bergmans, *Les peintres de la vie profonde: Cécile Cauterman et Rainer Maria Rilke* (Brussels, 1943), in which, however, there is no personal connection between the figures; Constant van Thorenburg-Mestdagh, *Histoire du commerce des vins à Gand, précédé d'une notice sur la maison Ct van Thorenburg-Mestdagh* (Ghent, 1904), with a picture of the family's 3-storey neo-classical mansion facing p. 7.

29. HGS, Madeleine van Thorenburg to George Sarton, 9 June 1951.

30. Johan de Smet, *Laethem-Saint-Martin: Un Siècle d'art flamand 1870–1970* (Tielt, Belgium, 2000), pp. 33–40.

Delvin was instrumental in founding the association of students and alumni of the academy, the Maatschappy van Leerlingen en Oud-Leerlingen der Koninklijke Academie van Schoone Kunsten in Ghent. Among his associates was his former pupil Julius de Praetere, a printer and designer who later became a teacher and friend of Mabel's in Zurich. In the association's *Jaarboek* for 1896–97, for example, Delvin writes in Flemish about the neo-romanesque temple to Charles Darwin, the British Museum (Natural History), in South Kensington. Delvin pledges allegiance to Pre-Raphaelites William Morris, Edward Coley Burne-Jones, and Dante Gabriel Rossetti. De Praetere, also in Flemish, writes about Symbolism, which he identifies as the art of the *Mystiekers*, or mystics.[31]

On 14 August 1906 Mabel's father dies in his office at the age of 65; he leaves virtually no assets.[32] Mabel, then visiting her parents, returns to her refuge at the Dangotte residence, where she continues giving lessons in English and in design. She also paints and studies.[33] Not least among the attractions of Ghent is her membership in a circle of young women with intellectual aspirations, through which she comes into contact with a young science student, George Sarton, to whom on 17 August 1906 she reports her father's death.[34]

*
* *

Mabel arrives in Ghent not only at the beginning of the Symbolist movement among young Flemish artists, in which Julius de Praetere is a leader, but also at a time when the new interior designs of Brussels architects Henry van de Velde and Victor *baron* Horta are attracting attention around the world. Her artistic sensibility is a fusion of both movements, which draw on socialism while depending on the patronage of the bourgeoisie. The Symbolist artists, supported by writers like Nobel-laureate Maurice Maeterlinck, are sympathetic to the trade-union movement among Flemish workers in the cotton and linen factories of Ghent, whose stupefying labor built fortunes for the mill owners.[35] The owners are at-

31. For De Praetere: Piet Boyens, *Flemish Art: Symbolism to Expressionism at Sint-Martens-Latem*, trans Tony Langham and Plym Peeters (Sint-Martens-Latem, 1992), p. 28; Paul Haesaerts, *Laethem-Saint-Martin: Le Village élu de l'art flamand* (Brussels, 1968), pp. 374–5. A copy of the *Jaarboek* is in the University Library at Ghent.

32. *Phoenix*, p. 50. In 1906 Mabel's brother Hugh Elwes, by then an engineer, marries Carmen Martínez in Mexico City. The couple have 2 children and then separate. The Sartons help support Hugh beginning in the 1920s, and George becomes his mainstay at the end of his life, in the 1950s.

33. "Scholar's Wife."

34. Agenda, 17 August 1906.

35. Herman Balthazar, "Genèse de la métropole industrielle," in *Gand: Apologie d'une ville rebelle*, ed. Johan Decavele (Antwerp, 1989), 155–83.

tracted to the designs of Horta, a Freemason and fellow-traveler of the Workers' Party, who is excluded from most public-works construction. Notwithstanding the origins of his income, Horta emphasizes in retrospect: "My friends and I were of the same mind...we were *reds*."[36]

Art and literature, especially in Flemish, evoke the life of workers and peasants, focusing both on mass movements and on traditional handwork.[37] The themes and images find their way into France and Germany, spreading along with Belgian commerce and industry. Flanders, the northern part of Belgium, was the center of the Industrial Revolution on the continent of Europe, and across the 19th century it harbors one of the most volatile brews of capitalists and workers. Wealth remains in the hands of French speakers, and French dominates the kingdom. Social reform inevitably becomes linked to self-determination in the Flemish language. Ethical tension is acute and problematic: Flemish socialists were not above collaborating with invading German armies during the first part of the 20th century.

At the turn of the century, people in Ghent look backward as well as forward. The landscape features splendid medieval structures, notably the *béguinages* or *begijnhofen*, religious communities dedicated to social works. The structures and the ideals that animated them excite the imagination of both Mabel and George. Along with architecture go painting and sculpture. The works of Flemish masters are on display in museums across Flanders, notably in the exquisite medieval town of Bruges, but also in Ghent. The Cathedral of St Baafs or St Bavon, a short walk from George's and Mabel's residences, displays paintings by Jan and Hubert van Eyck and Peter Paul Rubens. Works such as these are an accessible, fixed point of reference for people seeking a new way of viewing the world. The turn of the century saw renewed interest in late-medieval 'primitive' painters, notably those of the Low Countries.[38] George and Mabel often go out of their way to look into these images

36. Françoise Aubry, "Victor Horta and Brussels," in *Art Nouveau 1890–1914*, ed. Paul Greenhalgh (New York, 2000), pp. 275–85, on p. 276. Horta's views on socialism resemble those of the modernist Dutch architect Hendrik Petrus Berlage, who championed low-cost housing for workers while accepting commissions from rich capitalists. Iain Boyd White, "Introduction," in Hendrik Pertrus Berlage, *Thoughts on Style 1886–1909*, trans White and Wim de Wit (Santa Monica, Cal., 1996), pp. 1–81, on pp. 44–6.

37. In some trades, such as lace-making and tailoring, hand-work at home accounted for a substantial portion of production, a point made at the Universal Exhibition at Brussels in 1910. Emile Rossel, ed., *Le Livre d'or de l'Exposition universelle et international de Bruxelles en 1910* (n.p., n.d.), pp. 573–6.

38. For a study by a poet friend of George's and a member of the Laethem St-Martin artistic circle who later taught at the University of Ghent: Karel van de Woestijne, *De Vlaamsche Primiteven: Hoe ze waren te Brugge* (Ghent, 1903); *LM II, s.v.* Van de Woestijne.

from the past, as do the artists and writers of their generation. Flanders in 1900 is a place of both ferment and tradition.

All parts of Flanders are accessible to young George Sarton, and he absorbs them with relatively little guidance. George's father, Alfred Auguste Marie Joseph Sarton, was born in Poperingen in 1845. Alfred studied civil engineering at the University of Ghent and entered into the service of the Belgian railways, rising to become one of its directors and chief engineers.[39] Alfred's father Séverin Bonaventure Sarton had been a civil servant in the Belgian Treasury (May Sarton has him as the collector of city taxes).[40] Séverin was the only Walloon among George's grandparents; the others were Flemings. George's mother, Léonie Euphrasine Marie van Halmé, stemming from a prosperous family in Ostende, was born in 1860. Her family's impressive 5-storey house adjoined the modest 3-storey house of the Symbolist painter James Ensor.[41] Orphaned in childhood, Léonie was raised by a notary uncle, Hippolyte van Sieleghem, in Bruges; Van Sieleghem is George Sarton's greatly admired godfather. Léonie, fluent in German and knowledgeable about English, was a talented pianist.[42] She and the older Alfred marry in Bruges on 2 October 1883. Léonie's brother Carlos van Halmé writes in the late stages of his sister's pregnancy: "The child has not yet arrived, but everything has been taken care of, including the child's care. It will be named Georges or Suzanne. Léonie expects to raise the child in English fashion, that is, without diaper or bonnet."[43] Georges Alfred Léon Sarton is born on 31 August 1884. George's family on both sides are Catholic, but George's father is non-observant and a Freemason, and he is eventually buried without religious service. George is baptized for the sake of his mother, but he never receives Communion and is raised outside organized religion.[44]

George's mother Léonie dies on 6 March 1885, "of a hemorrhage... because she was too modest to call for help."[45] Her only child, George,

39. YMS, L. Motte to George Sarton, 8 November 1909.

40. *Phoenix*, p. 19.

41. YMS, annotated photograph.

42. Léonie interjects in a letter to her brother Carlos, for example, on 18 June 1880: "Dear me, pauvre dommelkoptje que je suis!" (NYPL). Her piano remained closed after her death, although it appears as if George kept it until 1919. When he liquidates his pillaged house after the First World War, George indicates to Mabel on 15 September 1919 that he supposes that she has written to a certain Vits about the piano, presumably his mother's.

43. YMS.

44. NYPL, George Sarton, "Autobiographical sketch prepared at the request of Herr Prof. Dr. Abderhalden, President of the Kaiserl. Leop. Carol. Deutsche Akademie der Naturforscher on August 25, 1935."

45. *Phoenix*, p. 13.

is a year-and-a-half old. Alfred Sarton travels a great deal in his work, and George recounts that he was "left alone and abandoned to the care of servants. Except for the little help which I received from my grandparents and from aunts and uncles, I had practically no familial education." His father watches him from a distance, literally, at dinner.[46] He recalls to Mabel on 4 July 1916 that "Uncle Arthur & Aunt Hélène have taken much care of me when I was young, in a way more than my father did." The house where he was born sat near the main rail line supervised by his father, but he grows up in his father's sombre townhouse located in the shadow of one of Ghent's great medieval churches, St Michel. It is not as grim as might be imagined. "My relatives love me a great deal, but they love me in the wrong way," George writes to Mabel on 6 December 1908, "and I must get beyond their affection." George is close to his grandmother, Anne Thérèse Sarton.[47] In a letter written by 8-year-old George to his grandmother, he remarks that his father informs him that she is not well. He looks forward to seeing her: "I will be really happy, because it has been such a long time, dear Mother, since I kissed you." We see here the salutation of respect, Mother, which George often uses when writing to his own wife and to his wife's mother, even before he married. In his letter, George reports that he dined for the first time with his father, and he mentions other relatives. He also mentions that Henry Verdeyen, a close friend of his, will come to dine on the morrow.[48]

The roots of George's rebellion and non-conformity derive more from an original mind than from emotional deprivation. He writes to Mabel about his childhood on 25 June 1915 in a letter sent from America to England: "There was no bad trick that I did not play. I was good only on Sundays and other legal holydays and on Thursday afternoons, when I called on my grandmother."[49] May Sarton reports that George was

---

46. "Scholar's Wife." It is possible that the servants were illiterate. In 1895, illiteracy in Belgium was about 15%, higher than the 9% in France, and far above the level in Germany, the Netherlands, and Scandinavia. Gilbert Lannoy, *La Renaissance de la communauté scientifique flamande* (M.Sc. thesis, University of Montreal, 1975), pp. 104–5, citing Julius MacLeod. In 1900, MacLeod estimates that a Flemish peasant knows 1200 words, or 600 concepts.

47. Possibly it is she whom he credits with remarking that 1811 was an exceptionally good *cuvée* for wine. Sarton, in *Isis, 8* (1926), 200. NYPL, George Sarton, "Autobiographical sketch prepared at the request of Herr Prof. Dr. Abderhalden," for the influence of his paternal grandmother.

48. NYPL, George Sarton to Anne Thérèse Sarton, 7 November 1892; NYPL, George Sarton to Mrs R. G. Elwes, 14 March 1911. In his agenda for 17 March 1904, George writes, "Entirely Henry Verdeyen."

49. Significantly, an early notebook of George's from 1903 (in HGS) bears the title, "Le Non-conformiste."

an unruly pupil in elementary school and into secondary school; we read George's plea for overlooking the vigorous spirit of young people, 20 years later, as a personal expiation: "One can never be too kind to the young; one can never be too tolerant of their faults, even of their intolerance."[50]

George attends the *athénées*, or state-supported classical schools in Ghent and Chimay, the latter from 1897 to 1902.[51] Chimay, tracing its pedigree to a religious college in the seventeenth century, was resuscitated during the Napoleonic occupation as a secondary school with Latin but not Greek, and it grew slowly under the United Kingdom of the Netherlands. It rose to prosperity when mathematician Adolphe Piret became director in 1852. By 1879 the *collège* inaugurated a scientific section. Two years later, it became a *royal*, secondary school, supported in part by the ministries of the Interior and Public Instruction. By way of contrast to classical secondary schools in Germany, France, and England, Belgian *athénées* featured tracks for humanities and for commercial and industrial studies; modern languages and natural sciences occupied as much time in the curriculum as Latin, Greek, and French. By the time that George attended Chimay, enrollment there was about 200 under the inspired leadership of a poet, Emile Valentin. George is most attracted to literary studies, in 1901 beginning a 3-act philosophical drama in German.[52] In his last year at Chimay, he does best in Latin, Greek, and French, less well in German and mathematics, and honorable mention in English, physics, and chemistry.[53] Although he does not become a *lauréat* in the national competitions, he is considered one of the school's successes.[54]

Late in the summer of 1902, George Sarton begins university at Ghent in philosophy, perhaps with a view to studying law. He takes the

---

50. *Phoenix*, pp. 15–17; George Sarton, "Evariste Galois,"*Scientific Monthly, 13* (1921), 363–75, p. 375.

51. A. M. Simon-Van der Meersch, *George Sarton's European Roots: Introduction to His Life and Times* [Exhibition organized by the Archives of the State University of Ghent and the Department of Communication and Cognition] (Ghent, 1984), p. 11. Chimay is in the south near the French border, one of the places in Belgium most remote from Ghent. It is possible that a *contretemps* with the director at the *athénée* at Ghent was behind the mutation. George's Aunt Elisa Sarton, sister of Arthur, Jules, and George's father and known as Mother Marie d'Agréda of the Société de Marie-Réparatrice, writes to him from Strasbourg on 12 March, around 1900: "I regretted that, almost at the end of your studies, you got into trouble with your Director." She comments that George reports that he wants to become a lawyer. NYPL.

52. YMS.

53. NYPL, diploma of George Sarton, 31 July 1902.

54. Jean Behen, *L'Athénée royal de Chimay: Notice historique* (Chimay, 1929), p. 49 for the curriculum, p. 68 for the apotheosis, p. 70 for his absence among the *lauréats*.

candidature examination in letters, in October 1902 and October 1903.[55] He enjoys famous teachers in the humanities, among them historians Henri Pirenne and Paul Fredericq, philologist Jozef Frederik Vercoullie, Buddhist scholar Louis de la Vallée Poussin, and Joseph Bidez, the last a classicist who turns to catalogue Greek alchemical texts during the 1920s, but after a year and a half, George grows disenchanted with philosophy and decides to study science.[56] On 18 February 1904 George notes in his journal that he will take science courses in the winter term "to connect better with life." His spirit, however, remains literary and humanistic. In 1903, according to his diary, he is working on a book, "Le Royaume des certitudes," as well as an introduction to metaphysics based on aesthetics. Although George plays no musical instrument (his mother's piano remains locked shut), he attends concerts whenever he can; during his early years in Cambridge, he purchases a record player, and recorded music becomes a solace for the rest of his life. At the winter concert at Ghent in 1903, he hears Pablo Casals, who, he writes in his diary, "gives a cello almost a human voice."

The theatre strongly attracts George. Early in 1903 he speaks to the Cercle des étudiants rationalistes on "theatre from the social point of view." According to the newspaper account of his talk, he begins by citing Octave Mirbeau: "As poor as he may be, a man doesn't live only for bread. He has a right to beauty, just as rich people do." Metaphysics seeks to make the leading ideas of intelligence common property; morality seeks to establish a common will, "'social' synergy." George summarizes the late Jean-Marie Guyau's *Art au point de vue sociologique* (Paris, 1901) and criticizes Hippolyte Taine's notions on art.[57] Art is, for George, the only common ground among people, for it establishes a unity of sentiment that goes beyond ideas. Art is also what links each person to the universe as a whole. George explores the history of drama, both from the reports of anthropologists on savage peoples and from Friedrich Nietzsche's account of the "glorious tragedy of Eros." Theatre

---

55. Simon-Van der Meersch, *George Sarton's European Roots*, p. 13. In the 4 Belgian universities before the First World War, academic progress is measured in terms of success on examinations. In each faculty, there are 2 *candidature* examinations, the second one corresponding roughly to a North-American bachelor's degree; there are subsequently 2 examinations paving the way to defense of a doctoral dissertation. According to the list of George's examinations provided in Van Oye, p. 157, George takes the first *candidature* twice in letters and philosophy, in 1902 and in 1903.

56. George first reads a scientific text in Greek—Hippocrates on epilepsy—in an anthology of Ulrich von Wilamowitz-Moellendorf's used by Bidez. George Sarton, *A History of Science: Ancient Science through the Golden Age of Greece* (Cambridge, Mass., 1952), p. xiv. Joseph Bidez, et al., eds, *Catalogue des manuscrits alchimiques grecs, 1: Les "Parisini"; Les Manuscrits des "Coeranides"* (Brussels, 1924).

57. Possibly, Taine, *Philosophie de l'art* (Paris, 1903), for a contemporary edition.

must appeal to the masses. It must be free or inexpensive. It should be supported by the government, traveling from place to place and financed by the host city.[58] In fact, George wants to organize a league for artistic propaganda to take charge of bringing art to the people.[59] George seeks to discipline his senses. In homage to Maurice Maeterlinck, George writes about an orchid exhibit. He is transported by the fragrances, and he has little patience with the orchid expert who is guiding him: "Doesn't science give rise to frequent perversions in aesthetic sense?" Technicians and experts, those seeking to create a new hybrid orchid, for example, are often cut off from the primal beauty of nature. Orchids give a model of the only nobility that counts—that of beauty."[60] Present in his earliest publications are themes that will reappear in his thoughts over the coming decade.

George writes in his earliest surviving diary, in 1903, that among his heros are Friedrich Nietzsche, Hippolyte Taine, Cesare Lombroso, Thomas Carlyle, and Richard Wagner. On 18 July 1903, the *Revue bleue* in Paris publishes George's thoughts on Wagnerian literature. In one of his very first serious writings, George demonstrates a predilection for bibliography and name dropping. He begins by observing that the German catalogue of publications about Wagner runs to about 10,000 entries, and it has been said that there are many omissions. Notwithstanding this enormous literature, George thinks that few people have seriously meditated on Wagner's theories and on the evolution of his thought. Thirteen critics, in his view, "form a sort of guard of honor that surrounds the temple and admits people into it." The list includes philosopher Hugo Dingler, composer Franz Liszt, and racist polemicist Houston Stewart Chamberlain. More names follow in profusion, poets and writers who have written on Wagner. George introduces a second theme that will occupy him over the coming decades: racial characteristics. He disputes Nietzsche's notion that Wagner is not a true German. From George's reading of Alfred Fouillée's *Esquisse psychologique des peuples européens* (Paris, 1903), "I see in Wagner all the traits of the German character, but alongside of these, the traits of genius, which have no nationality." Here is a third theme that will occupy George over the coming years: the nature of genius. Wagner's genius is cosmopolitan and, moreover, supremely optimistic. He concludes with a fourth theme that he carries to the end of his life, paying homage to great creators. He quotes Carlyle about the radiant illumination that comes

58. [George Sarton], "Cercle des étudiants rationalistes," *Journal de Gand*, First February 1903, p. 2.

59. George Sarton, "Le Théâtre gratuit," *Journal de Gand*, 5 February 1903, p. 2.

60. George Sarton, "Suggestion d'orchidées," *Journal de Gand*, 24 April 1903, p. 2.

from great men, adding: "More than ever, in this century of materialism, we must evoke the heroes of history and worship them. Their genius, which scholars cannot explain, must be the object of our veneration. Great men are the tripods of God"—that is to say, they support Him as if He were a Delphic oracle—"and the miracles of humanity."[61] In the *Revue bleue*, on 17 November 1904, under the pseudonym Dominique de Bray,[62] George writes about music. Hearing the symphony in D by César Franck evokes in him the same emotion as walking in the Flemish countryside around Ghent. The music of people and the music of God both derive from the same source of rhythm. Desire and music are the same thing, the young aesthete writes. Our soul is a Sleeping Beauty, to be awakened by the composer Prince. He cites Carlyle's opinion that a musical thought can be expressed only by someone who has explored deep mysteries; all deep thoughts are melodious.[63] Thomas Carlyle, the friend of Moncure Conway's, remains with George as a vital presence for over half a century.[64]

Wagner continues to draw George's attention. He comments later in 1903 on Wagner and on Beethoven. The heart is the only reliable critic, he begins, the heart is the soul of criticism. Critics pure of heart are Romain Rolland for Beethoven and André Suarès for Wagner, although Romain Rolland is poetic while Suarès is cerebral. Beethoven was supremely confident of his own ability; he suffered for the goddess of glory. Wagner waited longer for glory than Beethoven did, but his will was no less firm. Adversity brought out the best in both men, for as Benjamin Franklin observed, it is with men as with aromatic plants, "the more they are pounded, the more they exude their perfume."[65] George complements his writings on serious composers by commenting on Henri Bergson's notion of comedy. For Bergson, comedy is intrinsically human, it exists

---

61. George Sarton, "La Littérature wagnérienne," *Revue bleue, 20* (1903), 93–6.

62. Simon-Van der Meersch, *George Sarton's European Roots*, p. 29, following May Sarton, observes that George chose the name—which he uses repeatedly—from that of the hero of Eugène Fromentin's novel, *Dominique* (Paris, 1862). Fromentin's Dominique de Bray, an orphan, cannot marry the woman he loves; he resigns himself to a life in the countryside. The novel is celebrated for its psychological insight. Barbara Wright, *Eugène Fromentin, Dominique* (Glasgow, 2002), for a concise analysis. YMS, George Sarton to May Sarton, 21 June 1942, where George comments that reading *Dominique*, in his youth, calmed him and allowed him to dedicate himself to serious work.

63. Dominique de Bray [George Sarton], "Songerie sur la musique," *Revue bleue, 2* (1904), 380–4. George writes his pseudonym in his agenda on 15 March 1903.

64. Moncure Conway, *Thomas Carlyle* (New York, 1881), where Conway describes that his friendship with Carlyle began in 1863, through an introduction by Ralph Waldo Emerson.

65. George Sarton, "Beethoven et Wagner," *Flandre libérale*, 21 and 22 December 1903. Romain Rolland, *Beethoven* (Paris, 1903); André Suarès, *Wagner* (Paris, 1899).

outside emotion, and it carries social meaning. Bergson studies laughter at the expense of the person who laughs. Laughter occurs when our attention drops from a high topic to a lower one, or when the flesh gains on the soul. It is "the oil that lubricates the wheels of life." George concludes by recalling his arrival in Ghent during Mardi Gras celebrations, which he did not share in. He fled to his home, to be surrounded by books and engravings, by a beauty that had nothing to do with laughter. He implies that intellect and reason are as good as revelry.[66] George, from this time forward, has trouble with festivals. Bergson likes the review, as does George's classics professor Ernest Discailles.[67]

In November 1903 George visits the artistic Salon in Brussels. He is transported by the miniature portraits of Mary Lyttleton-Wyatt, and also by the "Head of John the Baptist" by Gustave van de Woestyne, one of the painters from the village of Laethem St-Martin, near Ghent. If he had the money, George would have purchased the painting. George knows Gustave's brother, the poet and writer Karel van de Woestijne. George also knows about the artists' colony in Laethem. He is impressed by Ferdinand Willaert's paintings at the Salon.[68] George will rediscover both the miniatures and the Laethem circle in Mabel Elwes. On 24 February 1904 he records in his diary the first part of Philippe Gaspard Gauckler's *Le Beau et son histoire* (Paris, 1873), the thoughts of a civil engineer on beauty, which he finds interesting for Gauckler's notions about "the universality of rhythm." He reads novels and a study of Gabriele d'Annunzio. He walks in the Flemish countryside. On 23 and 24 February he has read Eugène Fromentin's *Maîtres d'autrefois: Belgique-Holland* for the third time (the 13th edition appeared in 1904); it is a fine work of art criticism.[69] On 25 and 26 February he records reading the 21st volume of Romain Rolland's *Jean Christophe* (about which in a decade he will publish an essay) and also Francis Jammes's *Clara d'Ellébuse* (Paris, 1899); he writes to both authors.[70] He confides to his diary on 21 April 1904 that Jammes will be his guide. Over the next 8 years he returns several times to Jammes, a French author of bucolic spiritual novels.

In 1904, George's mind is ablaze with ideas. On 28 February he observes in his diary: "I believe that men can be divided into two main categories: those who have a unitary character and those who do not. A

66. George Sarton, "La Signification du comique," *Flandre libérale*, 30 November 1903.

67. HGS, Henri Bergson to George Sarton, 12 December 1903; Ernest Discailles to George Sarton, 12 September 1903.

68. Diary, 26 November 1903.

69. On Fromentin the artist, art critic, and novelist, Barbara Wright, *Eugène Fromentin: A Life in Art and Letters* (Berne, 2000).

70. Romain Rolland is amused with George's correspondence. Romain Rolland to Charles Péguy, 1904, cited in Simon-Van der Meersch, *George Sarton's European Roots*, p. 37.

chasm separates these two classes: the *unitary* is the restless man, the other is the quiet one." On 7 and 8 March 1904 he records afternoons with the landscape painter Ferdinand Willaert, director of the Académie de Termonde, whose painting impressed him in 1903 and about whom he wants to write.[71] On First March he reads a review of Thomas Carlyle. On 13 April, he extracts from Charles de Secondat *baron* de Montesquieu's *Persian Letters*; on 22 April he has finished Honoré de Balzac's *Curé de village*, which "dissipates all, or rather many of my criticisms against this *colossal man*." On 12 May 1904: "I have only one aim in life: each day to understand better, i.e., to *perceive* [sentir] better and better."

Writing is the path forward. At the invitation of Ernest Discailles, George publishes a short essay, *Songerie no. VII, notamment sur le bonheur et sur la gloire*, in the *Almanach des Etudiants libéraux de Gand*, and also as a separatum in 100 copies. The text is a meditation on Maeterlinck's *Sagesse et la destinée* (Paris, 1898). Happiness is a continual theme in Maeterlinck's writing, he contends. Happiness is accepting the world, not rebelling against it. The best of glory is posthumous, the reward of the genius that lies behind wisdom. Everything is vanity except for wisdom and glory. Our duty is to be ourselves, to continually improve, and in this way leave something behind after death. Here is already the guiding idea of the association he creates, Reiner Leven.[72] Karel van de Woestijne is also pleased with the *Songerie*.[73] At about this time, George considers abandoning university studies to devote himself to writing. His correspondents do their best to dissuade him. George's Uncle Arthur replies to a copy of George's essay with words of caution about false praise.[74] George's professor of classics at Chimay, Alphonse Cajot, is more circumspect about the *Songerie*. He urges George to continue with university studies, for time spent studying for examinations is not wasted: "Before being *someone*, be *something*!" An engineer, a doctor. Cajot cites artists and writers who have university diplomas. The diploma, in Cajot's view, is a passport in life.[75] Ernest Discailles, a friend of George's father, also urges George to continue with his studies.[76]

71. Haesaerts, *Laethem-Saint-Martin*, p. 401.

72. George Sarton, "Sur le bonheur et sur la gloire," Société générale des étudiants liberaux, *Almanach* (Ghent, 1904), pp. 182–98. HGS, Ernest Discailles to George Sarton, 19 January 1904, for the invitation. On Discailles: A. Breuer, ed., *Nos contemporains* (Brussels, 1904), pp. 216–18.

73. Karel van de Woestijne to George Sarton, thanking George for the *Almanach* version of the text, in Simon-Van der Meersch, *George Sarton's European Roots*, p. 29. The letter is dated 1903.

74. NYPL, Arthur Sarton to George Sarton, 29 March 1904.

75. HGS, Alphonse Cajot to George Sarton, 6 April 1904.

76. HGS, Ernest Discailles to George Sarton, 12 September 1904.

In September 1904 George completes the *Petite Amie*, also titled *Lettres d'un enfant sur la sagesse et la vie*, a work dedicated to Goethe.[77] It remains in manuscript, prose of youthful confusion about the large themes of life, soul, and desire, the narrator untroubled by money. Many thoughts and sentences reappear in his correspondence and in his early publications: Desire is "the conducting wire for the energy of the world." The narrator's favorite authors: Blaise Pascal, Ernest Hello, Eugène Fromentin (*Dominique*), Maurice Maeterlinck (*Sagesse et la destinée*), and Thomas à Kempis (*Imitation of Christ*). His motto: *Als ik kan*—roughly, such as I can, or I do my best—is borrowed from the motto signed by Jan van Eyck on several of his paintings. He approves of making haste slowly, and he thinks that the natural world is the source of all inspiration. "Nature, my dear friend, is worth more to me than all our abstractions."[78] The *Petite Amie* is not the only project that George does not complete. Publisher Félix Alcan rejects George's proposal to translate the *Onomastikon* of Julius Pollux, which he has never heard of, and advises George to establish a name before sending along a manuscript.[79]

By 1904 George has dedicated himself to nature. He takes the first and second *candidature* in natural sciences.[80] He belongs to the botanical society Kruidkundig Genootschap Dodonae, organized by Julius MacLeod, the Flemish-activist professor of botany, who in the years 1900–1904 believes, following Auguste Comte, that science points the path forward for social reform. MacLeod, in a publication of 1902 on "Science and History," is inspired by the German historian Karl Lamprecht to recommend that historians borrow the methods of scientists, notably biologists. George is attracted to MacLeod, who provides a preface to the Dutch edition (1904) of Peter Kropotkin's *Mutual Aid*, translated by MacLeod's wife, Fanny Maertens.[81] When George passes

77. A draft is located in George's agenda.

78. "Petite Amie," rough and fair copies inserted in Diary. The motto is also William Morris's, "If I can." Constance Howard, *Design for Embroidery from Traditional English Sources* (London, 1956), p. 95.

79. HGS, Félix Alcan to George Sarton, 7 September 1902.

80. Agenda, 13 July 1905 for the first part of the examination.

81. J. Gillis, *Paul Mansion en George Sarton* (Brussels, 1973) [Koninklijke Vlaamse Academie voor Wetenschappen, Letteren en Schone Kunsten van België, Klasse der Wetenschappen, Mededelingen, 35, no. 2], p. 9. Lannoy, *La Renaissance de la communauté scientifique flamande*, pp. 79–123, on Julius Macleod. Evidence from George's correspondence, diary, and agenda, appearing presently, suggests that MacLeod's son Andries MacLeod, a friend of George's, is inaccurate when he affirms: "It is true that my father exerted no influence on Sarton." Andries MacLeod, "Hoe is George Sarton ertoe gekomen zich met de geschiedenis der wetenschappen bezig te houden," *Scientiarum historia*, 1964, pp. 203–8.

the second examination in natural sciences, Ernest Discailles observes that he will be a scientist who is sensitive to poetry, but George continues to see himself more as a poet than a scientist.[82] The observation is accurate, for George's novella, *Une Vie de poète*, or *Poet's Life*, revised in December 1904, appears with the Ghent printer A. Vandeweghe. It bears an epigram from Eugène Fromentin's *Dominique*.

George's poet realizes, in his early twenties, that the course of human development sharpened primitive desire into gross and vile goals. The poet seeks to remain pure in his desire for beauty and justice. He swears that he will die poor. The poet sees life "as the accomplishment of a duty and the disinterested advancement toward an "ideal." He takes William the Silent's motto as his own: "To begin, I do not need to hope, nor success to persevere."[83] He does not care about bourgeois opinions. He "will write in the solitude of the pages in which I have seriously reflected." A great writer advises the poet:

> Do you know what a poet is? He's a gift that the heavens sometimes bestow on the earth. The poet is the supremely natural man [l'homme simple par excellence]. He's someone who feels, who loves, who suffers, and who has a precise and profound awareness of his love and his suffering. The poet is someone who loves completely; he's someone who gives himself. If there's a poet in you, rend your wounds with your nails and style, *your* style, will come out naturally from your veins, like water from a spring.

The poet decides that good writing does not need to be a torrent of words. He develops strong antipathy to the literary world. The poet falls in love and thinks that the only book worth reading is the book of life. The poet isolates himself with his family. He dies young, in obscurity.[84]

George sends out copies of the novella (to judge from chronology, he circulates it as proofs or manuscript), and he receives comments in return. The editor of the newspaper *La Flandre libérale*, Gustave Abel, sees it as a work of "social mysticism" that discusses large truths but

82. Simon-Van der Meersch, *George Sarton's European Roots*, p. 15; HGS, Ernest Discailles to George Sarton, 18 July 1906 for the second examination.

83. George writes on 24 March 1905 to Raymond Limbosch: "You would easily be able to find that William of Orange never spoke the words that I attribute to him." Van Oye, p. 37. Paul Fredericq, to whom George sends a copy of the novel, asks him about the source for the motto. George replies that it was in a book of Romain Rolland's. In subsequent correspondence, Romain Rolland cannot verify the matter. Paul van Oye, *In Memoriam Professor Dr G. Sarton, separatum* sent to May Sarton, 10 April 1957, pp. 234–40, on p. 238. YMS.

84. George Sarton, *Une Vie de poète* (Ghent, 1904), quotations on pp. 27, 11, 12, 15, 21, 27, 30.

offers no precise idea about them. It is all very well to speak about creating men for the society of tomorrow, but "there is more rhetoric than frankness in the social movement." Abel reads mention of God, but he wonders how the characters view the Deity; he thinks that the characters believe in a religion of beauty.[85] Childhood friend Henry Verdeyen thinks that it contains "profoundly sad ideas in an exquisite form." In Verdeyen's view, the novella hides a dark shadow: At the time of his death, the poet protagonist leaves a mistress and bastards who, following social convention, cannot carry his name.[86] The French biologist and historian of science Félix Le Dantec disapproves of the poet's "regrettable scorn for humankind"; in Le Dantec's view, scorn is inappropriate if one withdraws from society.[87] Beginning around 1905, George corresponds with Emile Masson, the political activist and translator of Thomas Carlyle from Pontivy in Brittany. Masson is struck by many of George's expressions in the novella, such as George's opposition to the advice of Professor Cajot: "I believe firmly that it is better to remain someone than to become something," which Masson expressed in his diary as: "The requirement to be something prevents one from being someone." George's book recalls Jean-Marie Guyau's *Irréligion de l'avenir*, "one of the most beautiful books that has ever been written or will ever be written."[88]

George's family has mixed reactions about the novella. Uncle Jules Sarton writes: "The book is above all for young people, and I would like to see it diffused widely." He promises to pass the book along to a friend with the hope for a review in a newspaper. But Jules thinks that the organization and the aim of the book are not clear enough. George does not yet appreciate Montesquieu and Plutarch, giving himself over entirely to Schopenhauer. His old uncle advises him: "Keep your ideal, cultivate

85. HGS, Gustave Abel to George Sarton, 13 July 1904; Chrysale, "Gustave Abel," *La Flamme* (Brussels), 4 January 1946.

86. HGS, Henry Verdeyen to George Sarton, Diary, 14 March 1904, for Verdeyen's death.

87. HGS, Félix Le Dantec to George Sarton, 28 May [1905].

88. HGS, Emile Masson to George Sarton, 19 March 1905. Jean-Marie Guyau, translator of the Stoic philosopher Epictetus and step-son of Alfred Fouillée, is known for *L'Irréligion de l'avenir* (Paris, 1887), heavily annotated by Nietzsche, and the posthumous *Art au point de vue sociologique*, admired by Tolstoy. *Irréligion* is by 1906 into an 11th edition. Half a year later, Masson writes to George repeating George's surprising comment that he knows "only citations" from Carlyle; Masson will send George his published collection of Carlyle's writings in French, if he has not already done so. HGS, Emile Masson to George Sarton, 18 July 1905 (the author is incorrectly identified as Père Yves Ernest Masson). Lewis Pyenson, "Ecrire et agir: Emile Masson, ami et correspondant de George Sarton," trans. Ph. Gury and S. Moal, in *Emile Masson, prophète et rebelle*, ed. J.-Didier Giraud and Marielle Giraud (Rennes, 2005), pp. 173–89.

your mind and above all your heart, be generous, devout, but never let yourself be impregnated with a distaste for the world." Instead of filling his mind with "ideas of revolt, replace it with pure, observational science, and especially work to establish a just balance between the mind and the flesh [le mens et la bête]."[89] Uncle Arthur Sarton sends along a detailed review. He thinks that the book is too intemperate in its criticism of rich people. The poet should have been led, by the logic of his reasoning, to action; instead he retreats from society. The poet is just the person whom Plato would have excluded from his Republic. But he concludes that the book is "well constructed, the parts are rigorously organized in it, and the author shows his talent." The author, Arthur suggests, should not retreat from life's realities and social obligations; he should avoid pessimism; he should not have, as professors have, "an immoderate love for the new, the unusual, and the unexpected." And he should mainly be concerned with using his talents "in the service of a just, useful, and productive cause."[90]

George's friend Raymond Limbosch also comments on the life of a poet. George receives the comments in a friendly spirit on 24 March 1905. He returns his own observations on Alexandra Myrial's anarchist tract, *Pour la vie* (Paris, 1899).[91] He remains unenthusiastic about it. Myrial seems to deny the goodness of people working for a common end. Myrial thinks that the best way of combatting absolutism is to ignore it, to declare it meaningless. In George's view, this is too positivistic. He prefers to approach the absolute with logic and mathematics. Whereas Myrial "is a positivist, I am *above all* a metaphysician." What pleases George most about *Une vie de poète* is its metaphysical, that is to say its religious, undercurrent. "Leave me my metaphysics, which is my happiness."[92] From the time of his young manhood, George engages positivism, although here we see that he does so in the spirit of a critic rather than a follower. George counsels Raymond Limbosch on 6 June 1905: "Keep pure, stay beautiful, remain happy." George struggles with all 3 ideals.[93]

Achieving purity is, for George, a worldly struggle. He recalls that the Dreyfus case made him aware of politics when he was about 14 years old. His own family divided over it. "My political awakening was the realization that justice cannot be obtained unless people are willing to fight for it and to risk themselves for it."[94] George lays awake on 17 October 1904

89. YMS, Jules Sarton to George Sarton, 14 March 1904. Previously, on 14 January 1904, George writes in his agenda: "I will no longer speak with my Uncle Jules."

90. NYPL.

91. Alexandra David used the pseudonym Alexandra Myrial. She married Philippe Néel in 1904.

92. Van Oye, pp. 37–9.

93. Van Oye, p. 41.

94. Diary, 28 February 1953.

thinking about "the horrors of the Russo-Japanese War, the horrors inherent in all wars....What is to be done to avoid such stupidity! Antimilitarist propaganda is the only remedy!" He continues: "So we must organize individual resistance against organized murder!" He imagines an international union for maintaining the peace, along with a pacifist oath taken by the members. As an epigraph on membership cards, he proposes: "Men are not born to kill each other *en masse* in the name of any sort of principle! But to love each other and to help each other."[95] George reaffirms his faith in socialism: "I have become once more, after two years of anxiety and illness, a socialist. I will do all I can for the socialist cause. *Because I believe that opinions not translated into acts are dead opinions, 'useless.'*" George sees socialism as a stage on the route to anarchism, for which people are not ready. "Socialism will lead them there, making men good, generous, and just."[96] Even though the bourgeois, liberal notion of universal suffrage "gives as much authority to imbeciles as to philosophers," by 1906 George cannot imagine anything better.[97]

George Sarton is an inveterate organizer, breathing life into a number of academies, societies, and committees. The impulse is associated with a tradition of student organizations in the Low Countries. On 30 December 1902 he drafts a proposal for establishing a Philosophical Society at Ghent.[98] Membership would cost F 20, and new members could enter according to a vote following a lecture to members. The society would have a bulletin for publishing the talks of members. Its library would be constituted by material received in exchange for its bulletin. George joins the liberal, anti-clerical student association 't Zal Wel Gaan (It Will Be Fine). The association makes a study trip to the Netherlands, where George sees Egyptian art for the first time.[99] One of the group's members, Irénée van der Ghinst, the son of a physician in Bruges, urges George to join him in forming the more radical association, Ter Waerheid (Toward the Truth), where George meets Henri de Man, the future leader of Belgian socialists and regretted collaborator of the Nazis, who had founded a socialist students' organization in 1905, and also the writer Paul Kenis.[100]

95. Diary 18 October 1904.

96. Diary, 18 January 1905.

97. HGS, Henri Fuss to George Sarton, 17 June [1906]. Fuss, an anarchist and a childhood acquaintance of Raymond Limbosch's, writes from prison and cites a letter of George's.

98. YMS, "Projet de fondation d'une Société de Philosophie de Gand."

99. Agenda, 27 September 1905.

100. Simon-Van der Meersch, *George Sarton's European Roots*, p. 19. Agenda, 28 September 1904, where George notes a letter he sends to 't Zal Wel Gaan: "Enthusiasm spoils nothing, comrades."

At about this time Ter Waerheid debates with the local socialist leader Edouard Anseele, with whom George had corresponded in 1901, about the best way to transform the University of Ghent into a Flemish institution, a goal of Flemish nationalists.[101] George recalls in his biography of Mabel:

> Anseele tried to direct my efforts in the direction of labor legislation, but he had misjudged me completely, for I was not a lawyer (or legal minded) and never had political ambitions. He believed that the "new society" would be created by the promulgation of better laws and that was true to some extent, and that was only one part of the story.

George comes progressively to embrace Anseele's reformist perspective. On 19 May 1907 he writes to Raymond Limbosch, an anarchist, that he himself is a social democrat, "that I have always been a *friend of laws*, and a *traditionalist* in all senses of the word. It is furthermore not a paradox to say that revolutionaries, the 'subversives,' are not those who want justice and truth but those who live in iniquity."[102]

From May through July 1905, George attends meetings at the Vooruit socialist cooperative in Ghent.[103] He lectures to striking workers in the flax industry "about past struggles of different peoples."[104] He hears debates on the question of Flemish liberation, and he takes a course in the Flemish language; his guide is Julius MacLeod.[105] People who were close to George Sarton recall that at a meeting organized by Anseele in favor of striking cotton workers, George meets Raymond Limbosch, anarchist son of a wealthy fashion designer in Brussels, who had taken a diploma as *agrégé* from the Ecole d'agriculture in Montargis, France, and then worked in the De Smet horticulture business.[106] In fact George meets Limbosch at the end of 1904 in the context of a lecture by the

---

101. Simon-Van der Meersch, *George Sarton's European Roots*, p. 39, for an extract from a letter of Anseele's of 1901. HGS, Edouard Anseele to George Sarton, 10 February 1901.

102. Van Oye, p. 47.

103. Carl Strikwerda, *A House Divided: Catholics, Socialists, and Flemish Nationalists in Nineteenth-Century Belgium* (Lanham, Md., 1997), pp. 78–80 on the Vooruit cooperative, whose meeting hall, Ons Huis, opened in 1902.

104. HGS, Jean Samijn to George Sarton, 27 June 1905. Samijn speaks for the Union of Flax Workers.

105. Agenda, 8 May 1905, 18 and 22 July 1905.

106. In his doctoral dissertation, *Een Sociale Geschiedenis van het schrijverschap tijdens de Belgische belle époque* (diss., University of Ghent, 2006), Christophe Verbruggen emphasizes the radical heritage of the Montargis school (where both Deng Xiaoping and Pol Pot studied).

anarchist Georges Thonar. In any case, George is captivated by this intellectual laborer.[107]

In September 1905, George along with Van der Ghinst and Limbosch create a socialist-leaning discussion group in philosophy, literature, and art, Reiner Leven (Purer Living). Reiner Leven takes its motto from the historian Jules Michelet, "To be strong, be pure," and George is the group's secretary.[108] Their sentiment inclines toward vegetarianism and teetotaling without, however, requiring adherence to these norms. The founders want especially to discourage prostitution and its patronage, and in their discussions they seek purity with Platonic zeal.[109]

When the group advertise for members in a local newspaper, Reiner Leven come into contact with a women's organization, the Flinke, or Bold Ones—young women studying at various trade and professional schools.[110] (At the time, Ghent was the center of feminist socialism in Belgium, symbolized by the cotton spinner and unmarried mother Emilie Claeys, who was prominent on the national level.[111]) Marthe Patyn, secretary of the Dangotte establishment, is the Flinke's organizer. Among its members is Véra Tordeur, the first woman student in mathematics and physics at the University of Ghent, who marries the engineer François

---

107. YMS, Céline Limbosch to May Sarton, 1957; Van Oye, p. 23; *Phoenix*, p. 42 for the Limbosch mansion in Rhode St-Genèse, a suburb of Brussels. Verbruggen, *Sociale Geschiedenis*, for the meeting of George Sarton and Raymond Limbosch in 1904.

108. HGS, Emile Masson to George Sarton, 20 November 1905, commenting "on your Reines Leven, and your motto from Michelet: 'To be strong, be pure.'"

109. In her illuminating discussion of George Sarton's years at Ghent, Denise de Weerdt suggests that Reiner Leven might be inspired by the Dutch movement Rein Leven, founded in 1901 along the lines of Leo Tolstoy's Christian-Anarchist thought, which emphasized toleration for free sexual relationships outside marriage. De Weerdt, *Free Love=Free Marriage? 'Free Love' and the Belgian Socialists, 1880–1940* (Amsterdam, 2000); De Weerdt elaborates on Sarton and Limbosch in "De Sartons en België," *Brood en kozen: Tijdschrift voor de geschiedenis van sociale bewegingen*, 1999, no. 1, pp. 53–71. In the present pages, I am indebted to De Weerdt's scholarship, which discusses Reiner Leven and also George Sarton's socialist activity, but my account departs from De Weerdt's chronology. On the late 19th-century purity movements, largely concerned with prostitution and the age of consent, Natalia Gerodetti, "'Lay Experts': Women's Social Purity Groups and the Politics of Sexuality in Switzerland, 1890–1915," *Women's History Review, 13* (2004), 585–610. Jan Romein, *The Watershed of Two Eras: Europe in 1900*, trans. Arnold Pomerans (Middletown, Conn., 1978), pp. 504–5 for the back-to-nature movement. Verbruggen, *Sociale Geschiedenis*, for the direct inspiration of Rein Leven and also George Sarton's early socialist activity.

110. *Phoenix*, p. 45, where the advertisement, signed by Van der Ghinst, attracted the Flinke because the women's group thought that the signatory, Irénée van der Ghinst, was a woman, by virtue of his first name. Agenda, 21 September 1905, where George notes that he must prepare Reiner Leven.

111. Strikwerda, *House Divided*, pp. 118–19.

Bouny, close to Raymond Limbosch. (The Bounys are life-long friends of the Sartons.) Also a member of the Flinke is Augusta de Taeye, who marries George's schoolteacher friend Michel Thiery in 1911. The feminist group include both intellectuals and working people, and they pride themselves on addressing each other by their family name, in masculine style. Marthe Patyn recruits Mabel Elwes and Céline Dangotte. Reiner Leven and the Flinke find themselves compatible, and in an unusual and egalitarian step, the women are invited to join the men's group while nevertheless retaining their separate institutional identity. The association raises more than eyebrows. Reiner Leven generates strong feelings among traditionalist Catholic students, and in 1906 one observer at a Catholic college in Antwerp, Auguste van Langendonck, condemns the group for its defense of free love. The criticism is typical of Catholic recriminations against socialists in Ghent,[112] and here it is not far off the mark, although the men of Reiner Leven are inclined to patronize the Flinke women. George writes in his agenda on 10 October 1904: "Women often fetter civilization" by seeking stability. Nevertheless, in his second and third manifestos for Reiner Leven, George writes about sharing social responsibility between the sexes.[113] According to George's agenda, Véra Tordeur is elected president of Reiner Leven on 15 May 1906.

George writes Reiner Leven's fourth manifesto, arguing for the 8-hour working day.[114] The text receives favorable comment in a publication of the Solvay Institute; Emile Waxweiler, director of the sociological side of the institute (whose writings provide an epigram for the manifesto and who provided the bibliography that George cites), thinks that George is one of the earliest writers in Belgium to appreciate the plans that Ernest Solvay and he propose for reforming industry.[115] Waxweiler, in sending along comments, nevertheless cautions George about his polemical

112. Strikwerda, *House Divided*, p. 132. Auguste van Langendonck, *Reiner Leven (Vivons plus pûr)* (Antwerp, 1906). Van Langendonck is a professor at the Collège Saint Jean Berchmans, a priest, also. I am grateful to Christophe Verbruggen for lending me a copy of the 24-page pamphlet.

113. Katharine Strelsky, "Bibliography of the Publications of George Sarton," *Isis, 48* (1952), 336–46, on p. 337; the second manifesto of 2 pages, *La Fraternité entre les sexes et l'éducation sociale* (Ghent, 1906), is a commentary on Eva (Emma) Pieczynska's notion of women's liberation through Christianity. Pieczynska, *L'Appel des femmes aux fonctions publiques* (Berne, 1898).

114. Reiner Leven, *La Journée de 8 heures* (Ghent, 1906), lent to me by Christophe Verbruggen.

115. H. Elkhadem, "George Sarton: Ses années de formation et ses réalisation académiques à travers sa correspondance avec Irénée van der Ghinst," *Academiae Analecta, Medelingen van de Koninklijke Academie voor Wetenschappen, Letteren en Schone Kunsten van België, Klasse der Wetenschappen, 47* (1985), 104–25, on p. 110; Simon-Van der Meersch, *George Sarton's European Roots*, pp. 21–3.

style—criticism that George sustains for the next 20 years: "Most of my changes are so as not to alienate a large part of the public: A focused and calm presentation makes the longest-lasting impression."[116] Jean Delville, an academic artist whom George admires, also congratulates George on his manifesto, which George sent along with writings about art (probably *Une Vie de poète*) and a poem. Delville comments on artistic vision, and his words stay with George over the coming decades. Delville contends that the values apparent in a painting reflect the mental state of an artist. "From Velasquez to Leonardo da Vinci, what degrees of mental and aesthetic sensibility!" Velázquez's portraits never penetrate into the soul in the way that Leonardo's portraits do. The distance between the artists is as great as that separating Frans Hals from Rembrandt. "Realists are the artists who are not developed enough and incapable of having an ideal. It is impossible to achieve great art through realism. The proof is that all realists restrict themselves to the limited representation of things, of *objects*." Great artists are philosophers, although a work of art is not philosophy. "Just as Wagner understood the philosophical and psychological value of sound, so idealistic painters understand the plastic and philosophical value of ideas and forms." The idealist sentiment finds an eager reader in George. "Great and good affinities unite us," Delville emphasizes. "You are ready for the study of *theosophy*."[117] Theosophy and its South-Asian roots in Helena Petrovna Blavatsky's infatuation with Swami Dayananda Saraswati will come to George later, but art increasingly draws his attention.[118] He wanders through Laethem St-Martin a number of times in 1905 and early 1906; on 11 April 1906 he visits an exhibit of the Laethem Symbolist artist Léon de Smet.

Politics and art unite George with Emile Masson. He offers condolence when Masson's child is stillborn. He learns that Masson is writing about the married life of the Carlyles, and also about Carlyle's biographer.[119] Emile Masson congratulates George on becoming secretary of Reiner Leven, but cautions: "There are so many illusions of action, meeting to talk, issuing pink, green, or blue cards, proclaiming 'We are the True, the Pure.' You go home proud of your country. I have gone through a bunch of 'revolutions' of this kind."[120] He counsels George, early in 1906, not to

116. HGS, Emile Waxweiler to George Sarton, 30 October 1906; Kaat Wils, "La Sociologie," in *Histoire des sciences en Belgique 1815–2000*, ed. Robert Halleux (Brussels, 2001), *1*, 305–322, Waxweiler on pp. 316–18.

117. HGS, Jean Delville to George Sarton, 20 September 1906.

118. Helena Petrovna Blavatsky's first important book, *Isis Unveiled*, appeared in 1877; later George does not mention it as an inspiration for the title of his own periodical, *Isis*.

119. HGS, Emile Masson to George Sarton, 29 December 1905, 25 February 1906, 7 May 1906.

120. HGS, Emile Masson to George Sarton, 20 November 1905.

move out of his father' house; he congratulates George on the anti-clerical sentiment of Reiner Leven; he is interested in Henri Poincaré's *La Science et l'hypothèse* (Paris, 1906) and Emile Boutroux's *De l'Idée de loi naturelle dans la science et la philosophie contemporaines* (Paris, 1895).[121] Masson describes how he and his wife Elsie toured Scotland, looking for traces of Thomas Carlyle and contemplating Carlyle's letters to his mother. Masson is pleased with George's third manifesto for Reiner Leven, "for which the heart must be in fact, 'The Fraternity of the Sexes.'"[122] Masson fervently hopes that "the good seed germinates on fertile land," but manifestos of this kind are not uncommon. "I would ask for less publicity and always more intimacy." And then he offers what many writers know: "The essence of publishing is hell! 'A white spot where there would be everything,' as the painter [Eugène] Carrière would have it."[123]

In a letter of 1905, Emile Masson describes how he met his wife Elsie. He needed information about John Ruskin for a popular piece he was writing. A woman correspondent in Strasbourg, whom he knew through the Union pour l'action morale, sent his letter to Elsie, then a French teacher in a school at Brighton, who gladly provided the information. "She and I, each of us separately, had, before our meeting, given up on 'happiness' and on selfish concerns. We had conceived of life as a religion of justice, requiring justice, even to the extent of self-condemnation. And we became engaged to fight on together."[124] Masson urges George to wait quietly for the woman of his life. The notion of common struggle is at the base of the Sartons' marriage.

On his own canvas, in his notebook, "Le Non-conformiste," George writes on 7 December 1903: "With women there is something I place above beauty. It is distinction." When he meets the distinctive Mabel Elwes, George is an irrepressible student of natural sciences, a published author, correspondent of Nobel laureates, and an omnivorous reader. He has acquired a firm sense of social justice and a clear code of personal conduct. He is insensitive in personal relations, but it would be a mistake to blame this character failing—as he, his wife, and his daughter subsequently affirm—on his having been raised as a motherless child in a loveless home, "as

121. HGS, Emile Masson to George Sarton, 7 May 1906.

122. The third manifesto extends the second manifesto, titled *La Fraternité entre les sexes et l'éducation sociale* (Ghent, 1906). The third manifesto of 13 pages, written by George but signed "A Christian," is titled *Au nom de: L'Education, la morale, la pureté, l'idéale (Histoire vraie)* (Ghent, 1906), and is a reply to the criticisms of Auguste van Langendonck. Christophe Verbruggen kindly lent me a copy of the texts.

123. HGS, Emile Masson to George Sarton, 18 November 1906. Robert James Bantens, *Eugène Carrière: His Work and His Influence* (Ann Arbor, 1983), for a general orientation.

124. HGS, Emile Masson to George Sarton, 2 April 1905.

bourgeois as a Balzac novel."[125] As George himself would have said, had he given disinterested thought to the matter, the counterexamples are numerous and striking. George grows up in an unusually dynamic setting, filled with many of the currents of industry and art that would dominate the 20th century, and he enjoys direct access to the people who animated these currents. The Low Countries, at the turn of the century, were at the crossroads of English, French, and German culture while retaining a special character with the Dutch language. Belgium especially, with tumult and controversy, had planted itself firmly in the Second Industrial Revolution and among the imperialist powers. It would be hard to imagine a better nursery for George Sarton's vision of science and history.

125. *Phoenix*, p. 12, quoted in Arnold Thackray and Robert K. Merton, "On Discipline Building: The Paradoxes of George Sarton," *Isis*, *63* (1972), 473–95, on p. 477.

CHAPTER 2

Purity

The Sartons' correspondence begins with a letter of Mabel's on 11 December 1905, thanking George for the honor of joining his society, Reiner Leven. Mabel notes that she and Céline Dangotte, in whose house she lives, would welcome a course given to their own society by someone with "elevated, healthy, and at the same time practical ideas," someone with deep knowledge and firm beliefs, for example, the author of the book that George had sent them, *De zedelyke Opvoeding onzer kinder*.[1] Did George know a suitable tutor? The pedagogical course could be an introduction to child psychology. Mabel, being English, has not read the Dutch book, but Céline spoke to her about it. Mabel returns several sheets that George left in the book.

Mabel writes again on 20 July 1906, in response to a letter that George sent to Céline, where George asks for advice in approaching a woman, Mélanie Lorein, a member of the Flinke. Mabel offers that they know each other sufficiently well so that he can receive her opinion that Mélanie is, in Mabel's view, young and impressionable, and that George would forcefully sweep her off her feet. "If you've smiled at the word 'forces' that I've used, do think however that it is a responsibility that you carry, this fact that you present the impression to others of being a 'strong person.'" She cautions him to act responsibly, to employ self-control, "for yourself and for those whom you know." George replies the next day. He prefers women who are "proud and free," not women who flatter men. The young woman, while not a mathematician, was nevertheless more than a mirror for his vanity. Mabel is mistaken about his "force"—at least in the "sexual" domain he is no lothario. He only knows that he pursues what he wants "obstinately and persistently," which gives the illusion of strength. George is glad that his letter to Céline has brought him into correspondence with Mabel. He appends Mélanie's most recent letter, which he has not yet answered. Liberated, respectful, reasonable, it is an inaugural exchange between intellectual equals.

George notes in his diary on 21 July 1906:

> I love Mélanie Loreyn, but I have promised my friends Céline Dangotte and Mabel Elwes not to tell her before October, lest I trouble this young soul. I promised, and she will know nothing about it. Is she the woman

1. Probably, Ågot Gjelms-Selmer, with S. Abas, *De Zedelijke Opvoeding onzer kinder* (Almelo, [1903]).

> whom I have loved all my life and who shares my existence? She attracts me, and charms me by her mystery and the tranquility of her gaze. Céline Dangotte and Mabel Elwes are two lovely souls, I am happy to know them. They make me a better person. Raymond Limbosch and Irénée van der Ghinst are also two lovely souls. I leave tomorrow for Holland to see Rembrandt!"[2]

His agenda indicates wide interests. On 30 September, George visits Laethem St-Martin to hear a Haydn concert; in coming years, he returns there periodically.[3] He indicates his libido by the astrological symbol for Mars, and he comments variously, "happy, disgusted." In a table for the month of October, he visits the Willemses on the 9th (Victor Willem is a biology professor at the University of Ghent), travels to Bruges and Brussels, notes Emile Waxweiler on the 16th and reads his *Sociologie* on the 29th,[4] and dines with the Garde civique—the local civil-defense unit—on the 28th.[5] Early in November he goes to Wondelgem with Mélanie Lorein; it is the village where he and Mabel will make their married home. Biologist Paul Pelseneer speaks to Reiner Leven on the 14th. In December, George again walks in Wondelgem. He is a regular visitor of the Bouny family. In February 1907 he is studying the paintings of Fra Angelico.

From October 1906 to July 1907 George keeps a record of the activity of Reiner Leven. There are trips to Zwynaarde and Wondelgem. The group hears Dutch poet and socialist Herman Gorter, Paul Pelseneer about the exploration of the deep sea, philosopher and historian Pierre Hoffmann about vegetarianism, a utopia for schooling, and "how to fight against envy and hatred,"[6] Brussels gymnastics professor Léon Dupont about physical education, classicist Paul Thomas about the Roman family,[7] Dr Joseph

---

2. YMS.

3. According to his agenda, George returns to Laethem on 4 August 1907 and 12 February 1908.

4. Emile Waxweiler, *Esquisse d'une sociologie* (Brussels, 1906).

5. George Sarton continues his association, irregularly, with the Garde civique for the next decade. It is evidence for his consistent belief, elaborated explicitly during the First World War, that pacifism is based on preparation for self-defense.

6. Agenda, 23 November 1904; HGS, Pierre Hoffmann to George Sarton, 10 February 1906. The sixth manifesto of Reiner Leven is Pierre Hoffmann's summary of a talk given on 3 April 1906, *L'Aspect moral du végétarisme* (Ghent, 1906). The fifth manifesto (apparently published out of sequence) is the summary of a talk at the University of Ghent on 26 March 1908 by Mlle Dr Józefa Ioteyko, chef de travaux at the Laboratory of Psycho-Physiology, University of Brussels, *Le Végétarisme au point de vue hygiénique* (Ghent, 1907). The seventh manifesto is a summary of a talk on 9 April 1908 at the University of Ghent by Fréd. Thalasso, vice president of the Belgian Vegetarian Society, *Le Végétarisme au point de vue social* (Ghent, 1908). Christophe Verbruggen kindly lent me copies of the texts.

7. Paul Thomas, *Moeurs romaines* (Brussels, 1899). *LM I* and *LM II, s.v.* Thomas.

Le Clément de Saint-Marcq on spiritism,[8] geometer Modeste Stuyvaert on mathematics, philologist Jozef Frederik Vercoullie on the evolution of language, historian Henri Pirenne on theories about the ancient Germans (George reads Pirenne's history of Belgium in October 1907), Victor Willem on ants, Waxweiler on social life, Jean van Impe on Donatello, chemist Frédéric Swarts on high temperature chemistry, and George's fellow students, biologist Paul van Oye on Heine and Paul Kenis on modern German poets. Other speakers or intended speakers include classicist Joseph Bidez, mathematician Paul Mansion, nutritionist Adolphe Miele, and painter Jean Delvin. Celine Dangotte speaks on Emile Verhaeren and Omar Khayyam, George on Maeterlinck and on popular universities in France and England, and Mabel on the Pre-Raphaelites. Speakers suggest colleagues. Waxweiler, for example, proposes Victor Willem and Edouard Anseele.[9] It would be a remarkable colloquium series today.[10]

In October 1906 begins a motif that runs through the life of the Sartons: Mabel's poor health, which frequently leaves her infirm and convalescent. Among the diagnoses, over the next decade, are a weak heart, curved spine, Graves disease, a thyroid condition, hormonal imbalance, and neurasthenia; among the treatments are surgery (appendectomy) and supplements of strychnine, phosphorus, arsenic, protein, and animal hormones. Although by 1938 Mabel had experienced heart attacks, it seems likely that her afflictions were in part psychological; it is probable that the intended cures did more harm than good.[11] Beginning in 1915 Mabel dispenses with them, and her health seems to improve. But she and George are moderns who, in their youth, place their faith in the fashions of experimental medicine.

In 1906 George writes with alarm about an illness of Mabel's. On 24 October he writes to Mabel that Irénéé van der Ghinst, who was friends with Céline, kept him informed about her illness. George suggests that Mabel read light works like Léopold Courouble's *Famille Kaekebroeck* (Brussels, 1901–). George enjoyed it so much that "he forgot to work for an entire day," but then he thinks the better of his suggestion (the novel is written in Brussels dialect): "Truly I think that you would not understand it. In the same way English novels have remained incomprehensible to me; they contain a way of thinking that is incommunicable—one only understands well the books of one's own race, because it is not only

8. Chevalier Le Clément de Saint-Marcq, *La Clef scientifique du spiritisme: Exposé populaire de la théorie noumenale* (Brussels, n.d.).

9. HGS, Emile Waxweiler to George Sarton, 15 November 1906.

10. Christophe Verbruggen, *Een Sociale Geschiedenis van het schrijverschap tijdens de Belgische belle époque* (diss., University of Ghent, 2006) for a reconstruction of the discussions and the lectures of Reiner Leven in 1907.

11. YMS, George Sarton to May Sarton, 3 July 1938, for Mabel's second heart attack.

with the brain that one understands but with one's blood and marrow." He recently read Rudyard Kipling's *Light That Failed*. He admired it abstractly, and he wanted to read it from cover to cover, but physically he found it so vexing that he wanted to pitch it out the window. He has begun *The Jungle*, but he doesn't know if he will finish it, for he has begun about ten books at once. "Am I foolish?" He imagines that Mabel doesn't think so, a consoling thought. George asks for her to return the copy of a text by Walter Pater which he copied out for her.[12] Racial thinking appears in George's writing, notably in his later flirtation with eugenics, but it breaks on George's increasing devotion to Asian science and his practical experience with the diversity of America.

Mabel answers immediately on 25 October with what George asked for: news. She wishes that she had no news to report, that she led a normal life. She contrasts her uncertain gait to her soul's energy. Like a small dog, her soul runs ahead of her physical self, scolding and laughing. She gives it the name of her dog in England, Brownie. There are times when she is filled with exaltation and imagines undertakings of significance and depth, and when strength comes to her she feels that she has enough to do "for two lives." She dreams about what she might accomplish if she did not need to earn money. George replies on First November that he is glad to know her situation. He reports that he was in her frame of mind years previously, when he was intellectually exhausted from studying metaphysics. When he was ill he took long walks in the company of an imaginary child (possibly the origin of his unpublished manuscript, *Lettres*).[13] He asks for the return of the book by Courouble, which he thinks might shock Mabel with its vulgarity. Sociologist and psychiatrist Frederik van Eeden's *De Kleine Johannes* (3 vols, Amsterdam, 1885–1906) would please Mabel better.[14] For Céline and also for Mabel, if she is interested, he sends along a text about Nietzsche.

George adds in a postscript that for the past years the only author he reads regularly is Ernest Hello, whose work *L'Homme* (Paris, 1872), into a seventh edition by 1905, he admires. The book is a collection of essays divided into the themes of life, science, and art from the perspective of Catholic theology. Hello, in particular, rejects Descartes's separation of spirit and matter. It may be imagined that George, a socialist attracted to the religious life, is particularly struck by Hello's emphasis on the synthetic union of nature and mind, and also, perhaps, by Hello's lives of the

12. *Phoenix*, p. 53 for a translation, which I do not follow.

13. YMS, George Sarton to May Sarton, 21 June 1941, for George's long walks as a young man in and around Ghent in the company of an imaginary child.

14. Jan Romein, *The Watershed of Two Eras: Europe in 1900*, trans. Arnold Pomerans (Middletown, Conn., 1978), pp. 533–4 for Van Eeden's anti-rationalist novel *De Kleine Johannes*.

saints. Hello's successor around 1900 is Francis Jammes, whom George also finds attractive.

The sense of unity guides George into middle life. He emphasizes to Mabel on 9 November 1924 that he is increasingly persuaded of the "unity of the world & my own utter insignificance in it." Furthermore,

> I deeply respect the religions of other people, however absurd its form, if it be genuine. I do not simply respect their religion, but I love it—through them, because of them. It is *their* form of the ideal beauty & justice & truth; however different in appearance, it is not essentially different. It tends towards the same direction.

For George, the Bible instructs that "the difference between the ignorant & the learned, the fool and the wise is not as wide as it may seem. If their hearts are right, they are almost entirely right, I am sure." In the course of his life the sentiment attracts him to a wide variety of universalist philosophies, from Transcendentalism and Theosophy to Buddhism and Christian Science. But as early as 1907 his friends remark on the wide and incompatible range of his literary masters in his search for "a discipline of faith." Emile Masson cautions: "Are you really sure that the false Persian [Charles Louis de Secondat, *baron* de Montesquieu] gets along well with his Jansenist friend [Blaise Pascal]? And that the admirers of the speeches of Diotima [Plato] would take seriously all those stories about symbolic crosses [Ernest Hello] that elaborate this mystery?" Masson remembers when he, too, "tried to assimilate all mystical, rational savants or cerebral barbarian products, whether insipid or acute."[15]

Mabel offers on 8 November 1906 that George has probably not read many English novels. *The Jungle Book* in her opinion is more typical of English writing than *The Light That Failed*. She grasps his point, that admiring a book while not liking it is to be in the situation of a cat having its fur rubbed the wrong way. She believes that she can write to George as a friend, and as she is often alone, she feels the need to communicate with someone. Mabel sends along a book of engravings by Burne-Jones, and also a print of his "Golden Stairs," which she invites him to hang on his wall. (Prints play a part in Mabel's and George's married life.) George finds time to reply on 15 November. He had a copy of the "Golden Stairs," but he gave it away, and he is glad to have received a replacement. The Burne-Jones prints allow him to put aside the work and struggle of a day and "purify" himself. Pre-Raphaelite work strikes him as "inferior art" with "too many artificial elements....imperfect beings, one part seduction, one part morbidity." Art up to the time of impressionism, in George's view, seeks "to get into and to reproduce nature correctly." The

15. HGS, Emile Masson to George Sarton, 28 June 1907.

reaction to this desire is an attempt to retreat from reality and create an idealistic art. He asks Mabel to consider Diego Velázquez, "the greatest of painters" who instructs: "There is no suffering, there is only harmony." Velázquez "doesn't distract us from life: he has us dive deeper in it to have us discover all its beauty and all its grandeur." George is reading English books, but he finds it hard going. In school he read only a bit of English literature. When he takes up the study of English again, "it will be to read the work of the great English physicists!" He had thought for years that only one language merited study, Greek. George addresses his own faults. He is not really lazy, but he has often wrestled with himself. To discipline his actions, since September, "I have written down everything I do, hour by hour, to be able to achieve complete control." He affirms to Mabel that it has worked, that his "guardian angel" has placed him on the right path. Here are themes for George's life: beauty, harmony, nature, physics, and list-making.

A month later, on 13 December, Mabel replies. She imagines that she has known George for a very long time, and she would like to converse with him at length. But she finds herself in a state, out of control, and a visit would not be wise. George's letter gave her pleasure, "I share *all* your ideas." Mabel will write out a list of English books for him. She now wonders if George was speaking about Kipling's *Jungle* or Upton Sinclair's. She passed by George's house in the afternoon. It had a slightly sad look, "the look of houses where there were never children." She has a picture "very clear and sad of you as a child, then a young man, struggling alone, in that dull house." She would like to see George, take his hand, and sense his sincerity, loyalty, and goodness. She is no fan of Belgian men generally, whom she thinks sink easily into vice. George and Irénée came as a godsend to her and Céline. Mabel then makes observations. First, she recalls that at one of meetings of Reiner Leven, George was brusque, and it wounded her. Second, she writes that she is glad of George's view of Mélanie Lorein because she is interested in him, although not too much. She wants him to know her faults. She sends along books by and about George Frederick Watts. If she continues to feel well, she will accept his invitation to visit.

George, on 21 December, reports that Jean Delville, who recently spoke at Ghent on the social rôle of art, wrote to George declaiming against the notion of realism. What did Mabel think about the topic? He urges her not to be morbidly introspective, as he has been in the past. His own sombre thoughts relate to his upbringing.

> You have clearly seen my childhood, my youth: Perhaps you have seen them too well, because I couldn't imagine anything sadder. Also, by contrast, my adolescence seems to me, despite all the disappointments, marvelously beautiful. Later I shall joyfully remember this lovely period in

> my life: I came to know beautiful souls and beautiful things. Previously, I lived in the shadows.

George shares Mabel's view of Belgian men. Between the ages of 10 and 20, he relates, he knew only base people. This is the reason, he explains, why he and Van der Ghinst founded Reiner Leven. The whole sad situation "is due in large measure to ignorance in education relative to the sexual question: There are only prejudices to fight in this area." George relates that he is trying hard to overcome a bad past. He "profoundly *hates* egoism, lies, and lack of simplicity." He loves "purity, chastity." He urges Mabel to be indulgent with prostitutes, who are merely victims. Where is she going in Germany, and when? (Mabel plans to travel there with Mme Dangotte on business.) He thinks to take a doctorate at a small German university some day, a place like Göttingen. He wants to see Mabel. She is right: In his previous correspondence, he spoke about Upton Sinclair's *Jungle*. On Christmas 1906 George sends Mabel a work by Ernest Hello, *"the most valuable book in my library."*

George writes to Marthe Patyn (the salutation is Patyn) on 17 January 1907 with encouragement to devote herself to serious study of social questions, because the sciences require preparatory skills.[16] Identifying style in art and literature is also difficult, but the moral and social ideas in the work are more easily appreciated. Which particular subject to study? George is interested himself in the relations between employers and their servants. He would be glad to work with her on the topic.[17] For sociology, the early years of the 20th century were a heroic time, when theory and practice were difficult to disengage, and when academic sociologists taught social workers, as much as they did journalists and revolutionaries. The line between scholarship and advocacy, emphasized by Max Weber, was only beginning to be drawn, and George's sociologist acquaintances like Waxweiler cross it frequently.

George writes to Mabel on 25 January 1907, returning the Watts biography, which he is glad to have read. He has mixed feelings about the Pre-Raphaelites. "I am more and more convinced that theirs is an inferior art, although the product of a superior mind." The proof lies in the fact that he tires of Burne-Jones. But he never tires of Velázquez, whose "premise is less artificial. Only things truly beautiful, profoundly, naturally, are *durable*. Everything else passes, following human caprice." George is sure that Hello would not have liked the Pre-Raphaelites. Their work appeared after a century of "prosperity and corruption." Watts occasionally attracts George because he allows us to see the goodness and the purity of our na-

---

16. May Sarton, *A World of Light: Portraits and Celebrations* (New York, 1976), p. 58, for men addressing Flinke women by their family name.

17. NYPL; *Phoenix*, pp. 47–8 for extracts.

ture, but George doesn't want to live in the Pre-Raphaelite "atmosphere of convalescence." He patronizes Mabel. "A substantial experience with art, having lost the innocence of early admiration, is required to admire these works *correctly.*" Mabel is wrong that one must first know the works of the "primitives." He has read Omar Khayyam's *Rubaiyat*, presented to Reiner Leven by Céline Dangotte. At the base, the philosophy is "not very noble." It is a matter of "frivolity, egoism, and libertinage." George apologizes if his lines annoy Mabel, but "he has increasingly decided to be categorical in matters of principle," and to leave scepticism behind. He thanks Céline for the return of the text by Nietzsche, but he has more need for the return of the book on vegetarianism.

Mabel, a student of design, proposes an emblem for Reiner Leven. George likes it, and she is glad, she reports on 26 January 1907. "I am absurdly sensitive about my designs when it is a question of someone whom I respect and like who condemns them, understands nothing about them, or remains completely indifferent to them." Her designs are like her children. She cautions George about judging the work of a sincere artist. Watts, in her view, was never a Pre-Raphaelite, even though in one work he seems like one, seeking to know "the highest beauty, the purest, noble and elevated ideas." The critics never write about how, in practice, artists execute their work. For this reason, their theories "are largely *myths*" for the untutored public, functioning to "obscure and falsify." Ruskin is one of the architects of this construction, fixing his own ideas in the works of the Pre-Raphaelites. The latter group, in Mabel's view, had a simple vision: "All their ideas, all their so-called theories, are summarized in the word *sincerity.* They found themselves revolted at artificial, banal, and *dead* art." A few young students, between 19 and 21 years old, discovered Italian primitives, and sent out a programmatic call, despite the fact that "they were ill-informed of those about whom they enthused." They carried a banner of fighting for the truth, and they were able to create a stream of clarity and purity in a country devoid of modern art. The Pre-Raphaelites "did not desire 'style' in the sense then current—they sought *the truth*—they did not *want artificial rules* imposed in the academies—they wanted *sincerity* as a rule."

Style is at the center of Mabel's métier as a designer. Mabel turns to the writings of Hello, whom George admires. "Don't reply with his sentence, 'The place of genius is the style.' For I will reply with his additional sentence on the 'style' of great men: 'their style is to be themselves,'" perhaps alluding to the notion of Georges-Louis Leclerc, *comte* de Buffon, that "style is the man himself."[18] George confuses simplicity with affection. What George calls intentional faults or artificiality in Burne-Jones, Mabel would

18. Buffon, *Discours sur le style, avec une notice biographique, une introduction et des notes explicatives,* ed. J. Pierre (Paris, 1896).

call simplification. George, in Mabel's view, is missing the beauty. To label the work of Rossetti, Watts, and Burne-Jones as *"inferior art"* raises Mabel's hackles. How dare he do so on the basis of a superficial understanding? George judges them inferior because they tire him. The works never tire Mabel. They have different sensibilities, George and Mabel. Velázquez, for her, is always admirable. In a few of his portraits,Velázquez goes beyond representing the soul of one child to show the child's soul in general—and those Mabel likes. Burne-Jones's visions of beauty are closer to the truth than the models that inspired him. In George's way of thinking, there was an "absolute suppression of imagination and personality." It is blasphemy for George to say that "to admire small things is to diminish large ones." This point of view denies appreciation of a flower underfoot or the fragrance of leaves because the small sensations prevent appreciation of "the large voice of the wind." Mabel admires Hello's grand words, but she draws the line at all his allusions to the Old Testament, which she cannot accept. George is off on the wrong track, and he should take care. Mabel's critical pen derives from a meeting at Reiner Leven, she writes in English: "when you 'snapped' at me—yes, yes you *did*—just like a little angry dog...and it did *not* sound a 'friendly' bark at all." George has wounded her. It takes him decades to anticipate her feelings.

George recalls late in life that he first read about Ruskin in October 1907, in a book by Robert de la Sizeranne.[19] This is, evidently, not quite accurate, but it is clear that early in 1907 George is uninformed about the Pre-Raphaelites. George writes back to Mabel on 30 January 1907 to affirm that their ways of looking at things are not really distinct. He acknowledges that he is too quick to judge the Pre-Raphaelites. His main point in writing is to affirm that the "little things" he spoke about, in his letter, belong not to nature but to artifice: a canvas of Rubens, the verse of Edmond Rostand, the music of Giuseppi Verdi. "You have me committing blasphemy, and my hair stands on end. I shudder to think that you were able to believe that I thought something like that. Remember that for nearly 4 years the subject most given to contemplation has been precisely the greatness and the poetry of ordinary life." Two months pass. On 27 March George writes again. He has reread Mabel's last letter. He allows that he spoke imprudently. He recalls the first letter Mabel wrote to him. She spoke about Mélanie Lorein. He dared dream, only to be let down. He does not deserve the happiness of her company: He isn't good enough or handsome enough. Since then he's made great progress, and it is due to Mélanie. He has met her and spoke briefly with her. She really is beautiful, and anyone who says the contrary is crazy or blind, although some days he finds her quite cold, like an inquisitor. George apologizes

19. Diary, 25 September 1949; Robert de La Sizeranne, *Ruskin et la religion de la beauté* (Paris, 1897; 1907), which George recalls acquiring around October 1907.

for writing to Mabel about Mélanie, but he must express his thoughts to someone, and it cannot be the woman in question. Since the time that he wrote Mélanie a letter that Céline Dangotte judged a love letter, he has sent her only banalities. George thinks that he is continually improving as he seeks his "ideal of knowledge and wisdom, his ideal of beauty, his revolutionary ideal." He wonders that questions about the future of society and natural law do not much concern Mabel. He invites Mabel to write to him in English.

Which Mabel promptly does, writing the next day at 5:00 p.m. upon her return from Brussels. She is glad that George—she uses the salutation "Sarton"— feels comfortable writing to her.

> When I read your letters and find things in them—nearly always things you most carefully *don't* say—odd gleams of courage, sunny bits of frankness, little revelations of your best self—know then that I feel something so deeply, quietly happy in me—that it gives the rare, deep sweet impression that I can never express but that is like a bird singing in the deep blue of the sky—something infinite, something reassuring—in short a little bit of real happiness!....Oh Sarton, Sarton, there is something in you that I like so much—I seem to be always arguing and disagreeing with you, but indeed in spite of that I feel that you are not such a horrible long way off from me as almost every-one is from every-one else.

Yes, Mélanie is beautiful, with "something of the real childish unconsciousness mixed with a certain womanly dignity." Mabel sees George clearly: "I do believe that you manage to think me even a little narrower in my interests & a little stupider than I really *am*." It is true, nonetheless, that Mabel has never studied things with method and perseverance. She loved to talk with her father about religion and sociology and other topics, and she read "a fair amount," but she was an irresponsible girl. She has always been, she thinks, too sensitive. What she means by his being off on the wrong track relates to his desire to "exaggerate" himself, as he wrote. She finds exaggeration in the writings of Hello, an unhealthy, hot-house fecundity. On the topic of prostitution, Mabel thinks, unlike George, that women are tempted by sensual pleasure just as men are. George tells her of his struggle for continence. Mabel has struggled too, and she struggles continually. She confides in George because she trusts him completely.[20]

The trusted correspondent replies on 3 April. He and Mabel think in much the same way. He wants to speak with her at length, rather than write. About matters sexual, he tends to agree with Ernest Renan that while women are taught to be discrete, men are taught to be indiscrete. And in the next sentence he launches into an indiscretion. He recalls

20. *Phoenix*, pp. 55–7 for a portion of the letter.

seeing a portrait of Mélanie done by Mabel. He asks if Mabel could lend him the portrait, for the preceding day he met Mélanie's younger brother. Mabel visits. After the visit, George acquires the latest book by Maurice Maeterlinck, the Symbolist resident of Ghent who was a student of natural history, *L'Intelligence des fleurs* (Paris, 1907), and he sends a fragment of it along for Mabel.[21] He read the book "with enthusiasm, *frenetically*," he writes to Mabel on 6 April. Maeterlinck's ideas are very similar to ideas he has had for a long time. "What a happy coincidence that I can send you this fine study several hours after speaking with you about the topic." George had seen Maeterlinck as an egoist; no longer. He returns to the topic of Mélanie, whom he hopes will some day appreciate him, though he doubts it. And if she will have him, she must accept his desire to be a *pariah*. "Loreyn knows that my wife will have to be prepared to be the wife of a pariah, a pariah herself; and that under no circumstances will I take care of my wife [en aucune façon je ne prendrai à ma charge l'entretien de ma femme]." George will send Mabel the portrait on Monday; evidently she brought it along when she visited; he does so on 9 April, and he regrets that he can't send Maeterlinck's book because he will speak on it before Reiner Leven. Here is the earliest announcement of George Sarton's plans for married life, plans when materialized that cause intense resentment in his daughter May.

Mabel then gives George wide berth, returning to her home in Ipswich, England, where she has a project to paint miniatures. On 10 June he writes to her there, "Ma chère Mabel,"copying out a dictionary entry about the town. Is Ipswich the town of her birth? George is having a hard time studying for his *candidature* examination, but he thinks it would help to have a copy of Donatello's St George before him. It is so much better than what the Pre-Raphaelites have produced. Soon after his examination, he will go to Paris to study at the Louvre, even though he is afraid of the iniquity of the big city. He hopes that Mabel is not annoyed by his presumptuous use of her given name (previously he wrote to her as his "dear friend"). Mabel replies on the 13th, describing her surroundings in English with emotion: "Oh I wish, I wish you were here & could see it all—all by yourself for you don't know me well enough nor I you to be able to feel your own thoughts undisturbed, if I were with you." George is, she thinks, acting like a child about Paris. Mabel writes again on 16 June (continuing a letter from the 15th) that a doctor has recommended complete rest for several months. But she already tried that the preceding year. The diagnosis disgusts her:

21. Presumably George bought the book unbound, possibly uncut, and was able to split its contents. He asks Mabel to return the fragment within a day because he wants to reread it.

> Months and years pass, and I make no progress. I learn almost nothing that will help me *achieve*, express the ideas that visit me on occasion. There is so much that I want to do—essays even—yes, of course, I know, I would never write them as I might like but now I don't even begin them. I waste my time doing foolishness often to earn enough to be independent, always in the hope of having one day enough to work seriously at my *craft*. What good is it to have these visions sometimes which exalt me. The underlying conviction that I would know how to produce Beauty one day, to see it in front of me, to sense life palpably around me, to know that other people do not see it, that I could reveal it, if only one single time. Tomorrow I will be old, and what will I have accomplished? Nothing? One must *wait*!! I don't know, I don't know. Wait, wait. There are times when I think that Hell is the condition of eternal waiting.

Mabel switches to English: "The sickening strain toward something which always escapes, an *actionless waiting*." She considers herself a happy person, but "I want to live *more* and *deeper*, and *wider*." Much of the time Mabel is content to do her best,

> the best that is in me each day, to understand the few people that I do meet, then that when I am old I shall *not* feel I have done "*nothing*" altho' I may never make my own particular dreams come true. But I can't give them up, cannot, must not and will not give them up, till I see they can do harm to anyone, and I don't think I could be content as I often am, if I had not got them at the bottom of my heart, where I feel them dimly stirring, often when I am doing the most material, practical, every-day things, they help me

Back to French to discuss matters of divinity: George closes his letter, "God be with you." But in Mabel's view, God has nothing to do with it. The notion of a personal god, containing something of immensity, is inexplicable. She never found an adequate explanation of God, but she is glad that George wishes her well in this way.

A letter from Ipswich to Ghent in one day (unthinkable today). George replies on 17 June 1907. He floats near her, like a wounded bird—the avian image that Mabel later appropriates as her own. He has just placed himself under Mabel's protection, and that may help cure her. He couldn't write to her sooner because he was studying for his examination, which he hopes to take in the second week of July, and also because Raymond Limbosch came to visit. He has had to deal with his relatives; he will be glad eventually to relocate far from home, even though his father is a good man. George falls asleep thinking of Mabel. He concludes with a peculiar Protestant thought: He asks God to cure her quickly, but she must also very strongly want to be cured. On the 19th Mabel apologizes for writing so late, for George's letter arrived on that morning. She urges him not to write, since he needs the time to study. George sends back a postcard on the 21st, recommending Alphonse Daudet's *Tartarin de Tarascon* (Paris, 1872), an ironic criticism of romanticism. On the 24th Mabel tells George that she was not born in the Ipswich house. Her father was a schoolboy there, his grandmother having a house nearby; he describes the house in a journal of his. Mabel's father was, she writes in English, "a serious-looking, brown-eyed boy, with spectacles, & his hair rather long, walking rather quickly in an absent-minded way."

George is grateful, on the 30th, for Mabel's gift of a print of St George, and he is surprised to learn that Ipswich has an electric tram. On 3 July he invites her to visit immediately upon her return to Ghent. Mabel lets George know that she is returning with a friend, Laura, who helped take care of her when she was ill. She has been producing designs for lace-work. She learned a great deal from the task, notably that the women who execute the design can produce only a few pieces in a lifetime—they take no enjoyment in it. Mabel would like to create designs based on plants that the women know so that they can have greater joy in their work.[22]

On 7 July 1907 Mabel writes that she is back in Ghent and would like to see George, but that her time is taken with her friend. Mabel wishes George well on his examination. She detests them. George has taken his examination, he writes on 10 July. But he performed poorly. Leaving the room, he told the jury that he would do better next year. He apologizes for not being home when Mabel came to call that morning. He is preparing to leave for a trip to Brussels and then Paris, but he would like to see Mabel before he leaves. There is no indication that the meeting takes place. Mabel informs George on 19 July that she is leaving for Enggistein, near Worb in the canton of Berne, Switzerland. Before she leaves, she is preparing some works for the artistic Salon. She writes at length about her emotions. She is very happy that George is her friend. She adds that Mélanie Lorein has been ill, although she is recovering with Céline's help. "Poor child."

George writes to Raymond Limbosch on 11 July 1907 that his "life has quite gone in one direction," that he has "a ministry [apostolat] to fulfill." Furthermore, "everything is subordinated to my purpose: my life, that of my wife, of my children, if I ever have any, that of my friends. Whoever hesitates in the face of the sacrifice is not worthy of undertaking great things. God has not chosen him." George prefers to avoid interacting directly with workers, choosing instead to "penetrate deeper into the spirit, the heart of the doctrine" of socialism.[23] George writes to Mabel from Paris, on 21 July, with thoughts about Mélanie, touring the Louvre, and

22. *Phoenix*, pp. 59–60 for an extract.

23. Van Oye, p. 49.

visiting Romain Rolland, with whom he talked "for a long time about music and literature."

On 28 July he writes from Ghent to thank Mabel for the fresh Swiss flowers that she sent to him in Paris, and for the English books sent to Ghent. He will read the books, although he is not sure that they really interest him. Mabel receives the letter on the morning of the 30th and promptly replies from the chalêt of Mme Meyer in Enggistein. The proprietress is about 70 years old, but very young at heart. Mabel is relieved to learn that the flowers arrived safely—the sending of flowers becomes a theme in George and Mabel's life together, as it does in the life of their daughter, May. Mabel would like to hold him prisoner in the forest until he tells her everything about Paris. She is suffering from toothache, and she uses George as, she writes in English, "a 'safety valve'—when I can contain myself no longer I 'let off steam' to you." Then she describes how Mélanie recovered, and how Céline, Mélanie, and Marthe Patyn traveled to the Netherlands on a fatiguing trip. About the books she sent, she hesitated over Alfred Edward Woodley Mason's *Four Feathers* (New York, 1902), although it interests her and she finds it typical of good English novels, and she thinks that they contrast to the French novels that she's read, notably Pierre Loti's. If George does not have much time, she urges him to read H. G. Wells's *In the Days of the Comet* (London, 1906).[24] She is expecting to receive other books shortly, in particular Ruskin's *Two Paths*.

Mabel repeats George's condemnation of Ruskin: "*His style lacks discipline.*" Then she gives a contrary opinion: "But there are qualities in his faults. He is spontaneous, generous, enthusiastic, personal, and sincere. He contradicts himself sublimely. He is mistaken sometimes, but he is never petty. He is *eloquent*. He has an imagination with an extraordinary richness and often a rare fineness." Ruskin leads Mabel to generalize:

> I need to see a true revolution in *not only self expression* but *in thinking and judging* in England. It was forbidden—by one of those unwritten laws that are nevertheless still enacted—to say frankly what one was thinking, above all one had to *diminish things*. One never said "*that is beautiful*" (still less could one one comment on the beauty in *magnitude* or *quality*), but one said "that's rather pretty." Enthusiasm, passion were "bad form." Originality also. The greatest crime was to "be noticed."

This pattern extends to the beauties of nature. The English have such a horror of *sentimentality* that they want to kill *sentiment*. George can find the pattern among the chauvinistic, narrow English, "who make fun of

24. In his agenda on 19 January 1906 George has already noted Wells's *Anticipations, ou de l'influence du progrès mécanique et scientifique sur la vie et la pensée humaines*, trans Henry-D. Davray and B. Kozakiewicz (Paris, 1904).

everything superficially, who *deny* heart, acts of surprising goodness, sublime devotion...and *unconscious*. The truth shines divinely at the bottom of the wells where it was intended to be interred." Mabel then describes the great beauty of the Swiss valleys and mountains. Mabel concludes with the thought that the death of her father has only now sunk in. George's words from last year are a comfort. Mabel is penetrating and aggressive in her letters from this time, while George seems superficial and distant.

If a student at Ghent sought science, as George did, he would have naturally been attracted to chemistry. The very word, around 1900, still carried the prestige and allure of August Kekulé von Stradonitz, who in the middle of the 19th century, at age 29, had come to the University of Ghent from Heidelberg as a full professor; there, at Ghent, Kekulé conceived his notion of carbon rings, pioneered aromatic chemistry, and started a family. The chemical tradition enjoyed spectacular success and attracted talented students, among whom was Leo Hendrik Baekeland, whose work on photographic paper, Velox, led to the research that produced the early plastic, Bakelite, and who becomes George's patron in America. George is attracted to the practical side of 19th-century science; in his first history-of-science courses at Harvard University, from 1916 to 1918, modern, useful inventions are much in evidence.

George writes on 3 August 1907 that he is spending 6 to 7 hours daily in a chemical laboratory, and he doesn't like it. "I really know, Mabel, that I will never be a laboratory man. I don't have the presence of mind, nor the very special ingenuity, nor above all the necessary *patience*. Because it takes the patience of an angel." George has been working without effect for several days to dry an apparatus and seal it hermetically. He shall have to begin again on Monday. And still he has a long time to spend in the laboratory, "to learn the theory." It is about oxidation and catalysis.[25] He is exasperated that he must devote so much time to the laboratory when there are so many other things to do, notably books to read (and he has not yet begun the ones Mabel sent). Since returning from Paris he has taken only one day off, at the Dangotte residence in the country with their friends. He wants to see Mabel and talk with her. The laboratory vexes him. "Tomorrow the professor leaves on vacation. I will be there alone with the assistant and a servant. So I will be a prisoner in this way for 2½ months!" The professor, Frédéric Swarts, the general chemist in the School of Civil Engineering, had lectured to Reiner Leven in 1906/07.[26] George does things on his own terms. Two days later he writes Mabel that Ruskin's *Sesame and Lilies*, which she sent him, pleases him greatly; Ruskin's work, written under the influence of Thomas Carlyle, is a discussion of good books

25. Diary, 31 August 1907.

26. Agenda, 13 June 1906.

and good women.[27] George reports that he quit the laboratory, which he worked in "with indifference, without zeal." George continues the next day, 6 August, that Ruskin will prepare him for visiting England. Yet he finds that Ruskin writes in a roundabout way: "His thought would be more beautiful if it were more condensed. But should I teach the eagle to fly? He knows better." Then he announces that he wants to ask her something important: "*You must give me your portrait.*"

Half a lifetime later, Mabel informs her friend Mary Bouton that she and George "both adored Ruskin in our youth & still admire him, & our first meeting ground (almost a battlefield) was a discussion of the P[re] R[aphaelite] B[rotherhood]."[28] So Mabel "heaved quite a sigh of relief" on 8 August 1907 (she uses the English phrase) when she learns that George likes Ruskin. She wants him to read more works by Ruskin, *The Crown of Wild Olive* along with *Unto This Last* (a work inspired by Thomas Carlyle), which Céline is sending him. She does not have the edition of Ruskin's collected works, only the first volumes of his *Modern Painters*. She will send him a portrait. Her mother is having a difficult time after her father's death. In English: "Do not be afraid. I am not *uselessly* sad. Out of that 'hurt' & sadness the light & air & sun are distilling sweetness & hope, and *patience*." Her father shared "awkwardness & clumsy *fingers*" with George, but he learned how to discipline his hands, and George will also be able "to do the most delicate microscopic work." She closes by asking him to translate, in the margins, all Ruskin's citations from Greek and Latin.

From the Swiss forest Mabel writes more importantly on 10 August about George and Mélanie Lorcin. She thinks, as she may have mentioned before, that Mélaine might be attracted to George and that it would be a dangerous liaison for both of them. George, she fears, is shocked and repulsed by the sexual side of love. He has not considered Mélanie's state of mind. She is quite ignorant of the facts of life. Mabel is sure of it from a sentence of Mélanie's about the birth of kittens. Mélanie needs to be informed, in the manner of a child. George finds that he wants to enlighten Mélanie, but that is not the same as love. George's attitude is the reason for Mélanie's distance from him. "George, my friend, might you in your life find the woman whom you will love not only with passion, but who will love you with miraculous tenderness for a lifetime." Mabel uses the word miraculous because she finds it expresses the divine quality of love.

27. Rosemary Ashton, *Thomas and Jane Carlyle: Portrait of a Marriage* (London, 2003), pp. 439–40. Emile Masson, on 20 October 1907 [HGS], reports to George that he anxiously awaits a copy of his translation of the letters of Carlyle to his mother, *Carlyle intime: Lettres de Thomas Carlyle à sa mère, dont plusieurs inédites, revues sur les originaux par M. Alexandre Carlyle*, trans. Emile Masson (Paris, 1907).

28. NYPL, Mabel Sarton to Mary Bouton, 25 June 1934.

She has seen it twice in her life. She recalls that he wrote that he "would not hesitate a moment to sacrifice the woman you love for your work, if necessary." But if he is loved, if he has children, how will he handle these responsibilities? "I do not understand! Unless with the intention, the great hope that the work undertaken at the expense of these people would benefit humanity (that would be perhaps a sublime thing, beyond ordinary judgments?)." Still, she thinks that passionate love precludes sacrifice:

> From the moment when it seems possible that your work can require this sacrifice *you cannot love anyone or elicit love from anyone* else. You may sacrifice *yourself*, perhaps you must do so, but you may *not* sacrifice other lives. Oh, don't you understand? You *want* to love Mélanie Loreyn, yet the thought of the possibility of sacrificing her to your work crosses your mind. I cannot express the disgust that this provokes in me.

George may not have connected the points, but he "does not have the right not to anticipate." Could reason and will prevent passion from arising? Only in part. What reason can do is to put them in their proper place. "Can a person develop without experiencing passion? (I do not say simply without giving in to it.) I doubt it. I have asked life to spare me passion because I hold that I could not give complete joy to a man. I have never been strong enough to have beautiful children."

George replies to Mabel's intensely personal insights with a description, on 14 August, of his day, listing his activities. These include returning to the laboratory in the afternoon—his exasperation with chemistry having been extinguished. George does not mention that he completed the first draft of his novella, *La Chaine d'or*, several days earlier.[29] He came up with the title and the general idea for it, a meditation on affection, while visiting Leiden during a study trip in 1905, when he noted "la chaine d'or qui entoure le monde."[30] He is not working hard, but he finds himself thinking intensely on a park bench in the afternoons. Mélanie has not understood what he said to her, which he regrets. But "the sexual instinct speaks by itself." Cats do not read books to learn about it. Then to the main point:

> You say that I may not sacrifice wife and children for my work. You forget that I announced my intentions to Loreyn before she would have fallen for me. The woman who loves me will accept this sacrifice *in advance*. I will not sacrifice her. She will accept the sacrifice herself, and if I hesitate, she will remind me of my duty, push me to the sacrifice. Is this not true for wives of sailors, miners, or chemists? Have they not

29. Diary, 11 August 1907.

30. Agenda, 21 September 1905.

> accepted, through marriage, the possible death of their husband, and of misery?
>
> Since I expect to disperse my fortune and to keep only what is necessary to live, I will always be on the edge of misery: My wife will accept misery *in advance*, then, for herself and for the children. If she loves me, she will at the same time love self-sacrifice.

After this prescient and egotistical passage, George continues that Mélanie will never love him, and that he thinks about her very little. He encloses a letter from Raymond Limbosch, to which he replied "that my aim, my ministry, has dominated my entire life, and that my friends, in agreeing to be my friends, had to accept the consequences."

George does not realize, perhaps does not care about, the effect of his words on Mabel, and she is too discrete to let on. Mabel receives George's letter the next day (was all Europe then accessible by overnight post?). She replies disingenuously that George's lines are filled with such a deep tenderness and goodness. In the Swiss forest, Mabel is distracted by a squirrel, and she writes about the beauty she sees. "Oh, don't destroy it with a mathematical definition." She sends along a photograph of herself in a hat and hopes that he enjoys his reveries in the park. Switzerland, the land claimed by British alpinists, exerts a strong attraction on Mabel. She writes to Marthe Patyn that there is something tragic about the distant Alps: "I have the feeling that *they have cleansed me, purified the soul and mind*."[31]

Thanking Mabel for the photograph and another portrait on 21 August, George reports that Mélanie returned to Ghent from a vacation. George asks Mabel's age. His laboratory work (he has returned) is advancing well, largely because Professor Swarts is absent.[32] (A photograph from this time shows his apparatus for measuring the speed of oxidation at constant pressure.) Céline proposes a day in the country, at which time he will plant an oak; in 50 years it will be splendid. By 28 August Mabel is back in Ghent but then goes directly out of town with Céline. In 50 years she will be 78, although she doubts that she will live so long. George can think only to send Mabel a graph of his moral and intellectual dedication in 1907, a "true portrait," on 31 August. He does this every year on the last day of August (he does not explain that it is his birthday). He went with his father, as he does every year on that day, to bring a bouquet of roses to his mother's grave. On his birthday

31. NYPL, Mabel Elwes to Marthe Patyn, 17 August 1907.

32. George, later in his career, is on good terms with Swarts. He asks the Carnegie Institution to send Swarts a copy of its publication on the atomic weight of boron and fluorine. CIW, George Sarton to Walter Gilbert, Assistant Secretary of the Carnegie Institution of Washington under Robert S. Woodward, 11 December 1919.

George summarizes his year in his diary. He notes that he read Henrik Ibsen's *A Doll's House*, presented by Céline Dangotte to Reiner Leven. He also notes, among the books that he has read over the past years, Stijn Streuvels's *Openlucht* (1905); Struvels, the pseudonym for Frank Lateur (nephew of Guido Gezelle), had his book *Lenteleven* printed in 1901 by Julius de Praetere.[33]

Mabel returns from Switzerland at the end of August and tries to visit George, but he is out; she is more comfortable writing to him than speaking with him, she writes on 3 September 1907, and she wishes to overcome her sense of embarrassment. She wants above all to thank him for his moral portrait. He provides a key to her thoughts, he helps her greatly. She very much hopes that he finds peace. On the 4th, George invites her to visit at a time that is good for them. Mabel is busy, she writes on 14 September, tutoring and also drawing.[34] Her miniatures were not included in the Salon exhibit at Ghent, but her drawings were. She thanks George on the 20th for the print of Auguste Rodin's *Thinker*, which has helped her through a recent bout of illness; he replies the same day that he too has been ill, with a toothache. Mabel will visit in several days, she writes on the 21st, when she returns from the country. On the 24th, Mabel writes that she is more ignorant than he supposes, but that she hopes the circumstance does not interfere with their friendship. She feels that sometimes she presents herself "under false colors. Understand that I know nothing, nothing, I learned nothing at all at school." Savant George writes from a café on 29 September:

> I don't understand your fears. Why would you want me to take you for a *savante*? I know that knowledge doesn't fall from the sky. Where would you look for it? Instruction of girls is, I know, extremely rudimentary. Worse, it is rudimentary while presenting the appearance of not being so. You certainly don't have to know many things if: 1) you don't have any illusions in this regard, as girls do; 2) you master thoroughly the few things that you must absolutely know, notably concerning your craft and things relating to social life. These are just a few things to keep in mind, and if your education is incomplete, there is no problem in fixing it: A little courage is all that is required.

---

33. YMS, George Sarton to May Sarton, 20 November 1935, where George calls Gezelle "the greatest poet of Flanders & one of the greatest of the whole world."

34. According to George in "Scholar's Wife," at about this time Mabel tutors Vladimir Peniakoff, the son of a Russian engineer then working in Belgium, who later has a successful career in North Africa. Peniakoff, *Popski's Private Army* (New York, 1950), p. iv for his unnamed private tutors. NYPL, Mabel Sarton to Marthe Patyn, Monday [before 1915], in French: "How is the Peniakoff girl doing—'petite Lumière'—I believe she was called? I would not have thought that she remembered me, although for my part I have a very clear memory of all 3 children."

On 8 October 1907 George reads Charles de Coster's *Thyl Ulenspiegel*, a paean to the Flemish past—notably Flemish courage in the face of Spanish occupation. The story stays with him for decades.[35]

Courage is Mabel. George has been revising his novella, and on 8 October Mabel writes that she has read the manuscript and she likes it very much. She thinks it due in large part to her affection for George, whom she sees in the book. But objectively, she finds "the whole unattractive—the ideas, philosophical, social and others, expressed (despite the form of conversation) in too *synthetic* a way to be *pleasing*. Also an *impression* of lack of *cohesion*." Perhaps it is because the characters are only substitutes for ideas, even if occasionally they really live. Still, the book might help some people, but only those "who are already rather far toward the *state of grace*—a small number." This evaluation, however, is not a reason for not publishing it. Mabel finds some of the pages beautiful, and none bore her. George agrees. He writes the next day: "Your impression is also mine. Regretfully, as you feel yourself, there is no way to correct that: *The whole thing would have to be redone*, in the clear light of day. But I can't think of it now, because I must get back to work." He is sending along the end of the manuscript, which he wrote out last night. "Same faults however as in the 1st part." He has written to a publisher, Pelletin, chosen at random. "There are 999 chances against me, that he will not want to run the risk of publishing this boring thing. In this case, I will place my manuscript quietly in my files. I am purged of it: That's the main thing." Upon reading the end of the manuscript, on 10 October, Mabel confides that it indeed has the same faults and merits as the first part. She is glad for his decision to contact the publisher. She thinks more than ever that there are good beginnings in the text. She learns that George had been to Ostende and so has seen the sea. She wants to be near it. She thinks that "a *part* of what I seek is *in the sky*. I am more and more persuaded that we don't look enough at the sky."

Mabel sends a note from Antwerp on 18 October, where she is at the home of a friend, answering an inquiry of George's about biographies of Ruskin. On the 22nd she is making good progress on a miniature of her young friend for her own pleasure, and the work is gratifying. She invites George and Raymond Limbosch to spend Christmas Day with her and the Flinke family, her usual celebration. She has had a photographic portrait taken by a local amateur, and she will send it to him. "It is really *me*, or I imagine that it is a true picture." She has sent a copy to her mother, and she is interested to compare George's impression with her mother's.

---

35. HGS, Emile Masson to George Sarton, 21 July 1914, where Masson reports that his wife Elsie is soon to give birth to their second child, and she plans to read "your sacred *Ulenspiegel* this time, as she did when she was carrying Yves." In 1916, George looks forward to writing about *Thyl Ulenspiegel*. Diary, 18 February 1916.

George walks in the country and returns to think only of Mabel, he writes on 24 October, wondering where she is, what she is doing, how she feels. He learned that he was quite naïve about publishing his book. The publisher did not reply, even though George sent a stamp for return mail. He wrote to Stock in Paris and Daman in Brussels, who answered that they have no time to read manuscripts. So he drew up a list of 15 publishers and he had typewritten letters prepared for them. (In the next to the last paragraph of the typed letter, he notes honestly: "I have no recommendations. I have written very little until now: 2 brochures and some articles.") Otherwise he's been very busy. He is attending courses in mathematics for the second *candidature*, he still works in the laboratory in the afternoon, and he also attends courses for the doctorate. He is becoming excited about the courses. He thanks Mabel for the invitation to spend Christmas with friends, but the holidays are for his family. Mabel replies the same day from Antwerp. She is happy for his academic work. She is interested to learn what the publishers will say.

George thanks her for the portrait on 27 October: "I like it because it familiarizes me better with your dear little head than the preceding ones [il me fait mieux connaître votre chère petite tête que les précédents]. But there is something that shocks me in the portrait, and I am vexed not to know what: perhaps it is simply the *tone*! I am strongly tempted to think so." He has heard from 2 publishers. One writes that he does not read manuscripts. The other, Ollendorf, says that he will gladly read the manuscript, but it takes him 4 to 5 weeks to do so; yet since George's manuscript is not a novel, there is no point in sending it. He has William Gershom Collingwood's study, apparently the best book on Ruskin.[36] "I have given up the idea of writing a *methodological* study of Ruskin: it would take too much time. Especially since reading and *annotating* his writings would take several months. But after my trip to London, Sheffield, and Oxford, I will perhaps write a short meditation on Ruskin. Let me know in any case your thoughts on Ruskin's life." Raymond Limbosch reproaches him for not corresponding more often, but George thinks that it serves no end to write letters for the sake of writing. He confides to Mabel that sometimes he spends half a day, or even a whole day, without thinking about her. Does that prove that he does not like her? "No, it only shows that I am an assiduous worker, and you should be glad for it. When my working day is over, I relax and I think contentedly [doucement] of you—it is a kind of reward." Mabel then sends along a note the next day about her copy of J. Marshall Mathers's *Ruskin, His Life and Teaching* (London, 1897), which has only one chapter on Ruskin's life; Ruskin's *Fors Clavigera: Letters to the Workmen and Labourers of Great Britain*, contains more information. Following a pattern in the correspondence, when George is thoughtless or selfish, she is curt and matter-of-fact.

36. William Gershom Collingwood, *Life of John Ruskin* (2 vols, London, 1893).

Two weeks pass, and George "begins to worry" in a postcard of 11 November, which bears instructions to be forwarded if necessary. He knows that he upset Mabel: "Have you received my latest letter, rather my mean note?" Mabel replies on the 12th that she is soon returning to Ghent where she will take a morning course at the Académie des beaux-arts. She would like to see George in the morning, before the course begins. Yes, she received his last letter, "which was *not* a mean note." She has been working hard, 6 miniatures in 5 weeks. George can't see her on the morning of her first class because he himself has a class then, he writes on 16 November. He is following lectures in analysis and geometry. He does very much want to speak with her. He is sending along books on Ruskin.

Mabel writes on 27 December to thank George for the Christmas gift, a book. She wants to hug him, as she hugged her mother and father. Céline had placed one of his books on her pillow, and it helped her rediscover the wonder of Christmas that she experienced as a girl. She writes with his stern picture in front of her, but she imagines him smiling. At this time, from short notes, Mabel is visiting George on occasion and the two of them take walks. In January 1908: "For half an hour this afternoon, I really wanted to see you, but I couldn't let you know and I also thought that it would disturb your work." George, perhaps on the 27th, regrets being out when Mabel called. He spent the weekend in Brussels, taking in art exhibits, an Italian play, and a function at the Solvay Institute; he encloses a ticket for the Sunday concert, which he can't attend, and he marks his signature with asterisks, presumably kisses. On 31 January she confides: "Did you know that you are the first male friend I've ever had?....I don't know why you have befriended me, but I hardly worry about it. And I am happy for it." She has an idea for a concert poster.

In addition to practising her craft, Mabel appears in the Flemish world of letters as an authority on art. She joins a group of artists and writers to propose a journal, *Iris: Vrij, algemeen nederlandsch maandschrif*, directed by Gerard Ceunis in Ghent. According to a prospectus, the journal is projected to appear in February 1908 and to focus on literature, art, sociology, and philosophy. The editorial board includes, remarkably, Emile Durkheim and Lester Ward (a pioneering American sociologist whose work George will eventually read), along with writers Frederik van Eeden, Paul Kenis, and Gustaaf D'Hondt; on the artistic board with Mabel are Jean Delville, Jean Delvin, Georges Buysse, Emile Claus, and Camille Lemonnier.[37] It is an association that would have honored even a senior artist or scholar.

---

37. NYPL, for prospectus. Christophe Verbruggen, "Tijdschriften en literaire strategieën na *Van Nu en Straks*. Carrièreplannen met *Iris* (1908) en *Nieuw Leven* (1907–1910)," *Revue belge de philologie et d'histoire, 83 (2005), 1261–86*, for *Iris* generally.

CHAPTER 3

Revolution

In the domestication of a martial image, we fall in love because we have been shot by Cupid's arrow. We are, however, inclined to receive the assault. When Cupid struck George Sarton, he was looking for a wife. Mabel Elwes was more than charmed by his fervor.

George writes to Mabel on 8 February 1908 as "Kaatje," a reference to Paul Spaak's play which they both admire, possibly read at Reiner Leven the day before.[1] Thanks to Mabel, he is once more going in the right direction. He will rewrite the *Chaine d'or* over the coming summer. He saw Mabel's portrait of Kaatje yesterday, which he very much likes. An aquarelle of Mabel's hangs over his bed. Upon waking, he gazed on it so long that he worried about arising. Honest George then breaks the spell: "Happily I remembered my obligations in time," and in a parenthetical remark: "The other two aquarelles appeal to me less." The cook is baking waffles for tomorrow morning, but if George works well he will treat himself to one in the evening. Although valedictions are notoriously difficult to gauge in French, George closes, for the first time: "Je vous embrasse comme je vous aime, avec tout mon coeur."

The *Chaine d'or* must really be recast, George's friend Emile Masson writes upon reading the manuscript. "It shows qualities that are still inchoate, qualities about beauty that have not yet found their proper expression. It provides evidence of a rich, refreshing, and noble youth that one day will emerge in a sovereign masterful style, provided that care is taken." George's hand is not yet strong enough to forge a durable text. Masson sends along detailed comments, pointing out infelicities and awkward metaphors. George's description of the Flemish countryside shows "no Flemish character" whatever; he overuses vague adjectives. He is mistaken to think that anarchism is something that precedes communism, at least according to Mikhail Aleksandrovich Bakunin and Peter Kropotkin. The characters move in an unrealistic setting. George's group of romantic revolutionaries, the Sympathiques, are not at all to Masson's taste, for they "have too great a fear of revolution." Masson would not follow any of the characters into the streets. He applauds certain of George's passages, for they mirror Carlyle

1. Paul Spaak, *Kaatje*, preface by Emile Verhaeren (Brussels, 1908), a lyric poem in four acts, presented as a play with music by Victor Buffin. The son of the author, Paul-Henri Spaak, became president the General Assembly of the United Nations in 1945. George notes the book in his diary on 8 February 1908; Reiner Leven met the preceding day.

and John Ruskin, even though he finds the narrative artificial. "Yes, I know, there is Tolstoy, the moralists in general, Ruskin, Maeterlinck, etc. But I suspect *that you must get out into the street*."

Masson reflects on the bloody nature of revolution:

> Murder is a crime if it is committed for a criminal end; but it is heroic if it is committed for a heroic end. I want to wipe out the ferocious beasts. We must have Hercules today, not Ignatius de Loyola. I don't want to break the sword, but to have it wielded by the Just, who will give up the crucifix. If there is a God, this is the example that He provides in all Nature and in all of history.

The problem, revolutionary socialist Masson thinks, is that George's novella is "too idyllic. It is too evident that your 'Reflections on life' are golden, but they are connected by a wire of inferior metal; the links are gold, and the chain is lead." Masson concludes in a postscript: "Remember that I have not only been severe, but severe without mercy." In a second postscript, written on the following day, Masson softens his comments by saying that George's sentiments are his own.[2] A writer of wavering conviction would have been devastated by these comments, but George replies immediately to Masson, who sits on the letter for a month.

Masson writes again in January 1908, to say that he supports anarchist Jean Grave, even though Grave's new book does not please him aesthetically. Masson is studying Nietzsche's *Gay Science* and has plunged back into the correspondence of Carlyle—whom he calls the Rhinoceros, borrowing from a tale of Nietzsche's. Masson wants to put together a book of anarchist biographies, *Porteurs de flammes*, and first of all will come Elisée Reclus (with whom George corresponds). Then to George's letter. George is right, that "whatever has no apparent use must be considered useless." Masson, different from George, thinks that "a logical unity to the universe does not seem undiscoverable [introuvable] to me, in the final analysis," even if there is still a hidden geometry that one must try to see.[3] The heart will win over the mind. George finds Bakunin and Kropotkin to be sentimental. But Eugène Carrière insists that sentiment must be given its place, perhaps first among all things.[4]

---

2. HGS, Emile Masson to George Sarton, 1 December 1907; the undated postscript associated with the letter by inference. In Lewis Pyenson, "Ecrire et agir: Emile Masson, ami et correspondant de George Sarton," trans. Ph. Gury and S. Moal, in *Emile Masson, prophète et rebelle*, ed. J.-Didier Giraud and Marielle Giraud (Rennes, 2005), pp. 173–89, the editors have inserted a number of notes that comment on my text. In their opinion (p. 181), Masson's appeal to violence here is atypical.

3. In "Ecrire et agir," the Girauds are correct that I mistakenly transcribed *introuvable* as *intournable*.

4. HGS, Emile Masson to George Sarton, 18 January 1908.

George Sarton's second novel (we shall see that it is a novel in its final version) is placed on the shelf while George focuses on science. He writes to Mabel when she is at the theater on Wednesday evening, 19 February 1908, explaining that the violets she gave him protected him while he worked: "You are my good guardian angel." Mabel travels to Paris at the end of February with Céline, and George prepares a vademecum, including directions to good vegetarian and temperance restaurants. According to his diary, George becomes a "strict" vegetarian on 16 February and also gives up alcohol, joining the Ordre international des Bons Templiers, a temperance organization promoted by Auguste Forel, the Swiss psychiatrist, entomologist, and sexologist. "My thoughts follow you to Paris," he writes on 22 February. Reiner Leven will henceforth spend a half hour reading from a work of philosophy. George will present Epictetus and also Marcus Aurelius. (Epictetus, the first-century Stoic, for whom reason can grasp nature and yet also be responsive to God's will, enjoyed appeal around 1900.) Two of the women members are enthusiastic about the new departure, "but I fear that it is, once more, only a girlish caprice." He wonders if the members of Reiner Leven have a 50% discount for a concert organized by the Flinke. On 24 March 1908 Mabel writes to her mother from Ghent about the results of the concert. "The theatre was *nearly full*."[5]

Mabel returns from Paris with a photograph of the "Crowning of the Virgin" (possibly by Fra Angelico, held by the Louvre) which George had recommended to her: "One of the most beautiful and greatest things made by humankind," he thanks her on 29 February 1908. The previous day he passed to see her but then decided that he had "not worked hard enough to have merited seeing you," and so he walked home. He wonders if Céline forgives him for the thoughtless lines he sent her in Paris. "I wrote that to annoy her, and now I am quite sad that it has come to pass. This really shows you my two sides." George is unable to read Marcus Aurelius to Reiner Leven because he must travel to Brussels. "Would you read this first part?" he asks Mabel on 4 March. He thinks fondly of the walk he took with her by the sea at Coq-sur-Mer, where Mabel had gone to paint and rest.

Their affection deepens. Mabel writes on 7 April from Coq-sur-Mer, where she is at the Dangottes' summer house, that she misses George, she "wants to hug him once, but with all her confidence and all her affection." George writes from Rhode St-Genèse, where he is with Raymond Limbosch, on 14 April. "If only you were actually here in flesh and bone with us." Mabel wants to visit him and Raymond, she writes on 16 April. George is leaving Rhode, but he "really wants to see you," he replies on 21 April. He is busy in the library and with his courses, but he urges her to come in the afternoon: "Come, come! You are urgently expected! I haven't

5. NYPL.

seen you for an eternity," writes "Vôtre George." Mabel visits immediately, that day, and George kisses her.[6] And a second letter that evening. "I will sleep with the angels tonight, thanks to you, O my friend. I have so often dreamed of kissing you." He was ecstatic: "I am so happy that I doubt my own happiness." He imagines that Mabel is "as ridiculously happy as I am. Thank you, Mabel!" She was leaving for 2 weeks. Could he see her tomorrow? "Excuse my penmanship: My hand trembles with joy." And for the first time, in closing, "A big kiss from your George." George writes to Raymond Limbosch on 22 April: "I have thought a great deal about what you said on the subject: Certainly her health gives me very great concern, but that should not be a serious obstacle to our marriage?"[7]

George enthusiastically writes to Emile Masson about Mabel and his love for her, and Masson wonders if he should dedicate an upcoming work, *Le Rouge-Gorge*, to "Mabelle."[8] Mabel, at Rhode, is circumspect on 23 April, commenting as usual and adding to her usual valediction "Je vous serre la main bien fort," for the first time "de toute mon affection." He continues, on the 27th, "as long as I live you will have in me a sincere friend, a friend for any obstacle." But he is not yet worthy of her. "If I become one day someone, Mabel, it is in large part to you that I will owe it." He has something very important to say to her. "You know that I am a sensitive person, that under my rough appearance I hide almost an unhealthy nervousness at certain times." Her last letter upset him. It closed with her offering him a good handshake. He concludes that he was too forward in kissing her (de vous embrasser), and in kissing her by letter, and she was calling him to his senses. If Mabel thinks "that two friends who love each other [qui s'aiment] can kiss without danger, you can easily let me know. *If not, I won't return to the topic*." He concludes by offering her "a good and firm handshake." In French correspondence, it is sometimes hard to know when lips touch and when love is intended. Here we have a clear indication.

George and Mabel meet. On 28 April 1908 George thanks Mabel for their conversation. He has not been as open as she was. He didn't have the courage to say everything he wanted to say, but he sensed that there were no secrets between them. He uses the second person formal: "For a rather long time I have dreamed of asking you to marry me, but it has only been for several weeks that I want it passionately, violently." Hence his state of mind.

> Because it is certain that if I would marry you it would be to have a family, *to have children with you*: Any other union seems to me against

6. Diary, 21 April 1908.

7. Van Oye, p. 51.

8. HGS, Emile Masson to George Sarton, 27 April 1908.

> nature, an immoral thing, an improbability too, given my temperament. However, you know what *terrible revolution* the conception of a child engenders in the body of a woman: It's a painful sickness that lasts 9 months & the convalescence. It is not surprising that I have thought a great deal about these things: You know, Mabel, that my mother died soon after my birth; my birth was the principal cause of her death. I believe that my mother was stronger than you. Also I think that to have a child irremediably compromises your health, if not worse. And I don't even think about the outcome of the child (which is however still more important), no, I think only of you. Also, after long reflection, I believe that a sexual union between us would be a bad course.
>
> I beseech you, Mabel, to pardon me for relating all these things. I ask pardon on two knees for making you suffer, you who have only ever done good. But truth demands it. I do not want any mysteries between us. And I swear to you that no matter what you may be going through, it is worse with me.
>
> Why has all this happened? Why do we have the same ideas, the same tastes, why do we love each other, and this obstacle comes between us? God knows. We must bow before destiny. I do not know what it will do for you. For me, not being with you, I will not be with anyone else. I will never marry.
>
> But who knows. Perhaps later, much later, when the little children who ask to be born will have ceased their cries to us, it will be possible for us to be united, to end life together. I look forward to this distant light to avert my eyes from the present.
>
> And now what must we do? I think along with you that we must avoid to do all the things of nature to raise our sentiments and to depress in us the desire that we cannot satisfy. To do otherwise would have us suffer uselessly. Here again you have been wiser than I. But nothing prevents our intellectual union from becoming closer from day to day. Let us strive to know each other well and to help each other. Do you want this, Mabel? I have no secrets from you. We will not know this deep solidarity that uniquely creates a common life, but at least will we not be able to have common interests like brother and sister?
>
> I am sure that I am expressing myself poorly, but perhaps your heart will understand mine, which suffers. Have I acted badly to tell you all this? I think that there cannot be mental reservations between us, but only *the clear light of sincerity, which purifies all.*

The two lovers meet on Wednesday morning, the 29th. George writes to Raymond Limbosch that Céline helped him understand Mabel: "We cannot marry, it would be ill-advised. But not being able to be hers, I will never be another's; I will never marry."[9]

Mabel writes to George the next day, also using second person formal:

9. NYPL.

I sensed that you understood that I could not speak to you in depth about what preoccupied both of us. I wanted desperately to come again to see you yesterday evening, but I would have troubled you by my own calm resolve, and it is also for this reason that I have not written. If you have expected a few words you will understand and excuse me, perhaps.

George, my George, all I would like to say and what I suffer from not knowing how to tell you, comes down, I think, to a few words. I love you with all my soul, but without passion. You cannot suffer under this restriction because you love me so much. Oh, George, George, it is possible that if it were not forbidden to me, I would love you in the same way, but since I *cannot* give you that, I will love you in a different way, so purely, so entirely, that *surely, surely*, I will give happiness to your life just as you will to mine.

George, why do you ask my pardon? No, no, *I* am the one who must ask pardon from you. Oh, I hurt, I *hurt* so much, that I do not know how to think clearly, as I should to know how to help you. That my own life is condemned to remain incomplete, I believed that I accepted that, but *because of me*, you, whom I love so deeply that every day I *want* you to be happy and strong, you, your life must be limited, your actions in life restricted [vos droits de vivre se couper], I do not know, I do not *know* how to accept it.

I have been culpably blind. I don't doubt that you can love me for all that (as only a short time ago). I feel so powerless, as if life played with me and handed me a role of abominable cruelty.

Listen, George, my friend. I promise you here to struggle against this revolt which most of the time covers me in a whirlpool where I no longer reason at all. Oh, have confidence in me still. Give me your hand. Yes, we will help each other walk in life. You don't hide your [illegible] suffering. I will never know how to relay the thanks that well up in me because you have now said everything.

Don't worry about hurting me. Look. I told you that I do not love you with passion. Something tempted me to do it, but I believe in you. Believe in me. I will deserve your confidence.

So often it is said, "It is not necessary to say that." "He knows it." Or, "That will uselessly hurt him." It is infinitely better to hurt *right away* and to explain.

Above this struggle and the hurt I feel is a light, [?] and sweetness that will never expire [Mabel changes to English], *that of having been greatly loved*. You have given me the greatest thing in life and you *ask my pardon*? No, no. I could thank God if I did not feel it had cost too much.

Raymond Limbosch writes to Mabel on 1 May 1908 about her attraction to George. He emphasizes that George continually talks to him about the hurt he causes her. In a second letter on the same day: "The form of affection that George displays for you must have given his conscience a rude shock; otherwise how could it continue? He thought about it over and over, as a thing that could not be impossible...." Raymond advises

Mabel to see George.[10] Mabel and George meet again. Mabel writes to George on 2 May:

> I just thought that you will have been able to see just now that I was crying. Yes, I cried yesterday in the evening, but good tears that left my soul refreshed and calm.
>
> You cannot think that I am sad or that I am unnerved with hurt [j'ai mal qui épuise]. It seems to me truly that yesterday *strength* [des sources] flowed in me and that my life marvelously grows and becomes refined because of you. I would have liked to tell you this, and when I was outside I only realized that I said nothing, nothing. I would have come back to say it rather than to write, but I was not feeling well.... Can I come by this afternoon? Just for a moment, on the way to Madeleine [van Thorenburg]'s? Yes, I will do that. I would like not to *write* when I *can* speak to you. If you are not there, I will leave these lines.

Raymond writes to Mabel on 7 May that he is "very glad" that Mabel sees the danger of responding to George's affection. "You are right: Don't flee from him, but don't *seek* him; don't be his consolation. It would be quite bad for him and dangerous for you, I believe. Yes, it is not love that he likes in you." George's letter shows that he has not yet calmed down. Raymond will interrogate George in a few months, to see his state of mind.[11]

George is close to losing his way, he writes to Mabel on 10 May, "I want *absolutely to be master of myself.*" They meet. He is glad that they both think the same way, he writes on 16 May: "In sum, first, I am firmly convinced, as you are, that the child creates the true conjugal union: this identity of interests and affection." It is perhaps from not having children that an acquaintance has suffered the most, notably her quasi-abandonment by her husband. George was very serious about asking Mabel to accompany him to Brittany. They could work together, he on his book, she on her painting. The question of money need not concern her: "You know that all that is mine is yours." George does the calculation down to the Belgian franc. He thinks that his father will give him F 400, and the sum will be sufficient for 5 or 6 weeks with Mabel. But Mabel has already replied on the same day:

> George, George, I believe that I very poorly expressed what I wanted to say about the friendship between man and woman. This is simply what I think: To have children together (between two people who also understand each other morally) *must* create a friendship, a connection, a communion that *nothing else* can give. Excuse me if I hurt you in recalling that. George, if I greatly enjoyed seeing the Carrières, I also greatly suffered. Many sing the praises of the infinite joy of motherhood [a theme

10. NYPL.

11. NYPL.

of painter Eugène Carrière's], and the divine tenderness of children, with a penetrating insistence that really hurts me, that I cannot forget, that my thoughts spin in a void, in confusion. This is why I am leaving for the country. I have a lot to do before I leave, and I know that there I will be able to regain my composure.

Nevertheless, I am deeply happy to have seen them, and I *want* you to see them, too. (There was a moment when I thought not to tell you to see them.)

I have an unreasonable desire, a "crazy desire," to go off with you to Brittany—you didn't say so seriously—oh, yes, I know that it would be hurtful, just as I for the moment do not *want* to be old, but with a heart young enough to want to go far, to seek the sea, to love you *well*, to make you happy (I am somewhat afraid to say it to you but I cannot hide it from you), and so, perhaps you understand. To be truly beautiful, I must be beautiful *inside*, above all. But I want to be so physically also. I always wanted to have a very beautiful body. I have really suffered from finding myself ugly. I want to lead a life as healthy as possible, so that, later, I will be at least in form and as beautiful as possible.

Don't laugh, George, and don't only think, "coquette." I truly believe that I am fine, and by telling you this, the "small fear" is only a kind of modesty, showing you a very intimate feeling, and not a shame.

Since my physical development became for me a duty and a habit, I have become less concerned with my imperfections. I felt that my arms were too long, and I don't know what else. It seemed to me that I appeared *ridiculous*. I can tell you that I no longer think so at all. I am still sometimes sad at not making more progress toward perfecting myself. I have days when I look in the mirror and find myself plain....

The letter breaks off, remaining a fragment. It is possible that the arrival of George's letter caused Mabel to stop writing. She begins another letter on the 17th:

My dear George, No, we *cannot* do that. The plan must be turned down. I was egotistical in showing you that it would have given me great pleasure, but in the end I should have known that for both of us it would be crazy. And it took me until the morning to arrive at this decision. I am sad that I don't help you more by being stronger, anticipating things better. Here is a good, firm handshake.

She begs to be excused for hurting him so. She would soon be off. George writes on the 17th. He is "out of his mind; I know that reason no longer guides me....I suffer more than I can say." He urges her not to decide anything yet. He wonders: "*Is madness not being careful enough or being too careful?*" The prudent course would be for him to flee from her, to detest her, "to destroy my heart." And on his envelope he scrawls, "You must ignore the word 'detest.'"

Céline writes to George on 21 May. She had hoped that George would leave Mabel alone, but his recent letter really shocked her. If George really loved Mabel, he would not have told her that she could not have a child. Mabel clearly told him that she did not love him; how dare he persist! Mabel suffers from angina. Does George have in mind waiting until Mabel is strong before marrying her? Does George know Mabel well enough to know if he can make her happy? In fact George knows little about Mabel; she knows him better than he knows her. George has for the moment lost his head. There is an enormous gap between his words and his conduct. Céline finds a superficiality in George's professions of love:

> You puff yourself up with words, you exalt yourself with words. You claim to be the injured party who scratches his wound and rejoices in it, forgetting that these hurts weigh on others. Remember that just a year ago you wrote letters to Mélanie Loreyn that some day would make you grumble like a child? Remember that you wrote to her that you would wait for her forever?

George really has to think carefully about his actions. Love requires time and patience. She will not press Mabel to ask Dr Nyssens what George wants, presumably information about contraception; Mabel is too upset. "If she wants to know about it, she will take care of it, either now or later." Dr Nyssens does not know her medical history.[12] Céline hopes that George will take her words in a constructive spirit. George, on 28 May, is drawn to read Arthur Symons, the English admirer of the Symbolists, who was notorious for his sexually charged poetry.[13]

At the end of May 1908, Mabel is in Ipswich. George sends her Biblical passages from Isaac and Matthew. Mabel is perplexed, she writes on 3 June. Why had he done so? Had he read them? Did he particularly like them? The verse "not all who say unto me Lord, Lord..." followed Mabel at the time that she lost faith in church religion. On Sunday evening, she and her mother attended the Unitarian service—her mother's church—and she found peace there. She wishes that he could have been next to

---

12. NYPL, Céline Dangotte to George Sarton, 21 May 1908. Probably Ernest Nyssens, whose notions on diet and teetotaling would have been congenial to Sarton and Dangotte. Ernest Nyssens, "Sur l'antialcoolism dans ses rapports avec le végétarisme," *Belgique médicale, 16* (27 June 1909), 303–6, where he advocates both vegetarianism and temperance; on p. 304 he cites Auguste Forel's view that alcohol use results in degenerate offspring. On a letter of Emile Waxweiler's of 30 October 1906 (HGS), George notes a publication by Nyssens on dietary reform. In his agenda, George visits Dr Nyssens on 7 March 1908. Van Oye, p. 30, gives Dr Nijssens-Verlijsen of Brussels, who later writes on spiritism. George's first cousin Marguerite Sarton, daughter of Jules Sarton and "almost the only friend of my childhood," married Emile Nyssens. Diary, 12 March 1941.

13. Diary, 28 May 1908.

her at the service. She describes the church and service in a number of pages, struck particularly by the sermon comparing Epictetus's version of Stoicism and Christianity. Mabel feels ill, discouraged, "every day dying a bit." She knows that she should pull herself together. "Don't judge me too hard, George, because sometimes I am afraid." She urges him to spend time in the Dangottes' garden at Assels, in the village of Drongen, or Tronchiennes, near Ghent. On 4 June, George sends Mabel the latest volume by Romain Rolland, which "he has just read with enthusiasm." He has read much of Goethe. "Now there has been a big change in me." Raymond writes to Mabel on 9 June 1908 about the change in George (he writes, "Georges"). George seems much less unhappy, "but one never knows with this precious boy!"[14]

At Mabel's suggestion, George travels to Mons to see the art galleries there. It was a fruitful trip, he writes to Mabel on Pentecost 1908, and he greatly admires the paintings of Paul Gauguin: "I like the violent harmonies of this primitive." He appreciates a magnificent canvas by Maurice Denis. He saw about 40 paintings by "Roig, a young Spanish painter still unknown who has an extremely original vision of life, but some of his canvasses are ghastly." The number may be exaggerated, but the reference is no doubt to Pablo Ruiz y Picasso. George visited Irénée van der Ghinst. "From the sexual point of view, I am much closer to him than to Raymond: He (V. d. G.) cannot be considered a sensual person. He wants to name me his heir to make sure that his estate goes for educational purpose"—a surprising decision, given Van der Ghinst's live-in girlfriend Marianne. Van der Ghinst does not share all George's political opinions (only one friend does, Maurice de Saegher).[15] George explains his decision to resign from the Bons Templiers and Reiner Leven. It follows

> an evolution in my thought from the sexual point of view. I cannot talk to you about these questions, which you could not understand or, worse, would misunderstand. I shall try to begin again to have a [illegible] with courage, a new life. As Maeterlinck, M[auri]ce De Saegher, Forel and others whom you do not know contend, it is bad, almost criminal to carry useless prejudices that we owe to Christian civilization. I shall begin my life again. I want to live. It is my right.

14. NYPL.

15. Maurice de Saegher is the first person to share all George's political ideas. Diary, 26 May 1908. Cécile Douard, an advocate for the blind who was herself blind, recounts a meeting with Maurice de Saegher (blind from birth) at an early 20th-century conference in Paris. De Saegher is *docteur en droit* and studing for the *agrégation*. They discuss supporting educational opportunities for the blind in Flanders. Jacqueline Aubenas, Suzanne van Rokeghem, and Jean Vercherai-Vervoort, *Des Femmes dans l'histoire de Belgique, depuis 1830* (Brussels, 2006), p. 121.

He begs Mabel not to judge him, but to speak to him only about herself. He signs himself "Your devoted friend, George Sarton."[16]

Mabel writes on 9 June. She implores him to remember that she disapproves of any sex between them:

> Oh George, George, if you knew how much I *hurt*, how much I am tormented by thinking of you, you will have pity. Perhaps it is true that I do not understand, but you have not *tried* to make me understand, fearing I understand poorly? There is something unjust in that. It seems to me that you treat me indifferently, that because I did not know how to love you completely and immediately you want me to take away a part of your confidence and your friendship.
>
> Yes, you do not *want* me to speak about *you*, I must speak *only* of me. Don't you understand that I love you enough for that to bother me terribly, to the point that I tremble with my entire body and my entire soul when you announce to me calmly, apparently, a radical change of mind for the conduct of your life, and that this change entails a comprehension, a conduct opposite to my ideals. And I must act—why?—to respect your individuality? For fear of my own ignorance of life? But if I see you in front of a precipice where you want to throw yourself, I should stop from preventing you because I am not *sure* that death would not be better for you?

She would not be able to do that.

> What you call living, what is that? Satisfy your sexual appetite? Oh, I also would like to see you *live* more, but not in this way. That seems to me, on the contrary, to be *dying*, to give in to instincts while imagining that it's *living*!! It's the term for weakness and lack of balance.
>
> George, be careful, listen, listen the words of *other* voices in you. They are there. Don't run to another *extreme* in your resolve. I should be calm and reasonable, and I don't know. I am so upset that I only have one thought, to appear before you on my knees and ask you, beseech you to promise me to *wait* for a few more months, George.
>
> And perhaps you are annoyed, impatient, you will say that I understand nothing. [Mabel switches to English] Indeed, indeed, I believe I understand a good deal—perhaps not with my *reason*, but with my

---

16. The Bons Templiers abstained from both alcohol and premarital sex. "If I were not Bon Templier, I would really prefer to drink several glasses of triple Uitzet" (a local beer) daily. Sarton to Raymond Limbosch, 17 May 1908, in Van Oye, p. 53. "The B.T. must themselves not only abstain from alcohol but also *be chaste*." Diary, 26 May 1908. According to Paul Pelseneer, the Bon Templiers in Ghent were friendly with the occupying German regime beginning in 1914. HGS, Paul Pelseneer to George Sarton, 17 December 1918. Pierre Hoffmann was a prominent Bon Templier. In 1916 he became rector of the University of Ghent, transformed from a French-speaking institution into a Flemish-speaking institution by the occupying German army. *LM II, s.v.* Peter Hoffmann.

imagination and feelings, and I do *not* "judge you." [Back to French] Yes, I know that you have a great need for tenderness, for affection, for caresses, that you have been deprived of them for so many years, that it is a hunger, a thirst in you, that you are weary of being always alone, of not being able to rest your head on someone you love. Oh, I understand, I understand, so well that you will never know how much I have suffered from not being able to give in to your desire and give you all the tenderness that flows over in me. It is a temptation that I felt continually to wrap you in my arms, to caress you sweetly as I would my child. (For a very long time it seems to me that I have loved you tenderly. When you left for Paris last year, I hugged you in my imagination to protect you, even knowing that it was a superstition and was "useless.") Why did I not give in? I believe because I felt vaguely that you would want me more; above all given that I knew that you loved me passionately. It would have been cruelty and not goodness to give you tenderness in place of love.

George, if *I* don't know how to make sense of things, neither should you always understand. When I came to speak with you the other day, I felt when I left that you had understood almost nothing about what I had *so much* wanted to tell you. I am going to try to make you understand once more, to say everything.

If *you* need tenderness—I need it also—and I have been afraid of loving *love* and not *you*, of undergoing the attraction of *being* loved (because I also thirst for it), yes, of being loved passionately, of being transported in a whirlpool of passion in which I will not be able to judge the future. It is a duty, not only to oneself, but also to others, an *essential* duty, *not* to do this. Unhappiness can result for both people, terrible suffering of a spoiled life. And you would tell me that I take things too tragically! (However, I've thought of that. It is somewhat true for me, and I am trying to correct it.) I really felt that there was the possibility of such a tragedy for us. I still feel it.

Mabel then writes about the importance of her own studies in design and painting, and her concern that married life would compromise her personal and artistic independence. She continues:

You ask yourself, George, why I keep returning to all this? Because I feel attracted to you almost irresistibly. For days I speak to you in a low voice, continually, I dream about you at night. I feel violently pulled in two directions. I am so weary, so weary.

Perhaps I should not show you this struggle or tell you all this. Nevertheless It seems to me that I treat you in the way that I would want you to treat me.

The reasons for my wanting to work are not the only ones that prevent me from marrying you. I believe that, despite certain ideas in common, we will never understand each other with regard to others. Please wait. I think that I often make you impatient. With Céline I had a life so wrapped in the tenderness of infinite patience, I was able to understand

> and be so close to her, that I feel that I will ask for much, just as I will be ready to give much. Do I sacrifice some needs? Do I have simple enough tastes to know how to live harmoniously in the middle when you dream. In sum, everything would be resolved if we loved each other beautifully and completely enough.
>
> I am ridiculously sensitive near people whom I greatly love. I too easily take small things "in a tragic way." I have been fighting against that for years and I made little progress. Happily, I also know how to fully enjoy, be profoundly happy with "small" courtesies, above all when they come from people dear to me. Are you not too "simple," perhaps too "important," to understand this? The other day when I spoke to you and it was so difficult, I felt you, despite everything, so far away. You told me rather seriously, "It isn't necessary to torture this letter like this." (I had folded it, perhaps, many times.) It struck me as a petty remark, dry and cold. Maybe it helped me, because it awakened my pride and I became calmer. But I found you a bit hard, a bit authoritarian (I relate this as an example.) Yes, I was calmed down when I left, but I still felt roughly treated. I saw that you understand many things differently from me. Oh, I no longer know who is right.
>
> One more thing. I am also not sure that you do not love the *love* in me, the possibility of love, and not *me*. Don't protest, George. I sense that alongside that you have had a true and beautiful friendship for me. But to me, it is not deep enough to guarantee us married happiness. Also, against my better judgment, I think about Mélanie often. You also loved her a great deal. (She, or *love* again, I can't tell.) Is it certain that you are not mistaken again? Will you not find a wife one day who would make you happier, complete your life better, than me?
>
> All this torments me, if only you knew how much! And the point of it, you will guess it from this letter if I do not say it, is that I began to love you passionately, how much I no longer want to acknowledge, because now it only brings me little joy and so much suffering and doubt. I hate myself for being someone who does not deserve life and love, does not deserve the great things that are simple and true.
>
> George, I have shown you all the words that I can, my naked soul. Please receive it dearly. In return, give it what it asks of you, a hold to your change of character and conduct. Write me something right away.

And she signs it: "I hug you directly, as I did a year ago, putting everything pure, loyal, and tender in it." Then she writes a long postscript. She can't go to bed without asking his pardon for the harsh words in her letter or if it seems as if she judges him. She wanted very much to have his reply, and she was also afraid of receiving it. "I love you without understanding you. I am even afraid of you, and not only afraid *for* you. I am afraid precisely of the unknown in you, of the incomprehensible, for me." She would rather have him angry at her, to receive harsh words from him, than for him to remain indifferent, or to "greet me as a child who cannot understand." She asks, if he had a son, would he teach the boy about his new view of life?

> To live is to work, sing, laugh, love humankind and the world, be very childlike while being completely a man or woman. To live is to *see* and not only to look, to live is to love and not only desire, to live is to grow every day in small ways, not only in grand ones, to live is to have immense joy, to live is to be able to suffer extremely and come out the winner, to live is to give completely and continually without calculating and without ever forgetting to give to oneself *forever* what is the most divine that one has, to live is to feel one in infinity, to feel intimately connected, inexpressibly, *inevitably*....

The next afternoon, 10 June, Mabel writes to say that if George thinks that she is not the one for him, or if it is not a marriage that will make him happy and he "would like to follow the course of Van der Ghinst," who is living with his lover Marianne, she asks only for his loyal and complete candor. She doesn't understand how Van der Ghinst acts, but for a long time she has thought that he was someone who could not live "freely" without prejudices. "I wonder what would happen if there were children, then if he thought to leave this woman one day. What *could* she become? And all the lies he would have to speak at home and to others. If something like that is done through *principle*, it must be done openly, and in this case I think it beautiful and glowing." But she finds the present arrangement mean. Did he sign the letter where he writes about these things "George Sarton" to distance himself?

George writes on 10 June, affirming that she knows all his thoughts and feelings. He shares her description of a true life: "It could not have been defined better." He no longer thinks "that chastity is *in itself* a virtue." Maeterlinck makes the point often. But he swears that he remains worthy of her. And again, in a second letter on the 10th: He wants to cry before Mabel on his knees. She misunderstands what he means when he says that he would sacrifice his wife to his ideals. Suppose that he were married with children, did not have much money but had a university appointment, he would have to chose between his appointment and his ideal of justice. He "would not hesitate to resign, which would be equivalent to sacrificing, for me, my family. Or following revolutionary writings and actions, I could be thrown in prison—my family would again be sacrificed." He would never exploit his wife, "as so many men have done, for their work, or their glory." In a household, man and wife should each have an occupation. They should share household tasks equally. They should have

> *a common mission: more than children*, I am convinced, that is the true cement of marriage. [Auguste] Forel, in his fine book, makes the point

> frequently. Notably on p. 511 he writes: "Marriage should be considered at once as a means of satisfying the sexual appetite and as *the moral and social school of life*, not only as the port or the refuge of egoism. Sharing of duties, absolute equality of rights and *social work in common* are done to solidify the sexual connection between spouses."[17]

If George's wife has the same views as he, his mission will also be hers. George is not surprised that Mabel does not understand him, adding, astonishingly: "If you loved me, you would understand me...Love is the kingdom of certainty." There is only one thing about Mabel that has remained hidden to George, "*your race*," that is, her nationality. He doesn't understand the English character very well, and so he misses a certain part of Mabel's constitution. Spending time in England for his studies would help overcome this "difficulty." If he was distant in their recent conversation, he sought to hide his own troubles. He doesn't understand what Mabel means when she writes about the love in her. He knows only that he loves her, with his entire being. A man loves a woman either because he needs love or because the woman attracts him. He could love another, but fate draws people together. If he had not met Mabel, he would perhaps have fallen in love with another woman. George does not want to analyze it. Mabel brings up Mélanie Lorein. If she had reciprocated, he might have still been in love with her. Love is an imperative of nature, and in this sense one loves love. "Have you thought about that, Mabel, *why* I love you? And you not?" Her letter gave him reason to hope. "*Whether you love him or not, your George is and will remain worthy of you. I swear it.*"

Mabel will not scrutinize George's love, she writes on 12 June 1908 from Hampstead in London, at the home of her friend Sophie Jackson. He knows that she loves him, a love that will govern her life, which she did not have before. She desires to walk with him by the sea, to feel that they will help each other in life. She senses that she loves him so deeply that it will bring him only pain. She needs time to see clearly, and she wants very

---

17. Auguste Forel, *Die sexuelle Frage: Eine naturwissenschaftliche, psychologische, hygienische und soziologische Studie für Gebildete* (Munich, 1905). On Forel generally, Willi Wottreng, *Hirnriss: Wie die Irrenärzte August Forel und Eugen Bleuler das Menschengeschlecht retten wollten* (Zurich, 1999). Mabel probably read a French edition; one appeared in Paris in 1906. Forel's temperance and his socialist sympathies would have attracted George and Mabel. Notably, Forel persuaded the Belgian socialist Emile Vandervelde to become a teetotaler. Emile Vandervelde, *Souvenirs d'un militant socialiste* (Paris, 1939), p. 136. Albert Einstein's path intersects that of the Sartons (Mabel was in Zurich when Einstein was a professor there); Albert Einstein's first wife Mileva Marić read and annotated *Die sexuelle Frage*, notably a passage in the section on Utopian ideas about ideal marriage in the future: Husband and wife "will have learnt to find the truest satisfaction in the accomplishment of their different duties, and in their work in common for the benefit of society." Michele Zackheim, *Einstein's Daughter: The Search for Lieserl* (New York, 1999), p. 270.

much to work. George is grateful for her lines, he says on 15 June. He will wait until after his examination to write to her. Raymond Limbosch writes to Céline Dangotte on 12 June that sexual conduct cannot be discussed in a general sense. If chastity bothers George more than sexual abandon, then he "is moral in taking a mistress. But how will he find someone who suits him and finds mutual satisfaction in the arrangement? What about the pleasures involved? What about the long-term effects? Chastity for its own sake is no virtue, as we see in the case of Catholic priests." Raymond believes that George has reached a "crisis of principles. We will never see him act like Irénée, live together with just any woman. In any case, that is not what he wants. When he speaks about a mistress, he means a woman with whom he has only a physical relationship." Raymond asks Céline why George has spoken to Mabel about all these matters, for he is sure that George has done so. Is it not because George "has the ardent desire to be loved"? Mabel wants to ease his pain, but she intensifies it.

> First, Mabel cannot marry him now. She must get better. Then, even though I don't know much about their relationship, it seems that if Mabel is suitable for George [again, Georges] (he claims so, but I am not sure, I don't know them well enough), George hardly seems suitable for Mabel. Does Mabel know George? Does she feel between them the spontaneous affinity that allows a deliberate and secure bond? She must recognize that George doesn't take account of his character which, although improvable, could lead to a life that he does not anticipate. She will have to anticipate things for two people, because George is transfixed, he is above common sense...or beneath it somewhere. She must think about the future objectively, *as if it did not concern her.*

Raymond is confident that Mabel will reflect in the right way, taking everything into account, knowing that she may say yes, no, or wait.[18]

On 15 June 1908 George writes to Raymond Limbosch at length. In matters of sentiment, George affirms, intuition reveals greater truths than reason. He urges Raymond not to analyze Mabel's feelings. When he reads Mabel's correspondence, he sees that she is attracted to him and also afraid of the unknown in him. This is his own fault. He has never spoken with her about his plans for the future, which he has had since he was 20. Explaining that to Mabel would help her rest easier. He refutes Raymond's objection that he should not have told Mabel about his change of convictions in sexual matters. He told her that he will be worthy of her, whatever she thinks of him, and, at any rate, "my evolution on sexual matters is much more theoretical than practical." He has come to believe that he is not much of a sensual person, but rather a normal person; that is, he is not dominated "excessively by sexual drives." His first

18. NYPL.

letter to Mabel after his examination will be about his future plans and how he understands her work. In a postscript: "Mabel's sexual experience is much wider than that of the Flinke; for she is older."[19]

Céline visits Mabel in England. They traveled to the small country town of Compton to visit the George Frederick Watts museum there, she writes to George on, she thinks, 18 June 1908. The friends saw Maud Allan dance, and Mabel thought Allan was less successful than Isadora Duncan, we learn from a postcard of George's of 21 June. George is studying assiduously, and also reading the memoirs of Marc van Vaernewyck.[20] He is unfamiliar with the failures of socialist Robert Owen, which Mabel mentioned in correspondence.

Mabel dreamed about visiting George, entering his room, seeing him working at his desk, she writes on 29 June. He rose, looking at her strangely without speaking. She had a terrible premonition. George took her by the hand and had her sit in a large armchair. Everything was still and the air hung heavy. Finally Mabel felt that she must speak, but no sound came out of her mouth. It was horrible. She still feels as if she committed a crime in her dream by not speaking. She asks him to hold her hand. She has ideas for lace designs. She is beginning to think about selling her work. She ordered a life of Robert Owen for 3 shillings. Could he give it to her? If he didn't have the money, Céline could lend it to them. She has not a penny with her. She concludes with a handshake, a hug and a kiss. "When we are old, whether I marry you or not, I would like to be able still to embrace you with the same kiss as now."

George has succeeded in the first part of his elementary studies, he writes on 2 July 1908. He is now a candidate in physical sciences and mathematics. Since 21 April his life has been dominated by bouts of mania and depression. He revisits a number of the points made in his letter to Raymond Limbosch. Mabel's fears result because she "does not know the aim and the unity of my life, you do not know exactly where I am going." He will tell her. The fears may also result because of his "untypical" life.

> Essentially I believe that I will struggle all my life to live in a healthy, harmonious, and joyful way. And all my life I will have more concerns than many men—and I will have a millstone around my neck like a convict (traine le boulet comme un galésien]. I will be the Wandering Jew who has no respite, who must continually go forward. It is in the hand of God. This is *a very real fear, I agree.* Because of this excessive labor, my

19. NYPL.

20. Probably, Marc van Vaernewyck, *Mémoires d'un patricien gantois du XVIe siècle: Troubles religieux en Flandre et dans les Pays-Bas au XVIe siècle, journal autographe*, trans. Hermann van Duyse (Ghent, 1905–06), originally appearing in Flemish in 1872–81.

life will perhaps be unhappy at times. And whoever accompanies me will perhaps suffer as much as I, if not more.

George affirms that in marrying him, Mabel would be especially able to carry on her work. He has given much thought to the material conditions of their marriage. He asks again if she will accompany him to Brittany.

Mabel writes on 4 July 1908 that she is coming with her mother to Ghent, and then to spend perhaps as much as a month with the Dangottes at Assels. In English: "I have not told her anything of you except that you are a great friend." Mabel on the 6th, once more declines to go to Brittany. She needs time to think clearly about what married life with him might mean. Mabel feared that she might have or develop angina pectoralis, but really she has never been strong. What would the "Wandering Jew" do with such a wife? Mabel would want to support the life of the man she marries. Mabel has said nothing of this to her mother. Mabel now thinks to spend 6 months or a year in Zurich, at the municipal school for design just confided to De Praat.[21] Her career is foremost in her mind: "I do not see how I could continue, or rather begin to work after my marriage, at least if there were no children, which neither of us wants, and you less than me." It would console her if "I could seriously work and give form to my ideas."

Upon passing his qualifying examination, George defends his thesis in Brussels.[22] It is a study in physical chemistry of negative self-catalysis in a heterogeneous system; on 29 July 1908 the 73-page, handwritten thesis is awarded first-prize for chemistry among all theses at the four Belgian universities, garnering a gold medal, and also a silver laurel branch from the city of Ghent. It is based on the work he did in Swarts's laboratory in August and September of 1906.[23] That George persisted in writing up research carried out years previously suggests strength of purpose and faith in the consistency of his thoughts. Throughout his life he appeals to notes written years, even decades earlier.

---

21. Mabel means Julius de Praetere, from 1905 director of the School of Decorative Arts and the Museum in Zurich, who in the years around 1900 was an engraver and printer in Laethem St-Martin, near Ghent. Paul Haesaerts, *Laethem-Saint-Martin: Le Village élu de l'art flamand* (Brussels, 1968), p. 375.

22. Van Oye, p. 53.

23. George Sarton, "Etude d'un phénomène d'autocatalyse négative en système hétérogène précédée d'une étude théorique sur la catalyse, l'autocatalyse et la catalyse négative," cited in a vita of July 1912; in 1935, in an autobiographical draft for the Leopoldina Academy in Germany, the thesis is cited as January 1908 and 80 pages. NYPL. In his diary entry for 12 October 1908, George indicates that he became candidate on First July and that the thesis has 75 pages, also the relation between the thesis and the time with Swarts.

By the end of July, George is off to Brittany, the land of Ernest Hello and Emile Masson, some of his favorite authors. He sends Mabel postcards, including a photograph of himself in Finistère on 10 August; he comments that he takes in the sun "like a lizard," but that he has been forced to break his vegetarian diet by eating a bit of meat. On 7 August, Mabel describes a long walk in the country mist with the eye of a painter. She is so taken with describing the scene that she atypically switches to English:

> And then a lark rose and poured down a very rain of trembling vibrating song, as if its heart had been absorbing sweetness till it was full and overflowing and now *must* out or it would die of ecstasy. And it went up & up as if drawn by some irresistible force, and I lost it, a tiny speck against long billowy grey clouds, and still now & then for a few moments longer, the air trembled into tiny waves of far-off song.

She needles George for failing to tell her that he passed his examination with flying colors, in first place. Her mother gave her miniatures of her grandfather and great grandfather, as a birthday present, along with a cashmere shawl; Céline Dangotte gave her an old oak chest for her atelier. George writes on 6 August (perhaps mistaking the day) that he did not tell Mabel about his success on the examination because she knew it already. He spent unforgettable days with his friends, notably André Spire, "the Jewish poet."[24] A high point of his trip to France, he informs Raymond Limbosch on 11 September 1908, was meeting Emile Masson and his English wife Elsie at Pontivy in Finistère. They "are just as I imagined, noble creatures."[25] He also meets the painter Jean-Julien Lemordant at Penmarc'h, Sainte-Guénolé.[26] Spire, the Massons, and Lemordant were friends. The Massons, especially, would have appealed to George. They were sympathetic to anarchism and its related doctrine of temperance, they admired Ruskin and Romain Rolland, and they translated Carlyle into French; their support of Breton and its use mirrors George's sympathy for more widespread use of Flemish in Ghent.[27]

---

24. Spire is known as a supporter of Zionism and socialism, and for his love poems. He corresponds with Raymond Limbosch and in 1930 writes an appreciation of Limbosch's poetry in the *Anthologie Limbosch*, a special edition of the periodical *La Nervie, 11*, no. 2 (1930). George notes Spire in his diary on 2 May 1908, after Masson writes enthusiastically about his poems, *Versets*. HGS, Emile Masson to George Sarton, 27 April 1908. Spire, a friend of historian of science Alexandre Koyré's, joins the New School in New York in 1941.

25. Van Oye, p. 54, for Masson and Lemordant.

26. Diary, 12 October 1908.

27. J. Didier Giraud and Marielle Giraud, *Emile Masson, professeur de liberté* (Chamalières, France, 1991), p. 160 for the visit of the "poet" Sarton in summer 1908.

Emile Masson recalls George at this time, a "big, thin adolescent, beardless, with a characteristic 'Germanic' earnestness and a stiffness."[28]

Mabel reports, on 13 August, that she likes the Finistère photograph, in which George seems a bit "Egyptian." Her strength is returning. In English:

> I feel that I have taken hold again firmly on life. At one time I felt as if I were losing my grip on it, as if the joy of life, of just *living* were slipping away from me. So many things, quite simple things were beyond my strength, and now that I can do those simple things once more, they are *intense* enjoyment. I want to laugh out loud all by myself sometimes just to feel the springiness of my feet against the ground.

Mabel sent George a picture of the Virgin of Memline, probably a painting by the medieval Flemish painter Hans Memling, whose works were displayed in Bruges; it is something she likes, and she asks him to keep it with him, even if, as he says, he feels a distance from it. Mabel reproaches George, on 21 August, for signing himself, "yours devotedly," signaling a distance, suggesting that he sees "prudence" in her refusal to marry him. She is sick with love, but she loves him "so deeply that, despite everything, my life glows." She is attracted to him physically, but "I do not *want* to allow you to build your future happiness on that and on our very real but very incomplete friendship." He closes his letters by saying, "God be with you," but Mabel, who invokes the Deity far less frequently than George, does not feel that God is near her. "I do not need him, it seems. I need *you*. But a you that escapes me, that I seek in vain, who just now doesn't seem to exist." She cannot come to join him in the south of France. "No, I cannot. Something *beyond* reason tells me that that would not end well."

On 24 August 1908 Mabel changes her atheistic credo. In English: "George please forget if you can, that I said that I don't feel the need of God being with me. That is not true, tho' I am not sure that I know yet just what you mean when you say it." Mme Dangotte has read all George's cards and asked Raymond a few questions about them. Mabel regrets it: "I love her very much, but she's a very indiscrete person who causes much mischief by her indiscretion," although the only mischief she can do for them is to make them more aware, a little ill at ease. George should really be discrete on his cards, as he was until most recently. Mabel spoke with Julius de Praetere about Zurich. "I would have to stay for at least a year in his school to learn how to work leather and to bind books, then there is lithography. The courses begin on 27 September." Would George return before then? She wants to see him before she leaves. She feels estranged from her mother, who has returned

28. Emile Masson, "George Sarton et le 'nouvel humanisme,'" *Les Humbles*, no. 10 (February 1919), 7–19, on p. 7.

to England. Her mother wrote to Mme Dangotte saying that Mabel is "no longer English, but *Belgian*." Mabel affirms that she loves Belgium, but she remains English. She loves George, and she worries that she will fall into his eyes when she sees him.

George is taken with Finistère. On 24 August he writes about swimming and being in the middle of "an epidemic of painters"; on 26 August that he wishes that Mabel "were here instead of me." On the 27th George confides:

> After much hope and illusion, I firmly believe that you will not love me enough to give me, *without hesitation, everything,* as I have done. You love me as a friend, a dear friend, for whom there will be no substitute in life. The conclusion has bruised me, not broken me, because neither that nor anything like it will break me. It has *transformed* me. A revolution has occurred in me, and I see everything differently from several months ago. It is as if I have aged 10 years. *I have accepted it,* because I am not one of those people who struggle stupidly against the inevitable. This result that I have come to—I will not tell you at what price—will perhaps make you very happy, and it simplifies everything. We remain two close friends without reservations, do not worry.

He gave himself to her, and it was an impossible gesture. He returns to Ghent around 10 September. There is no reason to be concerned about Mme Dangotte's meddling. "Whatever she knows about the nature of our relationship, there are only good things in it."

Mabel replies on 2 September in English that she does not want to write to him but to see him and speak quietly with him. "I too kiss you with all my heart and give you all my confidence—have I not already given it to you? Oh I want you, I want you—more than you believe." She continues: "I think you are quite mistaken to force a solution on yourself—but you must do what you think best and I must do the same, and we shall each go on loving the other in spite of all, in our own way—tell me some day *those* ways *meet*." Two days later Mabel reports that she will be operated on for a hernia. This means that she leaves for Zurich in the middle of October. She very much wants him to visit her in the hospital. George writes from Paris on 3 September that he is arriving in Ghent on the 5th and he looks forward to seeing Mabel on Sunday. There is no record of the meeting taking place. George writes on 24 September:

> Do not wait for me this afternoon. Do not expect me any longer at Korte Meer 26 [26, rue Courte du Marais, the Dangotte residence where Mabel lived]. I can't come. *Don't worry in any case.* I have changed a lot intellectually but my heart is the same as 4 years ago. Keep me informed about your health, because I dare not ask Marthe any longer.

He is very busy. In the morning, he works on physics. He has not seen Mélanie Lorein. "I can't come to see you, but I am near you *a good part* of the day."

In an undated letter from this period, Raymond Limbosch sends Céline Dangotte, who is with Mabel, thoughts about George and Mabel. It is possible that Mabel would have "a strong influence on G. and perhaps George would give up his past life," but "G. is too selfish [personnel] to change and he loves Mabel too poorly." Raymond wonders how Mabel can love George and yet "does not understand him." Perhaps her love is just an attraction or a sympathy that must still be put to the test. "But will this dangerous fool George accept Mabel; will Mabel discover the real George under George the lover? If I were Mabel, I would not hesitate to be severe with George." Raymond wonders how things would go if Mabel told George of her love. Raymond counsels Mabel to proceed carefully with George, hoping to discover if her affection is sincere. "It's terrible, a guy like that! Mabel alone could make him see reason by speaking to him openly, severely, as to a child who acts badly, because it is just that." Raymond has written to George to explain "the terrible exhaustion that he has produced in Mabel by his correspondence, and the great responsibility that results from it."[29]

Mabel is distraught on 26 September 1908. There was a contretemps between George and Céline, an unfortunate exchange of letters. His pride is the reason for it. He does not accept criticism easily.

> In large matters, you are already well along on the road toward enlightenment, but for smaller things in daily life, *you will agree in theory* that they are important, but you don't pay enough attention to them. I don't have to remind you that I think for *you yourself*, for your happiness in life, so that your days, each one of them, are lovely and smiling, these small things have as much importance as the main aim of your life. *You are quite unconscious of what you miss*, that I know, and could I not show it to you?

Perhaps repeating Raymond's advice, she writes to George that he is "still a child in many things (me too, and that is perhaps why I understand)." George's "usual generosity is a generosity that is too intellectual, too heavy, too filled with will, and not enough love." George writes that "you would like me to be a saint." Mabel: "Oh, no, I want you quite entirely and 'intimately' *human*, and you are one of the rare people whom I've met who is capable of it." Why does she write all this? "Because you *do not listen well*. You have not learned to *hear with generosity*, a very rare gift, and I know fewer men than women who have it. Do you know that often, to make you hear, to catch your attention, it is necessary to interrupt you

29. NYPL.

harshly?" In driving away her dear friend Céline, George hurts Mabel. She has the feeling that, in preaching to him, she is distant from him, and she hopes that he understands what she wants to say. Later on 26 September, Mabel receives George's card of the 24th. She replies that evening in English:

> You will not come to see me any more. Let it be so....George, will you remember always that if I have failed to give you all you needed, that I am and shall be still ready, and wanting to give you all that is in my power to give and that might bring you strength or happiness? Only one thing in return—your confidence. You asked me mine, but do you give me yours? Well even if not I can wait and what I have said is true just the same.

George answers Mabel's first letter, on 26 September, that it is not because of her that he refuses to visit, but, rather, he does not want to see Céline. When Mabel was in the hospital, Céline wrote him a nasty letter, saying that he thinks only about himself. He wants nothing more to do with her. He does not care what she thinks. She has acted "often like a mother-in-law (in the bad sense)." He writes that "men would judge in 40 or 50 years if I had been generous, i.e., devoted to others, or not." The appeal to posterity is characteristic. For the rest of his life, George imagines what a future generation will think of him.

George answers Mabel's second letter, also on 26 September. She is right, and he begs her apology. George feels, as he expressed to Raymond, that

> there is something hard, sad, and sombre in me, and I always thought that that derived from the fact that I was not naturally good, or, rather, that my childhood was not good. My childhood was really sad, Mabel, it's enough to make me cry. Entirely left to myself, reading bad books, doing bad things for the fun of it. As a young man I continued in my bad ways, knowing full well what I did, as a distraction, all alone.

Only at the age of 20 did he begin to reform, and only for the past 3 years does he know joy. So Mabel can understand how much Céline's accusations hurt him.

And Mabel does understand. She writes on 28 September that she will always stand by him. Her gift, she says in English, "is to love people." She tried to reconcile him to Céline. "Perhaps she *has* something of the jealous love of a mother for....do not forget that for years it is she who has nursed me through my illnesses, has fought for my health, has worked for my happiness and strength, far more tenderly, far more *closely* than most mothers." Céline has also acted unreasonably. In French: "Oh, my children, my two big, dear children, you have undergone hurt and isolation. It is sad, and I am unable to give in to it, to accept it. And after all, it's *stupid*, all this." And George writes on the 28th that he is ready to ask Céline's pardon. He will do what Mabel tells him to do. On the 29th Mabel urges George to come; Mme Dangotte is, in English, "very upset." George is completely contrite on the 30th, sending along a letter to show to Céline.

Raymond Limbosch did not take his *candidature* examination in the summer of 1908, when Céline and he were ill.[30] On First October 1908 George strongly advises Raymond not to change his university studies from engineering to commerce. "I am in any case convinced that it would be folly (except in the case of an absolute impossibility) now to give up mathematical studies." He urges Raymond to complete his *candidature* before beginning commerce.[31] The university is George's open door. George, in a diary entry on 12 October 1908, imagines a doctorate involving laboratory work. He passes his first doctoral qualifying examination on 27 October.[32] If he is able to land a university post in 4 years, he writes in his diary on 29 October, he will do what he has wanted to do for a long time:

> Write several biographies of great scientific geniuses. This work has never been carried out systematically, as few men are capable of it. Artistic geniuses have often been studied, while whose in science hardly at all, or in a manner too exclusive. It would be interesting to study the extra-scientific life of these men and the relationship of their life in human terms to their working life. To find behind the work the real force of their genius *in their heart*. These studies could be divided into 2 parts the work and the man: the second part being written for the wider public. It would be at once excellent moral propaganda and high scientific popularization.

Here we read the criticism leveled against George by the generation that succeeded him—criticism that surely would have amused the master. He was the last person to blur a distinction between scholarship and popularization. In his career he indeed identifies many of his writings as popularization or propaganda, and he knows that no discipline can survive without broad and general support.

After visiting with George, Mabel goes to the Dangotte country residence on 3 October. Céline and two other Flinke are beginning their evening course for the local children. "Mr Dangotte told me at lunch that I am much too 'masculine.' It seemed so funny that I really wanted to

30. Mabel Elwes to George Sarton, 13 August 1908.

31. Van Oye, p. 54.

32. A.M. Simon-Van der Meersch, *George Sarton's European Roots: Introduction to His Life and Times* [Exhibition organized by the Archives of the State Univeristy of Ghent and the Department of Communication and Cognition] (Ghent, 1984), p. 157.

laugh. It is a compliment that I do not deserve." George and Mabel find it difficult to meet. Their correspondence flags. On 16 October they walk in the dunes by the sea; it pleases George very much, he writes on 17 October. Mabel leaves with Céline for Zurich. She is 30 years old, and she strikes out on her own for the first time.

Part II

Apprenticeship of Innocence, 1908-1911

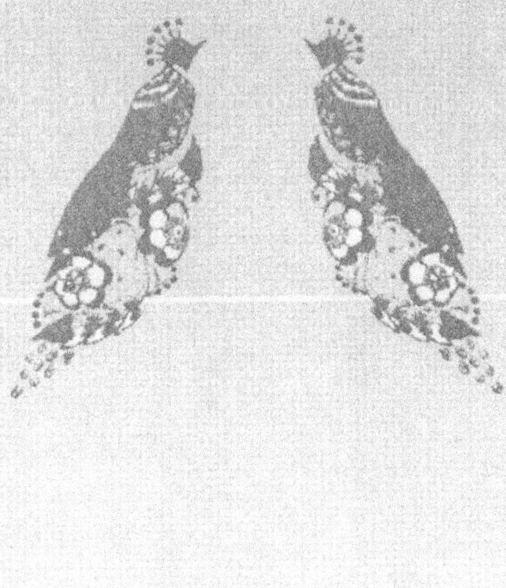

CHAPTER 4
Apart

No stranger to Switzerland, Mabel Elwes writes to George Sarton from Zurich on 25 October 1908. She reports that the school of design is marvelous, and the director, Julius de Praetere, is a fine man. De Praetere, born in Ledeberg and educated in Ghent, was a printer in Laethem St-Martin; from Belgium, De Praetere became, in May 1901, lecturer in applied arts at the Kunstgewerbeschule in Krefeld, Germany, returned to Ghent briefly, and then taught in Düsseldorf before being appointed at Zurich in 1905.[1] De Praetere, a man of about Mabel's own age, finds inspiration in William Morris (he claimed to have acquired Morris's typographical characters during a visit to England), and he is an advocate of renovating the medieval craft tradition. His interests range from engraving and printing to working with stained glass, pottery, and fabrics. Mabel has seen a magnificent furniture exhibit in the museum organized by De Praetere, but there were few pieces that she completely likes. She admires a reading and music room, "very beautiful, so quiet, a true 'atmosphere', all in black and white. Very few details, *space*, and furniture *logical* and simple without being cold." De Praetere relates that 100,000 visitors saw the exhibit. "What an education these exhibits must be for the public."

Mabel then reports on her visit to Dr Nyssens in Brussels, before she left. He found that her spine is quite straight, but her breathing is bad, due to angina. She came away with exercises and the admonishment not to wear constrictive clothing above her waist. She and Céline saw Eleonora Duse ("La Duse") in *Hedda Gabbler*. Mabel had seen her perform the play once before, but this time was better. She felt ill at ease watching the play, perhaps because she thought about Van der Ghinst and his girlfriend Marianne. In English: "I feel puzzled about it all. I don't understand." Noble Mabel does not mention that Ibsen's play is about intrigue, infidelity, and corruption in academia, where goodness goes unrewarded. To her mother Mabel reports on 27 October: "I am tired but satisfied with my day's work." On the same day to Marthe Patyn, "I love my work *so much*!" She will send her first bound book to

1. Piet Boyens, *Flemish Art: Symbolism to Expressionism at Sint-Martens-Latem*, trans. Tony Langham and Plym Peeters (Sint-Martens-Latem, 1992), p. 42; Paul Haesaerts, *Laethem-Saint-Martin: Le Village élu de l'art flamand* (Brussels, 1968), pp. 170–2; Johan de Smet, *Laethem-Saint-Martin: Un Siècle d'art flamand 1870–1970* (Tielt, Belgium, 2000), p. 153.

the Flinke. "I think that here I personify a bit *our collective soul*, that I work for *us all*."[2]

George on 26 October 1908 confides that during his last meeting with Mabel he spoke to her about many things that really did not concern her directly. It is his way, when he is involved in something, not to let it go, and this may be why people see him as a cold person. George writes on First November that he recently saw Céline, and he had a good talk with her. He thinks only good of her. "That proves that I truly love you, Mabel." George describes his activities on 8 November: toothache, assembly for the Garde civique, and teaching his course in geometry to the young women of the Flinke, which he does on Sunday mornings. The course is the result of his attempt to organize, with Michel Thiery an *athénée populaire*—plans that founder when it becomes clear that the government will not pay Thiery to teach in such an institution.[3] Mabel writes also on 8 November that the course she attends on the study of plants is more tiring than the one on the book. The professor for the latter is "a small German, extremely nervous and excitable with too many pupils, who adores his work but who *detests* teaching!! (He told me so.)" In her book course, Mabel learned how to color paper. She sent a batik design to the Flinke, and Mélanie can show it to George. She is quite taken with geometrical designs for her book jackets. She would like to bind a book for him. George, on 9 November remarks that Raymond and Céline will hear a talk by Albert-Louis du Chastain.[4] "Unfortunately I cannot attend, because I would prefer to die rather than to join the 'popularizers.'"

George is happy with his physics course at the university; his astronomy professor (Clément Dusausoy) is an idiot (*abruti*).[5] The future historian of science has not yet found his vocation: His course on history of mathematics is "quite interesting, but I consider it rather as a distraction." This is a rare reference to the lectures of Paul Mansion on the

---

2. NYPL.

3. HGS, Pierre Hoffmann to George Sarton, 24 August 1908, for governmental intransigence about the *athénée populaire*.

4. Albert-Louis du Chastain, in Brussels, was secretary of the Société française de bienfaisance and vice-president of the Société de moralité publique. He was a popularizer of French literature and also of the theory of evolution seen from the point of view of art and morality. A. Breuer, ed., *Nos contemporains* (Ixelles-Brussels, 1904), pp. 115–16. Despite his ambivalent feelings about popularization, George is not unfriendly to Du Chastain, to whom he sends greetings through Raymond Limbosch in 1914. George Sarton to Raymond Limbosch, 8 April 1914, in Van Oye, p. 60. Possibly George refers to another man of the same surname. Christophe Verbruggen has shown me a photograph of Raymond Limbosch with Jean du Chastain, dated 1915.

5. *LM I, s.v.* Dusasoy, for whom no publications are indicated.

history of mathematics. From 1908 to 1910 George follows Mansion's 2-year cycle of courses. Mansion pioneered the courses in 1888, when history of exact sciences became obligatory for doctoral students in physics and mathematics. Mansion emphasized an encyclopedic overview, along with special focus on selected topics, such as Euclid, the infinitesimal calculus, and Copernicus; cultural history was absent. Mansion followed Hieronymus Georg Zeuthen on Euclid, Moritz Cantor on rational mechanics, and Charles-Jean-Gustave-Nicolas *baron* de la Vallée Poussin on prime numbers, along with Mansion's own doctoral dissertation on the history of elliptical functions. George takes the last sequence of the course before Mansion retires in 1910, and Mansion has no formal rôle either in George's dissertation or on its examining committee.[6] Mansion is frosty with his most illustrious historian-of-science student. He declines to lecture to Reiner Leven because, "despite the manifestos, your program remains vague and it invites attack." Mansion shares George's orientation as a historian, although he points out shortcomings in his student. "As you write, to study the history of science, one must have the widest and most direct knowledge of the past, including the personal psychology of scientists." He observes that George employs the term Cantorism to mean studying history of science in the manner of Moritz Cantor, but Henri Poincaré uses it to mean "working through all the details rigorously, as is done now since Georg Cantor."[7] George's agenda reveals that he brings a number of his own ideas to Mansion's course. On 14 March 1904 George reads an article in the *Revue scientifique* for 26 December 1903 about the history of astronomy; on 29 December 1904 he reads French historian of biology Félix Le Dantec on malaria, the subject of Le Dantec's scientific competence; and on 8 November 1905 George's agenda records Paul Tannery's name.[8]

---

6. J. Gillis, *Paul Mansion en George Sarton* (Brussels, 1973) [Koninklijke Vlaamse Academie voor Wetenschappen, Letteren en Schone Kunsten van België, Klasse der Wetenschappen, *Mededelingen, 35,* 2], for a fine treatment of Mansion's career, with special reference to his work in history of science. Gillis identifies the lecturer in astronomy in 1907 as C. Dusausoy. Paul Bockstaele, "Benelux," in *Writing the History of Mathematics: Its Historical Development*, ed. Joseph W. Dauben and Christoph J. Scriba (Basle, 2002), pp. 45–59, p. 51 for the obligatory history course.

7. HGS, Paul Mansion to George Sarton, 7 November 1906 for Reiner Leven; 26 June 1910 for history of science.

8. Le Dantec's writings come up again in George's diary on 13 and 20 January 1905. Le Dantec inspired Belgian intellectuals. Théo van Rysselberghe's painting, *Lecture par Emile Verhaeren* (1903), features Le Dantec along with Verhaeren, Francis Viélé-Griffin, Henri-Edmond Cross, André Gide, Maurice Maeterlinck, Félix Fénéon, and Henri Ghéon. Robert Housee, "Peinture et sculpture (1800–1945)," in *Gand: Apologie d'une ville rebelle*, ed. Johan Decavele (Antwerp, 1989), pp. 267–91, on p. 180.

On 15 November 1908 George comments to Mabel about his courses: "Every professor (who need be concerned only with his specialty) tells us: You must study such and such book, you must read this paper, and all together it creates mountains of work. It is not surprising that I suffer on occasion from vertigo." He notes that Fernand Séverin has won the five-year prize in literature for his fine poem, "La solitude heureuse." It reminds George of the poetry Mabel forced him to read by the sea at Coq-sur-Mer. George saw the lines in English that Mabel sent to the Flinke. He quotes:

The shadows cling like soft blue veils among around
The trees and slip in trembling unchanging pools
And over all broods they far off upon the ground; deep still blue sky
Where like dream-feathers of a great wing spread across
The cloudlets lie.
A drowsy stillness every where
The trees, and slip down meekly in trembling pools upon the ground.

He likes the piece very much. Then he offers his opinion that the binding of a book must be above all practical, protecting the pages; the design was secondary. It does not make sense to bind a book with something that will show wear or dirt easily. George prefers marbled paper, which Mabel does not like. The books and bookcase that George gave to Reiner Leven never received much use; nor would they better serve the Union of Women of Ghent; the key to the bookcase is lost. On Thursday, he was too tired to work, and he read part of Rouse Ball's *History of Mathematics*;[9] Friday he read the *Merchant of Venice*. He comments that Raymond and Céline are attending the talk by Du Chastain, but, once more, he "would prefer to die under the worst conditions than to go to the 'popularizers.'" He counsels her, patronizingly, "Please listen, dear [petite] Mabel, don't try to do too many things at once." The rigorous distinction between knowers and popularizers remains with George for his entire career, even though scholars in the generation following his have given George's scholarship short shrift.

Mabel is very glad for George's letters, she writes on 17 November 1908, and she looks forward to returning to Ghent in a month. A Swiss doctor recommended by Dr Nyssens advises against vigorous physical activity. She is feeling a bit better, and her appetite has returned. She will no longer write postcards to him because the house matron is asking her when she will marry. She hopes that she can show her miniatures and perhaps sell them to pay for her expenses. She is making progress

9. Probably Walter William Rouse Ball, *Histoire des mathématiques*, trans. L. Freund, with additions by Roland Montessus de Ballore, (2 vols, Paris, 1906–7), noted in *Isis, 1* (1913), 562.

on design, although the practical side will take longer to master. "I cannot write to thou longer this evening, George—and I use informal 'you'—I did it above on occasion, I know, and now I begin to be afraid of seeing it." She spent a day with the De Praetere family, who live on the Zürichberg, and she read them "Kaatje," which they did not know. De Praetere dreams of creating a big cooperative factory in Flanders, surrounded by a garden village with furniture-making, pottery, lithography, bookbinding. It is his version of the decentralized future depicted in William Morris's *News from Nowhere*. De Praetere is organizing a fair for simple, inexpensive things made by local peasants, which he hopes to hold four times yearly. De Praetere, in the view of his critics, was an imposing teacher with a deep voice, who passed his ideas on to his best pupils.[10] Here we read the affinity between Mabel and her teacher in a common admiration of the arts and crafts movement and, generally, the ideals of socialism.

George, on 22 November, thinks "that we are better not to *tutoyer* as a matter of course," and he explains. His father does not yet know her, and he would disapprove for that reason. And to *tutoyer* between themselves while not doing so in front of him would be hypocritical. "But I will *tutoyer* you often in my dreams." He has been invited to join the local Masonic lodge Septentrion (in fact George joined the *loge* on the previous day, he notes in his diary). He welcomes the opportunity to have his opinions heard by significant minds.[11] George hopes that De Praetere can achieve his grand project.

A week later, on 28 November, George reports from the tavern of the Hotel Ganda in Ghent that he has not accomplished much. On the morrow there will be a big dinner celebrating his entry into the Masonic lodge, "Le Septentrion." Present will be the 3 Willems, the 4 Pelseneers, Raymond Limbosch, and Irénée van der Ghinst. He had long discussions with his professors. In several years he will leave for Germany. "The air is bad to breathe here. There is no scientific life, and the scientist is entirely isolated in our country: It is a universe without vigor." His 44-year-old physics professor, Edmond van Aubel (with whom he on occasion socialized), told him that if he were not married and head of a family, he would resign his professorship immediately to live in Germany as a

10. Haesaerts, *Laethem-Saint-Martin*, pp. 170–1, 375. Jan Pauwels and Boris Rousseeuw, "Julius de Praetere als Drukker en Illustrator van Guido Gezelle," *Gezelliana*, 2002, no. 1, 79–97, for more on De Praetere's personality.

11. NYPL, George Sarton to Raymond Limbosch, 15 June 1908, where George considers "entering the *loge*." George lectures twice on history of science to the Belgian Masons, on 3 February 1912 to the Ghent lodge and on 10 June 1912 to the Courtrai lodge. Diary, 5 May 1951. George renews his association with the Ghent lodge after the First World War. HGS, Paul Pelseneer to George Sarton, 5 March 1919.

Privatdozent.[12] George heard a fine lecture by the progressive priest Hugo Verriest. He alerted the Flinke, but most of them chose to hear the popularizer Du Chastain.

Mabel, on 3 December, is amused at George's description of the dinner party celebrating his becoming a Freemason. "It's odd, this practice of eating a great deal over a long time to show one's satisfaction. It's almost incomprehensible, isn't it, that intelligent and honest people still do that?" Mabel begs George to confide in her, not to hide the things that bother him. George explains, on 6 December, that at his Masonic dinner, he got along well with Paul Pelseneer, but not with his wife. He is glad that Mabel is happy with De Praetere, especially because "he is one of my countrymen [de ma race]." He was unhappy last week because he did not work much. This week went better: 37½ hours in profitable labor. Céline visited him, and it persuades him that she is no longer angry at him. Yes, he will go to Germany in a few years. Where precisely will depend on which part of physics he chooses. "Here in Ghent I am the *only one* who studies mathematical physics (I don't include a commandant in the general staff who is not worth much). So I can't speak with anyone about my work. There is another reason that *forcefully* makes me want to leave: My surroundings oppress me." George has had unhappy discussions with his physics professor, who,

> since his youth, has spent all his time on physics. *Nothing* interests him except science. He is married with children, but he has nothing to do with the life of either his wife or his children. He doesn't think that there is any intellectual need besides study. Thank God that I have an infinitely more complex character.

George congratulates Raymond Limbosch on 6 December about his engagement to Céline Dangotte and repeats some of what he wrote to Mabel, notably his plan to study in Germany. "Here the scientific air is bad to breathe. A few more years of 'world expansion'"—referring to Belgium's formal acquisition of the Congo as a colony—"and our country will fall to the level of Spain."[13] On the 8th Mabel reports that she has become friends with a girl who attends the Kunstgewerbeschule. In English: "She is kind-hearted, gay, and delightfully 'young' & open-minded. I am really fond of her." On the 16th, Mabel writes that she and her friend have found a pension. The work on lithography excites her. George writes on the 19th, inviting Mabel to meet his father for dinner on 31 December. On Christmas Eve he sends her his favorite book of the

12. J. E. Verschaffelt, "Notice sur Edmond von Aubel," *Annuaire de l'Académie royale de Belgique, 114* (1948), 48 pp. separatum.

13. Van Oye, p. 55.

year, Romain Rolland's *Michel Ange*. He inscribes it with the Flemish motto of Ghent, *love and fidelity*.

Mabel's friend in Zurich is Méta Budry, who comes from Vevey on Lake Geneva. De Praetere introduced Méta to Mabel and Céline in the fall of 1908. Mabel lodges with Méta in the pension. After a time, they move to a temperance house on the Zürichberg, after that to Zollikon, a nearby village on the lake, then back to Zurich. In addition to taking her courses, Mabel paints miniatures of children, including those of one of the scions of Nestlé and also the children of the family of Mathilde Wesendonk, Richard Wagner's mistress. Mabel creates her own group of Flinke at Zurich, who meet twice monthly, and they read Maeterlinck, Mabel's favorite writer. Mme Dangotte visits Mabel on several occasions, for it is she who pays Mabel's fees. Mabel's work is often featured in the exhibitions of the school. De Pratere does not share some of Mabel's social beliefs, and this leads to vigorous discussions.[14]

Early in 1909 Mabel breaks away from George. George asks, in a postcard of 10 January 1909 addressed to Mabel in Zurich, about her impressions of his father. In an undated letter, she writes a week after arriving in Switzerland. She is still a bit tired, and she ascribes it to the altitude and cold. She spends pages, in English, about descending the Zürichberg on a luge. She has taken up skating, and urges George also to learn. She describes her courses in bookbinding and brocade. She writes, in a postscript, "I smile at you from afar so that you won't be sad that I do not kiss you."

George replies to her on 17 January 1909. "Please, Mabel, tell me something or tell me nothing. But don't pose riddles to me like that. I am too nervous for that game." Why would she no longer kiss him? If she had issues with him, he begged her to raise them directly, and he would do his best to respond. He signaled a recent number of *L'Art et les artistes* (no. 46), where there are articles on Bronzino and Maurice Denis. "Maurice Denis devotes several pages himself to a great sculptor whom I do not know, Aristide Maillol." He sends along his photographic portrait. On the 17th, Mabel doubts herself. She writes "in haste." In English:

> I feel I have not yet really explained—let me try: I have been a little frightened at the passionate way you kissed me, especially the last time, frightened because I feel that could wake passion in me too, and because I am *certain* more & more certain that even if we loved equally passionately, we should *not* be able to be happy together. Do not ask me for reasons. I could give you many but I don't think they would satisfy you. I feel that in spite of the sympathy & affection that exists between us, that I am not a woman fit to be your wife.

---

14. NYPL, Méta Budry to George Sarton, 6 June 1952: "discussions extra violentes."

On the 19th, George receives her letter. He finally understands.

> I never wanted to force you to marry me. I know that there are great differences between us. I have thought *a great deal* about that this winter. A *day* has not passed when I did not think of us, in my imagination, as *married*. I am convinced that I could have made you happy. Of course, it would have been necessary for me to become used to your style of thinking and of *feeling*, and of you, yourself. We would both have had to *change*, in part: Is a marriage possible otherwise, without many reciprocal concessions? I had thought very seriously about all that. I had balanced the pros and the contras, and it seemed to me that we could make a good marriage, because we both greatly loved simplicity and sincerity. That was a solid and durable foundation.

He is terrified by her description of her fright at his passionate kiss. He sees that she has loved him only as a friend, "a bit less than Céline, or you have loved me at other moments like a mother loves a grown son." He feels as he felt the last summer, when he went off to Brittany. He did not give her his address intentionally, but she found out from a friend. Why, he asks, did she write to him in France? It was really to no end. Had she not done so, he could have greeted her upon his return as merely a friend. But back in Ghent, he loved her more than ever. He doesn't reproach her, although he feels anger. "I beg you to leave me with *rest and silence*. These are the only remedies I believe in." It was best if she did not write to him, at least for several years. If she does write, he will not reply. "You understand that I certainly don't want to remain miserable, but on the contrary I want to be *cured*." He will continue to think of her. "We have been good friends for several years, but now friendship is not possible between us, at least for me, who *loves* you." He signs the letter George Sarton, with a postscript, in English, "Silence and secrecy."

George confides to his diary on 21 January that he must study: "I am alone. Work, my friend!" He works, but he also plays. He socializes with Raymond Limbosch, Irénée van der Ghinst, and the Willems in February 1909.[15] It is possible that George takes up with prostitutes. "I am extremely sad today," he writes in his diary on 5 May 1909. "Yesterday, once more bad behavior; *diminished*. My sexual life is lamentable. One cannot say enough to young people how *amour de passage* is hardly love!" Léon Blum showed this in his essay of 1907, *Du Mariage*.[16] "But I must get used to thinking often about the *disgust* inspired by mercenary love."

In notes retrospectively dated both February and May 1909, Mabel writes that she can hardly stand the emotional pain. She wants all the

15. Van Oye, p. 56.

16. In *Du Mariage*, Blum applauds pre-marital sex.

happiness in the world to be his, she says using second-personal familiar. "Oh, George, don't you realize that I am too fragile to stand up to the joy of a lifetime and pass it on to another person who would issue from us? I have said it, but you did not hear." Mabel has always said that she is not in love with love. "I know that I love you not only as a very dear friend, but as a lover and a woman." Then: "If only there were not the question of a child, not to give into this terrible and sweet desire to hold a life in one's hands." And: "If I had given you my body, my devotion, if I had crossed with you the burning flame of passion that transfigures lovers and leaves them, at the end, with a beam of divine light." George has to find someone to complete his life, someone "deserving." She thanks him for having given her his love. Mabel switches to third person: "In many years, will I not be able to give him friendship? Can I not love his children? Can I not also love his wife because *he* loves her?" And in English: "Now I think the best way is to accept the love that has come to me and simply to try that, it shall sweeten all my life like a hidden flower that perfumes a whole garden. But what if he finds no one? What if he suffers, what if his life does not widen? What then?"

George continues on the course that he has plotted. Politically, he is a social democrat who hesitates over violent revolution and who rejects anarchism. He publishes an article in *Flandre libérale* on 10 June 1909 defending a Polish student whose extradition was demanded by Russia for an act of revolutionary violence—the theft of 660 rubles—committed in 1906.[17] George's affinity with Polish nationalists originates around 1904 in his friendship with Max Horwitz, a doctoral graduate of Ghent who had returned to visit and whose sister was arrested in Warsaw at a workers' meeting. Horwitz, too, landed in prison during the Revolution of 1905, and from prison he wrote to George eloquently about his activities. George transmits news of Horwitz's incarceration to the Massons in Pontivy, France, and Elsie Masson proposes to make him into a *cause-célèbre,* along the lines of the Massons' anarchist friend Gustave Hervé—a path that social-democratically-inclined George is not eager to follow.[18] In a fragment from this time, Masson observes that George "has decided for the program of [Karl] Kautsky & so you say, social democracy. Ah! A fat lot of good [la belle jambe]. You reproach me for being an 'anarchist.'" Masson quotes George: "It is only by a *powerful organization* that the proletariat will be saved." Masson

17. George Sarton, "Pour Micalchewski," *Flandre libérale*, 10 June 1909.

18. HGS, Emile Masson to George Sarton, 25 February 1906, postscript by Elsie Masson. Gilles Heuré, *Gustave Hervé, itinéraire d'un provacateur; De l'antipatriotisme au pétainisme* (Paris, 1997). Christophe Verbruggen, *Een Sociale Geschiedenis van het schrijverschap tijdens de Belgische belle époque* (diss., University of Ghent, 2006), gives Horwitz studying at Ghent from 1893 to 1898.

replies: "*Discipline*, that's just a word. Only one thing is necessary: that in the name of God or the devil, humanity—people—are clear to share the same horror of iniquity & the same passion for justice." The color of their uniform is of no significance.[19] Masson writes to George, who hesitates to see Hervé and anarchist Jean Grave when he visits Paris: "I clearly see that you are uneasy about Hervé. What's the problem! If he were a thousand times worse than the bourgeois press paints him, I would follow him." Emile Masson also urges George Sarton to consider the merits of Karl Liebknecht's socialist politics.[20]

Horwitz, a founding leader of what became the Polish Communist Party, was a trained mathematician.[21] In a letter from 1909, he asks George for information about Henri Poincaré's latest works, which George may obtain from the historian of mathematics Paul Mansion.[22] George, then following Mansion's course, writes in his diary in July 1909: "I pitifully passed my examination, the first step for the doctorate in physical sciences and mathematics." Of more significance than politics or mathematics is the appearance of his second novella.

On First July 1909, George sends an inscribed copy of *La Chaine d'or* to Romain Rolland. The book, a distillation in the form of a novella of George's intellectual progress over the preceding 7 years, is dated September 1907 and written under the pseudonym Dominique de Bray. It is dedicated to Romain Rolland and to Mabel Elwes. One of the characters is a man of 50, a writer and journalist who preaches revolution and works with a secret society of intellectuals, the Sympathiques. Another is a young assistant in physical chemistry at the university. The older man comments:

> It would be an admirable undertaking, worthy of a new Plutarch, to collect one day the biographies of all those who have dedicated their life to social regeneration. Their struggles, their misery, their failures, their glory, all would be described piously. It would be the golden book of revolutionary idealism, and to read it would stimulate us and com-

19. HGS, Emile Masson to George Sarton, undated fragment before the month of July, probably 1907.

20. HGS, Emile Masson to George Sarton, 20 October 1907.

21. Max (Maxymilian) Horwitz is known in the history of socialism as Henryk Walecki. In the 1930s, he was a member of the Balkan secretariat of the Comintern. In 1937 Stalin recalled him to the Soviet Union. Horwitz was one of many party members whom Stalin purged and murdered. Robert Conquest, *The Great Terror: A Reassessment* (Oxford, 1991), p. 406.

22. HGS, Max Horwitz to George Sarton, 24 July 1904, 12 December 1904 for the arrest of his sister, 25 January 1906 for his own arrest, 7 June 1906 on his way to Russian exile, 28 January 1909 for his epistolary activity and publications from prison, including a translation of Poincaré's *Science et l'hypothèse*.

> fort us in our hours of weakness. It is in this book that our children will learn to read.

The older man and his family are vegetarians and teetotalers (George cites ethical vegetarian treatises in a footnote). The man's donkey is named Francis, after Francis Jammes, "who portrayed them so well." The setting is indeed a transposition of Jammes's French countryside to Flanders, about which George waxes lyrical. Jammes's devout peasants are transformed into progressive socialists who drink deeply from the fountain of nature: "When you are troubled, my friend, you have only to question nature, as a child questions his mother." The questioning has to be deeply personal, for science without heart is "false science, charlatanry." The great sages—Plato, Epictetus, Pascal, Goethe, Carlyle, Emerson, Novalis, Hello, Ruskin, Tolstoy, and Maeterlinck, following George's reading over the past 5 years—are like links on a golden chain.

La Chaine d'or identifies Laethem St-Martin, "a village filled with poets and artists," as the suburb of Ghent that is closest to nature, where no mechanized factory disturbs the bucolic tranquility. It is one of the first characterizations of the famous Symbolist artists' colony. A poet at Laethem St-Martin was Karel van de Woestijne, an anarchist contemporary of George Sarton's at the *athénée* in Ghent who participated in industrial strikes; also present were sculptor George Minne, whom George knows, and painter Valerius de Saedeleer, who fell in behind the socialist Edouard Anseele. The artists and poets established a group, Open Wegen, in 1902, which heard talks from guest speakers on art and science (Julius de Praetere was a member of the group, as was the art connoisseur and chemist Hector van Houtte).[23]

In *La Chaine d'or*, George's young scientist affirms that he despises the ruling class, even if there are good people in it: "I hate philanthropy, in all its forms, because it is only a travesty of justice and goodness." A worker, part of the *ménage*, demands rights for household servants: a 10-hour, 6-day week, ending no later than 8:00 p.m.; decent food and lodging; 2 weeks of vacation per year. A manifesto is written, and a union is created. The scientist and the writer's daughter fall in love. The daughter and her mother spend the winter in San Remo. The scientist goes to Paris, takes in the museums, and returns to edify the family with talks about George's favorite artists. The scientist speaks about the deep joy in establishing the value of a constant of nature, paraphrasing the view of Alfred Nobel:

---

23. Boyens, *Flemish Art*, pp. 16–39. Els Burnez and Marc Somers, eds, *Latemse Memoires: Herinneringen van Mevrouw Karel van de Woestijne en Julius de Praetere* (Antwerp, 1995). George notes Minne's sculpture in his agenda on 14 April 1904, and he records his visit to Minne in Laethem St-Martin on 5 October 1904.

It leads to "accomplishing a work for eternity: for scientific theories pass one after the other and fall into oblivion, but a precise experiment has an eternal value." True scientists, he affirms, are suffused with patience and modesty. True liberty implies unusual self-discipline, in the spirit of Wagner's *Meistersinger* Hans Sachs. The scientist is injured in a laboratory explosion. His girlfriend reads him Hello's collection of aphorisms, *Le Siècle,* in the hospital; she understands nothing of them. The scientist is healed. He tells his lover:

> —Have you reflected, Béatrice, on the weighty mission that I must undertake? Have you thought about the dream of justice that I want to realize. We have often spoken about it together, and you know that it dominates my life. Know that I am ready to sacrifice you to my work.
>
> —I will be the happier, she says.
>
> —We will be persecuted; perhaps we will have to suffer a great deal. I do not offer you a life of pleasure. There is only one way to be just.

The couple is engaged and married, and they set off together "on the long road where no one saw the end."

The novella heralds George's willingness to deprive his family for his work, his ready acceptance of the status of a pariah, and his family's long wanderings.[24] The immediate reflection of the novella is the marriage, on 7 August 1909, of Céline Dangotte and Raymond Limbosch.[25]

On 31 August 1909 George summarizes his twenty-fifth year in his diary. He performed poorly on his examination. He is proud of the *Chaine d'or,* and he thinks that it may wake up young people. It is not, however, a great work of literature. He wrote it while also writing his thesis in chemistry. His next work will be a novel about an adolescent returning from London to Brittany, and the sea will have a large part in it. He feels largely alone, except for Van der Ghinst, the Willems, Maurice de Saegher in Paris, the Bounys in Mons, and the Massons in Pontivy. He has corresponded with Francis Jammes. He is friendly with the schoolteacher Michel Thiery. He met Jules van Biesbroeck, the sculptor who recently designed a monument to commemorate the anti-clerical law professor François Laurent. He likes Van Biesbroeck's artist father, Jules-Evariste, who resembles the father in *Chaine d'or*. George appreciates the English and their practical spirit, which he sees in the mechanical ether-theories of Lord Kelvin. He likes Kipling and H. G. Wells, and he wants understand Fabianism. He proposes to study Pierre Duhem, Ernst Mach, and Paul Appell to learn more about modern physics. He

24. Dominique de Bray, pseud. George Sarton, *La Chaine d'or* (Ghent, 1909), quotations on pp. 35, 13, 23, 24, 27, 34, 39, 56–7, 73, 78 in sequence.

25. Van Oye, p. 157.

wants to understand the development of mechanics, notably celestial mechanics, and he thinks that his doctoral dissertation will be in the area of "Hertzian mechanics," that is, electromagnetic waves. With regard to Mabel,

> The situation was hopeless. After having hoped and despaired thrice, I have become convinced that she would never marry me; she is too jealous of her independence; she lacks the sublime faith of Jane Welsh. Therefore I have broken off all relations with her as they had become impossible. That is for me an immense grief of which I shall be cured only when I love another woman. Alas! I can neither foresee that eventuality nor even hope for it. My heart is as if it were broken, but my life is not.[26]

Mabel, on 15 September 1909, informs her mother that she will be spending a week in a sanatorium near Zurich. The stay should allow her to rest. On 25 September Mabel writes to Marthe Patyn that she is feeling sad and alone, despite the presence of Méta Budry.[27] On this day appears the twentieth number of *Asselsch blad voor ouders en kind*, a biweekly magazine, one cent the issue, edited by Raymond Limbosch and sold by Marthe Pateyn; it is an educational effort of the Flinke. Mabel's traditional sketch of a mother and child, in the style of a copper engraving, graces the cover.[28]

George and his father visit London in the summer of 1909. There George reads about Fabian Socialism in the writings of George Bernard Shaw and H. G. Wells, and he rejects Karl Marx, whom he studied in 1908.[29] (George also finds H. G. Wells's *New Worlds for Old* [London, 1908], and from Wells he learns about samurai.[30]) George notices how his father's strength is diminished.[31] Early in the autumn, Alfred Sarton is dying. George writes to poet André Spire that his father is seriously ill. "His entire body is consumed: lungs, heart, stomach." George is exhausted with caring and with worry. He senses that if there is to be any improvement in his father's health, it will occur slowly.[32]

---

26. Translated in "Scholar's Wife"; alternative translation in *Phoenix*, p. 62.

27. NYPL.

28. The illustration reproduced in Verbruggen, *Sociale Geschiedenis*.

29. Diary, 12 March 1908. Denise De Weerdt, "De Sartons en België," *Brood en kozen: Tijdschrift voor de geschiedenis van sociale bewegingen*, 1999, no. 1, pp. 53–71, on p. 57, emphasizes that Mabel's father Gervase was one of the first Fabian socialists and a friend of Sidney and Beatrice Webb's. I have seen no document suggesting that Mabel wrote or spoke to George about Fabian socialism.

30. Diary, 1 April 1955. In his book *A Modern Utopia* (London, 1905), Wells imagines a voluntary class of leaders called Samurai, who endure periodical trials and hardships.

31. George Sarton to Mabel Elwes, 26 June 1910.

32. Paris, private papers of Marie-Brunette Spire, George Sarton to André Spire, 10 October 1909.

Alfred Sarton dies on 23 October 1909. Following his instructions of 1892, priests are forbidden to talk to him in his dying hours.[33] The funeral is attended by numerous Belgian dignitaries, among whom are the professors Henri Pirenne, Paul Fredericq, Pierre Hoffmann, and the physicist Henri Schoentjes.[34] George decides to liquidate his family house and much of its contents, including his father's wine cellar.[35] He moves into an apartment on the second floor at 3, Boulevard du Fort.[36] According to a meticulous drawing of George's, it is a small suite of 3 rooms.[37] Mabel, on 29 November 1909, writes to her mother that she has "worked well at the Academy" in Ghent.[38] On Tuesday she lectures to the Flinke on Rembrandt, and she still has to prepare her talk; she wants a nightgown and comb bag to go with her room, which is green.[39] Mabel Elwes and George Sarton live in separate worlds, and both are intensely unhappy.

33. NYPL.

34. "Les Funérailles de M. Alfred Sarton," *Flandre libérale*, 29 October 1909.

35. NYPL, Arthur Sarton to George Sarton, 24 November 1909, with instructions for the wine auction.

36. Geve, "Un Savant gantois aux Etats-Unis: Georges Sarton," *La Flandre libérale*, 15 December 1954, for the comments of George's landlord, M. J-F. Uyttendaele.

37. NYPL.

38. In fact Mabel continues to exhibit in the Cercle artistique et littéraire (1904, 1906, 1909). Karel Blondeel, *Vrouwelijke schilders in Gent (1880–1914), een socio-historische studie* (Licentiaat in de Geschiedenis, University of Ghent, 2003).

39. YMS.

CHAPTER 5

Renascence

On Mardi Gras 1910, a festival that has troubled him for many years, George enters plans for the future in his diary. He will work for a doctorate, studying among other things the history of mathematics in antiquity. When he has his doctorate, he will try to become assistant to Henri Poincaré, "the most intelligent man of our time." He hopes to devote his life to "philosophie naturelle." This "*living* history, the impassioned history of physical sciences and mathematics, is still to be achieved. Is that not truly history, the evolution of human intelligence, the development of human *grandeur*, also weaknesses?" He dreams about one day having a chair in history of physical sciences and mathematics and methodology. To accomplish that end, he must finish his "encyclopedic" education. George, who has spent 8 years in the university, has only begun his studies. And he knows that other young scholars aspire to similar ends. Emile Waxweiler, learning about George's career plans, informs him that a student at the Free University of Brussels, where Waxweiler teaches, is touring Germany and France. "It seems that a chair in the 'History of Sciences' will be created for him after his special doctoral thesis."[1] Waxweiler counsels George to become an actuary, because life is too uncertain to prepare for being a university instructor: It is discouraging that society does such a poor job in making use of someone with talents like George's but "adaptation to the environment remains the sovereign law of life."[2] Waxweiler likely refers to Théophile de Donder, who shortly obtains a lectureship in general mathematical physics and who takes over the courses in history of science given by Eugène Brand.[3]

Mabel sets down her feelings in notes written in pencil "after the Winter Concert" early in 1910, where she and George met by chance:

> Oh George, George, George, do you also hurt terribly so much this morning [Mabel uses second-person familiar], and last evening? Is is necessary

1. HGS, Emile Waxweiler to George Sarton, 6 December 1910.
2. HGS, Emile Waxweiler to George Sarton, 25 April 1910.
3. Paul Bockstaele, "Benelux," in *Writing the History of Mathematics: Its Historical Development*, ed. Joseph W. Dauben and Christoph J. Scriba (Basle, 2002), pp. 45–59, on p. 51.

for us to torture ourselves in this way? Was I mistaken yesterday evening, but it seemed that our two hearts had escaped us, and they were trying desperately to come back together. All my being leaned toward you, and words formed on my lips continually—always the same ones—simply that I love you, I love you, I *love you*. Oh, George, didn't you hear it? For me, the music itself was maddening because it seemed desperately to sing your sadness—your call to me. And I should have wanted to leave, but I couldn't any longer. I wanted to believe in a miracle, that despite your pain you knew that I am with you, and that it was a terrible and sad happiness because nevertheless I do not give myself. And I would have liked you to know, I really wanted that.

No, no, stay still. Don't tell me that it was cruel and selfish. Don't you understand that because your love has enriched me so, that sometimes I can believe that mine enriched you, even if it is not crowned with accomplishment?

George, my George, I am at the end of my rope—I can no longer think. I feel completely weak in the face of life and what it asks me to do. You are coming this morning, and I see in your eyes a sweet call, I would say to you, "Take me in your arms, help me, let me love you, for life wills it." George, I have struggled day by day, sometimes hour by hour in the night, to abandon your love. I believed that I wanted you to love another woman. By hurting you, I wanted however—so fervently wanted—to be *happy*. But it is such a slow request. I want to sense that you also will not stop loving. What will become of us? I will tell you.

George, I love you. Do you love me enough to wait for me? If I give myself to you and we have a son, he will be weak, and I also would be weak. You would have an invalid wife and also an invalid son. Let us help ourselves to wait, and I will concentrate on reconquering my health and strength. How would you answer?

Or, would we have the strength to live together without giving ourselves to each other completely? Sometimes I think that I could, but what if you couldn't? And I am not sure of myself—you know me—but do you see that once I open my arms to you, how passionately they will hold me? No, no, there is no doubt about it. I am sometimes afraid of myself. I do not recognize myself.

Tonight I said it all. What an inferior and weak person I must be. Why not confront the future? Would it not be better to go to bed together and if children are the result and they are not robust, to take care of them. They would be loved so securely that they would be able to face life. After all, physical strength is nothing compared with the resilience of the soul and the expanse of the heart.

But something cries in me, "No, no, *no*." I love my children too much to leave to them my own weaknesses. I have suffered *too much* to feel that I can never triumph, never work properly, to produce something that is worth while.

I am weak, we should realize it (*morally* also). *I dare not produce weak children.*

Madeleine van Thorenburg, an old friend of Mabel's and Céline's, advises Mabel to see George on the first day of spring.[4] George recalls in his manuscript biography of Mabel, "After very few words, the first we had exchanged for a year, we betrothed ourselves." George records the event in his diary on 26 March 1910. He and Mabel would wait to be married until her health improved and until he obtained a "stable position." George writes again about the event in his diary on 31 August 1910, quoting Thomas Carlyle's letter to Jane Welsh: "Has not a kind of Providence created us for one another? Have we not found each other? And might not both of us go round the Planet seeking vainly for a heart we could love so well?" Found, and found out. Mabel writes immediately to Mélanie Lorein to learn about George's past with her. Mélanie replies on 25 March 1910 that she never thought to marry George. She found herself attracted to him especially at the time of the death of his father, but she never revealed the sentiment.[5]

On 24 March 1910 George's Uncle Arthur answers the engagement announcement:

> I need hardly tell you how the announcement of the 23rd affected me: "Should one cry or should one laugh at such a solemn time?" It's from a bit of verse (Jean who cries and Jean who laughs) that my father and your grandfather sang, to the notes of Uncle Joseph Deschodt, of Douai.
>
> In truth, your fiancée's health is poor and she probably has no fortune. However, you are without a remarkable fortune. You will experience considerable difficulties in obtaining a post, because you are a free-thinker, a Mason, and President of the Socialist Students, all this in a Catholic country ruled, probably for a long time, by Catholics.
>
> Yet you know all this, as I do, and for that reason you are liquidating [convertissez en pièces de cents sous] all the movable goods of your father.
>
> So you are starting off on the gaming table of life without the two big assets which are the requirement of happiness: health and money....
>
> Love is a game, my dear George, and marriage is something else [une affaire], the most important thing in existence.

The letter reveals, first, that George has begun to sell off his father's furniture and, second, that George's socialist convictions are widely known.[6] George's socialist friend from Poland, Max Horwitz (who has a wife and at least 2 children), also congratulates George on his engagement and replies in detail to George's inquiry about an actuarial career.[7]

---

4. May Sarton, "Mabel Elwes Sarton," in May Sarton, *A World of Light: Portraits and Celebrations* (New York, 1976), pp. 49–68, on pp. 56–7; *Phoenix*, p. 63.
5. NYPL.
6. *Phoenix*, pp. 67–8 for another translation.
7. HGS, Max Horwitz to George Sarton, 14 April 1910. In a letter of 22 May 1912, Horwitz reports that he will take a position as actuary in Paris.

George invites Mabel, on 23 March, to dine with him and to work next to him, if she does not go to Zurich. All his things are hers, and she can borrow them without asking. He sends her his diary, so that she can judge how well he is working. Mabel replies, sending along the undated pencilled notes introduced previously. In English:

> From them you will know at least that I love you already almost as you love me, and yet differently, because no woman loves as any man. There is no need to ask my pardon, because my whole being has become dear to you. I am sad that I have not a *beautiful*—and strong—body to give you, but as it is, it is *yours*. It has been yours so long it seems to me, and would, and will be, *always* whether we marry or not.

George writes on 14 April 1910 that he is dining with Andries MacLeod, son of the biology professor at Ghent, Julius MacLeod.[8] He urges Mabel to remember that she is a citizen of England for only one more year. He would like to see her after lunch. In the evening, he must attend the meeting of the socialist students. Mabel, on 18 April, sends along important thoughts:

> For a very long time I have rebelled against the fact that the life of a wife is absorbed by the life of her husband, it is erased by him. Don't tell me that "it's natural," and so it is useless to rebel against it. I no longer rebel against it, and sometimes it seems a joy to be able to give in to his will, as for the place where we will live, and leave those whom I love for more than 10 years to follow you. I love to do it because it proves once again that you are the most dear thing in the world to me.

But she shall never sever her friendships. Mabel, gently, asks George to be a bit more understanding, to take her expressions of sentiment seriously. When she tells him that something bothers her, he should not reply that "It is silly to feel that way." She continues, "George, sometimes it seems that I could die if I felt a reproach from you, that minds too absolute could resent the thoughts of weaker minds."

Mabel insists on 13 May: "I would not want to accept to play the rôle of the wife who allows her husband to face the concerns of life all alone. I insist on my place in the struggle. *Please* let me, as I have never asked before, share, lighten to the best of my ability *all* the concerns, all the difficulties, all the troubles that confront you in the future." On 15 May Mabel writes what she wanted to say before she leaves for the country. She senses that sometimes, at night, George does not desire to caress her as ardently as she does him, so she begged him, "*never* embrace me out of habit, never, *even if you sense that I need to be held tight*, never take

8. Van Oye, pp. 104–8 on Andries MacLeod.

me in your arms out of pity, out of *habit*." She fears the commonplace in their marriage. Years ago she was concerned that one day George would wake up and sense that he no longer needed her. She implores him to reassure her by writing her a few lines. George's comportment is apparent for all to see. Céline's mother Léopoldine Dangotte is concerned about George's insensitivity to Mabel's needs, she writes to George. Mabel is sad that George is thoughtless in her presence. Mabel "loves you, in my view immensely," which is Léopoldine's own sentiment for Mabel: "I adore Mabel and regard her as a second daughter." George has said that he loves Mabel, but George must learn "to work with a person so delicate, so fragile and yet so strong."[9]

In the spring of 1910, Mabel begins collaborating with the architect Oscar van de Voorde on a modern cottage to be constructed for the Universal Exhibition at Brussels later that year. The collaboration says much about Mabel and her accomplishments. Van de Voorde worked with Jean Delvin and Julius de Praetere in 1896 on *Kunst en kennis*, the journal of the student and alumni association of the Fine Arts Academy at Ghent. Van de Voorde was in fact president of the association for a time, and he advocated interaction between architects, sculptors, and painters. In his contribution to *Kunst en kennis*, he expresses admiration for the Pre-Raphaelites, but he thinks that the present time deserves its own art, especially art for the masses. Van de Voorde will design the cottage, and Mabel will design many of its furnishings.

While Mabel receives an important commission, George becomes a professional astronomer. On 17 May 1910, George is appointed as an assistant at the Royal Observatory in Uccle, a suburb of Brussels, at an annual salary of F 2000, including a room with light and heat. His obligations are reduced to "a strict minimum, until he has passed his examinations."[10] George writes to Mabel from the Observatory on 23 May that he has reported for work there. He has a room in a miserable hotel. His office is comfortable, but it lacks curtains; the wallpaper has suffered from the humidity. The assistant astronomer George van Biesbroek, cousin of the sculptor, has been friendly. George prepared his lecture on celestial mechanics for the next day. His supervisor, the assistant astronomer Hector Philippot, established his schedule: 6 hours daily, or 36 hours per week, although he could work 8 hours daily if he wanted. His nightly work finishes at midnight, but Philippot suggests that he stay up later when the nights are clear. George has decided that he will keep with the 6 hours this close to his examination. Van Biesbroeck observed Halley's Comet, but it was not very impressive; he took great pleasure in showing George the big equatorial telescope. George will return to Ghent on Thursday

9. HGS, Léopoldine Dangotte to George Sarton, 24 June 1910.

10. NYPL, G. Lecointe to [Paul] Pelseneer, 17 May 1910.

and wants to see Mabel, although he must be back to work at 8; the next day he gives his course.

On 28 May George relates that he has given his resignation to Philippot, who along with the director Georges Lecointe urged him to stay. He is leaving on good terms with the astronomer. "I am conscious of having escaped a serious danger, and I quiver when I think that if the working conditions and the salary had been better, I would have been tempted to stay. I think that it is a weakness to have one's personal career placed on hold because of a need for security." George has "firmly decided to go directly toward the goal that I have defined, and I know that you, my beloved, would not think of preventing me from doing so." With his resignation from the observatory, he will devote himself entirely to his thesis. The *modus operandi* characterizes George's married life: He nearly always acts without consulting his wife. He writes in his diary on 30 May 1910 that to have remained at the observatory would have been to become "a gear without personality in a big and admirable machine." He insists that he will work toward his goals without a stable situation. "Mabel thinks like me on this point, and she will not, thank God! turn me away from my goal." Emile Masson, too, finds his "adventures at the Royal Observatory amusing," recalling his own beginning at the Sorbonne. In Masson's view, it would be difficult to remain human while working in such an oppressive environment. Yet George is right: "We are the Barbarians. We refine life too much, we want it to be too spiritual, too intellectual, too pure. Idiots! Reverse Barbarians, but Barbarians just the same."[11]

Mabel writes, on 12 June, that she has had nightmares. She is afraid of George and afraid of the future. She is ashamed to convey her feelings of weakness, and she is sure that she will overcome them. She wonders why George did not want to kiss her the previous day, when they walked in the garden. Is it because she was late? Is it something she said? She wants gentle understanding between them. In English:

> Let us try never to let the day *end* with a misunderstanding—a blank space between us. It is a dreadful feeling to go to sleep with that weighing on our hearts (indeed one generally does *not* go to sleep). When I was a small girl, I once or twice got up in the night to wake my brother and ask him to forgive me something done in the day because *if he were to die* he would then never have forgiven me, & I could not sleep. He was a little cross at being woken up, but he always gave me a nice sleepy kiss, however much I had wronged him. I don't think *any* human being who *tries* to do right and to love purely deserves the punishment of denial at bedtime. At least then, if it happens again, give me your hand loyally, even as to an enemy, and don't allow me to try to kiss you. I shall understand

11. HGS, Emile Masson to George Sarton, 6 August 1910.

without any words this next time, but oh I hope it won't come for a long, long time.

On 16 June in Ipswich, Mabel writes that she is "profoundly happy" to return to her country. She is working on bed covers for the cottage. She has read almost half of Auguste Forel, she writes on the 17th. "It is a beautiful book." She continues the next day:

> Forel's book brings up many things. I know very well that the sexual question was very complex and difficult to resolve, but I understand it 10 times better now. If the book made me calmer, it also raised a problem so agonizing that I must bring it up. I find myself calm because now I *know*. However, it shows that a man has sexual needs difficult to please, which must be satisfied before the age when he may normally marry. So he [Forel] easily admits that houses of prostitution are a necessity (in one form or another), and so there will always be women dedicated to this role? And he shows that the class of women for whom this is natural is very, very small (they are aberrant people). So these women will be eternally victims of the situation, *"inevitably."* This revolts me and saddens me. I do not see a way out. He gives no hope. *The two physical natures of man and woman are then not in harmony* on the level of humanity.
>
> I fear that what I say to you will seem rather futile, with your way of reasoning. Yet I must think about it more, before seeing in a clear way my own conclusions.

George writes on 16 June, explaining to Mabel that he recommended Forel "because the reading will greatly help you in *many* ways." George will be glad when his examination is behind him.

> I think that there is a strange contradiction in me. Basically I am lazy, there is no doubt about it, someone whose natural existence would be given over to dreaming. Yet an immense need to know and to understand more deeply pushes me, makes me continually begin new and broader undertakings. It would take me 20 lifetimes to achieve all I want to achieve. It would take me 50 to relax and reflect quietly. In 10 years I hope that it will be possible to ease up and to work only 2 or 3 hours daily, which is certainly enough for a philosopher. Ah, if I had been born in the time of Plato, I wouldn't have to learn so many things to penetrate the thought of our contemporaries. I wouldn't have this burden, this *joy*. Because the more I think about it, Mabel, the more it seems like we live in a marvelous time, one of the most splendid in history. Everything is open to question. I feel that from a great many points of view, humanity has arrived at a turning point in its evolution. The inventions of the past century have upset [bouleversé] daily life, changed its aspect. But engineers and physicists have more surprises in store for us. At the same time, a wind of criticism blows over science. Mathematical principles are now examined, and our doubts consolidate

> rather than shake our knowledge. Morality is being transformed, and our prejudices must cede to others. In the social sphere, the change is even more evident, and perhaps we will soon see the first application of modern socialism. Our generation is concerned much more with the beauty of things than the preceding one was. After a kind of slumber, the decorative arts are more popular than ever. And I could stretch out this list. Yes, Mabel, we have the good fortune to live at a critical time in human civilization.

From the heights of intellect George comments on the sensual. He has attended the Swedish exercises of Mabel's women friends. Mélanie Lorein was the best, he comments. George is struck by the effects of several months of gymnastics on her physique.

Mabel writes on 16 June, in English, "I see your face so clearly before me, and you have that look of pain that sometimes deepens on your face, and especially in your eyes, like a shadow." On 19 June, George writes to Mabel in Ipswich to say that during the 3 months of their engagement, his love for her has grown immeasurably. He has been working, though perhaps not so diligently.

> Basically I have a terribly undisciplined nature, and I don't know for which miraculous reasons I decided one day to discipline myself, for everything was in beginning. You say that my face often shows much sadness, and that doesn't surprise me. Most men are unconscious; I am someone filled with worries. You can't imagine how much each smile costs me. I live too little in the present, perhaps. My mind is almost always in the past or in the future, at least when I do pure mathematics, which is timeless [ne repète plus]. Later, when I am reading newspapers in your presence, you will see how often I become upset at far-off things. I take things to heart too much. I take life, mine and that of others, much too seriously, tragically. Is this bad? Basically, it is not as bad as finding life an immense comedy.

Then George moves to sex:

> I am not surprised that you anguish over the sexual question. It upset me *terribly* for 7 years. Now I look at it more objectively. I have almost never dared speak to you about it, because I wouldn't know how to end. I imagine a humanity much better than our present, where vile passions (alcoholism,...) will have practically disappeared, but I can't imagine a society where there would be other sexual liaisons than those approved by popular morality. However, I think that it is possible to diminish greatly the horrors of *forced* prostitution. Perhaps in the future, men and the women driven mad by desires of the flesh (either permanently or temporarily) will be able more easily to get together and satisfy themselves without thinking of selling and buying. All that will become clearer for you, but

> your great unease will remain. I won't say any more today. I am continuing to read Forel.

George asks Mabel to send him newspapers from Ipswich. He thinks that they reflect the "soul of a people." He adds, "It is truly an anguishing shock for the historian to think that all the newspapers, superficial and naive, are destined to turn to dust. This is terrible. They will disappear just when they have become really useful, i.e., when we have the necessary distance to read them with impartiality and to ferret out the essentials." George has followed the writings of historians for several years—Carlyle and also Bernard Pyne Grenfell on Egyptian papyrus—but here George speaks for the first time about his sensibility as a historian.[12]

George replies to Mabel's "jealous" thoughts—sent in pencil, where she wonders how she would respond if George fell in love with someone else, if she could respond as Forel advocates—by revisiting his relations with Mélanie Lorein. George loved her as much as he loved Mabel. Then he lost sight of her. After the death of his father, she was the only person that he was somewhat close to. When George moved into his present lodgings, she generally passed an evening weekly with him. She made tea and they talked. But he never spoke to her about love, not even indirectly. "I *never* kissed her." Before Mabel returned to his life, George realized that his affection for Mélanie Lorein had no solid basis. "Instinctively, I treated her as a child."

Mabel writes on 19 June that she has finished Forel. Despite the hopes that he expresses for the future, Mabel is left "with an enormous weight, thinking about all the misery and degradation that most of the book evokes." She can't accept his conclusion that women should marry between 17 and 19 so that men's sexual needs may be satisfied normally. Even if a woman is completely formed from the sexual standpoint at that age,

> from the moral and mental point of view, *she certainly is not*, in my view. I know no young girl of that age who is capable of choosing the man who will really complete her life, and above all I know none of that age capable of *raising children well*. I can imagine no future where they will be so. If it happens, it will be at the price of an *attenuated* childhood and adolescence, and I give great importance, on the contrary, to *a real*

12. According to his diary, early in April 1908 George attends a lecture by Grenfell on the discovery of Egyptian papyruses. Bernard Pyne Grenfell and Arthur S. Hunt, *Sayings of Our Lord from Early Greek Papyrus* (London, 1897), published for the Egypt Exploration Fund. George first discovers Egyptian antiquity at the Rijksmuseum in Leiden in 1905. George Sarton, "Why *Isis?*" *Isis, 44* (1953), 232–42, on p. 235.

> *adolescence*, long enough for a character to form joyfully, freely, slowly. He supposes perhaps that men carefully use *for a number of years* the contraceptive devices that he describes? I don't see it at all. The men who are capable of that will always be exceptional, I am sure of it. And then he often easily recognizes how difficult it would be to eliminate, even to improve the conditions of prostitution. It seems to me that in fact the conditions will change and improve, but the essential fact will remain. This is a sad, sad, sad thing.[13]

George writes on 21 June to Mabel in Ipswich that Mme Dangotte told him that he would not be able to make Mabel happy; the judgment hurt him, and he will try to disabuse her of the opinion. Raymond and Céline are building their house in Rhode St-Genèse, and they look fine. "I had to begin suddenly with a great many material details, organizing my living conditions, seeing to my fortune." He hasn't taken much physical exercise. When Mabel returns, "*you must make me* take walks and physical exercise. It's your responsibility." Madame de Saegher was called to Paris to see Maurice, who is gravely ill. But Maurice would not meet with her. It was awful for Mme de Saegher. "Better to kill oneself than to die slowly in this way." She considers Maurice "as a lost man." But George thinks that a resurrection can take place. The viscontéssa who took him into the "noble neighborhood" seems to be "only a bawd."

On 22 June, Mabel writes about a much-heralded exhibit of Japanese art in London, which George will see when he joins her. She felt that it was assembled, she says in English, "simply to flatter the taste—or want of taste." She thinks that it appeals to middle-class wants. She found much to admire, however, in the costumes from the past 600 years. She would have liked to spend all her time with the costumes, but she gave in to her mother, who pulled her forward, systematically scanning everything. "My father always gave into the wants, small and large, of Mother, and by this he did her no favors, for she did not realize what she made him do, and her character, at base generous, never broadened as it should have."

On the sexual question. Mabel thinks that she is a "normal woman," assuming that George doesn't go overboard, "which I don't imagine." She is ready to do everything to make their marriage a real union. "Never worry that because I hesitated to give myself, that I will not give myself now, completely, freely, spontaneously." Regarding Mélanie Lorein, George owes her no explanation. Her thought, which she feels "still on occasion, is a sharp regret that I know is '*wrong*,' based on a false notion of not having youth and beauty to bring to you. I don't feel the shadow of envy, that is to say hatred of her because she has these things. No, no,

13. Mabel's views seem close to those of Iwan Bloch, whose 1912 treatise on prostitution received a review by George in the first volume of his journal, *Isis*. George Sarton, "Iwan Bloch, *Die Prostitution*," *Isis, 1* (1913), 284–5.

don't think me petty for this jealousy, for I am not." Mabel then reflects on beauty. Everyone receives youth as a matter of course, but

> beauty seems to obey no law that we know. How shall I explain it? For example, I also occasionally imagine myself beautiful, and *that has no relation with my interior life*. It's like a cloak that is draped on the shoulders for several minutes and transforms one. The feeling comes at different times and doesn't *seem* to be a result of *my* life nor of my happiness or unhappiness, nor of my moral or intellectual victories nor my defeats. I know that many people believe that beauty derives from the soul, a center of beauty?

She thinks the better of all she has written so far, but sends it along anyway, urging George to read it "lying down." She is sad, too, about Mme Dangotte's warnings, which she received as well, but Madame has been very good to her, and Mabel loves her deeply. Finally, Mabel needs £5 and takes George up on his offer to send her money. On the 23rd, Mabel writes that she will indeed take care of George. In matters of his health, "I will become your dear mother, a dear mother who *must* be obeyed. So watch out, my child. You will see. I can be very, very demanding." George is coming to visit them in Ipswich. Mabel thinks that it will fatigue George, especially dealing with a language that he does not know.

Mabel visits the Carlyle house in Chelsea, London, she writes on 23 June, and finds George's name in the register there. She thinks that George resembles Thomas Carlyle, which would have greatly pleased her betrothed. George, also on the 23rd, says that he will not write about the "sexual question," preferring to speak to Mabel about it. He thinks that his "concerns are not those that exalt sensuality and eroticism." He has spoken with Céline about furnishing their future home, and she has given him her prices. In leaving for England on 26 June, George thinks back to his trip there the previous year when he was joined by his father, who "was so worn out, in part because of me, and the trip perhaps hastened his end." And then George is also sad because he is visiting "the country of birth of the one whom I thought lost forever." Mabel on the 25th writes from the Army and Navy Stores in London, where she has gone to purchase gifts.[14] Liberty's has nice armchairs for their new home. She is staying in London with her friend, Sophie Jackson. Mabel visited the National Gallery to see the Venus of Velázquez, among other paintings. She urges George to see a panel by Fra Angelico.

14. The Army and Navy Co-Operative Society, formed in 1871 by military officers, offered domestic items inexpensively. Mabel's parents were members, and Mabel continues to shop there.

The visit to London in June 1910 is fateful for George. He tours the exhibit of Japanese art at Shepherd's Bush. "It nearly took me off my feet," he recalled late in life. Then he visits the Indian Museum in South Kensington. Of greater significance, he meets Arthur Henry Fox Strangways, an expert on Hindu music and art, and Ananda Kentish Coomaraswamy, a specialist of Hindu and Indonesian art. Coomaraswamy returned to England from India with a large collection of Indian art; he and Fox Strangways share an interest in Indian music and belong to the India Society, where George probably met them. Both men remain his friends to the end of their life.[15] George is reunited with Mabel at the Round Church in Cambridge. They then spend 12 days by the sea in Cromer, Norfolk.[16]

The language of love goes beyond words, Mabel insists in her correspondence with George. Yet words tell us how they love. Until late summer 1910, George has written to Mabel only in French. On 16 September 1910, George writes, in English:

> I kiss you of all my heart. It is the first time, I think, that I write to you in your own language. You cann't realize how sad I am often, my dear, not to know better the tongue, which is and will ever remain, the most intimate to you. I am so sorrowful, for instance, when we read together, not to devine and to understand thoroughly the very words which trouth your soul. I will endeavour to know english; for our sake, I ought to know it to the bottom. I cannot prevent myself from thinking that you will only be mine when I shall perfectly understand the language with the help of which you were brought up, the language of your childhood, the very words which revealed to you the signification of things and the mysteries of life.
>
> I will endeavour to understand it, and this knowledge will yield to me the last sources of your soul. Ô dear! I'll do whatever lies in my power, but you will aid me? Will you?
>
> In my opinion, it is better for me not to speak too much, because it would be too annoying for you or for other people to correct me constantly, and to break me of my bad habits, and on another side to speak without being reproved would be rather dangerous. But I think it would be very useful for me to *hear* you speaking often the secret language—

15. George Sarton, "Why *Isis?*" *Isis, 44* (1953), 232–42, on p. 235 for the Japanese exhibition, Fox Strangways, and Coomaraswamy. James S. Crouch, *A Bibliography of Ananda Kentish Coomaraswamy* (New Delhi, 2002), for commentary on Coomaraswamy's enormous *oeuvre;* Roger Lipsey, *Coomaraswamy: His Life and Work* (Princeton, 1977), p. 91 for the India Society and the musical connection. "A. H. Fox Strangways," *Music and Letters, 29* (1948), 229–37, for a number of appreciations in the journal that he founded. On Shepherd's Bush: Ayako Hotta-Lister, *The Japan-British Exhibition of 1910: Gateway to the Island Empire of the East* (Richmond, Surrey, 1999).

16. "Scholar's Wife."

yours. I try to speak english with so many persons as I can, but often either the one or the other askes: Why don't you speak english with your fiancée?...Why? These questioning people are really intolerable.

I question whether it would not be better to read together (at the least, a question of an hour each day) than to speak? What do you think of it?

Don't occupy yourself too much with "our house" now. We have plenty of time. You must have more confidence in our luck, as Mrs Dangotte does. We don't find now the cozy home we look for, merely because we don't want it now. In 3 or 4 months, when we will really want it, the house will come: let it come and never try to offer violence to our good fortune....

Sleep soundly and softely...Don't be frightened: I come on tiptoe (like a sweetpea...) On your bedside, and kiss you once more, but so gently that you are not awakened. Sleep well, darling!

Mabel is touched. On the 17th in English:

Dear, (you are *so* dear, you know to have written me that letter) Is it not strange that this morning, on my way to you from the country-house, *I* was thinking about that very thing, our talking English together? I will try to do so...for it is true, I am never *quite, entirely* myself when I talk French, so you can only know me, and I can only be entirely yours when I have been able to talk English if naturally and easily with you. I sometimes long to do so, but it is such a painful feeling, to be afraid you don't quite understand what I say. I long to feel that when I talk to you, each word reaches your brain and heart *instantly*. It is a peculiar pain to me when that does not happen. Well, when I talk English, I feel the words arrive slowly, haltingly at your spirit's door, and then I send French ones running after them to get to you more quickly! And that is not a good plan. If I were to talk to you often, you would get more used to my ways of talking English, and little by little I should lose that fear that you don't quite understand.

Indeed, indeed, I will help you all I can. Could you find an hour four times a week to read together in? And to talk too sometimes? We must try to be very serious over it. Tomorrow we will decide...and listen! I will write to you sometimes in English. It is a good excuse to write to you. Some days I have a hundred things to say to you when you are not there, and then when I see you, the time is so filled up just with the happiness of being with you that I forget them. Now I shall write them down in *English* and send them. You are right, dearie mine about the house! I will trust to the little god who watches over our luck. See! I read and smile at him, and beg his pardon for my want of faith in him, and he seems to nod back at me and smile enigmatically.

She continues in English on 24 September with a decision about her career. She will not take an active part in the Dangottes' atelier, which has "almost been thrust upon me." She will contribute designs and turn up one day weekly, "but that is all." The enterprise "*is not in my character*."

She is not a good shop-keeper, and she needs time for herself over the coming year, "this not only for myself, but for you and for the children I hope to have one day." George remembers these words over the coming decades, continually discouraging Mabel from commercial undertakings. Mabel imagines: *"I shall not be really at peace till I am married to you."* She would like to feel like a small bird, heart beating wildly and held between George's hands.

Mabel apologizes, on 27 September, for being, she says in English, "irritable and unsympathetic all these last days." She did not know why. She wants desperately to see him, and then when he arrives,

> my voice sounds hard in my ears. I feel impatient with you and with myself, for the first minute, I am afraid for you to even touch me. I would most like to go away and leave you! Whatever you say or do irritates me, and all the time, something hurts and aches [illegible] in my heart, and I hate myself, and feel miserably ashamed that you should always be so gentle and kind and patient with me. Ah, George, I don't love you half enough.

She wants to show him only the good part of her, not her own shortcomings. She doesn't want to bother him with her small reversals. Her "debt of strain" began after her father's death and the year in Zurich and Céline's illness. "I still sometimes feel as if I were going to lose my reason, as if something in my brain would snap." George has given her happiness, but "I am only just beginning to lose the fear that that happiness would slip from my hands, a fear not definite, not even that of my incapability of bringing *you* happiness, something beyond even that, something vague but terrible." She recalls:

> I once sent you, in summer time, my heart brimming over with the joy of beauty seen, of streaming sun and dancing light. Tonight I send you my heart as it is, a poor, queer heart, with gray shadows and odd fears [illegible] in it, with useless tears running down on it. But I send it you all the same, because down in it, you will find stored your love and tenderness, you will see how they have kept it sweet, and how beside these flowers my love for you.

Mabel is recovering from a nervous breakdown. She and George go through a bad patch. George writes on 28 September to say that he can't come to see her because he has fallen behind in his studies.

On 5 October 1910, Mabel writes to Marthe Patyn, who is studying to become a nurse, asking for the address of the supplier of frames for her miniatures. She is working on wood-block engravings for the cover of Raymond Limbosch's book of poems, *L'Enclos*.[17] Mabel's design of

17. NYPL.

Raymond Limbosch's first books brings her to the heights of her teacher, Julius de Praetere. The poems originated after Raymond withdrew from the Faculty of Sciences at the Free University in 1908 to take care of Céline, who had fallen ill. Sitting up with her at night, he composed a book of verse.[18] *L'Enclos* (1910), and its successor written in Provence *Le Bois d'oliviers* (1911), are published in a square format in Antwerp using type by J.-E. Buschmann.[19] They appear as sister volumes. *L'Enclos* is organized around birds and flowers (an intimation of Mabel's by-name, "The Wild Bird"), and *Le Bois d'oliviers*, in blue and orange, features a butterfly-and-flower motif.

Mabel writes on 7 October in English that George's recent words have hurt her. She writes on 24 October that she needs him, she feels disoriented and homeless. She writes on 28 October, in English, that she thinks of George, for it is one year after his father's death.

> May I tell you very gently and quietly what thoughts helped *me* most after my father's death? It is so simple you may think that it does not need saying—still I will tell it to you. Is not the way to keep them living to make ourselves the continuators of all we loved best in them? When we are tempted to be intolerant, to make an effort to be as he was? To be kind as he was? To *cultivate* doubly for our own sake and his immortality all that was best in him? Then our children may inherit the fruits of our efforts & so the good be perpetuated.

She returns to their fireside talk and to one of the themes that will occupy George for the rest of his life:

> You know I do dimly feel that *all* great minds have a connecting link, whether they be artists or scientists or what, that they all trend the same way, because Beauty and Truth, if not *one same person*, are twin sisters. But I feel as if there is more *danger* for the scientific *average* man, to lead a narrow life, than for the average artist.

George replies immediately on the 28th. He feels sad about their recent talk, and her letter reinforces his sadness. Then he further reproaches Mabel, diminishing her insight:

> You find a common ground between the scientist and the artist, that is to say, between the two of us. Because this common ground exists between very great scientists & very great artists, it must then exist between us. You observe that simply, and for the first time after several months

18. YMS, Céline Limbosch to May Sarton, 1957.

19. Henri Liebrecht, "Le Poète et le livre d'art," in *Anthologie Limbosch* [*La Nervie, 11*, no. 2 (1930)], pp. 46–9. Buschmann was the favored printer for the Société des bibliophiles.

> of engagement. You can't understand all the details of my work, that is clear; but I had always hoped (and you never said the contrary) that at least you could become interested in the outcome of the work, of my efforts, that you could even like it as I like without really understanding the mysterious work of the potter or of the enameler.

George feels that she speaks of the scientific life as a foreign and hostile entity. What a thing to say to George, "who wants to devote his life to the inspired history of science!" When he had Mabel read the manuscript of *La Chaine d'or,* did he not ask if she agreed with the ideas that it advanced? She said that she had no objection to them. Did she misunderstand the book? Because in the book he "expressed, rather imperfectly, I agree, this perfect similarity between the undertaking of the scientist and that of the artist." Yes, he agrees that too few people see things this way. "But I am one of those who believes that great scientists and great artists, great thinkers and great musicians, great dreamers and great doers, live in the same atmosphere, *breathe the same air.* I am one of those who think that the enterprise of science is an enterprise of art. I do not think it only, *I sense it*, with all my being, all my nerves tell me that." When he reads a treatise on geometry, he hears an orchestra playing. "I know that art and science are identical, just as I know that I exist. The two, when they are beautiful enough, produce the same enthusiasm, the same frenzy, the same sacred intoxication, the same rhythms, the same thrills." If we are too insignificant to think in this way, George appeals to his masters, Plato, Leonardo, Dürer, and Goethe.

In old age George prefers practical Aristotle to idealistic Plato,[20] but in his youth: "This Platonic dream is held by the greatest scientists of our time—and also by many artists. And what feeling gave Jean Delville the idea of his great painting 'Le Banquet de Platon' [where Plato appears as Jesus under a wisteria vine, surrounded by naked acolytes, male and female] if not the deep intuition of the unity of human intelligence?" He continues:

> All great artists had a bit of this "scientific spirit" that you felt a short time ago as irreconcilable with the artistic spirit. I would add that without this *scientific spirit* (or something rather close to it) they would not have been great, or very great artists. The scientists who are the glory of humanity, *it is impossible* for them to conceive without imagination, lacking the qualities that are those of the artist.

---

20. George Sarton, *A History of Science: Ancient Science through the Golden Age of Greece* (Cambridge, Mass., 1952), p. 496 for the comparison between Plato and Aristotle, all to the benefit of Aristotle. But George notes in his Diary on 9 May 1954: "In spite of all my recriminations I am still at heart a Platonist."

Mabel can't imagine how upset George is. His "being seethes with anger and sadness," not at her but at "all those who say that the split exists." For if it existed, they could not understand and love each other.

> Despite the fact that all my time and "¾ of my life" are devoted to scientific labor, do I seem so far away from you? Do you think that I don't truly dream the things that you love? (Do not we love the same paintings in Old Art?) Or do you think that it is only out of consideration for you that I act in this way, and that when I attend the winter concert, it is not a real pleasure, a true interest of mine?

He will go back without fuss to his work as a scientist, as an artist.

> Later, in a hundred years, recognition will perhaps come, if I succeed to accomplish it, the artistic value of my work. And it requires as much imagination & inspiration to reconstitute the past, penetrate the thought of other times & write the history of human civilization as to write a sonnet or construct a poem. I tell you this so that I have in you (although you be on the other side of the divide) a close confidante whom I tell everything I think. But please don't tell anyone. One must not speak about something that has not been accomplished. But this clear light that I see there, far away, toward which I walk with confidence and joy and from which no one can turn me aside, I am guided by it. I know that it is the same star that appeared to so many men, and that it signals the crossroad where all minds meet, where all scientists are great artists and artists are scientists. I see it. The road is long, but I will remain strong. I understand that I will be criticized, misunderstood, fought against, scorned. I know that it will be necessary to sacrifice many enjoyments and leisure, and that life cannot be spent making excuses. I know all that, and I walk forward *intently.*
>
> And if I fall before arriving at the finish, I will have fought well.

George loves her truly, and wants to join hands across the divide, for if anyone else had said such things, he "would not have answered them; life is too short."

Mabel replies on the evening of 28 October, her second letter to George that day. She writes because if she comes to visit, she will be "stupidly emotional and not reply in the right way." She doesn't think that she would hurt him by saying that scientists are generally of a different temperament than artists. She does not consider herself to be an artist. Never in a hundred years would her work be viewed as the product of an artist. "I am a quite humble and insignificant artisan in the kingdom where artists are the kings." She does not even have the artistic temperament. If there is a distance between them, it is because George "*is sure that he advances toward the light*." He is so far above her. It is her "*pride of love to believe just now that I would have helped you to rise up to where I cannot follow you.*" She orders him to continue forward, and when it turns out that she becomes a burden years hence, she urges him to leave her.

> You must understand that you must do it for yourself, but you also owe it to me, for if I seem not to understand you well on occasion, I do understand you and I love you, and I admire you enough to realize very clearly that your life has an importance beyond mine, in any case, and it is possible that you will leave a trace for other people, which I shall not.

A long time ago she realized that she also has an end to accomplish, and she had the right to pursue it. But she no longer thinks so. She is driven to create, but she knows that she only produces small intimations of beauty. Her reward is the joy, "out of all proportion to its importance," of her work. The world also needs people of her qualities. She came to George "almost desperately when I felt that you needed me, and always with the fear that it was only your body and its [illegible] and its desire that called me." She surely misunderstands him sometimes. She wishes only continually to understand him better. She changes to English: "It seems to me, in spite of the end of your letter, that you have heaped on my head the bitterist of reproaches—that it is *you* that have *put me* on the other side of that abyss.[....] You ask me such strange questions—all your letter is so strange that I feel bewildered. I will try to answer each thing."

Mabel, returning to French, holds to her point that great scientists do seem different from great artists, although she does allow that an indistinct, divine inspiration unites them. She does not recall the *Chaine d'or* addressing the identity of scientist and artist. She and George view things with somewhat different sensibilities, neither better nor worse. When George speaks of his dedication to his work, he does so from pride, not from friendship. Although sometimes she annoys George, she "must not imagine being able to influence your life in its essential aim, not [she switches to English] for 'a hair's breadth.'" If, to prove that one loves, one must suffer long and hard, "know that I suffer, and that I have suffered more than I can say because of you." He has truly hurt her. "You fear that I will come and interfere with your work, tomorrow or Sunday. I will not. If you need to see me, you will come by bicycle."

Mabel writes in the evening of 29 October 1910 in pencil. She has forgotten much of what she wrote, and she no longer suffers under the "shadow of revolt." It is frightful that they wound each other. She is sad and contrite. The worst part of it is that the unpleasantness recalls past hurts:

> I have tried to chase them away, but they remain. They come intensely one instant and it's like wounds that open and only heal after causing a

> great deal of pain. I thought that I had forgotten many of these things, and I am afraid that they live still in my memory.

George, later on the evening of 29 October, writes that he was ashamed of his letter after having written it. But he could not do anything about it, for he had to set off for a meeting. Her reply is so good, so much better than his letter. He wants above all to draw closer to Mabel intellectually. He desires her, body and soul. He comes before her on his knees to beg pardon. He begs her to think that, "in his normal state," he is not proud, but rather humble. But internal struggles have produced the appearance of pride.

> I have asked myself for a long time if I had to find a position *of some kind* to earn enough money for making your material life quite comfortable, or if I could on the contrary give myself completely to my work...It is only recently that I have reached a conclusion: I know that I can and that I *must* think above all of the work that I must do. I will be a university professor, or I will not take a position.

Having arrived at this conclusion, Mabel has "given me a good lesson."

George, at the time of this epistolary marathon, has the Carlyles and their correspondence in mind: He writes to Emile Masson to recommend an article about Jane Welsh, and Masson sends him the new translation of the Carlyles' love letters which he and his wife just published.[21] The piece on Welsh is in a book by Saint-Georges de Bouhélier, which George sends to Masson. Masson, reading the book, thinks that the author takes the position of Thomas Carlyle's apologist James Anthony Froude. Saint-Georges de Bouhélier "did not know enough about the real living conditions of Carlyle, who suffered from human and superhuman sadness."[22]

Mabel continues with Welshian questioning. She still doesn't entirely understand what George has written. On Sunday she returns to George's lament. "What are the joys that you are forced to forego? And why do you think that you will be ridiculed, scorned, misunderstood?" Mabel thinks that he is mistaken. He will always have the respect of others in his field, "their sympathy also, if you meet them, almost certainly. Is it not one of the finest consequences of the work you have chosen that it brings you

---

21. HGS, Emile Masson to George Sarton, 30 October 1910, asking for a loan of the article by St-Georges de Bouhélier [Stéphane Georges de Bouhélier-Lepelletier]; Emile Masson to George Sarton, 23 October 1910, sending the book and signaling Elsie Masson's article on the same topic. *Carlyle intime: Lettres d'amour de Jane Welsh et de Thomas Carlyle*, trans. Elsie and Emile Masson (Paris, 1910).

22. HGS, Emile Masson to George Sarton, 1 December [1910]; St-Georges de Bouhélier, *L'Hiver en méditation, ou les passe-temps de Clarisse* (Paris, 1896).

closer to those men whom you admire the most?" George's friends would admire him, and the public at large has almost a "superstitious" respect for intellectuals—"It's that way in England, at least." George would triumph and would pay no attention to his detractors. But why worry about something that exists only in his imagination? George thinks that he must earn enough money to assure her comfort:

> I cannot tell you enough how that bothers me. Have I not convinced you that I do not want an ease that would cost you the time and the leisure necessary for your work?
>
> Oh George, George, you always doubt that I am able to lead a very simple life. Listen, I promise to do my part, loyally so that we never go beyond our budget. I will do it *joyfully, proudly*, I will not suffer. And if I must sacrifice some wants or whims, well, it *won't* be a sacrifice.

She would be glad to give up these things for him. George, who remembers everything, will keep this confession in mind.

On 5 November 1910 Mabel reflects from Rhode St-Genèse:

> Don't be sad, my love. If life is too short to do what you would like, the brevity makes it in part so terribly beautiful and dear. It doesn't stop me. It might be said, "you can have two lives, one for raising dear children and giving joy to the life of their father and to you, but afterwards, another, long, with a healthy and supple body, for working long and hard, for seeking and perfecting the power, the strength necessary to realize your dreams." I would jump with joy, I would feel *divine*, if this were given to me. But maybe I will work long, later? Perhaps I will grow old? Perhaps I will regain my strength? Perhaps the dreams will come back, persist, grow in real beauty, perhaps I will still have *the time to learn my craft*.
>
> You have told me about the work you dream about. Have I never told you what sleeps deep in the innermost recesses of my life, the ardent desire, quiet and persistent, to represent the human body in a truly beautiful way? I have seen them in a dream as I have never seen them represented, as we have sadly never been able to *see* them, in all the light of day and night. Do we know that bodies are divine and *luminous* against the blue night? Do we know that they dazzle with living joy in the dawn?
>
> Do you know that years ago I rose in the night and walked nude in my room, with a serious joy that I could not express in words, to see my body and my living movements in the shadow of a life so mysteriously beautiful that I cried? I am sure that I was enveloped in the beauty of my dream, but anyway, *I saw this beauty.* I can never forget it. For a long time I have despaired of never being able to capture it. I know with certainty that it is one of the many beauties, *rather than an essential part of the* ultimate beauty; perfect; necessary to life as air and sun. As long as people do not know how beautiful they are, can be, in body and soul, they will live only half aware.

Mabel regrets that she expresses herself so poorly. But "*this is like a wing at the core of my life*, something that vibrates and wakes up on occasion, something that I will never give up for anything, *anything* in the world. *Sometimes I believe that it is my reason for living*." Women are associated with birds and cages in Pre-Raphaelite art, and Mabel has appealed to avian imagery previously, but here she extends the image to a personal metaphor.[23]

On 4 November, George conveys that his work is going well, but there is a great deal to do: "I should for instance limit my studies to one period in history, i.e., the 17th century which especially interests me. And I sense that I will have to give up the study of Greek science, which however attracts me greatly! Alas, a human life is stupidly short! We are so weak and small." George writes on 6 November that he "has been distracted from his work by much correspondence and political preoccupation." The previous day, he ran around all morning trying to find information. "You see, I still don't know *exactly* what is going on in Argentina. I know that the reaction is very strong and that many socialist comrades have been persecuted, but I haven't been able to find out details. You know that we can't rely on the newspapers, which are informed by the agencies (Havas, Reuter...), all sold out to the capitalists [finance]." He still must go through his "voluminous" correspondence. He visited an art exhibit and attended a Flemish play. On the 10th, he relates that Irénée van der Ghinst has been appointed military physician in Brussels, a position that he wanted. George will look over an old house with a big garden near Madeleine van Thorenburg's residence.

Mabel travels to London on 9 November and writes from the train. Raymond purchased her a second-class ticket to Antwerp (typically she travels third class), and so "she is comfortably seated." She remembers that Andries MacLeod is visiting George that evening. At noon she continues from an Antwerp café. She worked hard at a printer's shop, where she was concerned about the right colors for a poster. "I was right to bring a large apron with me, for I am terribly dirty still. The blocks had to be cut much deeper than I had done. The errors could not be corrected." After lunch, she will sit quietly in the cathedral, for she can't return to the shop until 2 o'clock. She left her English money in her rain coat. Perhaps Raymond could drop it at the train station for her, or send it on to London. Mabel worries that George is not happy with her upon their parting. "Or was it simply the very wrong and inexplicable painful feeling that I often get when you go away in obscurity and leave me alone?"

23. Elaine Shefer, *Birds, Cages and Women in Victorian and Pre-Raphaelite Art* (New York, 1990), pp. 193–5, for the image of wounded birds used by William Holman Hunt and captured birds by Dante Gabriel Rossetti.

In London on 11 November, at her hotel in Bloomsbury, Mabel reports how she hates shopping, "going from one shop to another, taking in all the things that are stupid, or ugly or sadly tragically cheap, all the glaring lights, the noise, the hurry." She describes her day. She awoke late.

> I slept poorly. I was sleepwalking again, fortunately in my room, but I was quaking, cold as a rock, on my knees in front of the window, meddling with the blinds. I had no idea what I was doing. I can't tell you the horror of finding myself again in this condition, so foolishly, without knowing how long I was like that nor why (it was two steps from the bed) in the middle of the night. And so dark and cold. Suddenly came to my senses. I began to cry, in spite of myself, and little by little, warming up in the bed and crying for help, it passed, and I wound up sleeping soundly until this morning at 8!

She felt bad all morning, and she postponed seeing Thompson Buchanan's play, *A Woman's Way.* Mabel is trying her best to be frugal. She was sorely tempted by a perfume that she had spoken to George about, but she resisted. When they return to London together, she will have him buy it for her if she "deserved such a luxury that day!"

On 15 November 1910 Mabel reports from her hotel that she has been ill enough so that her companion Isabel Enke *née* Armstrong called a doctor. In part it is exhaustion, perhaps also something she ate. The doctor thinks that Mabel should rest completely for several months before her marriage. Indeed, the thought of designing nauseates Mabel. On the 16th Mabel writes that she is improving. Mabel wants to rest in Nice or in Switzerland. George, on the 15th, promises to pay for the cure; on the 16th, he thinks that she may have had an intestinal problem, but it is above all her nerves that need rest. She could go to Enggistein in Switzerland, where she spent a summer, or perhaps Nice, until their marriage. And after the marriage, she should not stress herself: "not only for the two of us, nor for our child, but also *for your work.*"

Mabel writes on 30 November, asking George to be a bit more gentle in expressing himself, less strident and self-centered. She begged him not to misunderstand. "I want to help you if possible not only in big things—am I capable of doing so? but for small things, these small things that are the bubbles rising to the surface but which often indicate what is below." She urges George also to correct her. She wants to sit quietly with George for a long time, before her operation. She signs herself, "The Wild Bird." The sobriquet is one that Mabel uses over the coming 20 years. Perhaps it is an allusion to Ibsen's *Wild Duck*, where Mabel might have associated herself with the invalid character Hedvig, although Mabel often uses images of birds and wings. George eventually assimilates the allusion. He writes after Mabel's death: "I am tired and my soul is black and blue all over, bruising her wings against the bars of her cage."[24]

Late in 1910 Mabel is at Eecloo, at the villa of the Enkes. (Peter Enke is a prosperous mill-owner.) George writes in English on First December 1910, a remarkable letter for someone who is uncertain of his command of the language. He addresses Mabel as "My dear wild bird":

> Is it not the best proof of love that you can give to me, than to try to make me better & gentler? But I ought to confess, that when you reprimand me with my faults, I cannot but prance and rear, at least in the very beginning. It lasts only a few seconds and then I accept it and I feel myself full of gratitude for the faithful friend, who helped me. But I think, it would often be better to reproach me with my faults by word of mouth, when I am near you and especially in those moments when we feel ourselves nearer than never, when you are so close to me indeed, that it seems to me as if there was only *one* heart, *one* breath, *one* soul in our both bodies; then it would be as if I grudged *myself*.

Mabel writes on 5 December, in English:

> Tomorrow is St Nicholas day. I am thinking of you as a little boy, when there was no-one to give you presents nor to watch your shining eyes when you got them nor to give you the best of all presents—a tender, loving kiss—and I feel stupidly sad at thinking of that little lonely body, and a want comes in me to make up for all those missing small joys—a want to give you the tenderest, softest kiss of all I have ever given you, a real mother's kiss as if you were my own dear son. And my thoughts turn softly to your mother and I say to her, with the heart-words that can't be written nor spoken, "Dear Mother of George, I will try to love him not only passionately truly, dearly as my husband, but also as you would have loved him, the way that good mothers love and that only the heart-words can even touch." And I imagine that she smiles back at me with kind, sad eyes (eyes that are very like yours are sometimes) and I seem to hear her say, "Yes, love him always, but you will only really know how to love him best of all, when you have a tiny boy of your own, held close in your arms."

We have seen Mabel's allusions to mothering George before. But here we see why he becomes so comfortable, later in marriage, calling her "Mother."

On 5 December, George reports on meeting with Julius MacLeod's wife, Fanny *née* Maertens, who showed him interesting embroidery from

24. NYPL, George Sarton to Mary Bouton, 21 November 1950.

25. The identification results from the fact that George's friend Andries MacLeod is not married at this time.

Russia.[25] He wants to read a part of *Hiawatha* with Mabel on Sunday. He sends her, by registered mail, 6 designs for propaganda postcards for the Etudiants socialistes. He thinks that it is difficult to sketch silhouettes of workers. He guesses that the postcards, engraved, will cost F 13.68 for the 6; printing would be F 4 per thousand. On St Nicholas, 6 December, George arranges for a Louis XIV commode to be sent to Raymond Limbosch.[26] George is assembling the furnishings for his bride's house, even if he has not yet located it.

Mabel writes to Marthe Patyn on 28 November that she is to be operated on for appendicitis. Her experience with the hernia was so positive "that it is a good sign." While she is waiting at the Dangotte residence, she is embroidering. She has ordered blue curtains for the bedroom of the Van de Voorde cottage, and if her client approves, she will have them made by the *béguine* orphanage. Mabel's embroideries will be featured in an upcoming number of the *Studio*:

> I had brought them to London to try to obtain orders from one or another of the large companies. Then I fell sick and despaired of not having accomplished anything. And then the idea occurred to send them to the publisher of *The Studio* with a letter simply describing them and saying that I thought that he would like to see them. He was more or less prepared, because Fernand K[h]nopff should have written to him, and an article on the cottage was projected. He asked to keep them for several days to photograph them!

Mabel casually records the intervention of the Belgian Symbolist painter Fernand Khnopff.[27] Then she adds that she will submit cushion designs for a competition announced by the *Studio*. She really hopes to win, for she needs the money. She has earned virtually nothing all year. Mabel doesn't want Mme Dangotte to pay for her contribution to the exhibition, especially so because she can never repay what Mme Dangotte spent to send her to Zurich. Mabel worries about costing her husband money, for the appendectomy, before she is married to him. Isabel and Hermann Enke have invited her to recover from the appendectomy at Eecloo. A copy of Raymond's book has arrived. The cover is fine, except that the orange color is too red. Mabel writes again to Marthe on 3 December. Raymond has returned from a trip to Germany with Mme Dangotte. On the train he read the correspondence between Thomas Carlyle and Jane Welsh, and in Weimar he found a lock of hair that Jane sent to Goethe.[28]

26. Van Oye, p. 56.

27. George writes Khnopff's name in his agenda on 23 March 1904.

28. NYPL.

Early in December 1910, Mabel undergoes the appendectomy by "two good and clever doctors," and she recovers at Eecloo. Mabel writes from Eecloo on 3 January 1911 that her mother accuses her of ingratitude. On 3 January George is sorry to learn that her mother has neurasthenia, the origin of her behavior. On 5 January, Mabel thinks that her mother has been suffering from it for some time. "It is the same sad state of mind that my father suffered atrociously from in the last years of his life." Mabel needs support from George, because her mother, "since the time I was very little, always doubted me *from time to time*, doubted my affection, my sincerity." She comments on the designs of a young socialist artist sent by George. They leave her completely cold. The artist would do well to practice more, to look again at the designs of Eugène Laermans, who portrayed peasants and workers. On 10 January Mabel writes that she has worked on designs for Christmas cards, which she—strangely to our eyes—hoped to sell at that time of year. She addresses George as Dominique, his pseudonym as a novelist. "How is the thesis going?" In English: "Has the first plunge in been taken and are you swimming along easily now? Or is there a counter current?"

Then about her independence. Mabel depends financially on George from this time forward. The situation continually rankles her: "There is something beyond my will, something wild rises up in me sometimes, and gives me such sharp pain and a bewildered feeling. Oh I don't think you can understand nor I explain, but you must be merciful to the Wild Bird who is wholly in your power now." George, on 11 January, hopes to finish his thesis in May. Four of the designs for the socialist students will be published, if possible. Mabel had wondered if the socialist designer could speak French. George replies that he speaks only Flemish. He is still a child. He made a gift to George of one of his sculptures.

George's thesis is going more slowly than he thought, he continues on 11 January. It is a question of efficient research. "I am often stopped by the shortcomings of my notes: It's a good lesson for the future." Systematic note-taking, humanists will affirm, is both a science and an art. George is a slow learner. We shall see later that much of his early research proves unusable for his general introduction to the history of science, and a great deal of the time he devoted to the history of Asian art never finds its way into print. George is ecstatic about Mabel's designs for cards. He urges Mabel to place a monogram on her designs and embroidery. On 13 January George thanks Mabel for her "good and reasonable" letter, in a letter where he omits a salutation. "You are sometimes so reasonable that I am amazed. And your letter proves to me once again that you are much, much better than I: I say this not to place you on a pedestal, nor to humiliate myself before you, but I say it because it is true." George comes back to identifying Mabel as his guiding light, his moral compass. She replies on the 13th that he must know that she loves him "as my friend, my husband, and *my son*." She does not belong on a pedestal. "I am a human creature like you. Nature cries out in me also, on occasion. I have suffered more than you will ever know." She has also struggled with her passion for him. George wrote about social convention. Mabel comments:

> There was never a question, I never thought about it, of social convention, at least for me. There is this. I am physically weakened and I think also morally. I would not be able to have a child. It is not about exciting the desires of sexual intercourse. These are, I know well, not only for that, but they have that as an effect. I imagine that sexual intercourse in itself would exhaust me.
>
> If I were completely healthy, if your work allowed it, and we longed for it in the same way, I swear to you—no, I need not swear, I tell you simply—there are no "social conventions" that would prevent me from sleeping with you. But if you would give in to your desire to take me now, the laws that you would violate are more fundamental. Not only would you desecrate me in what I hold to be most precious—my desire for harmony between my physical and spiritual life—you would betray *it*.
>
> You must not say "yes" because you think that I am better than you, nor because I am a woman, nor for any reason except that you know it to be *true*. Ah! I am quite a bit more "wild" than you think—the primitive, wild, sensual woman, who would *enjoy*, for a few moments, to feel dominated *by force*. That sleeps in me. But I want to tame it. Such a union seemed to me so nefarious, that I fear that I would hate the man who would have been the instrument of my betrayal to myself.
>
> And I do not fear that I will never give myself to you entirely, with the will *of my entire being. It will be*. I see it on the horizon of my life as a light that dazzles me as a divine light.

On 16 January 1911 George describes to "Bunny dear" the operation and recovery of his cat. He has sent Mabel a petition to circulate to the servants. Mabel replies the same day, in English: "*Poor* little Pussy."

Photographs of a number of Mabel's works appeared in the 14 January 1911 number of the *Studio*, accompanying an article written by Ferdinand Khnopff about the Brussels Exhibition which focuses on Van de Voorde's cottage. Khnopff deplores the fact that the modern school of Belgian art is unrepresented at the fair, especially because the German exhibitors in art are explicit about their debt to Belgian colleagues. The Belgian artists who exhibited, Gustave Serrurier-Bovy and Van de Voorde, did so at their own expense. Khnopff observes that Van de Voorde follows Serrurier-Bovy's notions about simplification in art (which stem from the ideas of John Ruskin, William Morris and Eugène-Emmanuel Viollet-le-Duc). "The architect was fortunate in obtaining the co-operation of Mme Dangotte, who chose and arranged artistically the various useful and ornamental articles; of Mlles Mabel Elwes and Méta Budry, who designed and executed the mural decorations and embroideries; and of M. [Victor] Acke, who made the furniture." Then follow images of 5 of Mabel's embroideries, produced both by machine and by hand. The motifs are flowers and a pair of peacocks.[29] Mabel's and Budry's embroideries attract the attention of visitors to the cottage, Léopoldine Dangotte observes in June 1910.[30] The exhibit and especially Khnopff's review mark Mabel's arrival as an artisan, at a time when George is still a largely self-published student.

29. F[erdinand] K[hnopff], "Studio Talk," *Studio, 51* (1910–11), 324–8. On 30 August 1910 Mabel writes to her mother that she must really do a watercolor of her cottage project for the *Studio:* "I do not feel up to it but can't put it off longer." NYPL.

30. HGS, Léopoldine Dangotte to George Sarton, 24 June 1910.

CHAPTER 6

Rules of Engagement

The Sartons exhibit many of the sensibilities and anxieties of intellectuals in the generation before the First World War. They are pacificist, they are sympathetic to socialism, and they experiment with vegetarian diet. George sees himself above all as an intellectual. Mabel sees herself as a designer. They are conscious about dressing correctly. They are attracted to beautiful expression in painting and music. They are omnivorous readers. They have confidence in modern medicine. They belong to no church. They carry 19th-century traits into their domestic life. First, they cannot manage to live together without the help of servants. Second, they exhibit a double standard about sex. It is likely that Mabel is a virgin bride. George, we shall see shortly, is not a virgin groom.

Mabel, recovering at the Enkes, writes to Marthe Patyn on 27 January 1911 with her hopes for her future children. Madame Dangotte continues with her inventory, and Adolphe Dangotte "is always impolite, even vulgar with her." A fissure in the Dangotte union opens into the wide rupture of divorce, with significant implications for Mabel. Given Céline's marriage and Mabel's long engagement, the Dangottes look forward to both women setting up their own nest. Domestic strife makes Mabel an undesirable houseguest. Mabel reports to Marthe that, during her illness, all her things were sent to the Enkes from the Dangotte residence. "It's *chaos*."[1]

Beginning on 16 February 1908, according to his diary, George becomes a "strict" vegetarian—that is, he does not eat meat and, if possible, avoids fowl and fish—and when he and Mabel travel they seek lodgings where vegetarian fare can be arranged, whether in a hotel or at a restaurant. George passes his second doctoral qualifying examination on 6 February 1911.[2] On 17 February 1911, Mabel leaves with Raymond and Céline to rest at a lacto-ovo-vegetarian retreat, Le Domaine de Créqui, located on the hills some 7 kilometers northwest of Nice. The cost is between 6 F

1. NYPL.

2. A. M. Simon-Van der Meersch, *George Sarton's European Roots: Introduction to His Life and Times* [Exhibition organized by the Archives of the State University of Ghent and the Department of Communication and Cognition] (Ghent, 1984), p. 17.

and 10 F daily. George accompanies them as far as Lille.[3] Mabel and the Limbosches have planned the trip since December 1910.[4] The idea is for Mabel to recover her health before marriage—rest is at the time prescribed for neurasthenia, Mabel's chronic fatigue. More to the point, she has no other place to stay.

Mabel, on 18 January 1911, writes upon attaining the shores of the Mediterranean, in English:

> Dear Heart, you will never think I am not feeling anything when I am going to leave you, if I talk and smile indifferently, and even kiss you as if I did not realize it was the last time? You will understand? I am too afraid of letting myself shew *too* much. I have to keep sharp command of my little army of body, heart and soul, and that is why, when you are gone, I feel really as if some part of my self has gone with you.

Céline spoke on feminism in Nice that afternoon. The house where she is staying is atrocious, very institutional. George writes to Mabel on 18 February 1911, from Ghent. He cites once more from Thomas Carlyle to Jane Welsh: "Has not a Kind Providence created us for one another? Have we not found each other? And might not both of us go round the Planet seeking vainly for a heart we could love so well?" George writes on 21 February that he is not looking forward to his weekly lecture to the socialist students at Mons. He dined with the MacLeods. Fanny MacLeod made a vegetarian meal; she "always worries that I will not have enough to eat and forces me to eat like several ogres."[5] On 24 February, George disputes Mabel's view that he is "so dreadfully reasonable." He couldn't work, so he looked into *Thaïs*, by Anatole France, a novel about a courtesan in ancient Egypt who comes to lead a pure life while her savior is destroyed, "one of the most beautiful I know."

On 23 February 1911, Mabel invokes the moral code that she expressed previously to George. She complains that Raymond and Céline have not taken her sensibilities into account when they sunbathe in front of her in the nude. On the first day she walked with them to the spot reserved for sunbathing, and then she returned to the house. "In the evening, they told me in a very friendly way that I was being foolish! Even that, if I wanted, I could draw them, and I really sensed that they didn't understand my scruples at all." She accepted Raymond's invitation to take in the sun with

3. Agenda.

4. HGS, Emile Masson to George Sarton, 26 December 1910, asking Mabel to write from Nice, when she arrives.

5. The home of Julius MacLeod. Andries MacLeod writes: "My mother says that you will always be welcome if you want to share our meals." HGS, Andries MacLeod to George Sarton, 11 April 1912.

him; she tried, but couldn't do it. "I would have thousands and thousands of times more difficulty to reveal myself nude to you!" It bothered her. "Are we wrong, or them? Or are we right? They do not love each other as we love each other." Mabel continues: "If I am shy about showing myself nude to you, is it not very simple and very natural? Am I bad because I ardently desire to appear beautiful to you? Is it not the fear of seeing in your eyes the recognition that I am not so?" Raymond sees her with a "perfect calmness, with a friendly indifference," but she cannot appear nude before him if she can't think of doing so before George.[6]

George replies on 25 February (2 days after Mabel writes) that Mabel's reflections on sunbathing worry him greatly, but he never thought of reproaching her. He is furious at Raymond for his lack of respect toward Mabel, that he "did not take account of your liberty of thinking & feeling. Myself, certainly, I would never act in this way toward you: to force you to act in a particular way." He sends along a letter for Raymond. George has difficulty answering Mabel's question, whether she should disrobe before him. It seemed to him "a kind of voluptuous madness.... It is a madness, a divine voluptuousness that God gives to us and that we must not reproach ourselves for, but rather *bless*. Why are we sometimes sad afterwards? Precisely because these sensations are so unusual, so extraordinary, so superhuman, that everything seems pale by comparison, and so one feels to have fallen in disgrace." George thinks that sensuality, when it is out of control, makes people into brutes; "perhaps sexual drunkenness is as destructive as that of alcohol and opium." They would help each other "not to abuse" what God has given them. He adds in a note: "Moreover, too much intellectual activity would compensate, if there were a danger of being overly sensual, which I think there is not." His letter to Raymond is frank and friendly: Raymond oversteps the bounds of common courtesy by imposing his will on Mabel. George affirms that "modesty, and all the delicacy of our sexual life, *all the mystery that we like to wrap our love in*," all these things are conventions. But they are "necessary conventions, introduced by civilization, which like morals advance stability and progress."

Mabel writes on 28 February, her father's birthday. In English: "Often when we lived too near London, he would take us by train out into the country and we would be away the whole day. I remember it so well! We were always a little disappointed because it was still so wintry." She returns to sunbathing, in French. She does not want to show George's letter to Raymond. Among the reasons for not doing so is that "I want to go out *by myself*. It is about me. I need your sympathy, even your advice,

6. Naturism early in the 20th century was associated with pure living. Jan Romein, *The Watershed of Two Eras: Europe in 1900*, trans. Arnold Pomerans (Middletown, Conn., 1978), p. 504.

but I want to act independently." She wants to speak to Raymond herself, "have the courage to tell him out loud that I want to be logical, and I do not find it logical to take part in *his* sun bath." She has, however, taken a sunbath with Céline. She describes her day: her breakfast is milk, decaffeinated coffee, bread, and butter; lunch: vegetables, salad, meat, potatoes or vegetables or pasta, fruits and nuts. She eats only half of what Céline eats. They have tea in the afternoon from small aluminum cups, along with cake or a biscuit, and a raw egg. The evening meal consists of vegetable dishes or pasta, and fruit compote or cheese. Mabel talks with other guests, Mr Das from India and Jean Philippe Vogel, a Dutch archaeologist who excavated sites in British India and whom Mabel met twice in Zurich; previously she sent Vogel photographs of her father's of Indian temples.[7]

George is sorry on 26 February 1911 that he is hard at work, and that most of his work involves making notes about Isaac Newton. "I hardly have the time to reflect on what I will write, to meditate deeply on his ideas. I must express myself quickly." He wants to express himself on the subject of the sunbaths because he "would never stand it if, later, Céline or R. entered our room repeatedly when we had the joy of being their hosts." George thinks that

> human love has been perhaps very much like what it was at the beginning. Every couple hides its familiarities, no progress is possible in this respect. Neither is it permitted to say that certain people love better or worse than others. Each loves in a special way. The main thing is to have agreement and harmony between the ways of loving, the temperaments of the two spouses.

Then he invokes Aphrodite, asking the goddess to maintain his youthful vigor. He has given a lecture in a beautiful auditorium at the Institut commercial du Hainaut. "The Cercle des étudiants socialistes at Mons is founded. But will it live?" If the F 300 that he gave her is not enough, he urges her to let him know. "You know that your dear husband would be very sad if you allowed yourself to lack for something." This sum would, according to the information Mabel provides, be enough to maintain her for a little more than a month. George, on 26 February 1911, writes wistfully from Ghent that he slowly walked by the river in the evening, "where the lights went on one by one, an atmosphere of beauty and innocence." He sees their marriage as a union for goodness. "Perhaps our restricted

7. It is possible that Mabel refers to Rakhal Das Banerji, who (while looking for traces of Alexander's passage in India) in 1922 discovered and excavated the Buddhist archaeological site Mohenjadaro and discounted the hypthesis of a pre-historic Aryan invasion of South Asia. Sudeshna Guha, "Negotiating the Evidence: History, Archaeology and the Indus Civilisation," *Modern Asian Studies, 39* (2005), 399–426, on pp. 405–6.

budget will not allow us to give each other rich gifts, but I hope nevertheless that we will be able to carry out much good around us. That will be our happiness, *and the dearest and the best of our secrets*."

Mabel writes in English on 28 February that she is not frightened of George,

> I am simply rather puzzled about *myself* and about the world in general. A thing that rings false to me is to say that a proof of non-sensuality is to have an absorbing intellectual employment. I don't believe that! Such work may indeed favor sensual exaltation & lead to its abuse, in its intervals, that would not be the *least* astonishing.

She is not pleased with his paean to Aphrodite. She salutes him the next day, "Good morning, Funny Dear!"

George writes on the same topic, on 28 February:

> I understand that a woman (or a man) who feels beautiful may want to show her body. In itself, there is only good in that. But then she must realize the desires that her body will evoke. *She cannot then reproach the man who desires her*; his desire is normal and healthy. Furthermore, if it is a man whose instincts are strongly conditioned by civilization and if he is also married, this desire will be able only to be ephemeral and without ardor. It's like a flame that he jumps through, but he will have to perhaps [illegible]. However this may be, she must not provoke him. That seems very clear to me.
>
> But I understand that a woman (or a man) can *enjoy* feeling desired? Is this desire not homage to beauty. And it may satisfy his sensuality. One thing results from these points. Either this woman is a cruel flirt, and we won't pursue this, or she is truly a woman with heart and so she has allowed herself the privilege of exciting this desire, and so will she not see it satisfied? I suppose furthermore that this desire may become so eloquent that it is transmitted to her. I don't say that all this will necessarily happen, but it certainly may happen.

Then George interpolates "Conclus," for we evidently read a draft. He continues that the woman can never be sure that the man before whom she disrobes does not desire her. She can never be sure that she is not beautiful or desirable. He ends "Destroy," dissatisfied with the draft, but not so dissatisfied that he does not send it to Mabel.

George writes an enthusiastic postcard on First March 1911: "I am happy that you like the *Roman du lièvre* as much as I do, because it is truly beautiful! I like Francis Jammes more than any other author in our time, & I was often shocked & saddened to feel that he was not entirely understood." George recommended previously, several times, that Mabel read the book aloud with Céline and Raymond. Jammes, the French poet and novelist, is by then the apostle of *naturisme,* where the writer focuses on common, everyday occurrences. He converted to Roman Catholicism in 1905 under the influence of Paul Claudel, and in 1911 the first volume of his novel about a Catholic peasant family, *Les Géorgiques chrétiennes,* appears. Here is the religious side to socialist George, a side that Mabel finds amusing. On 2 March, George is off to his lecture to the socialist students of Mons. He tells Mabel that before she left for Zurich, he loved her a great deal, but he did not desire her; he did not know her well enough to desire her. His desire for her grew when they became engaged, and it continues to grow. More on Jammes:

> I sense sometimes that my sensibility is so close to his. I feel that I understand so well what he writes, and what he does not write; all the nuances of thought that are hidden sometimes behind humorous words, or strangely rhyming verses, so prosaic. I like his simplicity, mixed with a sensibility so pure that it reminds me of the best parts of nature.

He also comments also on Anatole France's *Thaïs*:

> It is truly a complete work. So many books by Anatole France, his [*Fragments del'histoire contemporaine],* [1902], for ex., displease me with their fragmentary & anecdotal character. I like the details, but I do not like the whole. I think that this is decadent art, where beauty is sacrificed to elegance, and where the need to think carefully cedes to that of thinking with verve. And so this cunning scepticism shocks me. It is no longer great art, but rather dilettantism. On the other hand, *Thaïs* is a complete work, complete in itself.

George writes on 3 March about a lock-out at Eecloo. The bakery at the socialist cooperative Vooruit asked all its customers to contribute to the strikers. In one day they collected enough for F 4000 worth of bread.[8] George apologizes, on 4 March, for his clumsy thoughts about physical love. He clarifies: "We are both sensual people, but we both are repelled by artificial excess in sexuality." Sometimes he feels that his love could not be satisfied by millions of kisses and centuries of embraces, but the feeling is not right. "We must remain masters of our desire: so long as things are this way, we can give ourselves over to it without fear." He thinks she is right that an intellectual life, like his, leads to sensuality. But he thinks that intellectuals run less of a risk, in this regard, than a life dominated by bad theaters, clubs, and cafés, or "even than the life of a poet or of an artist." On 5 March, he has ordered jewelry, possibly wedding bands, for F 2300;

8. Since the late 19th century, the Vooruit was known for selling high-quality bread at reasonable prices. At the time, bread provided "at least 50 per cent of the total calorie intake of working-class families." Peter Scholliers, *Wages, Manufacturers and Workers in the Nineteenth-Century Factory: The Voortman Cotton Mill in Ghent* (Oxford, 1996), p. 167.

the Willems are making the arrangements. George very much admires the bedcovers Mabel made for Isabel Enke. He promises to begin canoeing (a vain promise) and also civil-guard exercises to get into shape. On the 6th, he has completed the manuscript of his thesis. It will run to about 300 printed pages. "There was no joy in finishing this work, because there was very little in writing it." The good consequence of the thesis is that it has taught George how to work. He devoted 6 months to it, although probably half that time was wasted. Still, it is a considerable accomplishment: "A new universe has opened before me. But this is a small thing, considering what remains to be done!" He calculates that he owes Céline and Raymond F 112.14. Should he send it along? He will begin to study for his examination. He sets out his work schedule for when they are married. The day stretches from 8 a.m. to 10:30 p.m., with time out for long walks and meals.

Mabel writes George a long, rambling letter on 5 March. It is about her feelings of intimacy. The theme turns round in her head. Mabel asserts her will in the letter. She wants him to be her equal, not for him to give in to her fancies.

> Another thing still, sometimes you have indicated to me that because I am virgin I cannot know anything about or enjoy sexual satisfaction. I would answer: *You are in the same situation*. Evidently you know a man's desire better than I, but I know better that of a woman, even if she is virgin. I will not comment on her satisfaction, but on *desire*, that I know as well as you.

Mabel knows, on the 6th, that George has not understood, and she regrets her last letter. She is trying hard to express herself. George should not address her as his "dear friend" when he sends "sentences filled with tacit or direct reproaches." Mabel is afraid to write in the future with what she feels or thinks. She replies to his request that she not keep his letters by saying that she has kept them all. She now feels better. She would like them not to speak any longer about their capacity for sensuality, nor about their desire. "For the former, I can't remember having spoken. But my head is so light and tired that I don't remember anything clearly this evening. I know that I've never felt worse than during these past days (yes, however). I know that I love you more and more. And this evening I sense how *good* and simple I can be."

She invokes him on 8 March: "Come close to me, my love, my child, my dear George." Mabel feels sad and lonely. She asks in English: "Is it just a physical reaction? I can't pretend [illegible] well. I have strange pains and queer feelings." More:

> Oh George, I feel overwhelmed. I feel almost as if I were losing my reason. Something is stretched so tight in my consciousness that I feel, one

> word more from some one I love, & that something *would snap*. George, help me, help me dear. For Heaven's sake, no for love's sake, don't just tell me I am exaggerated, am unreasonable, am hysterical. I am still trying very hard to get back to a better equilibrium, but something seems to have gone very wrong with me. I feel so worried and sad....

Mabel, wrestling with neurosis, has assumed too much about George, who is shocked into a confession. "I have just knelt near you," George writes on 8 March 1910 with an uncharacteristically clear hand. "Let me remain so." He has not the right to stand before her. He received her letters. He knows that he is beneath her, has he not often said so? But Mabel should remember where he was coming from. He is trying hard to improve, and perhaps he will never reach her heights. But at a certain point, do not efforts count above all? "One of the reasons why I fell in love with you so quickly, why I have been able to hope to marry you, is *the profound feeling that you really would be able to help me rise higher*." At the beginning, George grabbed her as a life saver. He came from so far away, in a place so sad and dirty, mired in the mud. She brought him light, and now this light is necessary for his life.

He must tell her some unpleasant facts about himself, and he focuses on his late father. As a student, his father had been very smart but he took things lightly. "He belonged to the rare category of students who had enough stamina, enough intelligence, and enough libido [assez peu de moralité] to put up a good front and do very well in his studies while living a dissipated life." He was one of the university's best, as a student and as a miscreant. From this time he acquired the habits of an alcoholic. "I should say that, however, my father was not at all what one commonly calls a 'chaser of women' and I think that he never deceived my mother, whom he adored and for whom he had an immense respect. But his sexual morals were extremely lax. He would not disapprove of my taking mistresses, provided that, of course, I treated them well." He continues:

> My father had the misfortune to obtain, when he was young, a very honored position that was completely secure, so that he did not continually have to try to find something better. He became lazy, and he liked to eat and drink. I never told you that after the death of my mother, he continued to be depressed and he began to drink again, just as when he was a student. When he grew older, he drank less, but as he was less tolerant of it, the effect was the same, and he wound up drunk more than once. These were terrible times for me. More than once as a child, I had to recall him to his senses. He suffered awfully, and despite everything, *there was so much goodness in his eyes*.

George believes that he inherited an alcoholic disposition, because his grandfather was also an alcoholic. But George also wants to relate his father's good qualities, which he thinks outweighed his father's faults. He was a good and simple man who continually helped others. "He was not only just, but I may say that he possessed this *consuming passion for justice* which I have inherited." He was for the most part tolerant. These qualities gave him many friends, even in his declining years. George and his father loved each other. In general ways he provided a standard of character, but he left George's moral education to chance. It was something that George had to discover by himself.

George's childhood and youth was extremely spoiled. It was only when he met Irénée van der Ghinst and they founded Reiner Leven that his eyes began to open, "and that I see the clarity of day. Before then, I was headed toward becoming, in my turn, the master of bad topics at the University of Ghent." What caused the transformation? It is due in part to the Flinke, but most of all it has been his contact with Mabel. "*You will never know how much you have helped me*." George has tried as hard as he can to "discipline himself." But he has often asked himself, "'Why has this austere soul been placed in a body with so many desires?"

One sentence of Mabel's recent letter makes George shiver: "'Again *in principle*, I do not want you to give in to me, nor give in to yourself, simply to make life easier for us, when it concerns an idea that is part of our personality.'" It is a fine ideal, beautiful, just, not to consider each other's faults, but he thinks that in practice it was a bad way to live. He hopes that jealousy will never come their way, but if it does, to follow what Mabel proposed would have a disastrous effect. It would drive George mad, or kill him, or worse. But if Mabel acted as she has acted until now, taking account of George's faults, the sky would be the limit. "I think that I will improve until the very end." The best way to make men better is to help them up. "Justice without goodness and indulgence, absolute justice that only follows 'principles,'" is odious to George. It is the kind of justice that makes children unruly. He thinks that Céline might have spoken to Mabel in this way, and it would not surprise him, "because I always thought that there was essentially something of the fanatic in her, something hard, as you yourself remarked not too long ago." Still, anyone who cannot give in without sacrificing his personality, such a person is "unfit for social life, much less for marriage."

George believes that the basis for education is kindness. He believes that people should be judged not only by what they are, but also by the effort they have made to become what they are. He does not think that, in proposing this, he is pleading for himself. "*It is not success that counts, but effort*." He gives an example. Neither of them abuses alcohol. Mabel does not like it, but George likes a good bottle of wine, and he has to watch himself not to drink too much. Is not his control more worthy than hers? It "enrages" him when Mabel writes, "'I sincerely believe that those who have come to feel like Céline, Raymond, and me have made a real step forward, toward the truth.' How do you know that it's a step forward? Have you been able to judge a lifetime? *It enrages me.*" George affirms, "Virtue lies not in the absence of desire, but in triumphal victory over desire." Would Mabel want George bereft of desire and vigor and in this way be "pure"? George has thought about this often, and it leads him to imagine suicide. He mocks: "Is there nothing more pure than death? O liberator, purifier, take me quickly, I call on you."

George feels far from Mabel. Is it that Mabel has distanced herself? "Your eyes, which would have me easily rise to the highest peaks, your clear and sweet eyes have looked away from me. I am a wretch, but you had faith in me, and I became good. Now you have abandoned me. But I know the proper path now!" He is not crying, but he feels hurt. He signs himself "ton ami, George." And then a very long postscript:

> P.S.: I must add this. I must make a terrible confession before you: I am not a virgin. But I hasten to add that it happened a long time ago and it occurred under such conditions that I have no memory of it. And so I did not lie, I swear it (I have never lied to you), when I said to you one day at the Bois de la Cambre, that the day when we sleep together for the first time I will be as upset, also as embarrassed as you.

George was wrong not to tell Mabel before. "The need to confess to you became much more urgent near the end of the past year, when our relations became so much more intimate." Chance had him read Tolstoy's memoirs. As an officer in the Russian army, Tolstoy led a dissipated life. When he became engaged to be married, he gave his fiancée his diary, where he recorded all the indiscretions of his youth. "His future bride received it as a terrible shock." The account gave George pause, and he delayed until the present time. He would tell Mabel the details, if she wanted, but he preferred not to do so. Nevertheless, he sends along a few lines about it.

In his first year of university studies, George went to class for only several months. He fell into bad company. He visited a café where students came with their mistresses—and what mistresses. George had stayed away from women. His companions thought that he was hiding his mistresses from them. George thought that this was the way of the world, especially since it was the way of his father. His father in fact thought that George was seeing women, because sometimes George would return very late at night. "The experience that I underwent was so brutal and quick that for a long time I was disgusted, a disgust joined however with desire—simple animal desire. I learned nothing from it, being rather the plaything of a woman."

George, however, was really not as debauched as his companions. He offers examples. He wanted to make a mistress of a girl who worked in a glove shop. After all, he wanted to be just like his companions. They went out together, and he kissed her twice. But just then the girl became sacred to him. He wanted her mother to know that he was squiring her, which astonished the girl. George traveled to Holland, and on his return he did not take up with the girl, because, although she was pretty, she was not intelligent, and he could not imagine marrying her. A second incident also shows George's essential goodness. Because he knew that he would never be able to seduce a girl for his mistress, he thought of finding someone among prostitutes. He found one in a café, and he asked her to take up with him. She accepted, and he rented a furnished apartment for her. The first evening that George came to her, "she had waited with such impatience that she was already in bed." George had delayed coming all day because he felt bad about the arrangement. The woman was dressed in fine linen. But at that moment, George no longer desired her. The woman wanted to embrace George, but all desire had left him. "I fled like a thief." He wonders what the woman thought of him. She lived in the apartment for a month, and George never came back to see her.

George believes that he should have revealed all these things the day of their engagement. He tells Mabel now, before their greatest intimacy. He would completely understand if Mabel recoiled from him in disgust. He considers the engagement held in abeyance. Even if Mabel now steps back from marriage, George offers to guarantee her material future with half of his income and the entire amount of any subsequent salary. "*You cannot refuse that from me* without offending me in a way that I do not deserve." He understands if she wants nothing more to do with him, but because he was once bad in one part of his life does not mean that his entire being should be rejected. "That is not true, it is not me. I am good now, *I am good*." He can write nothing more, and he dares not face her. His aunt often said that he would end up as a Trappist monk. "I am an unfortunate monk, of the kind of those who assassinate their father before taking the vows. And so, no—even if I were Christian, I would prefer to remain part of life, in the struggle." He asks only that Mabel not relate all that he has written to Raymond and Céline, although maybe she would ask their advice. "I know that the big gap in my education is not having been taught *self-respect,* which is inculcated so deeply in the heart of young English men. For me, that is the secret of the superiority of the English race over the French race."

Mabel writes on 10 March that she loves George. Just as he confesses to her, now she has a confession for him:

> I have also struggled against my sensations, just as you have. I was haunted by unhealthy desires for years and years. At first without knowing that

> it was bad, I believed that masturbation [l'abus de moi-même] was necessary to bring on my period. I know that that seems unbelievable, but it is so. Then I do not know how, instinctively I sensed that it was wrong. I began to struggle, but the attitude was established. Since then I have *almost* triumphed, but I still struggle with it.

So, Mabel offers, she is not the "pure being" that he imagines. Consequently, Mabel has immense sympathy for people who struggle in this way against all manner of aberration. She apologizes for not having understood George's prayer to Aphrodite. She is unsure about George's reliance on apodeictical reasoning: "Take the word *principle*. I don't want it. Principles unsettle me. They don't take account of individuality, the needs, faults, and qualities of the person who applies them." She sees herself, in perhaps the image of a Burne-Jones painting, as "a poor child from whom a warrior-king asks flowers." She urged: "Let us not destroy heart with the miserable words, like 'principle' or 'personality'—the key thing is to live, and to love better and better." In her last letter Mabel wrote that she, Céline, and Raymond have gone forward spiritually while George has stayed behind. She clarifies : "That doesn't mean that as a *whole* we would be better than you." It is merely "a step toward a state of greater goodness, *a healthier soul and body*, more harmony." And it has been a struggle to advance in this way, "you can't deny it!" Mabel thinks that George has "*not struggled* hard enough against the excess of desire." This is not simply a consequence of George's father. "I am sure that you are better, infinitely, if you master your desire in this way." She ends: "George, *I love you*. I am still your fiancée, your wife *for my entire life*."

George writes on 9 March to elaborate his thoughts. He thinks that it is not quite right to say that one should take account of a man's efforts at self-improvement, not merely at the results he obtains. Another equally just point of view is that only virtue, purity, and happiness matter. Someone who is corrupt, impure, or unhappy calls to us for elevation. If we only considered a person's efforts, it would be impossible to evaluate him. Is not the true nobility of soul where one does not have to make an effort to be pure, where one is pure from birth? George is not in a state of grace from birth. He has to work for it. George prefers people who are ardent and tormented, like Emile Verhaeren and Paul Verlaine. He returns to what he wrote about the methods of education. He favors modest methods of encouragement rather than inflexible methods of instilling principles. A good example of the method of discouragement was when Céline wrote to him, several years ago, that he was just a "vile egoist." But her letter only made George cross with her. George's second example comes from the question of sunbathing, which he can't get out of his head. He writes ironically: The three of them decided to take a step forward and expose themselves together; if George does not agree, it's because he lags behind. And the step must be taken without consulting him. The argument could also be used when discussing jealousy. George thinks that the method of discouragement, even using extreme force, is inappropriate. What Céline said to him did not help him at all; it stopped his forward progress. George wrote previously that if Mabel abandons him, his salvation would be in peril, but he no longer thinks that true: "I *am persuaded, on the contrary, that at the point where I am now, no obstacle can stop me from going forward.*" He continues: "Happy or unhappy, I will continue on to succeed in my small way. Nothing will distract me. Perhaps I will never get as far as you have, but at least I will have gone as far as I am able." George does not want Mabel to give herself to save him. "I do not love a guardian angel, but a woman." Is it not natural and healthy for George to desire Mabel physically? Is it not legitimate for a woman to be proud of her beauty?

George now considers himself at the same point as before their engagement. "I no longer consider myself your fiancé." George abused Mabel's confidence by not having told her all about his more distant past until recently. The best thing is for Mabel to live with the Limbosches. "The marriage will bring perhaps more disillusions. Certainly, I will shock you sometimes, despite all my good intentions (like any other husband). You will perhaps be glad to have a place of your own, to be mistress of your own home, but you will sustain all the inconveniences that result from a husband's authority, bad humor, or assertive desire, no matter how good he may be." He is sure that Céline and Raymond would accept Mabel into their household as one of their family. And she will be independent, for as George mentioned in his letter of the previous day, he will give her half of his income, or F 3000 annually. Alternatively, Mabel could go to live with Mme Dangotte. George thinks that, notwithstanding his confession about the folly of the time about his twentieth year, Mabel still has some shred of affection for him. But love is not enough to make a marriage. "One must be sure that a life in common is possible without continual affronts and discussions. The two spouses must be able to make reciprocal concessions and to *give in* to each other. If they see only their intangible personality, if they guard their independence too jealously, it is clear that they should not marry." If they were to marry, George would try to improve himself, but he would be "far from being perfect." If, "instead of helping each other and making concessions, we knock our heads together, it would be madness to begin life together." George writes quickly and to the point.

George adds a postscript later on 9 March 1911. He wants to make up with Mabel despite everything. Long-distance correspondence "causes much distress: Our letters continually cross in the mail." George writes that he wants desperately to beg Mabel's pardon, but he cannot do it. Still, he is desperately sad at hurting Mabel. He ends: "*Tell me what I must do to be pardoned and to receive your love*, and I will do it. I shall reestablish the discipline of my life and fight more energetically than ever to become worthy of you, *even if* you want nothing to do with me."

George writes on 10 March to ask if Mabel understands why he is generally grave and sad. "When I was a child, I played almost always alone in the garden, or I teased the servants and the dog, or I went away in my father's woods." By the age of 14 George had read a great deal, but mostly books or newspapers that poisoned his imagination. He feels now that "*the essential part of his nature is good and pure*." He spent the first part of his life struggling against his better nature. He had been convinced that a man of 20 should have mistresses. He could have gone in that way: He had the inclination and the money, and he was as free to do what he wanted. But his conscience saved him. What he related to Mabel proves that even when he wanted to act badly, he could not do so. As a young student, he indeed wanted to be the worst of the pack, and he did so out of pride. So when he proposes to act badly from pride, "please knock me on the head, throw me on the ground." George concludes that he changed his life through an act of personal will. He is so sorry to bring all this to her attention now, but he must. In her last letter, she underlined a sentence: "I should have replied, *that you are in the same condition* (i.e., that you are also virgin)."[9] If George had let this pass, he would have lied to her.

On the 11th Mabel writes about George's most recent letter, which makes her feel "extremely perplexed, extremely disoriented, undone." George wrote that sentiment is not enough. So she will "call up all my intelligence, all the reasoning that I am capable of," and reply to his points in detail. His letters were uncompromising, and so are hers. She wants only "*to help us both* find what is true, what is the best for our future." Mabel observes that George wrote, in a disgusted tone, "note well my dear friend, I don't want at all to meddle in your business, but only because I am upset—if I am upset, it concerns me a bit." Mabel thinks that George has not chosen his words carefully. In the future she hopes that he will think before he writes. George often wrote that he is inferior to her. It is wrong to make such comparisons:

> It is impossible to measure the relative value of two people of good will, as we are. You are different from me. I have undergone struggle different from yours (in addition to the ones I've confessed to you). I was not born 'in a state of grace.' It doesn't interest me if you are really superior to me. Let's struggle together, help ourselves in the struggle, and let us remain humble and proud at the same time, proud of our conquests, humble before all that remains to do.

9. George probably intercalates the parenthetical remark.

Nor should they feel superior to other people. "Certainly there are superior people in whom we see something divine. We should be ready to admire them without envy. They are rare. I feel much closer to human beings who are like me, but I thank God for having allowed our union to rise so high, it draws our attention toward the peaks and allows us to hope." George asks her pardon, but she does not know how to pardon. If he wants to make her happy, he should strive to be more just, and in small things. "Large sacrifices animated by the spirit of justice are on occasion easier to make than small ones, which involve a shoot of pride that spreads a long way, like a weed." George spoke in his letters about

> your sad childhood, your sombre youth, your struggles, your victories, *your pride at having arrived where you are.* You merit this pride. But don't mix passion in with it. You see? I believe that you are a passionate person who is blinded because of that. You do everything passionately, with exaltation. I don't think that is good. I find passion and exaltation necessary and beautiful, a vital part of life, so long as they do not dominate things. Even in the discipline that you want to enter, you do it with passion. So too in your self-remorse.

George must look at himself objectively, without exaggerating victories or defeats; "otherwise it is almost madness." Mabel knows that all this is easier for her—she is older and in some ways more experienced. Both George's strength and his weakness lie in his enthusiasm. To George's force, Mabel brings happiness and sometimes clear vision. George writes that he does not want them to give in to each other just to avoid controversy. "It is absurd to implement this idea, *which I never presented in an educational system*, to children." Such a "principle," in George's words, is unrelated to mutual help. Whenever George suffers a wrong or hurt, he believes that he deserves Mabel's consideration. Mabel thinks that this idea of George's is "simply based on the cult of personality." In fact, coming immediately to the moral aid of a person wronged is to reinforce "*a sense of defeat*, which, far from aiding him to recover, makes him despair." One really wants to help the injured person believe in his ability to triumph over his weakness. This view of hers, Mabel insists, is not a brutal view of life. "I believe that it is your pride as a man more than your desire, your passion, or your jealousy that prevents you from recognizing this and that drives you to write entire pages of reasoning absolutely unrelated to the question." Mabel has tried out her ideas on children, with good effect. "You see, George, that educating children with kindness and good [feelings?] alone is a false ideal. The child has no naïve soul." Mabel thinks that, exceptionally, one has to give in, but he has no reason to imagine that if she were to retain her views with dignity, she would as a consequence cause rancor in the marriage: "What do you base yourself on daring to imagine *an instant* that I would bring unhappiness, discord, or even much worse? *Where is your spirit of justice?*"

Mabel is furious. George has made no effort to see how she thinks, to imagine her moral progress. He has thought only of himself. He does not seem to value what Mabel, Raymond, and Céline think about passion: "The incontestable fact that disordered, torturing desire, which is an obsession for most men, having them degrade and victimize women, *this desire must be mastered.*" The dominating desire, Mabel holds, follows in part from "the fact that we hide the mysteries of the body from one to another." If people saw themselves naturally, children especially would be spared many of the moral struggles that Mabel and George experienced. "I know that I myself am not abnormal, that in any case I am capable of desire. However I am able to appear nude before a man without being troubled." Whatever the solution to this question, George erred in his manner of judging and discussing. "I am much more afraid for our happiness from this than because of everything that you have written about your youth."

Mabel recalls how she came to take *bains d'air* together with Céline. Even after having lived with Céline for several years, they had never seen each other nude. Mabel told Céline that she was unhappy with this false modesty, and so they began to walk and wash in full view of each other. It cost an effort, at least for Mabel, who did not think herself beautiful. Later, returning from Germany one day, Raymond spoke about *bains d'air,* and that it would be so nice and good if people were to walk nude, without any fear or desire. Raymond's argument impressed Mabel. When Mabel cared for Céline during her illness, she often saw Céline nude in the presence of Raymond, and that without thinking about it. One day she entered the bathroom where Raymond was taking his *bain d'air* nude. A few days later, he entered Céline's room where Mabel was taking her *bain d'air,* and he brusquely offered that he found it absurd and silly that we dare not appear nude to each other. Mabel found Raymond's argument persuasive, but she did not follow up on it. "Why? I really don't know! Did I vaguely perceive in my subconscious [sub-conscient] that you would not approve?" Still, she did not undress in front of Raymond. And she forgot about the matter until Céline and Raymond wrote to George about taking *bains d'air* together. She really couldn't imagine that George would be upset about it. Mabel looks forward to life with George: "I love you. I want to become your wife, the mother of your children, your companion, your friend. I have weighed everything for a very long time and I know in advance what I risk. *I accept these risks.*" Mabel continues on 13 March. George is wrong to think that he has failed her from the sexual point of view. "I think that you remain extraordinarily pure." On their wedding night, Mabel does not really imagine that she will be embarrassed.

On 12 March George writes that he will work every day to be worthy of her love and her pardon. Once they were married, they would express their thoughts freely. He thanks Mabel for her confession. George wrote to Mme Dangotte with the main lines of what he related to Mabel, so Mabel should not be surprised if it comes up in conversation. The next day George sends a deliberate and reasoned reply to Mabel's concerns. He reiterates that he will try to moderate his impulsive nature. He talked intimately with Dr Nyssens, who helped calm his uncertainties. He wants Mabel "to feel and to understand that she is completely independent, and for this reason I was severe." George believes that "account must be taken of *effort and grace*. Good will is necessary, but in it alone nothing is achieved. And to be able to judge people, one must often gain distance from them." George thinks that they should marry as soon as possible, but not the very day of his examination. "Yes, when I saw you at the Winter Concert, you were truly so beautiful and your eyes were so beautiful, it seemed to me that they pulled me in. When we return to the Concert, we will be married."

George continues on 14 March 1911 that Mabel's notion of coming to live in his apartment "raises several objections." The apartment is dirty, ugly, and uncomfortable; he doesn't like the people he rents from. "Finally, it would be very difficult for me to work here with you, since the rooms all connect with each other, and your presence would distract me from my work." He wonders if Céline and Raymond could rent them their house when they are at Coq-sur-Mer. George would find an apartment to work in, and they would take their meals together at the home of Van der Ghinst. They would need domestic help. When they have found a house, he will install his office there, and so give up his apartment. Mabel will handle their domestic help, not George. She should not forget that they "will have rather little money."

George outlines their means. They will have F 6000 yearly. Rent would be F 1000, food F 1800, Mabel's and George's personal expenses F 1200 (for "clothes, incidentals, including theatre, small trips, small tokens of affection, etc."), laundry F 360, domestic help F 360, coal, gas and water F 240. This leaves about F 1000 for "intellectual expenses" such as books, societies, journals, travel, charitable support (frais de solidarité), and medical and dental expenses. He knows that one cannot predict medical expenses, so they will have to be very frugal. George is troubled about the servant, but he thinks that it may be the best practical solution. "What is above all necessary is that we both accomplish our mission." They must arrange things to promote Mabel's health and George's work. When she returns, if he has passed his examination, they should hire a servant girl, get married, and honeymoon in the Netherlands for 10 days. When they are installed in their lodgings, they will travel to England.

George continues on 15 March, that he proposes their honeymoon in the Netherlands because he loves "the simple and *austere* beauty of the country." In an advertisement in the *Flandre libérale*, he asked for a house with a kitchen on the main floor, "because this seems to me so much nicer than a basement kitchen," common at the time in Belgium. George comes back to the question of sensuality. He confesses that he also masturbated as a child. He will attend the lecture of Emile Vinck, a Freemason and member of the Workers' Party, on "responsible procreation."[10] He concludes by hugging her tight, "Bunny dear." Mme De Saegher, after visiting Nice, reports that Mabel's room is not healthy. George will pay for a room higher up in the hotel.

On 17 March 1911 George recalls that they became engaged on 21 March 1910. Mabel came into the small study that he had rented in the former Dominican Monastery, less than a minute's walk from his father's house. He remembers the room perfectly, and Mabel also. Mabel was wearing a fur coat. She sat some way from George. They became engaged very quickly. In the afternoon, they went rowing on the Lys. The next day they went to Coq-sur-Mer. He thought that they would marry on the anniversary of their engagement, that is, the first day of spring. If Mabel is recovering and relaxing, then she should stay at Nice or elsewhere. But if not, "it would be better to return, it is a waste of energy and money." When Céline and Raymond leave, "since you are not vegetarian," there is no reason to stay on in the same place. The lecture by Vinck on "responsible procreation" was "quite miserable. I was really ashamed, mostly because young girls had come to hear."

Mabel feels "sad and upset" on 15 March. She thinks "that your confession has altered your feelings a bit for me." Mabel thinks that they became engaged on 18 March 1910; the 21st was when George asked if they could marry after a year. On 17 March, Mabel writes to say that George has helped her immensely in her struggle against self-abuse; his belief in her purity gave her the strength to become what he imagined. She has completely bared her soul to George. Céline plans to speak in Nice before the union of women in the cigar and vermicelli factories; "her topic is suffrage—the vote." On 20 March, Mabel writes that she is sending design drawings to Mme Dangotte. She will also send George a collection of

10. Emile Vinck, *Procréation conscient* (Ghent, 1909), citing Auguste Forel on pp. 14-15. In *La Réduction des heures de travail et la journée de huit heures* (Ghent, 1904), Vinck begins with Karl Marx on p. 5. Vinck, whose income derives from the legal world, lived in a house designed in 1903 by Victor *baron* Horta. Françoise Aubry, "Victor Horta and Brussels," in *Art Nouveau 1890-1914*, ed. Paul Greenhalgh (New York, 2000), pp. 275–85, p. 276. Denise De Weerdt, "De Sartons en België," *Brood en kozen: Tijdschrift voor de geschiedenis van sociale bewegingen*, 1999, no. 1, pp. 53–71, on p. 60, for Vinck as an opponent of traditional Catholic views on sexual activity.

terra-cotta figures for a Christmas crèche, *santons* traditional in the Nice area. In a second letter on the 20th, Mabel thanks George for the money, just received. She apologizes for having been so trying. In English: "I am *your* wild, wild bird, you know that, don't you? Though so wild sometimes, yet truly *yours*."

On 21 March, George writes that he has looked at several houses and begun a book on astronomy by Guillaume Bigourdan. George comments on England's attempt to sign a treaty with the United States on arbitration. He doubts that it is British initiative. "The English people would do much more to attain the unity of the empire of world peace by giving home-rule to Ireland, than by international treaties. They should work for peace and justice at home first!" The secretary of the section for arts of the Cercle artistique wants to host an exhibit of decorative art. "I think that you would do better not to show *too quickly*, but to wait until you have *many* things, and varied things, to show." One of the Flinke attended the lecture by Vinck and borrowed his copy of Forel. "The Union des femmes gentoises has furthermore just bought the book." Could she send him a lock of her hair "without thinking that I am a child?" On 22 March, George writes that he welcomes Mabel's *santons* because "I am very interested in all kinds of popular art, and also they connect with older traditions." On 23 March, George dreams about their house. He wants all his books in a library, not dispersed throughout the rooms. Two new vagabond cats arrived at his place, and his pet cat Snouke "is jealous beyond words."

From the time of their courtship, the Sartons are drawn to cats, and even when they live on the verge of destitution, cats are part of their *ménage,* their psyche, and their correspondence. On 31 March, George reconsiders Mabel's observation that he unconsciously compares her to his cat. George thinks this is because Mabel is the first woman he really knows, and "it is the manner of little boys who have no sister and never knew their mother." Mabel writes from her hotel near Nice on 14 April: "George, it is horrible that people with a bit of education [un certain développement] dare to speak about love, fraternity, raising the mind and the soul, that these people disparage the ordinary life, intense, beautiful, mysterious, of a beast which finds refuge with them." Mabel loves animals. She witnessed the shooting death of a cat that had adopted their hotel, and it upset her a great deal. One of the other guests advised her not to interefere, quoting Epictetus about staying out of other people's business. Mabel will find a home for one of the remaining cats, and try to heal another, which is sick. On 18 April, George writes that he concurs about animal rights: "It is quite strange that human beings do not feel the suffering of animals at all—human beings who are highly educated, perhaps Theosophists!"

On 24 March, Mabel challenges George's opinions. She writes about the British Isles:

> You are convinced that home rule for Ireland is desirable, necessary? Even if she governs herself *poorly*. The Irish, "even individually," govern very poorly in general, they don't have any self-control. But perhaps that is better than being governed harshly by someone else. Their hopes cannot be refused. It wouldn't thrill me if the peace treaty were signed. (I don't think that home-rule for Ireland would do much for *world* peace.)

Mabel will not contribute works for the exhibit in Ghent. What she really wants to do is to organize an exhibit of men and women from different countries. "I would be happy to show that women, so inferior in what is called high art, are a great force now in decorative art." Mabel has finished Oscar Wilde's *De Profundis,* which included some letters and "The Ballad of Reading Gaol." "It is one of the *very beautiful* books I have ever known."[11]

On 23 March, George's pity for Mme Dangotte's husband has changed to disgust. "What he did vis-à-vis Madame is truly ignominious." On the 24th George reports that he spent all morning at court with Mme Dangotte. "Very *hard* day for her. Finally, the worst part is over! It wore me out." But George is neglectful of details. So on First April Mabel berates George for having only alluded to the Dangottes' marital woes; because of her close relations with the family, she merits knowing precisely what is going on, although not the salacious details. "It is so extraordinary that you falter so easily when dealing with me, so quickly after we came so close to each other. It's a small thing, and I feel that I am quite wrong to allow myself to be upset again in this way." Further, "You are quite unconscious of having hurt me, and it seems quite futile to tell you right now." Mabel thinks that this incident shows that "I am morally—or nervously?—still quite weak." She suffers tachycardial episodes.

Mabel writes on 3 April 1911 that her mother is ill and that she dreams about her. Her mother feels alone, certain that she will die. "I really must see her before that. I really must make up with her." She will leave either for Corsica or for Cavalaire, a small coastal village near St Raphaël, which has "a beach with pretty white sand and large white heaths ending in the sea." On 6 April Mabel writes from a nearby flower festival, where she has traveled with a male guest at her hotel. She is distressed that George remains insensitive to her feelings, scolding her for rising late or for a mistaken apprehension. "I know so well afterwards that I

11. Mabel's binding of *De Profundis* in F. O. Bailey Antiquarians, *A Fine Auction of the Estate of May Sarton, Sunday, May 19, 1996* (Portland, Me., 1996), p. 21. On 28 April 1911 George comments to Mabel that he hardly knows Oscar Wilde.

am wrong to be upset in this way. I really don't have much self-control." She assures George that she scolds herself more harshly than he imagines. "Why all this doesn't get better for me, as with a physical wound? I wonder! Do you know that I never receive a word from you, or from Ray. or Cé. even, without feeling an *absurd* distress. I can't suppress it." The distress is instantaneous. Mabel changes to English: "I feel as if I were some instrument tuned to its highest pitch—the last touch sets wild, throbbing vibrations through it. Sometimes I feel as if some cords would break if a further strain were put on it." On 9 April Mabel writes to confess to George that she "acted badly" during her recent excursion. "It is difficult for me to make a complete confession because I would betray someone else. Now I only feel one thing: I know myself better, and I will no longer let myself be defeated." She lets George perhaps infer that her male traveling companion kissed her. Whatever may be the reality behind Mabel's lines of spring 1911, it seems clear that she carries a heavy burden of guilt into her impending marriage.

On 27 March, George thanks Mabel for her lock of hair. Raymond and Céline brought him the *santons* from Nice, "very curious and original: I like this naive and simple art. These 'santons' have the colors of sugar figures placed on spice bread in Flanders." He still thinks "that England would do much for world peace by giving autonomy to Ireland, just as a man improves life by being just and good in his *internal* life." On 30 March George introduces the brief on adultery filed by Mr Dangotte. "The details of the brief are absolutely scandalous, *repugnant*." Mme Dangotte was called before the judge, along with Mr VdV (whom we infer is her lover) and witnesses; she was confronted by the witnesses.

The professor charged with examining George's dissertation strongly suggests that he change a few passages. George is not sure if he will do so. George's "thesis" on chemistry unfortunately criticizes the instruction in chemistry given to future doctors, offering the impression that he wants to argue with his professors.

> I introduce these criticisms because I am irresistibly led to them by the nature of the things themselves. I am not to blame if all science instruction needs to be completely restructured. In the first part of my thesis, my memoir on "history of science," I regretfully present still other criticisms. I think a great deal of this work. It has been considerably enlarged in my mind since the idea first came to me, and as I understand it, it will take several months to flesh it out. It was only after that that I began my "Newton"! I must do the former work first, because it is in truth *an entire program*. And since it will be impossible for me to complete by myself the historical work that I have imagined, especially given that I am not really an assiduous worker and that I will never be an *érudit* in the proper sense of the word (this is too much opposed to my impulsive and impassioned nature), in developing my ideas on the

> way that history of science should be studied, I will have to hope to be followed by other savants better suited to the task and more patient than I ever will be. Furthermore, something as extensive as this can be achieved only with the collaboration of many men.

Mabel is sympathetic. On First April she understands why George wants to publish his work. "My dear, that you may accomplish most of what you want to do! That I might bring you happiness and peace and help you by creating a sympathetic environment, that one day, when we are old, despite all we have not accomplished, at least looking back at these plans, our ambitions, we may find *a beautiful* effort."

George writes from Brussels on 2 April that he is studying for his doctoral defense and also attending the matinee of *L'Africaine,* Giacomo Meyerbeer's opera about Vasco da Gama. George returns to his sensibility as a scholar. He finds that he is torn between the desire to study certain precise questions carefully and the desire to know everything that is happening, "to enjoy all the intellectual joys of our time." He is happy not to have specialized too early, as the Germans do. He wants to read fine books and see beautiful things *"as recreation,* as reward. Otherwise, it is not a man's life, but a slave's life. It is *with you* above all that I want to experience these joys." On 3 April, he is studying furiously, but he must attend a meeting the next afternoon on pacifism; in response to a speaker on liberal pacifism, George will reply from the point of view of a socialist. He is organizing a meeting with the Liberal Students against the centrist government of Frans Schollaert. "I must march because no one else will do it in the way I want." And on 4 April George writes anxiously to say that he never thought for a moment that Mabel wanted to know the lurid details about the Dangottes, that she had an "unhealthy curiosity," as she writes. Evidently George is reticent to describe the charges in detail, because the affair reflects poorly on Mme Dangotte, yet Mabel needs to know the circumstances because she is scheduled to be a witness in the divorce proceedings. She writes to her mother on 21 April 1911 that the hearing has been deferred for 6 months and so she need not return to Belgium to testify.[12]

On 8 April George suspects that he has never given Mabel either *Une Vie de poète* or *La Chaîne d'or*. He is sending them to her, along with reprints from his callow youth. The novel he had been planning for many years is related to these early publications, but he will only write it when he has the leisure "at the age of 80." On 9 April, George is ready to compromise about their future living conditions. "I think that we should settle for one of those banal, modern houses, you know the kind. We will work to make it as nice and as welcoming as possible. I say modern house,

12. NYPL.

because above all the house must be dry and able to be comfortable: gas, water, bath, kitchen on the main floor." George regrets that he can't do better. He is more upset with the idea of the deprivation of other, more deserving people, than that of their own deprivation. "We are poor among the rich, but we are still more rich among the immense mass of poor." George comments on Mabel's mother. "She belongs to the unfortunate race of those who 'gnaw' on themselves." George knows, because he is like that sometimes, too, when a disagreeable idea comes to dominate his thinking.

On 10 April George offers, in reply to Mabel's letter of the 9th, that he is in disgrace. He won't talk about it, but he emphasizes that he was completely enveloped in love for Mabel. On First May, he will be with the socialist students in Liège, "a trip I am glad to make!" He agrees with Mabel that the most important thing is their intentions, and not their actions, "although we must do what is necessary so that our actions are the best possible." Regarding Mabel's confession of an indiscretion, "you tell me too much or too little, because I can't prevent my mad imagination from seeing 36 things instead of one, which is true." He wants them to speak about it no more. On 13 April he writes: "We are organizing for the end of the month a big anticlerical meeting to defend the neutrality of the schools and the freedom of the child. It is troublesome to me (especially now) but it is necessary." He agrees: "Baudelaire is one of our greatest poets."

George is a dilettante, and he makes unreasonable commitments: beginning married life, completing a doctorate, and playing at socialism. On 17 April 1911 George writes to a friend, Maxime, at Antwerp where he is spending Easter with his family, to outline his socialist point of view:

> I believe that the exploited everywhere will never improve their lot unless they demand it, unless they rebel and drive fear into others. It is not in the nature of humanity for privileged people to abandon their privileges spontaneously. Or if it happens, and we have an example during the French Revolution, it can only be by wisdom and generosity in the last instance. No doubt there are generous people, and you are one of them, but will not these people always be a small minority of eccentrics?[13]

Mabel writes from her vegetarian hotel near Nice on 14 April that she very much wants a nice house with a garden, but she will settle for something less. She agrees with George that they are privileged, but "it weighs on me less than on you. I do not feel that I must renounce what we possess." They had the right to be happy, "*one of the first duties of existing.*"

13. YMS. Possibly Max Horwitz.

Mabel, on 16 April, applauds George's support of the "freedom of the child." She loves him especially for his "social work." She sent her mother a small gift, and that has reconciled them. She wants to return to Ghent before the middle of May, but she "has no home now." She wants to know when they will be married. She thinks about the house at quai des Tillents. Would the people make concessions? Would they put in water on the first floor? "Gas is a lesser question. Lamps properly managed are not bad....I am afraid the kitchen is a great draw-back—small, dark and *down-stairs*," that is, in the basement. "Let us try to find a house at least a bit warm and *open*. The upper part of the house on the qu. des Tillents is *very* open, and the view is splendid." She still suffers pangs of worry: "Sometimes a dreadful anxiety flashes across me like a stroke of lightning, for a second."

On 18 April, George finally decides not to give away his fortune (a choice made by Bertrand Russell at this time) but to use it prudently, for good cause. "You know how I hate personal luxury, because I cannot prevent myself from seeing the blood and the tears that it hides." He thinks that Mabel is less frugal than he: "I sense that you would allow yourself rather easily to spend a great deal of money, if you had the occasion to do so." He wants to live simply with Mabel. "Luxury is never necessary." George consoles himself that he does not have silk wall hangings and Persian carpets with the thought that he does not have paintings by Diego Velázquez or Eugène Carrière. He thinks that Mabel has spent F 100, given to her as a wedding present, foolishly. "I really think that you spend a great deal." For Mabel's stay in the south of France, George gave her F 350 monthly, or F 12 daily, which is more than they calculated necessary; Mme Dangotte gave her money, and he imagines that Céline and Raymond paid for some of her expenses. When George traveled in England, he lived in grand style for F 400–450, including travel. "I would never want you to be deprived of anything *necessary* (food that you need, for ex.), but you should be more economical, *more sparing*. Remember that the accounts of our household will be much more complex than your small accounts while traveling, and you cannot constantly ask me for more money, but rather work with what I give you." George is annoyed if Mabel asks for more money before the end of the month. She must learn not to satisfy all her desires. George, the Fabian socialist, can imagine only a traditional bourgeois household, where the wife depends entirely on a husband's financial calculations.

George writes in the same letter that to love beauty and justice is to hate what is ugly, bad, and unjust. "I hate injustice and perfidy in all their forms. I hate ugly things above all when they are pretentious and rich. I hate evil, but not always enough." George finds himself too self-indulgent, and this leads to moral complicity. "In the sexual domain, for ex.,

I am always ready to understand or excuse too easily, and that proves to me that I have things to wrestle with still." He includes F 350. "I hope very much that you will not need more money. *And above all do not return several days earlier in order not to ask for more*. But arrange things to remain there until 13 May with the money you have." Money is at the root of misfortune in many marriages, but here we sense that other, less tangible matters are also at issue, notably the idea of beauty. George is still smitten with Eugène Carrière, one of the most popular of Belgian artists, who kept a cordial distance from Paul Verlaine and the Symbolists, while Mabel's masters and collaborators are the Belgian Symbolists. George continues to admire Carrière even after having dedicated himself to Leonardo for several years. When in 1919 he comments on Carrière's absence from a book by his friend Emile Masson, Masson exclaims: "Don't speak to me about Carrière, who has always seemed to me—notwithstanding his incredible heroisms—a complete conventionalist [un parfait pompier]."[14]

On 23 April, Mabel is in Cavalaire in Var, and she is depressed. In a second letter around this time, perhaps on 23 April, Mabel reports that she walked in her sleep. She has thought more about the question of expenses. "I think that my *greatest* temptation is to spend in order to *give* and not to spend as a personal luxury, as you seem always to fear." She knows that "giving is not right when one has only a certain sum of money for a special purpose, but giving is *my personal luxury*." She repeats: "I find it bad to give when it is not right, when finally what is given does not belong to you. My mind has told me that for a long time. But I do not know what small devil, subtle in sentimental reasoning, has always suggested to me that giving is *as* necessary as eating and drinking." Mabel then presents a precise accounting of her expenses, along with rebellion against her financial dependence.

> I must confess that I harbored yesterday a very painful feeling of impotence because I cannot pay you back for these expenses, because in the future, despite everything, I depend materially on you. When I "waste," it is not even the money that I earned, as in the past, that if with effort I earned 1000 or 2000, that this is absorbed in the total. I want it that way. But I lack a certain generosity. It really bothers me now to accept that all the money that we spend comes from you. I would have preferred to supply half of it, or even more!

At this time Mabel is designing the cover of Raymond's book of poems, using olive trees as her inspiration. Mabel worries that she will not succeed.

14. HGS, Emile Masson to George Sarton, 28 September 1919. Robert James Bantens, *Eugène Carrière: His Work and His Influence* (Ann Arbor, 1983), pp. 49–51; in 1890, Carrière executed a portrait of Paul Verlaine; the poet then dedicated a poem to the painter.

The worry is groundless. The design of *Le Bois d'oliviers* is strong and clear. Raymond dedicates the book to Mabel and George.

On the 24th Mabel has received George's F 350. Mme Dangotte has sent her no money at all since she arrived in the south of France, and Céline and Raymond have paid for almost none of her expenses. She observes that she lived more cheaply in Zurich than George lived in Brittany. She asks George pointedly about money:

> *Would you tell me please if you consider a luxury any object that is beautiful but expensive*? And how do you decide what is "expensive"? Simply *anything* that costs more than a certain amount? Or anything that is *comparatively* expensive, to what the same object, simpler or less beautiful, would cost? For example, if I can buy a *pretty* hatpin, in silver, Dutch, for 1.50, is this a luxury, since I can find a simpler one with a black head for 20 centimes?

On 26 April Mabel reflects on her future life with George: "I see on the road ahead many difficulties and struggles (I have seen that you will advance more intrepidly than I in things relating to social principles), but I've also seen that there will be many joys, and that at the end we will be very close to each other, and to our ideal." She returns to reflect on the question of money. She is concerned about how the budget shall be established, its limits and its fairness. Mabel believes that even in the best of all possible worlds, there would exist differences among people. This is because some people would need to surround themselves with beautiful things for their own happiness, and some people would intentionally restrain their desires for accumulating things. George says that one can have beautiful things without spending much money. Mabel wonders. She begins a meditation that diverges to the theme of artistic creation. By art, she thinks not of unique works of art that should belong to the nation.

> Artists would live how? From the generous pocket of the state, if they work. Ah! It is still quite complicated. How would one say, "that person is a true artist"?? This one produces in his life only 10 works, and they are nevertheless divine! The *apparent* work involved in the art is so minimal sometimes compared to the true work carried on in the soul. Perhaps it would be the finest proof of the true artist for him to remain poor, for him to work at a trade to earn his living, but for him to work at his art without the hope of material recognition. And then in "good times" would one not want *better* and more, beauty?

Everyone would like to have the time to visit the house of artists, to see works of art in the making. But in them one can find nothing except flowers or beautiful things of inspiration. Mabel wants artists to know that the value of the things that inspire them is insignificant compared to what the artist gives, "a ray of their luminous thought that revealed the beauty hidden in that matter." She repeats,

> All beautiful things in private homes...are for me a national treasure. Each house organized with love of the beautiful becomes a small center, a source of light, different from the larger centers of museums. Each object *lives* if it is in its proper place. Each object becomes more or less beautiful by its surroundings and its *use* and its isolation. Despite *everything*, paintings or works of art in a museum always have a *colder* beauty, and *always* suffer because of their having been "collected."

In homes it is possible to create a center of beauty in harmony with human sensibility. "So I emphasize that *I* am allowed to want to be surrounded by beautiful things." We know that Mabel has seen the Carlyles' house in London, and that both she and George admire how, in their mind, it nourishes the spirit of the historian and his wife. The discussion of household finances calls to mind Jane Welsh Carlyle's eloquent, 3000-word letter to her husband about increasing her allowance.[15] Jane gets more than Mabel does.

Céline's doctors think that Céline's house is too damp and insalubrious for Mabel to live there, George reports on 22 April, but he offers that it will be fine in summer. On 23 April George writes that the anti-clerical demonstration for 21 May in Ghent is taking very large proportions; 10,000 people are expected. George, on 26 April, answers Mabel's discussion of finances. He writes that his consternation at Mabel's spending is reasonable, especially the matter of planning ahead. George does not always have money readily available. To send the most recent amount to Mabel, for example, he had to alert the bank 15 days in advance. George does not in any way want to deprive Mabel of necessities. If she needs more money, he asks only that she warn him in advance. George is tempted to spoil Mabel, but then he thinks about her parents. "You know better than I how your mother was harmed by having been spoiled, and you know how your father was hurt by it." It seems that something concrete came from Mabel's discussion: George has decided not to go to Berlin for a week to study union and socialist organizations because the expense is unjustified. George continues about frugality on 27 April. "You must get used to not spending more than you are permitted to do. That is the first principle to follow in managing a household, and the main point of the question." For their marriage, Mabel must obtain a copy of her birth certificate from the clerk of the municipality where she was born, and to have it notarized by the Belgian consul. Then it has to be translated by the legal translator in Ghent. They can marry 15 days after registering at

15. John Stewart Collis, *The Carlyles: A Biography of Thomas and Jane Carlyle* (New York, 1971), p. 132.

the city hall. George is quite in agreement with Mabel in what she wrote about the artist's house and a warm house. George continues in a second letter of 27 April: *"You do not depend financially on me."* His capital belongs only to Mabel, and in the event of divorce, half is hers.

Mabel writes on 29 April that her relatives want to help pay for her operation, F 640.20. She understood the proposed trip to Berlin as something to "help the general organization of the movement for justice," not as something to help a particular person or group, and so it is "difficult to judge *in advance* if the result of the trip would be worth the F 150." Mabel thanks George for his thoughts on luxury, but she will not raise them again in writing. "I fear again that I have engaged you in 'dialogues.'" Mabel is sad at her temptations to spend money, and she regrets that she spent too much in France, but "I am not angry at myself as you believe I am for the main thing that you reproach me for. I was hurt and ridiculously upset (as I still am sometimes and as I fear will happen again for some time) because you suspect that I hid what I received from Madame and Céline." She adds in a footnote that if they ever divorce, she will not accept half of his income. In that case, "*I know* more than ever that it would be necessary to be independent, to earn my living." Only if she were ill could she accept anything. She is glad on 30 April that George was not upset by her thoughts about homes. She is asking her mother to obtain the birth certificate. On First May, George indicates that he has paid for her operation; he prefers that Mabel's family apply what they would have paid for the operation toward a wedding gift. On 5 May 1911 George writes that his aunt has given them a gift of F 200.

The bride is blushing. Mabel, on 4 May 1911, cannot remember how old she is, whether she was born in 1878 (she was) or 1879, although she knows that she was not born in 1880. If in 1879, she calculates that she is six years his senior (George was born in 1884). She has never realized this. And now she worries that George will have an old wife. Mabel, on 5 May, has reread the *La Chaîne d'or,* which George sent her. On 8 May Mabel grows nervous. She returns to Ghent via Paris about 15 May.

"G. passed his examination this afternoon with high honors," Céline Limbosch writes to Marthe Patyn on 15 May. In the same letter, Mabel elaborates. At his examination, George talked a great deal about celestial mechanics. He dealt with the question of whether the earth circles the sun or vice-versa. "It is a bit delicate because the Catholics do not accept one of the theories, precisely the one George defended. The 'jury' was split, and the *majority* were against him, yet he replied so well that they were *forced* to give him highest honors." The information comes from Victor Willem, who attended the examination along with other

16. NYPL. The *Journal de Gand* announces the defense on 12 May 1911; it did not report the result in its issue of the 16th.

professors.[16] Mabel's remark about heliocentrism may be taken as an assertion of her artistic sensibility, reinforced by Dante Gabriel Rossetti's feigned indifference about the relative motion of sun and earth, but it may also reflect Céline's and Raymond's disaffection with science. Whatever the secular truth behind Mabel's remark in Céline's letter, George's jury lacked the professor best qualified to comment on history of science, Paul Mansion, who had retired the previous year. Mansion requested permission to continue teaching his history of science course, but the authorities appointed a younger man, Arthur Claeys, who sat in judgment of George's dissertation.[17] Emile Masson learns about the successful defense from Mabel, with whom he and his wife Elsie correspond, and congratulates George. "I imagine that, if they talk about you in Belgium with as much humility as they use with your doctoral colleagues in Germany, this must have the effect on you of what happens when a girl becomes a married woman."[18]

George's dissertation, "Les Principes de la mécanique de Newton," is a manuscript of 130 pages. In a vita dated July 1912, he writes that he would like to publish the thesis "after having entirely rewritten it, notably in relation to the principle of relativity."[19] Albert Einstein's relativity came to George shortly after his thesis defense, in a letter from Henri Lauer, a Polish student in Paris preparing for the *licence*. Lauer explains that George's friend Max Horwitz asked him to send George information about the principle of relativity. Lauer advises George to begin with Einstein's articles of 1905 "On the Electrodynamics of Moving Bodies" and "Does the Inertia of a Body Depend upon Its Energy Content?" The former article is, Lauer emphasizes, "remarkable." Then Lauer suggests reading about Hermann Minkowski's reinterpretation of relativity in terms of 4-dimensional space-time. Finally, he suggests Arnold Sommerfeld's and Max Born's elaborations of relativity in 1910 and 1911, along with Max Planck's discussion of inertial and gravitational mass in 1908.[20] Lauer is precisely the right informant. From 1908 to 1910 he studied mathematics at the Zurich Institute of Technology; during the First World War he was assistant to Einstein's mathematician collaborator on general relativity,

17. J. Gillis, *Paul Mansion en George Sarton* (Brussels, 1973) [Koninklijke Vlaamse Academie voor Wetenschappen, Letteren en Schone Kunsten van België, Klasse der Wetenschappen, *Mededelingen, 35,* 2], p. 10; Paul Bockstaele, "Benelux," in *Writing the History of Mathematics: Its Historical Development,* ed. Joseph W. Dauben and Christoph J. Scriba (Basle, 2002), pp. 45–59.

18. HGS, Emile Masson to George Sarton, 23 June [1911].

19. NYPL.

20. HGS, Henri Lauer to George Sarton, 20 June 1911. The connection between Lauer and Horwitz is intriguing, because at about this time Horwitz visits with Joseph Stalin in Vienna (where he has met Russian revolutionaries before). HGS, Max Horwitz to George Sarton, 13 June 1913.

Hermann Grossmann, and in 1919 he defended a doctorate at the Zurich institute under the direction of Einstein's old professor Adolf Hurwitz (Hermann Weyl is co-director).[21] George slowly makes progress with Einstein. On 16 February 1912 he notes in his diary that knowing the science of his own time begins with learning the relativity principle. In one of his first substantial publications in history of science, in 1913, George refers to Einstein's relativity; and Einstein receives more attention from him before the war begins.[22] George's work schedule, outlined in his diary on 16 February 1912, includes Newton and relativity, but George never publishes his dissertation, and although he lectures on Newton in his courses, George publishes very little about him.

George writes to Céline on 29 May 1911 when he learns from Mabel that she will be hired by Mme Dangotte to design embroidery. George disputes the matter. "I am very happy to see Mabel working, but I do not enjoy seeing her 'commercialize' her work." He would much prefer Mabel providing designs without charge to avoid unpleasant discussion about money between employer and worker, which would be "much more delicate" if undertaken between friends. Besides this, George holds all commerce in very low esteem, and he hopes that life will allow him to realize his dream of showing how to improve social relations and "rehabilitate the commercial class." We can be sure about the bristling effect of the letter on Céline, who loves Mabel as a sister and who is committed to her mother's enterprise.

Not only is Mabel's career jeopardized by marriage, but Mabel's very identity is also called into question. Mabel writes from Céline and Raymond's house on 18 June 1911 to Marthe Patyn. She learned "suddenly, without *any* warning, that I was an illegitimate child." She is horribly upset because the circumstances were never revealed by her mother. She and George must hide the truth, "because of Mr Dangotte" who may have decided, notwithstanding his divorce, to support Mabel as an adopted daughter.[23] The engagement announcement in one of the newspapers of "Mabel Cole," giving her mother's maiden name, makes the

21. Henri Lauer, *Sur la réduction des formes positives d'Hermite* (diss., Zurich Institute of Technology, 1919), vita.

22. George Sarton, "Henri Poincaré (1854–1912)," *Ciel et terre, 34* (1913), 1–11, 37–48, on p. 42. The last *Isis* Critical Bibliography for 1914 , no. 5, notes Einstein and Marcel Grossmann's *separatum* of 1913 on general relativity, along with Max Laue's publications on the principle of relativity, and no. 6, prepared in 1914 but not appearing until 1919, extracts from Einstein's general treatment of relativity in *Scientia* in 1914. *Isis, 2* (1914), 287; *2* (1919), 454.

23. Léopoldine Dangotte writes a codicil to her will on First February 1912, to the effect that if Céline inherits F 300,000, she give F 50,000 to Mabel, "whom I consider as a second child." Mabel will also receive a dozen machine embroideries, which she may choose. NYPL. Here it is possible to imagine intended compensation for the change of heart of her ex-husband.

truth evident. But there is good news. Mabel and George have bought a house in Wondelgem. "It must be transformed a bit, the façade is *very* ugly now."[24]

How the Sartons bought the Wondelgem house has been related by May Sarton and Roger Calcoen. George's father left a well-provided wine cellar. George, who had flirted with teetotaling, would not entertain on the scale of his father. He put the wine up for sale with the house. His father's friends, knowing the quality of the wine and happy to come away with good memories of Alfred Sarton, were generous with their purse, and the money was sufficient for George and Mabel's new home. George emphasizes in his unpublished biography of Mabel: "The house of Wondelgem was purchased with the money obtained from it [selling the wine cellar]."[25] Even if the cellar went for F 25,000, the cost of the unimproved house, chronology suggests that the story is an exaggeration. First, George enjoyed a substantial inheritance, F 50,000, from his godfather Hippolyte van Sieleghem, who had raised his mother when she was orphaned and who died in 1901.[26] Second, Alfred Sarton died a year and a half before George and Mabel married, and George liquidated his father's affairs early in 1910, more than a year before the purchase of the house. Five years afterwards, on the prairie of Illinois, George makes light of the matter: "I will have a great time, when I go back to Belgium, and shall explain to my old uncles why I sold my father's cellar. I did not want my wine to be drunk by the Huns."[27] Until the Wondelgem house can be renovated, the newlyweds will reside at 4, Rempart de la Biloque in Ghent, near the canal known as the Coupure, in George's biography of Mabel the "queer dilapidated but charming house" of Céline and Raymond Limbosch.

Two commentators have contended that George Sarton recognized "his own historical naïveté and lack of training" only in early middle age.[28] In notes for an autobiography written at the end of his life, George does emphasize: "It took me an awful long time to learn the

24. NYPL.

25. *Phoenix*, pp. 65–6; Roger Calcoen, "George Sarton," *Handelingen der Maatschappij voor Geschiedenis en Oudheidkunde te Gent, 34* (Ghent, 1980), 181–91; M. J-F. Uyttendaele, George's landlord in 1910–11, recalls that the wine cellar was "the best investment" of George's father. Geve, "Un savant gantois aux Etat-Unis," *La Flandre libérale*, 15 December 1954. George sold the contents, along with the family jewels and furniture. The recollections are unclear if the proceeds of the sale went directly for the house at Wondelgem.

26. YMS. Berthe [probably Van Halmé, George's first cousin] to May Sarton, 18 July 1957, with a copy of the will.

27. NYPL, George Sarton to Mabel Sarton, First July 1915.

28. Arnold Thackray and Robert K. Merton, "On Discipline Building: The Paradoxes of George Sarton," *Isis, 63* (1972), 473–95, on p. 488.

fundamentals of historical method."[29] History, however, is game for mature minds. Works of historical genius seldom appear in the springtime of life. A balanced assessment would conclude that, in his youth, George armed himself remarkably well for his career. He attended university courses for 8 years, and these included the full spectrum of the sciences, in addition to philosophy and classics; he concluded his studies with distinction. His professors were, in sum, unusually accomplished. But a synthetic urge dominated his academic trajectory, as he recalls in 1930, instead of an increasingly narrow focus on one specialty. "How could one explain otherwise that as soon as he had understood a new theory or mastered a new technique, instead of applying it, he proceeded to study something else? That strange conduct puzzled his teachers very much; they thought him shiftless and capricious because they were not able to see what he was driving at."[30] One of his professors is more generous in retrospect: "We didn't want to complain about you because you had *le feu sacré*."[31]

By the time of his marriage, George was completely conversant in politics and art history, as well as the letters of his own day. He knew the sciences, from natural history to mathematics. He had followed systematic lectures in his chosen field by a competent, if unexciting, professor. He had traveled to major museums and libraries in Western Europe. His bride was equally well prepared for her career. She had studied for years with excellent Belgian, English, and Swiss artists and designers. She had learned a vast range of techniques in painting and crafts, and she also had traveled widely. Her work had appeared in print and had been sold commercially. It had been the subject of eloquent praise by avant-garde artists. Both George and Mabel Sarton were well-connected in the intellectual world of early 20th-century Belgium. With a sense of optimism, on 22 June 1911 the innocents—one *docteur en sciences* and the other *artiste peintre*—join hands in the Ghent Town Hall to confront the world together.[32]

29. YMS. George Sarton, "My Gratitude."

30. NYPL, "Notes written in answer to Dr Kidder's request of Sept. 4th, 1930," 10 September 1930.

31. HGS, Paul Thomas to George Sarton, 11 January 1924.

32. *Journal de Gand*, 23 June 1911, for the occupations.

Fig. 1. Mabel Elwes, by Francis R. Elwell.

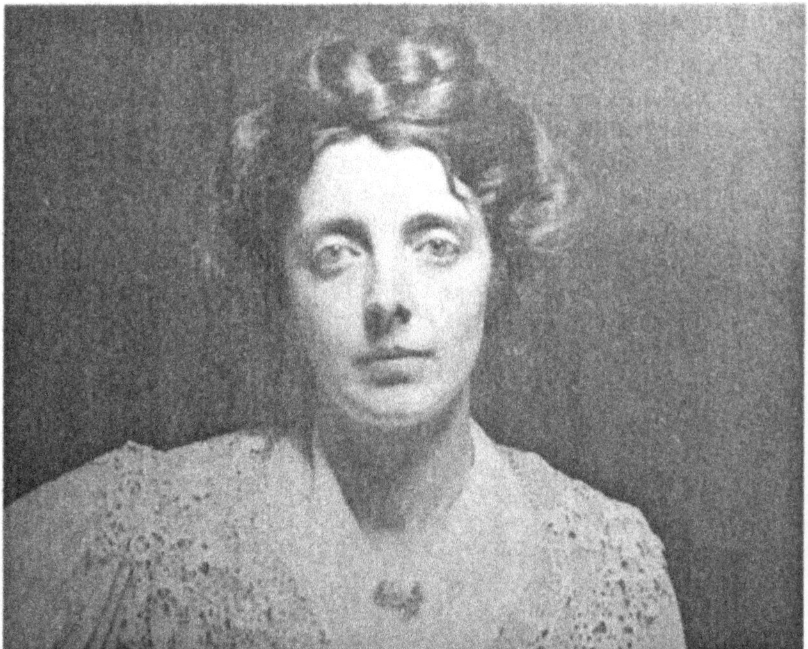

Fig. 2. Mabel Elwes, Antwerp, October 1907

Fig. 3. Mabel Elwes, circa 1914.

Fig. 4. Mabel Elwes, circa 1920, in a dress of her own design.

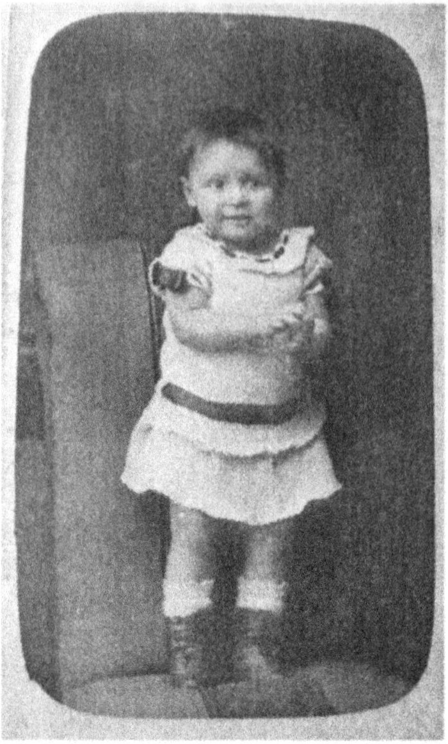

Fig. 5. George Sarton as a toddler.

Fig. 6. Young George Sarton, titled "Les deux amis" and presented by George to Mabel Elwes on 14 January 1909.

Fig. 7. George Sarton, circa 1900.

Fig. 8. George Sarton, circa 1911.

Fig. 9. George Sarton at work, Widener Library.

Fig. 10. George Sarton, circa 1918.

Fig. 11. George Sarton at home in Cambridge, Massachusetts, presumably 24 Agassiz Street. The oak sideboard and the portrait, of the *seigneur* Joseph Duvet de la Tour (1708–1803) are likely items given to the Sartons by George's Aunt Hélène, as recorded in George's diary on 15 December 1919.

Fig. 12. George Sarton working at home in Cambridge, Massachusetts. Cats are the Sartons' constant companions.

Fig. 13. George Sarton, sketched by Lucy Stanton. Possibly a study for the pastel of George in ALS. George mentions the sketch in his agenda on 1 September 1922.

Fig. 14. Dining room in Wondelgem, circa 1914, May Sarton squirming in the high-chair.

Fig. 15. May Sarton sketched by Mabel Sarton, 1916.

Fig. 16. May Sarton in a dress designed and executed by Mabel Sarton. May Sarton presented the photograph and the dress to Susan Sherman on 29 October 1993.

Fig. 17. Céline Dangotte, 19 January 1910, sketched in pencil by Mabel Elwes.

Fig. 18. The Flinke *circa* 1908. Left, Marie Mees, Alice van Damme, Marie Praet, Esther Delahaye, Marthe Patyn; right standing, Victoire Ledewyn, Lucie Boulanger; right seated, Mélanie Lorein, Augusta de Taeye. Their 19th-century attire may be contrasted to Mabel Sarton's bold *couture*.

Part III

Alliance of Singularities, 1911-1915

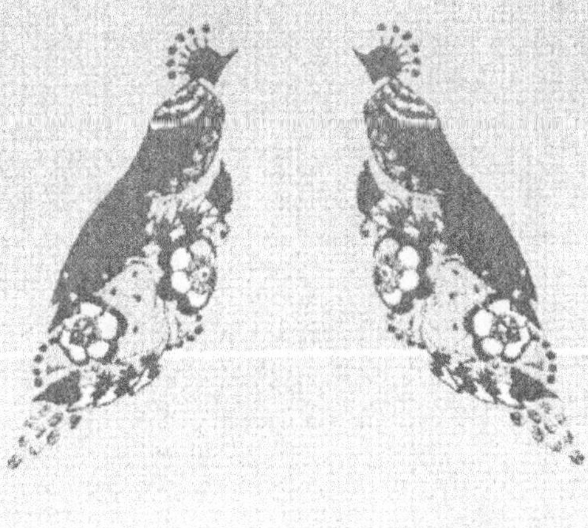

CHAPTER 7

Arcadia, Ego, and Isis

Modern unions of love often entail common striving as well as compromise. Family and career, work and expectations, necessarily depart from tradition and respond to changing circumstances. Passion of many kinds is disciplined. The early years of the marriage of George and Mabel Sarton, a long time in coming, became a stage-set for the assertion of the independence of husband and wife. George launched his periodical, *Isis*, around which eventually coalesced a scholarly discipline; notwithstanding bouts with depression, Mabel expanded her career as a designer.

Their honeymoon in the Netherlands set the initial tone. Years later Mabel recalls to George

> that view of wide grey & brown water & big sky that we saw from the windows of our first resting place (was it not Dordrecht?) I remember wondering if our lives would be something like that. I felt that it suited *you*. It had dignity & reserve & a sort of big-ness, but I felt a strange little fear & awe for myself as if I did not "match" you & it![1]

They take in art museums in The Hague, Haarlem, and Amsterdam. At Haarlem, they see an exhibit of batik. Generally, they "spent part of the day in different museums, then returned together to the same ones in order to compare our impression." Reaching conclusions independently, George recalls later, is "a rule of our lives: We must have each his own independence and solitude."[2]

The village of Wondelgem is a 45-minute walk from the center of Ghent. On the eve of the First World War it featured villas constructed by owners of the textile factories in the northern suburbs and owners of the industries lining the Terneuzen Canal. Camping near the rich and powerful appeals to university people—consider the professors who lived in Charlottenburg just outside Berlin or on Hampstead Heath in London—but it was an audacious choice for a private scholar on a budget.

The Sartons' house is "situated at the end of a magnificent garden, with such beautiful trees, so well spaced, that it looked like a park and was so well designed that it seemed much larger than it actually was."[3] The

1. Mabel Sarton to George Sarton, 22 June 1922.
2. "Scholar's Wife."
3. "Scholar's Wife."

garden at Wondelgem is often recalled by Mabel and George, and it is in the center of May Sarton's autobiographical writings.[4] Perhaps inspired by his gardener friend Raymond Limbosch, George also tries his hand at cultivating the natural world. He plants a herb garden at Wondelgem in 1912.[5] May Sarton writes in retrospect that it completely failed, although George just says, in his biography of Mabel, that he lost interest in it and was inept at gardening. Several years later, writing to Mabel from America, George hopes to give Mabel a garden and teach May gardening. "I have great hope that she will soon be, if a small, at least a good gardener."[6] Mabel loves to watch things grow. She writes to George on 6 February 1925: "A seed catalogue is still a magic source of comfort & solace to me!!"

At Wondelgem, the Sartons are indeed "poor among the rich," as George writes to Mabel on 9 April 1911. The approximate cost of renovating the house, indicated by George in a letter to Mabel of 4 July 1912, is nearly F 35,000. Mason, gardener, plumber, electrician, painter, and motor and water pump come to F 9070; their house itself, unfurnished, costs F 10,000; the land (presumably with the unimproved house) costs F 15,000. The money comes from George's inheritance. George describes how he hopes the house will appear when renovated:

> Very simple, very humble. The passer-by should notice nothing out of place, and yet should be astonished. Its height should follow the proportion of its lines.
> Made of local material: bricks whitened with chalk.
> Shelter for two bicycles near the entrance door.
> Windows much closer to the surface of the façade than at our place, with iron and wooden braces for placing flower holders.

The lower windows would be guillotine style, the upper ones would open toward the exterior, as in a *béguinage* he saw. He wants an English-style entrance door, covered in copper, and a letter-box opening inside the house. The house would be surrounded by gardens. On the interior, there would be a dumb waiter and a slide, along with a large washroom; in his study, high windows over book-cases. There would be flatware and linen as in Céline's house. Mabel adds her own comments. She wants a red tile roof and hardwood trim. She proposes dormer windows opening to the interior. (A photograph of the house shows that she prevails.)[7] Electrification

4. *Phoenix*, p. 66.

5. NYPL, design for herb garden.

6. George Sarton to Mabel Sarton, 26 April 1915.

7. Roger Calcoen, "George Sarton," *Handelingen der Maatschappij voor Geschiedenis en Oudheidkunde te Gent, 34* (Ghent, 1980), 181–91, on p. 185.

and plumbing follow the advice of their industrialist friend Peter Enke; fireplaces are imported from England.[8] Mabel designs furniture and introduces themes from nature—flowers and birds. Until a bed can be delivered, the Sartons sleep on the floor. When the house is furnished, George and Mabel have separate beds, and Mabel has a workroom of her own.

Two weeks after the marriage, George writes to Mabel's good friend Madeleine van Thorenburg, asking about Mabel. Van Thorenburg replies that in her opinion George can make Mabel happy, but that he has an unrealistic appreciation of her: "Oh, George, I believe that you are still a big child who feels from time to time that he must hold others very high, very high, very high above him."[9] George is not alone in coming to terms with new realities. Mabel writes to Méta Budry, in English, that she "felt deadly tired & weak all the first fortnight or so after I was married."[10]

George writes in his diary on 1 September 1911 that he is much concerned with construction at Wondelgem. He thinks that Mabel will find rest there, and that he will be able to work. He worries about Mabel's health and his "*material* future." He has decided "to devote my life to the history of science." He believes that he must learn about the art of Asia. His light comes from the east: "The Orient—all intellectual manifestations of the Orient, art and science—attract me. I feel that I am a bit Oriental. Perhaps I will study later the science of Egypt and Chaldea more closely." He dreams about a museum for history of science.

With his marriage, George opens outward in the manner of his projected Wondelgem windows, toward his calling. Mabel turns inward. She writes to George on 9 September 1911, from the home of a friend where she is pregnant and resting, about George's "cry of revolt" that he must have a servant in the house so that he can carry out his work. Has she not managed to arrange things for him? If she had not been pregnant, she would have worked harder at the task. The pregnancy had been discouraged by their physician, Adolphe Miele, a friend George's and collaborator of Victor Willem's on nutrition and infant development who had lectured to Reiner Leven in 1906.[11] George recalls in his biography

8. NYPL, Mabel Sarton to Eleanor Sarton, 23 October 1911: "All sorts of troubles with our English fireplaces chiefly—pieces have been broken on the way."

9. HGS, Madeleine van Thorenburg to George Sarton, 6 July 1911.

10. NYPL, Mabel Sarton to Méta Budry, 19 July 1911.

11. Adolphe Miele and Victor Willem, "Contribution à l'étude des causes et du traitement de l'atrophie infantile," *Revue d'hygiène et de médecine infantiles*, 3 (1904), 3–33. At the time Miele was an assistant in human anatomy and Willem was chef des travaux pratiques in zoology at the University of Ghent. George notes Miele in his agenda on 22 October 1905, around the time George mentions Willem (3 November and 30 November 1905); he socializes with Willem and Miele on 22 March 1906 and 4 April 1906. Miele's talk: "La Puériculture scientifique."

of Mabel: "Soon after our marriage our Ghent doctor, a very kind man, Dr Miele, had advised us that she might not be able to bear the pains of pregnancy and labor and that we should avoid impregnation. We tried to avoid it but were too impetuous and maladroit and the child grew up in the womb." Mabel writes to George on 4 November 1911, when the couple move into the Wondelgem house: "We are starting on a new bit of life in our own home—the last bit of the material part of our independent lives will disappear." The theme of bringing children into the world appears frequently in the first 20 years of the Sartons' relationship—Mabel conceives a number of times—and it is never without controversy.

So, too, the matter of servants, whom Mabel often employs and seldom enjoys. Mabel writes to Marthe Patyn that the Limbosches, Mme Dangotte, George, and she discussed the question of the happiness of servants over dinner. "George spoke about the 'happiness' of passive women, and I cried out in revolt, that's not at all reasonable, such happiness is false." George defined happiness as the balance between a person's needs and his means of satisfying them. "There is a truth in that and in his definition but it is a very small truth for me, so remote from life, so unilluminated." Raymond took her to task for criticizing George without proposing an alternative, which she could not do. "I felt so horribly far from everyone, as if there were a wall of incomprehension between us."[12]

In an undated letter before May's birth, Mabel expresses the concern that George is oppressed by the responsibilities of wife and family. In English: "Do you know that last night the thought of a child being born to us, was a dread to me?" She worried that he thinks that "through your marriage you have taken a burden of forced work on your shoulders" which will deprive him of "undisturbed work that you had dreamed of (that I feel you were destined to do). Can't you understand that *that* is what hurts the most, far, far the most?" She continues, "Can you really believe that I am trying to lead you away from that 'vie d'étude' that you have marked out for yourself, so as to have you for *my*self?" Mabel thinks that George has made large concessions "since the time that you told me that your wife must be sacrificed to your work, and must find her happiness in the honour of being sacrificed." But Mabel accepts George's focus. "As a whole, I have absolutely accepted the fact that your work must come first of all in our lives." Mabel reminds George that "I learn lessons *slowly*, painfully, haltingly."

About this time George joins an Italian and a German society for the history of science. He acquires a second-hand set of the first nine volumes of Karl Sudhoff's *Mitteilungen zur Geschichte der Medizin und der Naturwissenschaften*, and he takes systematic notes on them.[13] In his

12. NYPL, Mabel Sarton to Marthe Patyn, 2 November 1912.

13. "Scholar's Wife."

diary on 7 January 1912, George writes: "I have now a much clearer view of the history of science: my dream is much more precise. I can see now distinctly what my task will be. There is an enormous amount of work to accomplish. I will do as much of it as possible." He continues, "My life will be truly one of slavery, but a happy slavery." He proceeds by annotating secondary works, notably Moritz Cantor's 4-volume history of mathematics, entering the corrections suggested by Gustaf Eneström and his collaborators in *Bibliotheca mathematica*.[14] In time he will study science as it is done, and for that he will start with Henri Poincaré's book about cosmogenical theories.

George Sarton is known today as the founder of *Isis*, the leading journal in the history of science. The journal takes shape during the first months of his marriage, and George edits it through Mabel's death in 1950. One senses that, among all George's schemes and dreams, this project is distinct. Early in 1912 George contacts French mathematician Henri Poincaré and German chemist Wilhelm Ostwald—distinguished scientists with wide interests in philosophy and history—about plans for a periodical.[15] "Above all, I must not neglect any personal effort, nor any initiative, to create the 'revue générale de l'histoire des sciences pures et appliquées, de *l'histoire de la science*,'" he confides to his diary on 2 March 1912. He will name it *Isis*, the shortest name that he knew for an Egyptian deity, according to Roger Calcoen.[16] George finds support among scholars who will later come to his aid, notably David Eugene Smith, professor of mathematics at Teachers' College of Columbia University. Smith is glad to figure on George's committee of patronage, provided that *Isis* is published according to high standards and provided that George has funding sufficient to see the journal through 2 or 3 years.[17] Smith doubts that George can hope for financial support from American universities; he imagines that contributors to early volumes will not require payment. Smith offers to provide an article on Japanese mathematics of the Seki school or on Hindu

---

14. George Sarton, "Les Tendances actuelles de l'histoire des mathématiques," *Isis, 1* (1913), 577–89, p. 586 for the importance of Eneström's corrections to Cantor's history; also Wilhelm Lorey, "Gustav Eneström: In memoriam," *Isis*, *8* (1926), 313–20, on p. 318.

15. Diary, 12 January 1912.

16. Calcoen, "Sarton." The enthusiastic support of Poincaré and Ostwald in Sarton, "Chronique et correspondance," *Isis, 1* (1913), 96. George writes in a plea for funding, appearing in *Isis*, that he chose the name "because it was short, as I might have chosen Minerva, Athene, Hermes or Clio if those names had not been preëmpted." George Sarton, "The New Humanism," *Isis*, *6* (1924), 9–34, following which, as Appendix I, "The Publication of *Isis* (An Urgent Appeal to American Scholars)," pp. 35–39.

17. HGS, David Eugene Smith to George Sarton, 12 April 1912.

geometry.[18] Smith, who becomes one of George's firmest supporters, then volunteers to send out sample copies of *Isis* at his own expense, along with a covering letter that he writes.[19]

In 1912 George Sarton is a private scholar with wide interests and contacts but lacking in accomplishments. Can he obtain support for his project? Some academics accept when asked by George to appear on the cover of the early numbers of his journal, to review for it, or at least to commend his project. Others, like Pierre Duhem, the polymath French outsider widely recognized as an authority in the history of medieval science, decline the invitation. "The orientation of your review is not perhaps in the same direction as my own interests," Duhem writes to George, "and my name on the cover could perhaps, one day, embarrass both you and me."[20] George's old professor of history of science, Paul Mansion, writes: "You are quite right to think that I do not want to support *Isis*, for your point of view is absolutely opposed to mine."[21] Philosopher Abel Rey supports George's journal while emphasizing his own project for a series of "primary works leading to a synthetic history of science."[22] Henri Bergson and Karl Pearson decline politely.[23]

To a 21st-century reader, for whom George's project carries an air of fantasy, it may be recalled that independent scholars enjoyed an honored status early in the 20th century. Writers whom George admired like Maurice Maeterlinck and Romain Rolland were in this category, and the pattern continued for decades with seminal figures like Abby Warburg and Walter Benjamin. At the time of the Sartons' marriage, one private scholar in particular would have been familiar to intellectuals in Ghent. He is the distinguished classicist Franz Cumont. In 1910 Cumont, *professeur ordinaire*, became the unanimous choice of the university faculty

18. HGS, David Eugene Smith to George Sarton, 25 October 1912. Probably Smith means Seki Takakazu, a key figure in the formulation of the Japanese style of mathematics known as wasan. Seki and wasan are treated by Smith in *A History of Japanese Mathematics* (Chicago, 1914). Smith's close colleague was Mikami Yoshio, also later a correspondent of George's. Mikami, *The Development of Mathematics in China and Japan* (Leipzig, 1912).

19. HGS, David Eugene Smith to George Sarton, 14 November 1912.

20. HGS, Pierre Duhem to George Sarton, 11 April 1912, printed in George Sarton, "Acta atque agenda," *Archives internationales d'histoire des sciences*, *4* (1951), 323–56, p. 354.

21. HGS, Paul Mansion to George Sarton, 12 February 1912. George reflects a lifetime later that Mansion, a devout Catholic, may have associated the word Isis with Freemasonry—Mansion would have known about George's Freemason father and also about George's socialist activity. George Sarton, "Why *Isis*?" *Isis*, *44* (1953), 232–42.

22. HGS, Abel Rey to George Sarton, 14 September 1912.

23. HGS, Henri Bergson to George Sarton, 12 April 1912; Karl Pearson to George Sarton, 4 April 1912.

for a chair in Roman history. On political grounds, however, the Catholic minister in charge of education refused to support the faculty, whereupon Cumont resigned from the university. It took the minister more than a year to accept the resignation. Soon afterwards Cumont, in his early forties, left Belgium for Paris and Rome, where he lived on his inheritance.[24] Franz Cumont was one of the earliest supporters of *Isis,* appearing as the only Belgian member of its first *comité de patronage,* although he advised George that he would remain in the background.[25]

Notwithstanding the attraction of an independent life, George is conscientious about seeking a teaching post, which he needs to supplement his income. Late in the winter of 1912, he writes to ask for a position from the head of educational affairs in Brussels. He reminds the head that they met briefly when George defended his chemistry thesis in 1908.

> My principal aim is to devote my life to the study of the *history of science.* It is in large part a new discipline, not yet constituted or presented in a systematic way, but for this reason more interesting. Is there nothing more captivating, more exciting, than to work in an intellectual domain that is still virgin, where almost everything is to be accomplished or to be reworked, where the methodology is still to be perfected? The history of science is the history of human intelligence in the highest sense.

To undertake this kind of work, one must have a "sense of history," as well as an understanding of psychology and sociology, in addition to a wide knowledge of science. "One must be at once an historian, a sociologist, and a scientist." If he were rich, he would devote himself to his studies completely, but he is forced to earn his living. He has a wife and may become a father. "I must think of providing my wife and my children with a decent life, free from worries." He understands that the professor of astronomy, Clément Dusausoy, has retired. Dusausoy's successor has possibly not chosen his pedagogical assistant, or *répétiteur*. George would like to be considered for the post. When he has accomplished part of his life's work, he would like to imagine applying for a chair in history of science at the university, if one were to be created. George asks the director how to proceed, emphasizing that he is writing at the suggestion of Ernest Discailles, whom he met recently at the Royal Library in Brussels. The request, supported by his friend Discailles, brings no result.[26]

---

24. *LM I* and *LM II, s.v.* Cumont. HGS, Franz Cumont to George Sarton, 27 January 1913, where Cumont writes in French: "Following the death of my father, I have decided to leave Brussels."

25. HGS, Franz Cumont to George Sarton, 19 May 1912.

26. NYPL, George Sarton to Head of Secondary Education, 14 March 1912; HGS, Ernest Discailles to George Sarton, 16 March 1912.

Modern science and scholarship are founded on the union of teaching and research promoted at 19th-century German universities, but George knows from his correspondent Emile Masson how routine teaching saps creative work. Masson wrote to George in 1906: "Ach Gott! Time is short....The damned classes where I labor for nothing, scarcely stirring the long ears of these mules, eat up so much time and energy! You know, I don't even read the newspapers and periodicals I subscribe to!"[27] George accordingly solicits advice about a teaching career from his friend, biologist Paul Pelseneer. Pelseneer recalls George's desire to study what interests him rather than to teach, but Pelseneer thinks that teaching in the normal school is a prudent choice, for it does produce income for a commitment of 8-9 hours weekly and it really does not impair research. Pelseneer urges George to see mathematics professor Modeste Stuyvaert, who taught in various secondary schools for many years, about the course left by Dusausoy.[28] In any case, George must apply for a position in the schools through the director of secondary education in Brussels, who "unfortunately is clerical and hardly welcoming." The director, however, is also a Flemish nationalist, who would be sympathetic to George's offer to teach in Flemish. The director of university affairs is, however, a rare, liberal civil servant who would treat George fairly. The civil engineer Jan-Florimond Vanderlinden controls the posts of *répétiteur* in the Faculty of Sciences.[29] George asks Stuyvaert about the normal-school position. Stuyvaert indicates that he has withdrawn his own application for it, and he urges George to apply; the course pays F 600 for 3 hours per week of teaching. But there is another candidate, a teacher at the *athénée*.[30] George writes again on 17 July. Again, no result.[31]

In his biography of Mabel, George recalls walking with their maid, Valentine van Walle, to a socialist country retreat on the eve of May's birth, in his memory a shameful and thoughtless excursion. May, or as a birth document reads "Eleonore Marie," is born on 3 May 1912, after a painful, unassisted labor. Mabel writes to Méta Budry on 10 May: Her daughter, sleeping in the garden, "has lots of very long, very fine dark hair—dark blue eyes & the sweetest, healthiest pink skin I ever saw on any baby!" Again to Budry on 15 May: "The May-Baby & I are both deliciously well and happy." Two weeks after May's birth, Mabel reports to Marthe Pateyn that she has a great deal of milk, and that she hopes to

27. HGS, Emile Masson to George Sarton, 7 May 1906.

28. *LM I* and *LM II*, *s.v.* Stuyvaert.

29. HGS, Paul Pelseneer to George Sarton, 13 June 1912 and 10 July 1912. LM I, *s.v.* Vanderlinden.

30. HGS, Modeste Stuyvaert to George Sarton, 18 June 1912.

31. Van Oye, pp. 20–1.

breast-feed May through the summer.[32] On 18 June Mabel informs Méta that May suffered from "a sort of poisoning through the milk" after she was weaned. In English:

> May-Baby begins to *look*—not only to see, but to *look*! And her eyes are two soft, bright miracles, & so clear, so clear, nothing in the world it seems to me is so clear. When you look in them, it is as if they were two bleu-grey flowers with black centres, down under deep clear water. May-Baby is growing fast, fast bodily and spiritually. All sorts of little doors in her brain that were fast locked and sealed when she was born, are already *ajar*...some are open![33]

On 22 June 1912 Mabel writes to George from Knokke, on the North Sea, where she is resting with newborn May, the maid Marthe, and Céline and Raymond. She wants to rejoin George, who will not spare a weekend to visit them. She asks him to take a vacation when she returns, for him to work only in the morning. She needs him during the day, especially after they have made love the night before. Sometimes she wonders if George is right,

> that the end & aim of human beings is purely intellectual, that all the rest is simply "marginal." And by "intellectual" I mean the acquirement of knowledge, tending to exact truth. In which case the things that simply "ornament" life are very secondary, so secondary that personally I should choose to "stop" my present way of living. I should either try to help, in however small a detail, towards that "acquisition of knowledge" or else stop altogether and die voluntarily.
>
> But perhaps the child might help, might do a small share of the essential work? If so, by bringing her up, indirectly I help too. And as for the measure of doubt contained in that, all intellectual workers undergo the same conditions. I suppose none is *sure* his work will contain a new grain of knowledge.
>
> Don't say to me that Love and Beauty are in themselves "knowledge."' Life understood as work towards exact knowledge, would mean that the only important thing would be to know exactly what constitutes Love and Beauty—probably both would be pronounced nonexistent—vague words covering an "ensemble" of facts not in the least resembling the current meaning given to those words.

On 24 June 1912 Mabel writes from Knokke that she is sad that George, then in Paris, does not celebrate their anniversary. They shall be away from each other for several more weeks.

32. NYPL.
33. NYPL.

George left his wife and new-born daughter for a celebration of the summer solstice sponsored by the French Astronomical Society. He writes to "chère petite maman" from Paris on 23 June 1912. He asks Mabel if she remembers their honeymoon trip in the Netherlands. When he first visited the Netherlands with his father 10 years previously, he did not know about Jan van Goyen and Johannes Vermeer. The artists made a deep impression on him then. George finds more affinity with the Netherlands than with France. The French shock him because they talk too much and because they are too undisciplined. French democracy is ill. Only the Swiss in Europe are really inspired by democratic principles; after the Swiss come the English. At his banquet the preceding night (on the first floor of the Eiffel Tower) there were 400 or 500 people. George sat in a brilliant constellation at the head table with astronomer Camille Flammarion, engineer Gustave Alexandre Eiffel, and physicists Pierre Puiseux and Charles Edouard Guillaume. George sat between Georges Claude, the industrialist who manufactured liquid air and neon lights, and Mlle Philibert, sister of an admiral. Next to her was the speaker Guillaume and a Danish singer. Guillaume is, like Paul Tannery, interesting for his character as well as his work. On the morrow George will visit the publishers Alcan and Hermann; that afternoon he tours the Conservatoire des arts et métiers.

George has Mabel's letter on 24 June 1912. He feels rebuked. He promises to manage his time better. But he must continue to work hard.

> The career that I have chosen (I was not entirely free to choose another one, for one must follow one's aptitudes) is perhaps the one that requires the most intense work for the meanest remuneration. If I had been a merchant or an industrialist, I would have been able to give you more time, and also more comfort. But the Gods gave me at birth such a thirst for knowledge, for justice, and for the ideal, that I am committed to the work of a slave.

He reminds Mabel that he warned her, before their engagement, that he could never belong to her completely, for he must follow the lines of a mysterious force. But now that he knows love, he is doubly a slave—to his mission and to her. He writes: "You are my best friend, my confidante, my councilor, my wife, my lover, and the mother of little May."

George liked to think that 2 children were born to him in 1912, May and *Isis*. He honors Mabel as their mother. "In our earnest innocence," he writes in his biography of Mabel, "we had conceived something immortal,... which deserved to bear the name of an eternal goddess."[34] The first number of *Isis* is dated March 1913, but George observes that he circulated a prospectus of 50 pages somewhat earlier, and his programmatic announcement of the journal appeared in November 1912. "To call May and *Isis* 'twins' is thus inaccurate though they were separated by an interval (six months) too short for ordinary sisters." George recalls that he and Mabel sent out thousands of copies of the specimen number to scholars identified from the pages of the annual academic directory, *Minerva: Jahrbuch der gelehrten Welt*. "I wrote the addresses, Mabel did the wrapping and added the stamps."[35] Years later, Mabel designs a bookplate for George, and elements from it are incorporated by George's successors on the masthead of *Isis* and on the letterhead of the History of Science Society. Her art is carried by the discipline to the end of the 20th century.

George's more serious purpose in traveling to Paris in summer 1912 is to persuade publishers to take on *Isis* at their own expense. On 26 June he writes to Mabel, "gros chat," that his best efforts with publishers in Paris have failed. George presents his notion too quickly, in 5 minutes. Then he explains all the problems associated with it before he comes to the advantages, but by then it is too late. When he returns to Ghent, he wants to make a few more inquiries on behalf of *Isis*, notably with the French patron of science Prince Roland Bonaparte and another person near Ernest Solvay. "If these initiatives bear no result, I will give up the project." The direction and publication of *Isis* "would take enormous time, &—I have no illusion of [illegible]—would bring me nothing." But he reports that he learned a lot in Paris. He visited the Greek collection of the Louvre, and also Arts & Métiers, which he finds superb. Ernest Renan is right to speak about the Greek miracle. George truly regrets that Mabel cannot be with him to drink in the museum culture. George really likes Bronzino's "Portrait of a Sculptor" (also known as "Portrait of a Young Man"). It is not as good as Titian's portraits, but still it is lovely, and he refers to it in his *Chaine d'or*. George emphasizes that he pursues universal interests, to his disadvantage: "I don't run a large risk of becoming a specialist. I suffer sometimes too

34. In his memoir of Mabel, George writes that the name for his journal was chosen by him and Mabel together. In helping to choose a name, Mabel may have thought back to a similar-sounding journal for which she was artistic consultant in 1908: *Iris: Vrij, algemeen nederlandsch maandschrif*; prospectus in NYPL. George elaborates various precedents in "Why *Isis*?" *Isis*, *44* (1953), 232–42. The most persuasive is a book by James Teackle Dennis, *The Burden of Isis, Being the Laments of Isis and Nepohythys* (London, 1910), which George bought in Ipswich during a happy visit in 1910 with Mabel. Diary, 25 February 1953, for Dennis.

35. "Scholar's Wife."

much from feeling so terribly complex, so curious about everything; from the scientific point of view, it is a great weakness." He has taken a resolution to devote only half his time to things he truly enjoys; the other half will be for earning money. The resolution, taken many times early in his career, is never kept.

With the family's installation at Wondelgem, George avoids the Limbosches, while Mabel gravitates to them. Céline writes to Mabel in English before the birth of May, that there is an estrangement between Raymond and George. The Limbosches, she offers, "have been considered more or less as 'undesirable'" in Wondelgem. But they will come when the baby arrives, if they are needed.[36] The estrangement between George and Raymond grows. At base it relates to George's success in science and to Raymond's rejection of it. George believes that science is a light for the world, while Raymond finds illumination primarily in the truths of mysticism.

Mabel, at Knokke with the Limbosches, writes on 25 June that George's dedication to his work stimulates her to imagine similar dedication, but as a mother and housewife she is engaged in "'nameless' work, which can't be discussed or measured, which begins again each day, which requires only an inessential part of one's talents." If she designed, she would feel fulfilled. She is not yet organized well enough to find the time for that creative work, and the servants Marthe and Valentine are not independent enough to allow her the freedom, but she must find the time, if only 2 hours daily. She also wants a larger block of time for herself, a half-day free every week. She hopes that she helps him to achieve his dreams, but helping George is not enough for Mabel. "I must work, too. I *will work*. It is a conviction, a need, a hope, a happiness that is *always* in me. I must make it happen."[37] May smiles and laughs in her sleep. In English: "I am sure of it. She dreams of other things than *eating* now!!" On the 26th Mabel writes in English that she wants George all the more because she is

36. NYPL.

37. At about this time, following the death of her mother, Céline Dangotte establishes her own interior furnishings company in Brussels, Art décoratif Céline Dangotte, or ADCD. NYPL, Mabel Sarton to Marthe Patyn, [Spring 1912], where Mabel anticipates the birth of David or May and where Mabel is sad that Mme Dangotte's new life has been brutally cut short, and so she shall not see Mabel's child. Mabel provides Céline with designs and embroideries. Céline's artistic director is Albert Van huffel [sic], a former member of the Kunst en Kennis artistic group in Ghent whose botanical motifs owe much to Julius MacLeod and whose early furniture followed the designs of Charles Rennie Mackintosh. Elisabeth Dhaniens, "Les Arts appliquées depuis 1800," in *Gand: Apologie d'une ville rebelle*, ed. Johan Decavele (Antwerp, 1989), 349–71, on pp. 358–9, notably a photograph of the furniture Van huffel designed for Céline Dangotte's residence in 1918. Jane Turner, ed., *Grove Dictionary of Art* (London, 1996), *s.v.* Van Huffel.

"feeling very far away from Ray. & Céline. They puzzle me and I feel sad about them." Raymond does not like May, does not like small children at all. The next day, the 27th, Mabel writes in English to thank George for the portrait of Tolstoy: "There is something the matter with me—all sorts of things in Raymond & Céline shock and hurt me, so I can't long be happy with them." She is having a grey dress made from "*the remnants* of our grey curtains in our room!!" Mabel closes: "Best, best love to you, Petit Rat du Petit Chat et le Chaton."

On 29 June 1912 George writes from Paris that he spent six hours at the Bureau international de poids et mesures, where the director, Guillaume, gave him a tour. Following his discovery of invar, a nickel alloy for which he will receive a Nobel prize, Guillaume "has revolutionized chronometry and geodesy." At Sèvres George met the Brussels astronomer Paul Stroobant with his young wife. The next day, George visited the Egyptian collection in the Louvre. Commentators have framed George as a positivist, but here we see him recognizing the integrity of pre-classical art, a non-positivist sensibility:

> I "discovered" today that the stylization of primitive works of art is not at all as I thought a stylization of ignorance, a "first approximation"—something analogous to what happens in the domain of science, where one finds first, with the help of crude instruments, the very simple laws, which subsequently become more and more complex depending on the progress of technique. I attentively examined some of the main pieces in the Egyptian section, and I am persuaded that the artists that produced the works must have had very precise anatomical information.

George thinks that "*it is not primitive art*, but on the contrary a mature art, I would say an art at its height." The artists of 40 centuries ago "understood marvelously well the beauty of the human body." George now appreciates primitive Greek art better. George often thinks about Mabel's work. "Essentially, your freedom depends not on me, but on the servants. With two assistants, two good assistants, it seems clear to me that 2 hours a day are enough for taking care of 'domestic' chores." George will arrive in Knokke on Tuesday. On Monday he will have to take care of his correspondence at Wondelgem.

The young husband has second thoughts about his role in the family. George's diary on 1 July 1912 records:

> I shall try to devote every day more time to my beloved, who has accepted to be my life companion, and who is indeed a faithful companion, my mainstay, my adviser, my inspiration, my friend, my beloved, and the mother of our dear little May. I feel very guilty toward her. No doubt she knows how much I love her, she knows that she can rely upon me with full confidence. Yet I have often neglected her, I was too deeply immersed

> in my work, my eyes were fixed upon a distant ideal and I forgot to accomplish my nearest duty. I remained too much alone near her....
>
> Thanks once more, my dear wife, my beloved Mabel, whose conscience is so clear and so straight that mine seems gloomy and tortuous in comparison, thanks my beloved for helping me so well to live with kindness and justice: it is for that very reason that I am most grateful to you.[38]

George laments that his trip to Paris was a failure. He will stop devoting himself exclusively to history of science and find a remunerative position. He will fight for social justice. He wants to publish a piece on household servants with the Solvay Institute. It is a theme discussed years previously among his friends in Reiner Leven, who perceived that domestic labor would become an important ethical issue in the 20th century.

At the height of the summer in 1912, the Sartons celebrate their house and daughter by inviting George's relatives for a feast. George's uncle, Arthur Sarton, thanks the couple: "Your beginnings, dear George and dear Mabel, which have just taken place, are spectacular and impressive [coups de théatre et de vrais coups de maître]." George and Mabel are vegetarians, but they fed everyone extremely well with meat, fowl, and fish. They have established themselves in "a small terrestrial paradise, far from annoyances and the thousand-and-one occasions for being subject to formality and social obligations." He likes their cottage, with its large openings to the light, its "furniture, perhaps a bit too rectangular, but still original, in interesting wood, carefully put together, and set with medallions painted by the mistress of the house," its hot and cold running water, its electric lights. He finds it a small part of England in Belgium. He is attracted to the "unique garden," with its trees.[39]

During their time in Wondelgem, George and Mabel see themselves as Fabian socialists sympathetic to the Belgian Labor Party. George records in his diary: "As my friend Victor Willem tells me, I am condemned by my ideas and by my way of acting to the base of Society"—George translates the phrase a lifetime later in his biography of Mabel: "Your ideas and your manners have banned you from Society."[40] The base of society is grateful. In 1912 George's socialist friend Max Horwitz writes from exile in Vienna to ask if he can travel to Russian Poland as George Sarton, and he asks to borrow George's passport, validated for man and wife.[41] George sends Horwitz his passport with a Russian visa, and Horwitz uses it on

38. Diary, 1 July 1912, George's translation in "Scholar's Wife."

39. NYPL, Arthur Sarton to George and Mabel Sarton, 16 July 1912; *Phoenix*, pp. 72–4, for a complete translation, which I have not followed.

40. Diary, September 1912.

41. HGS, Max Horwitz to George Sarton, 22 May 1912.

a number of occasions to visit Warsaw.[42] Also in 1912, closer to home, the president of the Society of Socialist Students in Ghent, H. Chaudon, thanks George, the retiring president, for his 2 years of service. George has been a model of impartiality, and he has gone beyond his duty by helping the organization both morally and financially. Chaudon asks George how to go about inviting speakers, and whether a third manifesto should be published. The socialist leader Henri de Man, Chaudon notes, has promised not to charge for his talk of last year.[43] George continues his association with the socialist students, speaking to them late in 1913.[44]

George was one of the founders of the socialist society on 10 March 1910, which joined Belgian students with foreign students. He wrote manifestos for it.[45] The first manifesto, appearing in April 1910, sets out the society's constitution and aims. The second manifesto, dated 19 October 1910, reaffirms that the society is repelled that the mass of citizens are enslaved by a small minority of rich men. "We find it impossible to reconcile individual liberty with economic slavery." If a reader shares this view, he is invited to join the society, which seeks to complete the socialist education of its members through study and lectures, as well as collaborating with the working class. George observes that many students thought only to advance their career and did not want to defend the oppressed. He cites Peter Kropotkin, calling them to the cause. Students belonging to the middle class are subject to the vicissitudes of the financial markets, making them the victims, rather than the allies, of capitalism. Students might be especially interested in reformulating the administration of society, in the manner suggested by Sidney and Beatrice Webb, including the creation of cooperatives, a minimum wage, a reduction in the hours worked, a guarantee for savings accounts, and the prevention of unemployment. As George Bernard Shaw emphasizes, the essential point is for ordinary people to take control of their own destiny. The student society is "clearly revolutionary." It is anticlerical and antimilitarist. "We hate the army." The people do not yet have a country of their own, for they are forced to serve the interests of the ruling class, but once they had one, they would gladly die for it. The socialist students would not bow before the ruling class. "We are ready to fight with all our resources, and to extract from them bit by bit the reforms required for advancing the general well-being and improving human life." Their children would succeed if they did not. George notes

42. HGS, Max Horwitz to George Sarton, 13 June 1913, returning the passport "with numerous seals on it" and asking for a recertified document.

43. YMS, H. Chaudon to George Sarton, 7 December 1912.

44. Agenda, 22 December 1913.

45. Diary, 31 August 1910.

that in 1910 the student society heard lectures by Emile Vandervelde, Louis de Brouckère, Edouard Anseele, and Auguste Debunne, the last in Flemish. The members of the society look forward to the appearance of sister societies and to a national organization. By naming their invited speakers, not all of whom shared the same principles, the society clearly indicated its non-sectarian nature.[46]

George writes a third manifesto in October 1911, but it remains unpublished. Titled "Socialism and Individualism," George recalls in the manuscript that the aim of the organization is above all self-improvement. They hold that society is organized in a "criminal and odious" fashion, where the mass of people are slaves to a privileged few. They do not demand equality for everyone, an "absurd and chimerical" condition, but they do require "equality at the starting line," or the possibility for people to develop freely. George addresses an introductory paragraph to students who think that they will be able to speak about social questions when their career is firmly established. But by then, "their spine will be irremediably curved, and when they finally dare to speak freely, they will lack the strength to do so."

He begins his main topic, liberty and the state, by examining Herbert Spencer's notion of liberalism. In Spencer's view, the state exists to safeguard liberty, to prevent exploitative excess, and it has models for doing so: free cooperation in a laissez-faire regime, which he favors, or forced cooperation under a military regime. But George observes that there are an infinity of possibilities for state intervention. Liberals early in the 20th century resemble their Spencerian antecedents in name only, for they have compromised their principles by introducing obligatory schooling. The liberals have in fact followed the lead of the socialists. Everyone recognizes that liberty must be organized, and the liberals in Belgium are hardly in a position to criticize the socialists for compromising liberty. George agrees with Spencer's call for eliminating useless regulations and streamlining the legal system. The bureaucracy has become a plague in the modern state, and George sees capitalism at the root of the problem. Demoralization leads people to look for, precisely, a secure position with the state. George thinks that there is a practical balance between centralization and local control of resources.

For individualism, or personal liberty, to have any sense, George continues, people must have a certain degree of dignity and security. George thinks that it is equally dangerous to imagine that this condition can be satisfied by a single dictator, on the one hand, or by the undifferentiated mass of workers, on the other hand. He imagines that an intellectual élite will, in

46. Carl Strikwerda, *A House Divided: Catholics, Socialists, and Flemish Nationalists in Nineteenth-Century Belgium* (Lanham, Md., 1997), pp. 288–9 for the speakers; the first 2 published manifestos in YMS.

the future, make the best decisions for society. Such an élite is in the best position to respect the qualities of all people, for he is resolutely against a Darwinian struggle for existence, where the strong defeat the weak. In addition to the economic side of socialism, there is a moral side. The moral side is based on "the feeling of human solidarity, combined with a sort of pride, self-respect, at the joy of being useful, and of doing good work." The feeling will become something like a religious ideal. People

> will profoundly feel that their conduct, whatever it may be, will have distant consequences after their death; they will feel also that they will be able to pay back the debt they owe to all who have created their civilization, to pay off the goodness that they have inherited, only by continuing and bettering the great work undertaken by humanity.

This sentiment is impossible in George's own time. The entire capitalist structure is "rotten with lies, injustice, and mistrust." Only in a collectivist society will people be free, and liberation can take place only when people are infused with the moral ideal of socialism, for capitalism is characterized precisely by its "lack of a guiding principle, its lack of an idea and of faith. Capitalism is essentially centrifugal, anarchic." Socialism seeks to "replace a soup of disorganization, anarchy, and decomposition, the hunger for gold, by progress, order, beauty, and synthesis. And it seems that this is the goal of a long evolution of humanity." Socialism will connect the solitary workman, conscientious in his task, "by countless and eternal links with other men." Under socialism, if it wins the day around the world, nations will encourage science and the arts and will compete to honor those among their citizens with greatest aptitude in these areas. The socialist state will rehabilitate its citizens, "transform all its bad citizens into useful people." The richest nation will be seen not as the one with the most bankers, but as the one with the highest standard of living and with the most creative people. George imagines a socialist world where nationalism assumes a nobility lacking under capitalism. He concludes that there is no real contradiction between socialism and individualism. Rather, individualism is irreconcilable with egoism. The heralds of the new society—George names Carlyle, Emerson, Ruskin, Tolstoy, and Walt Whitman—perceive that there is an essential human solidarity. It is apposite to recall Auguste Comte's view, that the laws of the universe take precedence over human laws: "There is no freedom of conscience in astronomy, in physics, in chemistry, in physiology." Laws are instituted in society because people are imperfect, and many of the laws relate simply to conventions, such as driving on the left or right side of the road.

"All human society can be considered an immense organism, continually tending toward more self-awareness and more beauty." The self-awareness is the conscience of its citizens. The greater the diversity among its citizens, the more complex and perfect the organism. The individual is just a motor for converting chemical energy into intellectual energy. "The progress of humanity lies in increasing the efficiency of the motor. This is what modern socialism seeks, with all its strength." George extracts a personal lesson. By seeking to become better individuals, "by becoming more refined, more sensitive, and more noble, we act in the best way for promoting progress." Contributing in this way is how our separate personalities can achieve meaning:

> The man who is proud of having accomplished his task to the best of his ability, who is aware of the useful rôle that he plays in society, humble as it may be, who is aware of the value of his life, who knows himself and is in command of himself, who senses himself to be a fine human being, free and disciplined—he is the pure individualist, and he is also a good socialist. Such a man builds the happy citadel of tomorrow.

In that future age, people will divide their time between manual and intellectual labor. A worker's effort will produce joy comparable to that of the artist when he creates beautiful things. "The artisan will become an artist." The allusions come from George's personal education. They appear in his correspondence with Mabel and in his artistic manifesto, *La Chaîne d'or*. These socialist sentiments provide a guide for much of his career.[47]

George summarizes his past year in his diary on 8 September 1912. Mabel's health is not good; she suffers from *névralgies* or neurasthenia—the poorly-defined afflictions of headaches and fatigue that at the time seemed related to the nervous system.[48] His family's material situation is troublesome. He earned a small sum from a number of articles in the *Revue générale des sciences* and from teaching at the Athénée de jeunes filles, an extension of the science courses he gave on Sundays to young women.[49] He projects giving up his "speculative and completely disinterested studies," although he has worked his way

47. Manifesto in YMS.

48. Frank R. Starkey, *The Glands of Life: A Contribution to the Study of Organotherapy* (Philadelphia, 1914), p. 91: "The opinion is becoming prevalent, especially among laymen, that neurasthenia has no physical basis. This is a very grave error; but it must be admitted that mental or emotional stimuli, either in themselves abnormal, or excessive in degree, though normal in kind, may cause physical symptoms." Starkey continues that the surest trigger for the disease comes from "bad education,...perniciously wrong training in two most important things: sexual functions and relation of personal rights to social duties." The "rest cure" is commonly prescribed (p. 100).

49. HGS, Andries MacLeod to George Sarton, 14 March 1912, observing the beginning of George's course at the Athénée des jeunes filles, which is not a state institution; Pierre Hoffmann to George Sarton, 24 August 1908, for governmental intransigence about recognizing the *athénée populaire* that George sought to establish with his friend Michel Thiery.

through secondary works in history of science appearing over the past 10 years. May gives him much pleasure, even though she "speaks like the birds."

The articles in the *Revue générale des sciences pures et appliquées*, a journal in Paris that published writings by Paul Tannery, Pierre Duhem, and Gaston Milhaud, begin with an éloge of German chemist Albert Ladenburg. George discusses Ladenburg's chemical discoveries and alludes to personal struggles in his life. He focuses on Ladenburg's published lectures on the history of chemistry during the 18th and 19th centuries, which have great historical and methodological value, although they are somewhat stiff and not at all synthetic: "The author has not taken sufficiently into account external circumstances that influenced the development of chemistry." A significant omission concerns industrial chemistry: "A history of science obtains its full importance only if it is continually confronted with the history of civilization or, at least, with the history of technological progress."[50] With his second article in the *Revue*, George inaugurates a column on "history of science" in the sense of Tannery—the interactions of all the various scientific disciplines in connection with their social and cultural influences. In addition to Tannery, George names Moritz Cantor and Pierre Duhem as writers who are leading the way forward, and he gives favorable attention to Karl Lamprecht's institute devoted to the history of civilization in Leipzig. He anticipates that in 10 years "history of science will become itself a new science."[51] In his third article, George uses a bibliographical work about Blaise Pascal to emphasize that the only useful bibliographies are critical ones. In the review, he chides scholarly antagonists—Pierre Duhem for his weak appreciation of library research and Paul Mansion for a weak publication in history of mathematics.[52] It is erroneous to imagine that science present, given its progressive nature, itself preserves the best of science past, George announces in his fourth article. Because science advances, science in the future may well focus on past discoveries held to have little significance today. Hence the necessity for studying the history of science. Wilhelm Ostwald recognizes the value of history of science, and he has dedicated a large series to republishing scientific classics.[53] The bicentennial of

50. George Sarton, "Albert Ladenburg," *Revue générale des sciences pures et appliquées*, *22* (1911), 937–8.

51. George Sarton, "L'Histoire générale de la science," *Revue générale des sciences pures et appliquées*, *23* (1912), 93–4.

52. George Sarton, "La Bibliographie de l'histoire de la science," *Revue générale des sciences pures et appliquées*, *23* (1912), 131–2; Albert Maire, *L'Oeuvre scientifique de Blaise Pascal* (Paris, 1912), with a preface by Pierre Duhem; Paul Mansion, "Programme du cours d'histoire des mathématiques de l'Université de Gand," *Biblioteca mathematica*, *1* (1900), 232–6.

the birth of Mikhail Vasilievich Lomonosov offers George the occasion to elaborate on the underappreciated innovations of an 18th-century savant—a precursor of quantitative chemistry and the mechanical theory of heat.[54] How, then, should one establish a chronology for scientific discoveries? In his next column, George introduces a recent discussion by Gustaf Eneström about the difficulties of establishing a chronology in history of mathematics, for one must consider not only the emergence of a concept but also its publication and, more important still, its acceptance. Which dates, for example, should be assigned to discoveries by Evariste Galois, Niels Hendrik Abel, Hermann Grassmann, and Gregor Mendel? George looks forward to a synoptic presentation of the facts of history of science without regard to discipline or nation, which will prepare the way for a synthetic narrative.[55]

There is already such a chronology, however, and in his next installment George discusses it. Ludwig Darmstaedter, in collaboration with many colleagues, is the author of a 1200-page chronological table of scientific discoveries, which George finds useful—notwithstanding the absence of bibliographical sources. Franz Feldhaus has published a parallel volume in history of technology, and Feldhaus has set up a private institute in Berlin to publish a more detailed history, based on 150,000 facts organized by 20,000 subject headings, in addition to thousands of illustrations. George is impressed with Feldhaus's enterprise, as much by its cooperative nature as by its commercial character. Funding is assured from industrialists, periodicals, technical journals, and museums.[56] For his next column, George chooses the 250th anniversary of the creation of the Royal Society of London. He uses the occasion to write an institutional history, and all Europe is his tableau. George emphasizes the society's interest in direct experimentation, in corresponding with other savants in Europe, and in publishing new facts. A quarter-century later, George publishes the epoch-making study by Robert King Merton about the early years of the Royal Society, and in it Merton focuses on the soci-

53. George Sarton, "Les Classiques de la science," *Revue générale des sciences pures et appliquées, 23* (1912), 217.

54. George Sarton, "Le Bicentenaire de la naissance de M. V. Lomonosoff," *Revue générale des sciences pures et appliquées, 23* (1912), 300–1.

55. George Sarton, "La Chronologie de l'histoire de la science," *Revue générale des sciences pures et appliquées, 23* (1912), 311–12. Gustaf Eneström, "Ueber die Bedeutung von Quellenstudien bei mathematischer Geschichtsschreibung," *Biblioteca mathematica, 12* (1912), 1–20.

56. George Sarton, "L'Histoire de la technologie," *Revue générale des sciences pures et appliquées, 23* (1912), 421. Ludwig Darmstaedter, *Handbuch zur Geschichte der Naturwissenschaften und der Technik* (Berlin, 1908), undertaken in collaboration with René du Bois-Reymond and Carl Schaefer and extending an earlier edition of 1904; Franz Feldhaus, *Ruhmesblätter der Technik* (Leipzig, 1910).

ety's Dissenting founders (and its Pietist ethos) as well as on the society's interest in technology relating to military affairs; both themes are absent from George's publication in 1912.[57] The last of George's columns is a short report on the Italian Society for the Critical History of Medical and Natural Sciences, founded in 1907, which publishes a journal. George uses his column to condemn the Italian government for ignoring the society and suppressing the only Italian chair in history of medicine, at Rome.[58] George abandons writing for the French journal, but many of the ideas he publishes recur in the decade to come. Feldhaus's private institute and the life of scientific societies, in particular, find expression as George makes his way in the intellectual wilderness of America.

George Sarton does not travel light as he sets off on the road to posterity. On All Saints' Day (First November) 1912, he formulates ambitious plans to write general books about philosophy of science and history of science. He comments in his diary:

> This is somewhat like trying to redo the work of Comte and that of Spencer. Yes! But one must bear in mind 1° that my views are in many respects different from those of Comte and Spencer, 2° that I know *science* better than Spencer and the *history of science* better than Comte. An undertaking such as the one of which I am contemplating the realization is not truly philosophical unless it be solidly established upon the *history* of science.[59]

George's history of science will take 20 years; after that, he can begin the philosophy of science. He projects a work schedule, "because neither much work nor good work can be accomplished without order, and because I always worry about overtaxing myself due to lack of physical exercise." One morning weekly will be for science, the other mornings for history of science. Sunday is for administrative tasks regarding *Isis* and for reading.[60] Uncle Jules wonders about George's future. He has seen the first number of *Isis*. Jules remembers that his favorite authors, around 1870, were John Stuart Mill and Auguste Comte. He twice cautions George that his project will take perseverance and money. "Do you have all that?"[61]

57. George Sarton, "Le 250e Anniversaire de la Société royale de Londres," *Revue générale des sciences pures et appliquées*, *23* (1912), 306–12.

58. George Sarton, "La Société italienne d'histoire des sciences médicales et naturelles," *Revue générale des sciences pures et appliquées*, *23* (1912), 537.

59. Translated in "Scholar's Wife."

60. Diary, 21 November 1912.

61. YMS, Jules Sarton to George Sarton, 4 January 1913. Presently NYPL.

At the moment George imagines that he does, and even if his enterprise in editing and publishing a journal fails, it will have contributed to his ultimate goals. In *Isis,* George often introduces questions and research projects, and the main section becomes a place for circulating his own programmatic statements about the field of study. He uses its extensive survey of literature—the bibliography and reviews occupy more than half of the journal's space—to extend his grasp over an enormous variety of sources, inviting active scholars to contribute reviews and articles in their preferred language. He casts his net wide. The bibliography includes the expected studies of great scientists and their discoveries, along with past mechanisms, gadgets, diseases, and plagues. It also features science on the research frontier—genetics, relativity, and quantum physics—along with sociology and peace studies, concessions to George's commitment as a social reformer and internationalist. Finally, the bibliography introduces titles in the history of art—George's common bond with Mabel—which he understands as a complement to the history of science. By analogy with present practice, George is elaborating a vast post-doctoral research program, and *Isis* is the means of achieving it from the unlikely location in a suburb of Ghent. In a short period of time, the appeal to posterity makes the journal more than a means to an end. By May 1914, George plans to publish a number of copies of *Isis* on fine Van Gelder paper, for "it may be—we can't really say—that *Isis* will be considered by historians in the future as a vital source for the study of scientific thought in the 20th century."[62] For George, a scholar who has no scholarly book to his credit before he reaches the age of 43, *Isis* is at once his glory and his burden.

George is clear about the aim of *Isis*. At the beginning of the first number, in a text completed in November 1912 and too long to be nailed to the masthead, he outlines his hopes and his vision. At the present time, he observes, science is enjoying a synthetic renascence, from electromagnetism to atomism to notions of heredity and evolution. This rebirth has been prepared by synthetic thinkers, like Hermann von Helmholtz, Claude Bernard, and Marcellin Berthelot, and by positivist philosophers; a reaction by spiritual, humanist, and pragmatist thinkers has set in. It is George's idea to connect philosophers and scientists by focusing on the general parts of scientific knowledge through studying the history of science. Auguste Comte pointed the way forward, Paul Tannery was his foremost successor, and a number of scholars are active in the field. He proposes to take the matter further by focusing on the development of scientific ideas. "All natural, psychological, or economic phenomena that have influenced and modified the evolution of scientific phenomena shall be studied in my review, not for their own

62. George Sarton, "Notes sur la revue *Isis*," *Isis*, 2 (1914), 162–3.

value but rather as aids and only as explanatory devices." History of science is, then, the demonstration laboratory for philosophy of science. It must be studied as a whole, by epochs, not divided "by country, by specialty, or in any other manner."

Five scholarly areas complement the history of science. First is the history of civilization. General circumstances can explain special developments. Mesopotamian predilection for state archives seems to relate to the emergence of systematic astronomy; the culture of scribes in medieval Europe slowed publication of Latin translations of Greek authors; the metric system was as much a scientific reform as a rejection of royalist authority. Second is the history of technology. The demand for vacuum pumps in Leiden led directly to the Dutch precision-instrument trade; introduction of Arabic numerals in the 13th century and decimal fractions in the 16th century, connected with trade, had a profound impact on science. Third is the history of religion. Religious strife has had a deep impact on science, whether by suppressing it or by causing the exodus of wrongly-oriented believers. Fourth is the history of art, a favorite topic for George. There is much science in the techniques of art. Despite fundamental differences between artists and scientists, they both deal with the dreams of humanity, and in certain individuals, like Leonardo da Vinci, Albrecht Dürer, or Bernard Palissy, the dreams become manifest simultaneously. To understand the history of chemistry, one must study "the evolution of the decorative arts: the art of the ceramist, the glazier, the engraver, and," in homage to Mabel, "the miniaturist, the painter, and the enameler." The history of the representation of nature in art helps clarify the diffusion of plants and animals. History of science is in a certain measure the history of taste, and many great scientists were also great stylists. Finally, the fifth complementary area consists of archaeology, anthropology, and ethnology. Here one finds notions like the cyclical theory of culture, then much in the news. Archaeology promises to provide information about the development of technology, and there is much to learn from anthropology about popular notions of science.

History is also useful for scientists, George continues. The errors of the past are instructive, and history is "the guide, the indispensable catalogue for the construction of syntheses and for choices made from new viewpoints." Old science is filled with overlooked wisdom. Some scientific innovations, the result of "abrupt discontinuities," derive from genius, and nothing is more exciting than returning to an original source, which is pregnant with heuristic insight. George deplores the "exclusive religion of positive facts," which has banished interest in past science "to the worst of metaphysics, scientific idolatry." There are times, like the present, when "paradoxical and arresting discoveries" make it necessary to inventory the state of science, and for that one must appeal to history. History of science is useful in pedagogy, where science is best presented, as Ernst Mach does, "in the order of its historical development," and where science past serves to inspire scientists future. Finally, history of science is related to questions of individual and group psychology, to the possibility of general laws for the intellectual development of humanity. The path forward is by comparison.

In an anticipation of the academic debate over internal and external history of science in the 1960s and 1970s, comparison limited to the domain of science is denoted by "internal comparison"; comparison between scientific phenomena and other intellectual and economic phenomena is called "external comparison."[63] He notes generalizations such as James George Frazer's categorization of agrarian ceremonies and Karl Lamprecht's search for whether the development of children's perception recapitulates the perception of past epochs in human history. Individual psychology relates to the genius of technological invention; it finds an application in classifications of intelligence, like Wilhelm Ostwald's distinction between the romantic and the classical temperament, or more rigorously in Francis Galton's and Alphonse de Candolle's studies of genius. *Isis* is the place to consider families of scientist, which is a branch of eugenics.[64]

George holds in conclusion that science is the great pacifier. "It is the cement that unites the highest and most inclusive minds of all nations, all races, all beliefs. All peoples benefit immediately from discoveries made by particular peoples." *Isis* will present notions of tolerance, denouncing "imperialist tendencies that some scientists try to fix to the science of their country or their race." George calls on everyone sharing his vision to help him achieve his goal.[65] The goal, so all-embracing that it seems to be a

---

63. For example, Roy Porter, "The History of Science and the History of Society," in *Companion to the History of Modern Science*, eds R. C. Olby, G. N. Cantor, J. R. R. Christie, and M. J. S. Hodge (London, 1990), pp. 32–46, on p. 39. The once-vital controversy between internalist and externalist approaches to history of science is now seen as an artifact of times past. Unlike postmodernism, it is absent from the index to a prominent successor to Olby's volume, *The Oxford Companion to the History of Modern Science*, ed. John L. Heilbron (Oxford, 2003).

64. George wrote to de Candolle's son, who replied that Alphonse de Candolle avoided all metaphysics, seeking a purely objective view of the world, and that Alphonse's favorite writer was Montaigne. HGS, Casimir de Candolle to George Sarton, 4 May 1913.

65. George Sarton, "L'Histoire de la science," *Isis, 1* (1913), 3–46. Among many critical discussions of Sarton's early views: Arnold Thackray and Robert K. Merton, "On Discipline Building: The Paradoxes of George Sarton,"*Isis, 63* (1972). 473–95; Tore Frängsmyr, "Science or History: George Sarton and the Positivist Tradition in the History of Science," *Lychnos*, 1973/74, pp. 104–44; Hosam Elkhadem and Annette André-Félix, "L'Humanisme selon George Sarton, historien de la science," *Mémoires et publications de la Société des sciences, des arts et des lettres du Hainaut, 86* (1975), 9–26; Lewis Pyenson, "What Is the Good of History of Science?" *History of Science, 27* (1989), 353–89; Lewis Pyenson, "El Fin de la ilustración: Reflexiones próximas y lejanas sobre la Historia de la Ciencia," *Arbor* [Madrid], *142*, nos. 558/559/560 (1992), 69–91; Lewis Pyenson, "Inventory as a Route to Understanding: Sarton, Neugebauer, and Sources," *History of Science, 33* (1995), 253–82.

scientific counterpart to Nobel-laureate Rudolf Christoph von Eucken's advocacy of spirit over nature, reverberates with an antinomy. George claims universality for his endeavor, and he rejects approaching it through exclusive studies of scientific disciplines, but he demands nonetheless a place for history of science among the disciplines represented in the academic world. Europe, where specialization had eroded general education, would not grant his appeal. America, fixed on the notion of civilizing masses of university undergraduates, might be expected to show greater sympathy.

George's long statement, breathtaking in its scope, apparently does not produce the desired effect in his readers, so George constructs a shorter statement for the second number of *Isis*. He insists that "history is only a means...not an end. The end is the philosophy of science, acquiring a more perfect familiarity with the nature of humanity." History is liberating, just as travel dispels prejudices. "Only history can give us a clear and complete grasp of the spectrum of modern science and allow us to appreciate its real significance. It can best help scientists to find new paths where their work will be more fruitful. George is not able to study history for its own sake. He also "wants to understand the better to predict, to act with more precision and wisdom." *Isis* is more a periodical in philosophy and sociology than in historical erudition; it is a tool for achieving *synthesis*. To make his point, he changes the subtitle of the periodical to: *Review Devoted to the History and to the Organization of Science*. By the word organization he means mostly the "internal organization" of science, its principles and theories and the evidence brought to bear on them. He also means "external organization," including institutions, schools, and bibliographies. George wants to keep his readers informed about the latest syntheses in science. "What is to be realized is today's scientific philosophy (or tomorrow's), not yesterday's."[66] George emphasizes his program everywhere he can. For example, he notes in the third number of *Isis*, in a review of Maurice Vernes's *Histoire sociale des religions* (Paris, 1911), that *Isis* is directly concerned with the questions raised by Vernes. "We anticipate preparing a *Social History of Science*."[67] Also in the third number of *Isis*, George reports on the Twentieth Peace Congress, held at The Hague: "Peace is the first condition necessary to guarantee the existence and the development of science. And for science to expand to its full potential, not only must peace exist, but the heel of military might must not be placed too harshly on ordinary people." Studying the organization of science naturally leads one, "morally and logically," to become a pacifist and to affiliate with other pacifists.[68]

66. George Sarton, "Le But d'*Isis*," *Isis*, *1* (1913), 193–6.

67. George Sarton, "Maurice Vernes, *Histoire sociale des religions*," *Isis*, *1* (1913), 538–9.

68. George Sarton, "XXe Congrès de la paix," *Isis*, *1* (1913), 483–4.

Does George in fact have the elements of a discipline under control? Certainly, he affirms later in 1913, when he discusses work in the history of mathematics. Léon Brunschvicg's history of mathematical philosophy and Pierre Boutroux's principles of mathematical analysis receive pages of measured praise, even though neither book is really a work in history of mathematics. Notwithstanding recent scholarship, there is, in fact, no book in mathematics comparable to Ernst Mach's history of mechanics. Still, the history of mathematics has achieved a sophistication far in advance of the history of medicine, despite the "immense and terrifying literature and all the periodical publications devoted to the history of medicine." He asks whether the explanation for this asymmetry lies in the intellectual culture of physicians, which is too superficial and empirical, or in the field of medicine itself. Here is a question worthy of extended thought.[69]

George's discussion of the history of mathematics is essentially a literature review. Indeed, as he is still furiously taking notes for his own syntheses, bibliography has a central place in his scholarship. George elaborates the bibliographical end of *Isis* in May 1914. Periodicals are multiplying at an alarming rate, he observes, due to relentless specialization on the one hand, and the collectivization of intellectual work, on the other hand. It has become vital to control the secondary sources in history of science. Libraries face the dilemma of not being able to acquire every periodical on the market. To be well informed, librarians require rational bibliography. He reviews 62 periodicals and series relevant to the history of science, and he concludes that *Isis* is unique in its focus on the history and organization of science.[70] In George's inventory, periodicals in the history of medicine are more numerous than for any other area of science, broadly speaking. Setting out his Critical Bibliography in 1913, George wants to be catholic rather than protestant, including works by "dilettanti" to the extent that they are useful while carefully avoiding vacuous accounts by famous scientists; indeed, history of science as a hobby, rather than as a discipline, "is the greatest danger facing us." The worthless, "more or less spiritual" publications of dilettantes are especially prominent in medical periodicals.[71] Physicians, indeed, are George's rivals from the beginning. He writes in May 1914:

> It is somewhat remarkable to observe that the various societies for history of science, which now exist in many countries, are almost always

69. George Sarton, "Les Tendances actuelles de l'histoire des mathématiques," *Isis, 1* (1913), 577–89.

70. George Sarton, "Bibliographie synthétique des revues et des collections de livres," *Isis, 2* (1914), 125–61.

71. George Sarton, "Bibliographie analytique des publications relatives à l'histoire de la science parues depuis le 1er janvier 1912," *Isis, 1* (1913), 136–90, on pp. 140–1.

> of medical origin and concern themselves principally with the history of medicine, that is, with the history of an applied science. This phenomenon is interesting, but it is in certain respects annoying, because this completely unjustified hegemony cheapens the point of view of the historian of science, lowers his horizon, and weakens the impact and the value of his scientific and philosophical works.[72]

History of science first—let the medicos, with their special point of view, stand in line. George maintains this perspective across his entire career, notably when he crosses swords with the Institute for the History of Medicine at Johns Hopkins University in 1935.[73]

Mabel, too, wonders about George's calling. She writes from Rhode on 8 January 1913, as if she were Henrik Ibsen's Nora. She thinks more than ever that if she is to be cured of her ailments she

> must work, *not as an amusement* without a goal but regularly, conscientiously. I want to have the time and the quiet necessary for it. I feel in me that it is my fundamental duty. I don't want to talk about it any more. If you agree, it will help me; if you disapprove, I will do it anyway (I don't see any way out of this dilemma, dear, sorry to say).
>
> You are absorbed increasingly in your work. I must live in my own way as solitary as you live in yours. I *must* have an occupation that satisfies not only my heart but my mind, if not, I will slowly die. Elenore [sic] May respectfully must not absorb all my energies (I agree completely) and *at her age* the care that I give her (even if it is not *perfect*) does not constitute, for me, a satisfying task.
>
> For you the year 1912 was very satisfying. For me, it was one of the worst from *almost* all essential points of view. I must reestablish myself. I must fight against myself and against circumstances, even against you, if it were necessary. My will has been subordinate to yours in a lamentable way. I certainly know that in part it is a physical effect. I know that my physical stamina in itself has been sapped.
>
> You know, I am surprised that I *have not dared* to think for myself completely independently, without letting myself be influenced *in advance* by your way of thinking and reasoning. This is not good! I don't want to continue like this! I prefer to be imperfect but *to be myself*. Ah well, I believe that for a year, perhaps more, a part of me has died, something that was a strength (because I believe that a small part was lost which had to be lost). I want to rediscover this strength, and I will find it only through working, by suppressing the constant revolt of a part of me which has been deactivated *by force*.

---

72. George Sarton, "Un Nouvelle Société d'histoire des sciences," *Isis, 1* (1913), 250.

73. George Sarton, "Second Preface to Volume XXIII: The History of Science versus the History of Medicine," *Isis, 23* (1935), 313–20, where George contends that history of medicine is ancillary to history of science.

Mabel is more modest than George. Instead of dedicating only 2 hours daily for her household tasks, she wants to dedicate 2 hours for her work. Céline's situation differs from Mabel's. Céline has no child, does not wash, purchases her vegetables instead of growing them, and has efficient maids. Mabel asserts her right to be happy: "Another thing that I fear will shock you but that I feel intuitively. *We are not here only to work,* but to love, and love includes many things other than your work and mine. Even to have a bit of fun, if you know what I mean." Mabel adds in a postscript that she won't ask George again to leave the house. She will go to England, either alone or with May. George will join her if it is necessary for his work—otherwise she prefers him to stay at home. "You must put up courageously with the solitude that you have imposed on yourself, but you may not impose it on me (at least, not as completely)."

It is a marriage on the rocks. There is, in the modern age, an inevitable tension between husband and wife when children appear. Parents with intellectual pretensions seek to make their name in the world, to create. Creating children and a home is insufficient. *Otherness*, the abstract beauty formulated by the Pre-Raphaelites and Symbolists, becomes a mantra that consumes the thinkers. George could have realized his goals while teaching school, as his friend Emile Masson did in Brittany; Mabel could have created while working in the Dangotte establishment, as she did in 1910. But both husband and wife seek abstraction from the world and its requirements. The search entails a lifetime of material deprivation.

Mabel, alone at Wondelgem with May, writes in English in an undated letter early in 1913 (18 months after marriage, she indicates) that she desperately wants George. "I have thought of what last year has been for me. I feel it has been a moral failure. I have rarely felt ashamed of myself as I do now. I am not thinking of external things accomplished, but of what I have gained *in myself*. Well, I have not gained much if anything. This new year *must* be a better one." She wants George to help her take up her work again. In another fragment, in French, "I hurt so horribly, so deeply. It seems to me that I will never be able to be happy." Her life is a failure. "*I should not have come back to you.* That is the truth. They were right who said that I should not marry." Mabel does not cry for herself, but for George. "I cannot give you a son, our little daughter alone. I cannot give you a normal sexual life. I cannot follow your work." And she cannot even keep house for him. She feels a weakness that she does not understand. George has conceded her a second servant. "I *detest* your 'concessions.'" George believes that Mabel does not stay within her budget. She tries hard, but sometimes she cannot do so. On 27 February 1913 Mabel continues in English that she is glad that she did not send along the previous fragment. She really wants George to spend time with her and to confide in her. Mabel has decided to take the house in hand, not to modify her management of expenses but to *"revolutionise"* them. Mabel needs to know from George about what "honest wages" would be for their servants—she thinks that they are not paid enough. About a maid for May: "If that person is doing *the work she can best do*, is properly paid & has a just amount of leisure, then it is not concession to *our* principles, *mine* as yours." She thinks that she will be able to contribute financially to the cost of the household by working. "If I had had entire charge of May Baby since she was born, I am perfectly sure, I should be worn out now & incapable of doing other work, incapable of doing *that* work as it should be done." Mabel does not want to feel that George, in French, "'sacrificed' your principles to my weakness" by having servants. But the house entails activity and expenses of all sorts. "I love this house enormously (despite its drawbacks!). I love it above all for its *peace*. I also love the garden more than I can say. But if you were to sacrifice it to the product of your intelligence, to subjugate it, to assure these joys to me and little May, *I could not accept it*." Regarding May, Mabel thinks that they must prepare her to earn her own living, but also give May a small income so that she can live independently of them. Mabel hopes, in English that she and George will no longer hurt each other by using "edged words."

George wonders in his agenda on 4 November 1912 if Mabel is pregnant. She is. Mabel, pregnant, travels to Bordighera, Italy, early in February 1913, to relax. Mme Dangotte has been driven to "the point of a break-down" by her lawsuit.[74] Mabel, too, is not doing well. She delivers a baby boy (whom she alludes to in the undated fragment introduced previously), a blue-baby who fails to draw breath.[75] Marthe Patyn helps her recover. Mabel writes to Marthe that the doctor finds her uterus displaced and ovary inflamed. She will have to be operated on in 2 weeks. She has almost stopped bleeding.[76] George writes to Marthe on 12 February to thank her for her help. Later Mabel informs Marthe that the operation will not take place, and that she is healing: "I feel all my energy returning. My head is filled with ideas for designs and embroidery. I feel the wellsprings of my will reestablished. Life has become, once more, incredibly precious to me, despite its pain and struggle and difficulty."[77]

Shortly after the death of his infant son, the first number of George's literary child *Isis* is finally in the mail. George's faithful correspondent Emile Masson, who is gripped with intestinal illness and consumed by a campaign to preserve the Breton language, sends comments. Masson is impressed with the magnitude of George's task, and he invokes Carlyle to

74. NYPL, Mabel Sarton to Eleanor Elwes, 4 February 1913, from Bordighera.

wish George good fortune. Ignorant of science, he nevertheless offers that science seems dominated by the pragmatists whom George denounces, not the positivists. Masson would like science to serve humanity.

> I believe that a scientific conception, yes, purely scientific, of life, is wrong, is ineffective, is destined finally to death. A conception of life that is heroic, "musical," something like what Nietzsche calls "disaggregated,"must soon replace this [illegible] Socratic conception that we endure in the rudest tortures carried out in the infamous prisons we call "houses of education."

Masson wants fewer books and more air. He is in favor of biographies, but only those "of *real workers* for the consolation of humanity and for the actions of humankind."[78] Masson, who also corresponds with Raymond Limbosch, seems much closer to the poet's sensibility than to George's.

Emile Masson's friendship with George Sarton is grounded in their common literary striving, their young families, and their social values. Differences in principles—Masson was attracted to anarchism and humanism while George was a social democrat with scientistic tendencies—mattered less than a commitment to the Enlightenment abstraction of Humanity. Over the course of the 20th century, sectarian violence and cultural pessimism compromised the abstraction's appeal, but in the years before the First World War, a confraternity of European intellectuals shared an optimistic vision of human betterment through purposeful political action. The diverse rebel angels—from bomb-throwing anarchists to pastoral pacifists—residing in Paris in the 1914 novel by Anatole France, *La Révolte des anges*, come close to the mark.

Early in April, George prepares for a Belgian general strike scheduled later in the month; he travels to the Netherlands to disseminate socialist propaganda.[79] The strike has been called by socialists for universal suffrage, following their failure to form a government in the parliamentary elections of 1912.[80] George sends Mabel, "Bunny dear," a note in English from the Solvay Institute in Brussels on 15 April 1913, for the first time using the letterhead of *Isis*. He comments that the general strike has begun well, and that Maeterlinck has promised to assist the strikers. He

75. Margot Peters, *May Sarton: A Biography* (New York, 1997), p. 12, for the second confinement, based on a letter of May Sarton's, probably following the recollection of Céline Limbosch.

76. NYPL, without date.

77. NYPL, Mabel Sarton to Marthe Patyn, 24 February 1913.

78. HGS, Emile Masson to George Sarton, 26 March [1913].

79. Agenda, 1 & 6 April 1913.

80. Herman Balthazar, "Genèse de la métropole industrielle," in *Gand: Apologie d'une ville rebelle*, pp. 155–83, on p. 178.

sends kisses and signs himself "Ye old rat." The moniker, which he uses with pride over the coming years, signals his thriftiness or, in the traditional French sense, his miserliness. On the 16th he writes that he is returning to help her. Mabel writes on 21 April to Méta Budry, complaining of exhaustion. Cécile Cauterman and her mother have been visiting. They leave soon, and Mabel, in English, is "dreading new people—& worse dread still, not having anyone at all! We went for a fortnight to Céline's but somehow I was more tired when I came back than when I went." Mabel admires Méta's embroideries at the Dangottes' establishment in Brussels. On 3 June, Mabel urges Méta and her sister Berthy to visit.[81] Again in English: "Our garden is marvellously beautiful just now—or so it seems to me—so rich and such various greens & so quiet! Sometimes the patient look of the trees is like a quite [*sic*], cool hand put on one's forehead, on a feverish day." She has begun to receive the phospholipid lecithin (at the time derived from egg yolk)—one injection every 2 days—on the advice of Dr Miele.[82]

In the summer of 1913, Adolphe Miele records nervous afflictions and various aches and pains. In reply to Miele's questions of 9 July 1913, Mabel indicates that sometimes thoughts rain down all around her, out of control, and that she fears losing her mind. She is not depressed for long periods of time, however; she has not become obsessed with her child; her emotional attachments have not oscillated greatly. Miele prescribes phosphorus, strychnine, and beginning on 27 September 1913 opozones—preparations of the newly-discovered hormones.[83] On 29 October Miele finds a urogenital-tract infection.[84] On 29 December, Mabel has discharge from her ovaries. Opozones are once more prescribed. On 28 January 1914 Mabel is afflicted with severe neuralgia. In the first decade of the 20th century, dried ovarian substance was commercially available and used to treat nervous and emotional disorders, including delusional insanity, associated with conditions diagnosed as "ovarian" in

81. NYPL, Méta Budry to Mabel Sarton, 25 August 1925, on the death of Budry's sister Berthy.

82. NYPL. In the middle of the 19th century, phosphorus was held to be essential for mental functions. The contention found decreasing support by the eve of the First World War. Miele's prescription of lecithin for mental imbalance is a conservative, even outdated therapy. Theodore L. Sourkes, "An Element of Thought: Phosphorus and Mental Philosophy in the Nineteenth Century," *Journal of the History of the Neurosciences*, 7 (1998), 108–24; Sourkes, "The Discovery of Lecithin, the First Phospholipid," *Bulletin for the History of Chemistry*, 29 (2004), 9–15.

83. One form of sex-hormone treatment at the time consisted of powdered corpus luteum, derived from domestic animals, taken orally with meals. It was indicated for neurasthenia as well as for menstrual bleeding. B. A. Thomas and R. H. Ivy, *Applied Immunology* (Philadelphia, 1915), p. 316.

84. NYPL, Mabel Sarton's health records.

origin.[85] There is hubris in Miele's treatment, but before the notion of controlled substances and medication, patients had few rights and doctors few restrictions. Mabel suffers through several cures for her chronic illnesses. Miele notes her progress with elaborate charts (a concession to his earlier career in research), but the regimes are unnecessary if not ineffective. With her exile to England in 1914, Mabel does not do worse without them.

On May's first birthday, 3 May 1913, Mabel writes to Marthe Patyn that May is in Mons with the Bounys, and she is sad. George took May there by train on 28 April. They decided to send May away because their maid suddenly quit to take a job as cook. Even changing May once tires Mabel, and when May cries, Mabel cries also. Mabel's time away from May did not begin well, for she cried a lot. She has a house-guest, her friend Lilly Gull from Zurich.[86] The separation is hard on May. "I did not know either that I myself would be so strongly attached." May remains with the Bounys for 4 weeks.[87]

Mabel then travels to the Limbosches in Rhode, leaving May with George. Raymond Limbosch writes to her on 18 June 1913 that they will welcome her, and that they did not understand her state of mind when she wrote to them from Paris, where she apparently fled from her family. "Certainly we must accept our fate, but the problem is to find it among the facts, the possibilities, the dreams that are all part of our life! I believe that the best way is not to reason too much about it."[88] On 25 June George informs Mabel about domestic arrangements: He will not eat with May. He thinks that May is left-handed. Paul Carus's journal *Monist* spontaneously reproduced the announcement for *Isis* that George sent it; he transmits the latest number of the *Monist*'s sister journal, the *Open Court*, with an interesting article on a Buddhist sect that George has told Mabel about previously. On the 26th, George changes his mind and eats with May. They go out in the garden to pick flowers. "May and I are a pair of old friends who understand each other very well." He is writing an article, and it is difficult: "I am elaborating a point of view rather special and new, our ideas about heredity. I am doing something rather bold: I relocate the atomic theory of chemistry in biology, in the most mysterious part of biology, the evolution of living things." In his valediction, he bids goodbye to "petit coq," a nickname signifying courage and resilience which Mabel

85. H. Batty Shaw, *Organotherapy, or Treatment by Means of Preparations of Various Organs* (Chicago, 1905), pp. 203–4.

86. NYPL. According to George in "Scholars Wife": "Lilly belonged to one of the most eminent families of that ancient city and later she married the economist and philanthropist Carl Horber." The Horbers sustained *Isis: Isis, 19* (1933), 197–9.

87. NYPL, Mabel Sarton to Méta Budry, 3 June 1913.

88. NYPL.

will retain. George relates on 27 June that May "is exceptionally good, to the point that I cannot refrain sometimes from telling her so (she understands nothing though)."

Mabel on 14 June 1913 asks George to forgive her for hurting him. In English: "Is it not terrible & strange that we can each do such things & hurt each other so cruelly once more? How can *I* have said anything that could make you speak sarcastically of my love & kindness to you? How could *you* say those words in that tone?" Mabel has been struggling with premonitions for weeks, "even had the feeling of being truly a little *mad, of thinking things which were not of my real mind.*" Thoughts come to Mabel and create apprehension and foreboding: "As if something in my brain (?) were at such high tension, that a small shock, would break it, *and I should lose my whole self control.*"

Over the next weeks Mabel remains apart from George and recovers her composure. On 26 June 1913 Mabel tries out colored inks for her designs. She is relaxed, and "I feel a kind of healing each time I realize *that I no longer have responsibilities.*" Mabel thinks that this kind of rest is just what she needs, and it will help her when she comes home. On 27 June Mabel reports, in English, about her maids: Madeleine's "state of health—moral & physical—the poor girl seems really pursued by sorrows of all sorts & is so very depressed that Marthe even was afraid one night she would try to poison herself." On 29 June George reports that May spends part of the morning and afternoon in his room. George gave May some rose petals to play with, and she soiled her nightgown. He can't get the spots out. "Can you do it?" May must be vaccinated. George is working to publish no. 2 of *Isis*. The journal takes an enormous amount of time. It provides him with "enough work for *twenty* years." George thinks that Raymond Poincaré's visit to England and the French-English alliance "is really a great event." It "will be a great element of peace in the world: Because it is certain that France and England *sincerely & deeply* desire peace."

For George Sarton, the world in the large differs from the world of intimacy. He devotes hours to international peace and social justice, but he demands domestic peace as a condition of life, rather than as an endeavor of devotion. He writes to Emile Masson in May 1913 that Mabel "worries me enormously." Masson delivers good advice: "Friend, why not leave off *Isis* for Mabel, now and then? Remember Carlyle. Remember Jane Welsh, notably *Reminiscences*. What a lesson—*grande & terrible.*"[89]

89. HGS, Emile Masson to George Sarton, 17 July 1913, referring to George's letter of May. Charles Eliot Norton, *Carlyle intime: Jane Welsh Carlyle, Reminiscences*, trans. Elsie and Emile Masson (Paris, 1913). From the beginning of his correspondence with George, Emile Masson warned George about Carlyle's single-minded devotion to his work, to the detriment of his humanity. HGS, Emile Masson to George Sarton, 11 July 1905 (the correspondent is mistakenly identified as Père Yves Ernest Masson).

The Massons have just published a translation of Welsh's memoirs and, as they do with all their books, they send a copy to George. At about this time Masson writes in a similar vein about George's self-imposed lifework, an "immense, immense task, possibly too vast for one person."[90] George receives similar advice from Irénée van der Ghinst, who questions the life that he projects. George rejects the advice, although he admits that Mabel and he sometimes wonder "if I am not wrong to act in this way," and he avows that he will not succeed unless he attracts a patron.[91] The resulting pattern attaches to George for much of his married life. He worries, regrets, vocalizes, and then continues his enterprise.

On First July 1913 George writes that his evenings are completely devoted to administrative tasks: "After supper I become my own secretary." He wants no. 2 of *Isis* to appear before the end of July. In a second letter on First July, George writes that he understands Mabel's internal struggle. "Essentially, the disease that afflicts you is to have a soul and a body arranged badly. You are too fragile to keep all the worries, all the disorientations, on the inside. I have perhaps as many concerns and passions as you, but my shell is harder." Mabel does best to rest and recover her vocation of being "a human being, artist, lover, and mother." George, in charge of May during Mabel's absence, inadvertently provides an illustration of his expectation for domestic harmony. He reports that May was quiet all morning in his room. Then at noon she became excited. George finally slammed his fist on the table, causing May to cry profusely.

Mabel is in a "sad mood" on 30 July [June?] 1913. In English: "I had a miserable dream of you, where you said cruel things—all mixed up with *real* circumstances." When she tries to imagine George's face, he appears angry. She cannot put aside "the remembrance of different reproaches you have made me—all my faults pile themselves up and crush me." In the past, Mabel has been afraid of going mad and has felt so tired and mentally sluggish, she switches to French, "that I had wanted to die, and it was an active wish" deactivated only by her "vague instinct" that told her it would be a sign of weakness to do so. It was not the fact that she is physically ill or weak, or incapable of serving George and May. "*It is not being master of my own destiny*. What drives someone mad is being the slave of circumstances." But order is returning to her, slowly, as she undertakes things one at a time. Mabel wants George to understand her struggle, which she has never spoken about before. On First July Mabel

90. HGS, Emile Masson to George Sarton, 22 November [without year].

91. H. Elkhadem, "George Sarton: Ses années de formation et ses réalisation académiques à travers sa correspondance avec Irénée van der Ghinst," *Academiae Analecta, Medelingen van de Koninklijke Academie voor Wetenschappen, Letteren en Schone Kunsten van België, Klasse der Wetenschappen*, 47 (1985), 104–25, on pp. 113–14, citing George Sarton to Irénée van der Ghinst, 15 July 1913.

reports that she spoke about the Franco-English alliance at dinner, and "they," presumably Céline and Raymond, "made fun of me." On 3 July Mabel feels better. She anticipates each letter of George's with a mixture of, in English, "joy and fear and love." She wonders if she could not become George's secretary, for her attempts at designing have not been proceeding well. "I believe that it is possible that I will be *less* fatigued and depressed having worked as secretary than having done nothing except housework and caring for the child." She proposes beginning from 3 to 5 in the afternoon, 3 times per week. As is usual, she will rest after the midday meal, and they will have tea at 5. She thinks that she could turn out to be better than a professional secretary, although the disadvantage is that she hardly knows German and Italian.

In 1913, Mabel spends about 7 months in Rhode St-Genèse and Knokke without May, for, as she writes to Marthe Patyn on 20 September, May wears her out, "even when she is good."[92] On 8 July Mabel writes that Madeleine van Thorenburg has given her George's letter and journal. Mr Coppieters came and spoke to them about the revindications of Flemish nationalists. In a few days there is to be a concert by the Société chorale socialiste of Paris, attended by 2000 people from Ghent. "If the Flamingants dare to interrupt it, 'they will be buried alive,'" a friend reports.[93] Mabel designed a cushion for Céline, who has asked that she revise a number of others done previously. She dreams of tigers and lions. Méta Budry asked Céline for the source of her silk fabric, and Céline thinks not to tell her, because Méta is her competitor (notwithstanding their collaboration in 1910 on the Brussels cottage). Mabel has explained to Céline that she wants to show her own things at an upcoming fair in Leipzig, which Céline sees as a kind of betrayal. Mabel, on 10 July, passes along Dr Miele's evaluation of her health. "In any case, just now I am tempted to cry 'it is my head, my brain that is sick.'"

George's head is wrapped around writing. He publishes a long éloge of his hero, Henri Poincaré, for the Belgian astronomical journal *Ciel et terre*. George is especially taken with the French mathematician's philosophy of science in his book, *La Valeur de la science*, which George characterizes not as pragmatist but rather as idealist. Reality, for Poincaré, consists of the mathematical laws of nature recognized by people everywhere. This harmony is the only truth that we may attain. Civilizations merit con-

92. Marc Hooghe, "May Sarton (1912–1995)," *Ghendtsche Tydinghen*, *28* (1999), 54–63; the letter in NYPL.

93. The linguistic controversy between Walloons and Flemings divided the socialist movement in Ghent. Socialist leaders like Edouard Anseele viewed French as an essential tool for self-improvement, and they allied themselves with the liberal, anticlerical, francophone bourgeoisie. Catholic activists as a result found fertile ground among Flemish speakers. Strikwerda, *A House Divided*, p. 128. Mabel refers to the comments of socialist Emile Coppieters, brother of Léopoldine Dangotte.

sideration only by their science and their art. Thought, "a brief flash in a long night," exists for itself alone.[94] George, a believer in heroic revelation of beauty and truth, imagines that the advancement of civilization depends on the inspired products of thought, works of genius. From his student years, George is attracted to "responsible procreation"—meaning eugenics—and the heroic, meaning exceptional intellect. Inspired by the Fabians in England, he falls in with elitist social democrats of the time. He mostly speaks French in a Flemish community, with whose aspirations he sympathizes. And he marries an English woman. Due in part to these circumstances, he devotes an entire section of the second number of *Isis* to the topic of genius and race, trying to sort out the debt owed by exceptional talent to heredity and to environment.

In the long section, chemist Nobel-laureate and history aficionado Wilhelm Ostwald, an original member of the patronage committee for *Isis*, expands on his book *Grosse Männer* (Leipzig, 1909) to discuss the inheritance of genius, expressing uncertainty about whether Hugo de Vries's mutation or Charles Darwin's variation is a better explanation for the appearance of talent.[95] The eugenicists William Cecil Dampier-Whetham and his wife Catherine Durning Dampier-Whetham, authors of a history of science emphasizing that "creation is usually the outcome of one race working out its own salvation," elaborate the notion that natural knowledge "brings with it that extension of man's dominion over the forces of Nature which Francis Bacon foresaw and desired for the human race, as distinguished from technical craftsmanship and emotional expression." This science "has been almost overwhelmingly the achievement of men of Northern descent." The Dampier-Whethams emphasize that purity of blood-line predisposes a nation to excellence, while times "of the biological intermingling of races...are seldom periods of creative power and high intellectual achievement."[96]

George wonders whether class differences within a nation are more extreme than differences between intellectuals across nations. He offers that civilization transcends nations, existing "wherever there is knowledge and light, in the country and in the most distant provinces as well as in the largest cities." In effect, he affirms that Wondelgem is as civilized as Berlin or London. In reviewing a note in the *Bulletin* of the Solvay

94. George Sarton, "Henri Poincaré (1854–1912)," *Ciel et terre, 34* (1913), 1–11, 37–48, on pp. 41–2.

95. Wilhelm Ostwald, "Genie und Vererbung," *Isis, 1* (1913), 208–14.

96. W. C. Dampier-Whetham and C. D. Dampier-Whetham, "Three English Men of Science," *Isis, 1* (1913), 215–18; their book: Dampier-Whetham and Dampier-Whetham, *Science and the Human Mind: A Critical and Historical Account of the Development of Natural Knowledge* (London, 1912), sympathetically reviewed by George in *Isis, 1* (1913), 125–32.

Institute, George asks about the relation of racial purity and racial fusion in the development of science.[97] George discusses the October 1912 issue of *La Revue anthropologique*, devoted to ethnic expansion, which concludes that no generalizations about race are possible, given the lack of reliable data.[98] He goes further in 1913, rejecting Houston Stewart Chamberlain's racist method in a scathing review of Chamberlain's *Goethe* (Munich, 1912): "I detest his intellectual method, which he uses once again in this monstrous work." Chamberlain's book is digressive and pompous, and its siren call will not appeal to readers of *Isis*.[99]

George sets out his own thoughts on nature and nurture in the article mentioned in his letter of 26 June 1913 to Mabel, which he publishes, after the article of the Dampier-Whethams, as a meditation on promoting conditions for the creation of geniuses. George is inspired by Ralph Waldo Emerson's view that the entire history of humanity is summarized in the biography of a few great personalities. He investigates the nature of genius, following works by criminologist Cesare Lombroso, mathematician Henri Fehr, anthropometrician Francis Galton, biologist Alphonse de Candolle, and his admired Thomas Carlyle. Great men accomplish their work through great effort. George thinks that genius "certainly in large measure" depends on race—George's name for the atoms of nationality that compose the international fabric of the world. To examine the question scientifically, he considers the debate between Neo-Darwinians and Neo-Lamarckians. The Neo-Darwinians "ascribe more importance to the race, to the connection of the individual in his milieu." The Neo-Lamarckians, who are materialist, give determining influence to the milieu itself. George then devotes a great deal of space to reviewing current notions of heredity, concluding against the inheritance of acquired characteristics and against any "creative" direction on the part of the environment. We may count on our fingers the number of French-language biologists who rejected Lamarck in favor of Darwin.[100] Here George proclaims himself a Neo-Darwinian. His concern, however, remains racial adaptation in changed environments, and for this he looks to physicist-anthropologist Franz Boas's study of immigrants in the melting-pot of American New York. How does all this relate to genius? Exceptional people will always arise randomly, George asserts. Somehow genius emerges in one place or another, largely, in his view, from the books that people

97. George Sarton, "Le Rôle de la fusion des races dans l'origine du christianisme," *Isis, 1* (1913), 104–6.

98. George Sarton, "Croisements ethniques," *Isis, 1* (1913), 106–7.

99. George Sarton, "Houston Stewart Chamberlain, *Goethe*," *Isis, 1* (1913), 519–21.

100. Robert E. Stebbins, "France," in *The Comparative Reception of Darwinism*, ed. Thomas Glick (Austin, 1972), pp. 117–63, p. 162: "Félix Le Dantec, perhaps the leading French writer on transformism from 1895 to 1910, followed Lamarckian lines."

read. The habits of reading, "the way in which they choose their food for thought, is infinitely complex and random." Suddenly, in a lightning-bolt, the reader's vision is transformed. George concludes: This notion of genius "fits well with the individualistic and heroic concept for which I have, in some measure, provided a scientific justification." It is also a justification for George's own, long program of reading, which after 11 years brings him, at age 29, to create *Isis*.[101]

Another set of publications in 1913 concerns scientific internationalism and bibliography. As a student, George chafed under the inadequate catalogue of the library at the University of Ghent. In 1908 he wrote to the authorities about the absence of a catalogue, following a letter on the same topic 6 years previously.[102] Paul Thomas, in charge of the library, agreed with George, but Thomas suggested that the only way to achieve the goal would be through a private donation.[103] The question of cataloguing returns in 1913, since half of *Isis* deals with reviews and bibliographical notices. George contributes programmatic thoughts to the periodical *La Vie internationale*, edited by Henri Lafontaine at Ernest Solvay's International Institute of Bibliography and published by the Union of International Associations with the financial support of the Carnegie Endowment for International Peace. The venue derives from George's admiration for the social policies promoted by Solvay and especially for the Solvay Institute's commitment to pacifism. In George's view, Lafontaine's companion annual publication, which gives a list of international associations and also features the work of the union to promote them, "provides the best demonstration of practical internationalism."[104]

George's first publication for Lafontaine relates history of science, as he sees it, to the promotion of internationalism. He emphasizes that the new career of international organizer first made an appearance in science, where cooperation is required for understanding the natural world. Scientists are the architects of internationalism and, moreover, the elucidators of worldviews: The science of a historical period largely determines its "ideology," whether consciously or unconsciously. With time, scientific apprehensions are increasingly available to educated citizens, becoming part of their common heritage. For this reason, the

---

101. George Sarton, "Comment augmenter le rendement intellectual de l'humanité?" *Isis, 1* (1913), 219–42, 416–73.

102. Agenda; in Marc De Mey, "Sarton's Earliest Ambitions at the University of Ghent," *Isis, 75* (1984), 39–45, George's letter of 4 November 1902 to Ferdinand Vander Haeghen.

103. HGS, Paul Thomas to George Sarton, 14 March 1908.

104. George Sarton, "Annuaire de la vie internationale," *Isis, 1* (1913), 289–90. George notes Henri Lafontaine's work as a pacifist and socialist, in connection with the Committee for the Defense of National Interests and International Civilization, in his agenda on 17 May 1905.

history of science is the history of human progress. It supplies the idealistic, motive force for studying the human past. "Humanity at large invents and moves forward." And simultaneous discovery (the investigation of which will drive forward the great innovators in history of science during the 1950s) shows that there is a partial, social determinism of new ideas; in history of technology, we can see an evolution of objects similar to evolution in the natural world. In sum, the task of history of science fits well with the internationalist task of Lafontaine's institute, for it makes no sense in science to speak about a "national culture." There is neither German nor French science, and there is no special style for science written in one or another language: "Style, in its purest form [dans son acception la plus haute] is independent of the transmitting language." He adapts Buffon: "Style is not a literary quality, it is a quality of the soul."[105] Here we read the political justification for the change in the subtitle of *Isis*, made to include the "history and the organization of science."

George Sarton delivers a pacifist and internationalist message to Lafontaine's readers. Science belongs to no nation. It is open to everyone. George maintains this view through the nationalist rhetoric of the First World War, the ostracism of German and Austrian scientists by the victors in the 1920s, and the imposition of political ideologies upon scientists in Germany and the Soviet Union during the 1930s. Sarton's view evaporated during the last third of the 20th century, when scholars with "national science" as a field of specialization make large claims for how special patterns of education and funding determined the course of scientific research in various nations. The matter remains controversial, but most historians of science would affirm, along with Anton Chekhov, that there are no national multiplication tables.[106]

In a second article for *La Vie internationale* (written according to a footnote in January 1914) George examines the world of journalism, in which the daily press makes no attempt at impartiality. George calls for a moral force in journalism, so that the press may be an agent of truth and peace. He dreams about a press that would provide reliable information about the economic, moral, and social state of the world. George reviews a proposal by pacifist Alfred H. Fried for an international telegraphic

---

105. George Sarton, "L'Histoire de la science et l'organisation internationale," *La Vie internationale*, 4 (1913), 27–40, on p. 37. Lewis Pyenson and Susan Sheets-Pyenson, *Servants of Nature: A History of Scientific Institutions, Enterprises and Sensibilities* (New York, 1999), pp. 14–15 on the importance of simultaneous discovery for Robert K. Merton, Derek J. de Solla Price, and Thomas S. Kuhn in the 1950s.

106. Lewis Pyenson, "National Culture and Styles," in *Oxford Companion to the History of Modern Science*, ed. John Heilbron (Oxford, 2003), pp. 557–9; Pyenson, "An End to National Science: The Meaning and the Extension of Local Knowledge," *History of Science*, 40 (2002), 251–90.

agency, which could act to reduce misunderstandings in international conflicts.[107] In this piece, George penetrates into the nature of popular writing. His analysis will serve him during the war, when he finds work as an occasional journalist and when he writes articles for popular literary periodicals.

George's noble sentiments about publishing find a practical referent. From the beginning of *Isis*, George the organizer elaborated a plan to produce an encyclopedia on cards from the book reviews, printed only on the recto.[108] He offers the product in May 1914. The resulting card index will be a boon to scholars and an enterprise of "*idealist propaganda*, a work of high solidarity, seeking to respond to both human intelligence and the human soul."[109] The project requires setting a number of conventions. First, the printed text of the encyclopedia must conform to certain card dimensions. As he notes in an unpublished essay, there are four standards in circulation: the one used for international postcards; the one chosen by the Library of Congress; and 2 idiosyncratic possibilities derived by Wilhelm Ostwald, who in 1911 founded an organization for undertaking projects such as interinstitutional bibliographies. The best idea, George thinks, is to arrange things for index cards of the first 2 sizes.[110] Second, there is the general question of organizing the encyclopedia by subject, and this relates to a taxonomy for the sciences. In 1913 George engages Wilhelm Ostwald and Henri Lafontaine on the question. Ostwald is a confirmed believer in the Comtean hierarchy of knowledge and a skeptic of the procrustean and culturally limited decimal system originally devised by Melvil Dewey at Amherst College in the 1870s. Auguste Comte's hierarchy of the sciences would be valid into the foreseeable future, Ostwald believes.[111] It is an open question if George could have brought the project to fruition in the absence of a European war, but bibliographer Henri Stein writes in 1917: "We can scarcely imagine what end is pursued in reaching for a domain so large and so poorly defined, and how it will

107. George Sarton, "L'Organisation scientifique du journalisme," *La Vie internationale*, *5* (1913), 391–428, on pp. 401, 421.

108. George Sarton, "Bibliographie analytique des publications relatives à l'histoire de la science parues depuis le 1er janvier 1912," *Isis*, *1* (1913), 136–90, on pp. 141.

109. George Sarton, *L'Encyclopédie sur fiches* [Brussels, 1914], 2-pp. broadside tipped into bound copies of *Isis*, described without extravagant rhetoric in *Isis*, *2* (1914), 249–50. In volume 2, each entry in the Critical Bibliography is marked by the word "ISIS," flush right, to facilitate the scheme.

110. YMS, George Sarton, "L'Uniformisation des formats" (1914). Ostwald's organization is titled, Die Brücke, Zur Organisierung der geistigen Arbeit.

111. HGS, Wilhelm Ostwald to George Sarton, 12 December 1913; Henri Lafontaine to George Sarton, 23 December 1913. We shall see that George visits Melvil Dewey during the summer of 1915. NYPL, George Sarton to Mabel Sarton, 27 July 1915.

be possible to carry out this extraordinary program."[112] In the middle 1920s, the John Crerar Library in Chicago and the Library of Congress reproduce some of the content of *Isis* on index cards, but when George is offered the possibility of resuscitating the card scheme in 1935, he resolutely refuses to consider it.[113]

George, on 7 July 1913, hopes that Mabel is feeling better. He acquired a history of Egyptian civilization and a governmental study of daylight-saving time. George is considering a typewriter. He wonders if Raymond thinks that the machine can be repaired, "an important point." George guesses that a typewriter would last 10 years. So he favors a durable and more expensive one, say, for F 600 rather than for F 200. On 8 July George sends along a prescription for phosphorus from Adolphe Miele to help Mabel. May walks quite well by herself. On 9 July 1913, George types a letter to Mabel on a Remington, delivered to Wondelgem for his trial. The machine costs F 600 and is not as pretty as Raymond's, although it is more durable. The keyboard is rather different from the one at Rhode that he is accustomed to using. Mabel, replying to the letter, thinks that F 600 is rather steep.

On 11 July, George, still typing, says that he has reread Dr Miele's letter, which attributes Mabel's disease to "a lack of the internal secretion of an organ." Does this not indicate a serum treatment of some kind? He asks Mabel to question Dr Miele when she sees him next. He will be making preserves from cherries and, later, strawberries. He played with May in the garden, and she is covered with mosquito bites. On 12 July Mabel offers that she "could not reasonably act as your regular secretary," but she is making progress with Raymond's typewriter, and her skills might really help George. She thinks that her indisposition is due to her period, which is now over—menstruation is frequently ascribed by Mabel as a cause for feeling ill. Mabel is interested to see what will result from the Leipzig fair where she exhibited, although she is not optimistic. Later on the 12th, Mabel types her first letter to George. She is glad that he finds his work rewarding, but, in English: "don't overwork your thinking-machine or its creative power may break down." On 14 July George types to "mon cher petit coq" that he has read John McFarland Kennedy's *Satakas, or*

112. Unsigned note in *Bibliographie moderne, 18* (1916/17), 109, transcribed and sent to George Sarton by Arnold C. Klebs as an appendix to a letter of Easter Sunday 1920, in HGS. Klebs attributes the note to Stein, the journal's editor. Klebs, who admires the plan, adds: "The work involved would hardly pay."

113. George Sarton, "Index Cards for *Isis*," *Isis*, *6* (1924), 44. HGS, Solomon Gandz to George Sarton, 22 February 1935: "Have you ever given thought to the plan for a Union Catalog on cards for the special collection of History of Science, which would be subdivided according to subjects and authors/ It seems to me that this would be the logical continuation of your *Introduction* and *Isis*, which, in a sense, laid the foundation for the plan."

Wise Sayings of Bhartriharis (London, 1913), which helps him integrate the austerity of Indian thought with its exuberant sensuality. The book reawakens George's feelings for the Orient, "but this time just as my desire awakened and roared, I felt glad that I was not born there."[114] George went into Brussels in the morning to look over another brand of typewriter. George knows that typewriting is the way of the future, for in May 1914 he advises contributors to *Isis* about how their typewritten manuscripts should be submitted (big margins and wide spacing) and urges them not to give in to typists, whose preference is for long lines to avoid changing sheets frequently.[115] The interest is ephemeral. George writes with a pen for the rest of his life.

Mabel notes on 16 July 1913 that George, when visiting Brussels, was very close to her at Rhode, but that he chose not to see her, returning immediately to May. Mabel is interested to read the *Satakas*. She imagines that the topic could sustain a substantial work of scholarship in India, but that Hindus would not undertake it: "Hindus in general are not inclined to that, but rather to meditation, & rather *passive* work." Mabel returns to Wondelgem. By late August she is depressed again. She writes in English on 26 August:

> Husband, I wish I could give you the key to my heart & my brains so you would see that if I do & say wrong things I am bitterly punished. I see quite well & clearly that I am *very* irritable & impatient now, & I try very hard not to be so (it is harder than you can guess perhaps).
>
> I feel it is after all rather foolish to want to explain to you. What is there to explain? I am so tired I can not even think of it all. There is a deep, gnawing pain in me & I am tired, tired, tired & I want to *rest*, & I can't.

Having failed to land a teaching position, George recognizes the need for money. He writes in his diary for First July 1913:

> In any case I *must* not abstain from involvement with the material, financial results of my work. It is sad, but at the same time it is *good* and *right*, and I should be glad for it. Life is beginning. Up to now, material concerns have not concerned me, often to my shame. The privilege that I have enjoyed weighed on me as an insult.

George summarizes the past year in his diary on 31 August 1913. About one-third of his time has been taken with administrative tasks, "not requiring great intellectual effort, I mean setting up *Isis*. The realization of

---

114. Reviewed by P. Masson-Oursel in *Isis, 1* (1913), 512–13, who finds Bhartriharis, in certain aspects, a moralist comparable to Horace.

115. George Sarton, "Notes sur la revue *Isis*," *Isis, 2* (1914), 162–3.

this project has exceeded my hopes. The journal is truly very good and will be useful. If I can keep it going—I have only 85 subscribers, and I need at least 200 to cover my expenses!" (George knows that the *Bulletin* of Emile Waxweiler's sociological institute in Brussels has about 800 subscribers.[116]) He has not yet made money, but he thinks that he can do so by writing. "I don't want to do it yet, for I am too afraid of prostituting myself to write in this way, *to* earn money."

Late in summer 1913, Mabel still lives apart from George and May. She writes on 13 September, in English: "The house did not seem too empty & sad when you got home this evening. It is not the same as if there were no little May. I feel happier when I leave you *because she is there*. Though she *is* only such an 'incomplete' little person, still she prevents the house from feeling too quiet!" On 15 September 1913 Mabel watches the Limbosches play tennis and wonders if she shall ever be strong enough again to do it with pleasure. George writes to "petit coq" on 14 September that May asks for her. May "has become truly a little civilized being." George played with her in the garden and showed her his book on animals for children. May was able to distinguish the duck from other birds. George offers that Raymond is crazy to ride a motorcycle, but George nevertheless suggests that Mabel travel with Raymond on it to the Palais du Cinquantenaire to see the head of Isis in the library of the Egyptian room. George asks Mabel to sketch the head, for he will need it in January. On 16 September, George reports how he cured May of cholic. May tried not to bother George with the affliction, and George was touched. In the evening he gave her stomach sharp blows with his fist, solving the problem. He has taught May to distinguish the sun from the moon.

Céline is Mabel's adoptive sister, and Mabel feels at home in the Limbosch household. George, however, feels threatened by his rich friends and resents them. He writes on 19 October 1913 that he is against the plan to move to Rhode, near Brussels. It will not allow Mabel the opportunity to rest, and it will probably result in estranging Mabel from the Limbosches. He thinks in any case that Mabel is more tired when she visits Rhode than when she is home at Wondelgem. Furthermore, their maid would probably not follow them, and she has been very good with May. George says all this even though he has made arrangements for the move, obtaining an office at the Solvay Institute. He urges Mabel to think about the more distant future: "You cannot remain indefinitely at Rhode; you must come home one day to your family [foyer conjugal]." George has abstracted from discussion his need for Mabel, for if sentiment comes into the picture, "would it not eventually have you think that you have become indifferent to me?"

116. HGS, Emile Waxweiler, to George Sarton, 28 October 1912.

George's conscience, Emile Masson, wonders on 22 November 1913 if George's task is too great for one person, placing his family at risk. Again he recalls Carlyle:

> Ô my friend, think of Carlyle! What a lesson, what a terrible lesson! As far as I am concerned, when I am tempted, I think about him, about his terrible *Reminiscences*, and I become calm. I walk over to the next room, that of Mother and the children. Five minutes are enough to straighten me out....We continue to think that despite everything the world (in particular the world of Science and even of Literature) is moonshine, and that *real life* is, above all serious, fecund, *divine*, provided that it is lived with piety, that is, with confidence and joy. From this comes our age-old scorn for Writing, for Paper, for the *world of paper.*

It increasingly seems as if humanity is doomed to place ink on paper. Masson implies that George should consider replacing paper with real life.[117]

The sentiment is shared by French writer Romain Rolland, who is struck by George's literary piece on Malwida Rivalier von Meysenbug in the Brussels periodical, *Flamberge*. Meysenbug, who died in 1903, was a radical feminist who encouraged Rolland's earliest literary efforts. In his article, George begins by commenting on the essential goodness of Rolland's work. Rolland's subjects—Beethoven and Michelangelo—have unhappy and turbulent lives. Their goodness is purified by suffering. Their mettle is hardened by misfortune. Then he invokes Rolland's angel, Malwida von Meysenbug, born in 1817, whom Rolland met when, as a young man, he studied in Rome, and with whom he became close friends. At the end of her life, Meysenbug continued to champion radical causes, notably the "heroic defense of the Boers."[118] George attracts the attention of the master. Romain Rolland writes to his friend Jean-Richard Bloch that the piece, signed Dominique de Bray, is really written by George Sarton, "one of the most remarkable minds of the young generation." Romain Rolland has known George for 15 years. "He is essentially a scientist; but he is interested in everything with a magnificent passion. I was afraid lately that he peaked too soon. To hold up, he has to be strong."[119]

117. HGS, Emile Masson to George Sarton, 22 November [1913], although the year may be 1912.

118. Dominique de Bray, pseud. George Sarton, "Romain Rolland et Malwida von Meysenbug," *Flamberge*, no. 11 (1913), 503–9, George's article dated November 1912. The issue is devoted to Romain Rolland. A copy of this issue is available in the Bibliothèque nationale de France (Tolbiac, Rez-de-jardin), Paris. In his book, *L'Enclos* (Antwerp, 1910), Raymond Limbosch dedicates a poem to Malwida von Meysenbug.

119. Romain Rolland to Jean-Richard Bloch, 17 December 1913, in *Deux hommes se rencontrent: Correspondance entre Jean-Richard Bloch et Romain Rolland (1910–1918)*, ed. Mme Jean-Richard Bloch and Mme Romain Rolland (Paris, 1964), p. 223. Marie-Brunette Spire kindly alerted me to this passage.

While trying to keep his family intact, George seeks outside funding for *Isis*. After the appearance of the first number, Emile Waxweiler delivers a small grant from the Solvay Institute, and he urges George to contact Henri Lafontaine about support from America.[120] In December 1913 George applies for the Prix Binoux of the Paris Academy of Sciences; he also writes to the Carnegie Endowment, to the Carnegie Institution in Washington, and to the Solvay Institute. The applications eventually bear fruit: He will receive the Prix Binoux in 1915, the Carnegie Institution underwrites his career in America, and the Solvay Institute provides him with an office in 1919.[121] George applies to the government for a grant, and he lines up Henri Pirenne as a supporter.[122] The amount is far below his needs.[123]

Mabel writes on 5 January 1914, where she is at Rhode with May, about her marriage. George seems distant:

> I had a sad feeling last night because it seemed as if you *were already gone*—your *spirit*—as if you had not time nor energy to give me. I had all sorts of things I wanted to tell you, & I wanted you to put your arms round me *tight* just a moment & you just said good-night in a great hurry & somehow then I *dare not* come to you.

Mabel wonders if she is a bit deranged because of the opozones. She hopes that 1914 will be better for them. "Last year for *me* has left a feeling of disgrace & failure & emptiness." Mabel writes without salutation on 9 January:

> My function and my social responsibility? And don't you even consider *at all* my work? And you leave completely aside that I consider it as great a duty to use the talent that I have to raise my child? Oh, not that I see myself as a "great artist"—far from it! I feel humility about my work, and this never goes away. But I am *committed* to give what I have, as little as it may be. Do you believe that I am blinded by the enthusiasm of Céline? But I have only to think about her coldness of execution....

120. HGS, Emile Waxweiler to George Sarton, 30 March 1912, 17 January 1913 for the grant of F 300.

121. George previously sent the Carnegie Institution an announcement about *Isis*. Note written on HGS, David Eugene Smith to George Sarton, 18 February 1913. The Belgian consul in Edinburgh advised George that the Carnegie Institution is the best place to look for American support, and that the Carnegie trustee to enlist for *Isis* is Andrew Dickson White at Cornell University. HGS, Charles Saroléa to George Sarton, 20 November 1912.

122. HGS, Henri Pirenne to George Sarton, 22 February 1914.

123. CIW, George Sarton to Robert Simpson Woodward, 26 September 1914, for mention of an "annual grant (small indeed)" from the Belgian government; HGS, Emile Masson to George Sarton, 21 July 1914, for allusion to a subscription to *Isis* by Albert I, King of the Belgians.

Mabel reminds George about how Mme Dangotte supported her. Mabel thinks that she is less talented as a mother than as an artist. "In any case the duties of being a mother exhaust me infinitely more than the other duties." Still, Mabel does not want to see herself as weaker than she is.

Mabel details her day. Breakfast at 8:30, chores from 9 to 11. Walk or work in the garden with May, 11 to 12:30. Dinner at 12:30. Rest from 1:30 to 3. Mabel's artistic work from 3 to 5, made possible by the afternoon maid. Tea at 5 and May's bath and meal, then needlework from 6 to 7:30 with May again. May's bedtime 7:30 to 8, with her meal. Evening meal 8 to 8:30. Then Mabel can read and write. There are visits and shopping. May can play by herself a part of the time, but Mabel must help her on the potty, and when May does regress to soiling herself, it provides Mabel with even more work. George writes that Mabel lacks discipline. But Mabel thinks that it is not merely that—"for you it is infinitely simpler." Mabel, in an undated letter, gives the cost of servants: F 1.50 for a cleaning woman one day per week; a young girl in the morning would go for F 1 daily (including F 0.20 for the tram).

George writes on 9 January 1914 that Mabel's recent letters have upset him. George repeats that he is willing to help Mabel raise May. As long as they are without servants, he will put in several hours a day at it. He thinks that raising young children is mostly for the mother. His "isolation" is normal: He does what any man does who earns his bread and carries out his calling. He is following his destiny. His task is great, and it requires much work. Mabel knows all this. It seems to him that he cannot be distant, on the one hand, and overbearing, on the other hand, as Mabel claims. He thinks that if he is guilty of anything, "it is weakness rather than tyranny." He will help Mabel realize her vocation, but for that to happen she must "rest, cure herself, and *discipline herself*." If 1913 has been a bad year for Mabel, it is because she was sick, not because of anything George did. "I believe that the causes preventing the development of your personality are *internal causes*. You will be free only to the extent that you are disciplined: I hope that you don't come to this realization too late!" George does not impose his solitude on Mabel: The choice to live in Wondelgem is more hers than his. He asks her to return home, and he gives her complete liberty to travel, with or without May. "I will take good care of our home, but if you want it to remain bright and warm, don't desert it too often." George concludes by hugging both Mabel and May "the other tyrant." George writes in another letter on the same day that he never denied that Mabel's work "is also a large part of your reason for being," but he thinks that her health is the important thing. And Mabel's maternal responsibilities are most important now. "How can you think and write to me of such things. It's truly as if you had to defend yourself against me, as if I had become a dangerous character to guard against."

Then he gives expression to his suspicions about the Limbosches. "It's really as if other people have turned you against me." George is laboring hard, all alone, and he endures Mabel's criticism. Why do her comments always materialize when George is not around? "Why write to me about all this when it would be so much simpler *to talk about it*?"

Mabel writes to George on 11 January 1914 that she wandered around Brussels, aimlessly, because she could not stay at Rhode. She worries that May will catch cold again, for the house in Rhode is heated only in the kitchen and great room. She is sending along May's laundry and books, which George can place "on my bed (in *our* room)." Here we see that George and Mabel did not share a bed. George's last letter shows that he has understood nothing of what Mabel has said. If Mabel has been in George's shadow, it is because she allowed herself to be placed in this situation. She has been happy to help him achieve his work, and he sees her help as a sacrifice. Mabel needs to speak with George, but she always hesitates to do so: "When I feel that you are really busy, I am paralyzed and I put off talking with you. Of course I know that you told me that I can when it is necessary, but I ask myself always is it truly necessary." Now George says that he doesn't want to ask for her help because "it would have the worst effect on my personality!!" George's words have settled profoundly on Mabel's heart and soul. On 14 January Mabel is very much heartened by George's letter. Mabel wants Adolphe Miele to see if radiography helps explain why she is so thin.

George, on 13 January, feels better after reading Mabel's letter. He wants them to talk about their problems in the future, rather than to write about them. He thinks that it is easier for him to be bound by marriage than for Mabel, and he sees her "revolting" against this bond. When she returns, he hopes they will discuss her criticisms in English. But in the middle of April 1914, Mabel is again at Rhode with May. George misses them, on 10 April: "I am much more affected than last year. Perhaps it is simply that little May has become a much more bustling, boisterous, and demanding person, whose presence and absence are more strongly felt." George has used the silence in the house to arrange papers in his study, which have lain untouched since their marriage. The shower was installed that morning; he will insulate the tank and the pipes later. Mabel writes on 17 April 1914 that May is afflicted with inflamed ear tubes. Mabel is taking care of May by herself, and so she has no free time.

Mabel thanks Méta Budry on 3 May 1914 for the gift of dresses for May, who is chattering away, adding Flemish diminutives to French words, mixing in English ones as well. They plan to travel to England in June. Mabel is designing furniture and decorations for a model cottage in the Brussels salon in June, which will have a section in *art décoratif* for the first time. Mabel has chosen the woods and the yellow silk wall covering. The grey carpet is divided into squares by a black-and-white design with a bright bouquet in each square. Her garden is a "fairy-tale" of flowers. Mr Dangotte has married Mme Van der Noot. The event is without importance, but "they are arranging things better than ever to disinherit Céline." George is growing fatter and working harder than ever. "I often worry that he will exhaust himself."[124]

The family is apart again in early June 1914. Mabel is in Rhode without May. George writes to Rhode on 4 June that he is nearly finished with no. 5 of *Isis*. His time is spent between history of science and Peter Kropotkin's memoirs, *Autour d'une vie*, which greatly impress him: The book is "at once melancholic and joyous, sad and heroic." He looks forward to the day, 15 years in the future, when he may read the book to May. George was not sure about their intended move to the Limbosch house in Rhode in the winter, but after having seen it "I no longer hesitate an instant: It is still much too humid, and it would be unwise for us to spend the bad season there." In a postscript:

> The only reason that we would have to live there this winter would be to help R. & C. I have thought about this side of the question. It is truly impossible. I am ready to risk my health for *their* health, but I absolutely refuse to do it for their house. And when I say my health, I mean also that of little May and you. If you must be in Brussels this winter, there is no lack of other places to stay.

On 7 June George writes that he had to spend as much time correcting the proofs of his analytical bibliography for *Isis* as he spent writing it, and he is furious with the printer. "I am extremely exhausted. And yet I must carry out an infinite number of tasks. And for what result?" On 17 June, he hopes that the Salon has not been too upsetting for Mabel. "Little May is very good and proper. (Are you really sure that she is not always like this? It is at least possible?)" On 19 June, his work is going well, "Little May is very good. I care for her like a mother." George is glad that Mabel is working energetically. Mabel is lucky that the exhibition has been postponed due to the international situation. He really wants to see her installation of furniture. He is glad that a number of artists like her work.

Notwithstanding the impending war, George and Mabel go to England with May in the middle of July 1914, George to London, mother and child to Ipswich. George, on 11 July, wonders what Mabel's mother thinks of *Isis*. (Mabel and her mother are barely on speaking terms, she writes to George on 16 July.) He is going to work in the India Office, and to the British Museum or National Portrait Gallery in the afternoon. To May in English: "Daddy expects that she will do her duty." George

124. NYPL.

meets again with Ananda Kentish Coomaraswamy, who shall become one of George's closest advisers on Asian topics. Coomaraswamy, the son of a Tamil father and English mother, was first director of the Geological Survey of Ceylon. By the eve of the First World War, he had become an Indian nationalist and his interest turned toward Indian philosophy and art, which he collected. George indicates that Miss D. M. Larcher will show Coomaraswamy's miniature embroidery to Mabel. Coomaraswamy's wife Alice Richardson (known as Ratan Devi), whom he married in 1912, is English, and she has written a book on Hindu music with a preface by Rabindranath Tagore.[125] For no. 6 of *Isis*, which George wants to devote exclusively to science in India, he prepares a review of Coomaraswamy's book about Indian arts and artisans. The review, finally appearing in 1919, serves to reprimand Western ignorance—its "immense egoism"—of Eastern civilizations. Indian art is situated at a pole removed from Western, individualistic art; but Indian crafts and decorative arts draw strength from the caste system, which keeps trades within families. George draws lessons, notably that a traditional genre of art contains the seeds of its own decline, for it has no strength to resist foreign corruption.[126] The war aborts George's plans. Had it not intervened, it is possible that he would have devoted himself to Sanscrit and, perhaps, the art of South Asia, instead of to Arabic and Islam.[127]

One evening in London, George visits John de la Vallette, a Dutch scholar of Hindu art. On 14 July 1914, George reports on De la Vallette, who is really Javan; his father was governor of Java. He is agent for a large navigation company in London. He and his wife dressed formally to receive George; "I wore my suit of blue wool which, said in passing, requires urgent repairs; furthermore, in the heat, I had left my handkerchiefs at the hotel. I didn't feel embarrassed for a minute." On the 12th, George hopes that Mabel can rest by herself for part of the day, if her mother and May get along well. George's experience with the India Office "convinces me that the English bureaucracy is not worth more than that of the Continent." The Office receives all Hindu publications, but the people there can't be bothered to unwrap them quickly. The most recent volumes unwrapped are from 1911.

---

125. *Thirty Songs from the Panjab and Kashmir*, recorded by Ratan Devi with introduction and translations by Ananda K. Coomaraswamy and a foreword by Rabindranath Tagore (London, 1913). On Coomaraswamy as a supporter of the arts & crafts movement and on the inspiration of his first wife Ethel Partridge (later Ethel Mairet), an expert on spinning and weaving, for Mohandas Gandhi: Tanya Harrod, *The Crafts in Britain in the 20th Century* (New Haven, 1999), pp. 177–9.

126. George Sarton, "Ananda K. Coomaraswamy, *The Arts and Crafts of India and Ceylon*," *Isis*, *2* (1919), 404–7.

127. George Sarton, "Why *Isis*?" *Isis*, *44* (1953), 232–42, on p. 242.

George recalls in his biography of Mabel that they "were living in a kind of dreamland" in Wondelgem, for there was not enough money to sustain their home along with his projects. When he compliments Raymond Limbosch on the appearance of his book of poems, *Les Faunesques*, George sends along a balance sheet for *Isis*. It has cost him F 7489.63, with F 4116.97 for printing and F 1554.11 for postage. Taking into account his subsidies of F 1300 (probably from the Belgian government) and subscriptions of F 2654.18, he has lost F 3336.95. But since he forecast a deficit of F 4000, he calculates that he is doing better than expected.[128] Expected perhaps, but nonetheless ruinous.

Dreams are ambiguous, as even George, a sceptic of psychoanalysis, knows. The tocsin of August 1914 wakes the Sartons. George is 30 years old, and Mabel is 6 years his senior. In common with other intellectuals of their generation, they have a wistful apprehension of the world. The writers and artists they admire are uncertain about the Industrial Age, seeking to recover parts of the human spirit excluded—they would say left behind—by impersonal machines and exploitive social structures. The idealism receives a rude shock from the horror of the first great capitalist war.

128 Van Oye, p. 59, George Sarton to Raymond Limbosch, 15 March 1914.

CHAPTER 8

Conflagration

On the morning of 4 August 1914, when German troops invade Belgium, the Sartons are in Wondelgem. The violation of Belgian neutrality places a heavy strain on their pacificist beliefs. In an undated letter from Brussels, George writes that he does not think that he will be drafted for the army, but he will go if called. The central office of the Red Cross is "absolutely in ruins." On 5 August George writes that he went to the Institut de physiologie at Brussels, which asked him to return on Saturday. The Red Cross wants only professional healers. On 7 August, George is still trying to find useful work in Brussels, "sad atmosphere here."

Mabel writes to George on 8 August that Raymond Limbosch spent part of the night transporting wounded soldiers. George might help Céline and Raymond; they have 120 people working for them. On 9 August, George replies that there is hardly any point in helping the Limbosches: They spent all night in their automobile to ferry only 2 wounded, while George took care of only one when he was on guard duty. He thinks that half the people working with ambulances and nursing stations should go back to their usual employment ("supposing that these are useful"). He writes around this time that most of the men he sees in Brussels are exhausted, rather than wounded. He proposes to return to Wondelgem on Monday and resume his own work. Travel is difficult. "I have already been accused of espionage several times."

On 9 or 10 August Mabel reports that she has volunteered at the local ambulance, and asked to do anything possible. She indicated that they could house 6 wounded people. She is collecting everything they do not absolutely need—linen and clothes—for the wounded. She has been to the Institut de physiologie, and she is glad that she does not have to work there with some people she knows, including Mr Dangotte: "atmosphere of bourgeois snobs too content with themselves." Adolphe Miele is in Brussels. His wife and children are "without money and without news, and she is half crazy." Mabel has ordered more opozones, but she does not know if she can obtain them. Although she has taken 3 injections daily, she thinks that she doesn't need them any more. May has been very difficult. Whenever the doorbell rings May asks if it is George.

At about this time Mabel informs Marthe Patyn that they have almost no money. They are thinking of closing the house and going to live with the Limbosches. "My salon for which I worked so hard seems in vain."[1] Céline writes to Mabel on 10 August that she is sending F 200.[2] On 10 August Mabel writes to George in English that "my 'outer' calm & energy have come back," and that she is upset only about little things. She went to the post office for the F 200, for which she received an invoice in the mail, but she must still wait a day. Mabel writes to Méta Budry on 13 August 1914, agreeing to postpone Budry's visit to Wondelgem. Life is returning to normal since the Belgians are defending their homeland; "one is becoming used to *everything*, even this horrible state of waiting."[3]

In his diary on 5 September 1914 George observes: "What unheard of madness inspired the handful of men who were the only artisans of this a thousand-times damned war?" He thinks it happened because of defects in political organization. "May this childhood disease be the last, and Democracy emerge from the test revivified and strengthened." He continues:

> At this time work is very painful for me because my thought is continually distracted by the obsession of this war, this universal madness. It is as if suddenly all the machines in a workshop were turned upside down, against their correct use. And in fact civilization now turns against itself, and the good becomes bad, the bad good. Where will this lead us? In any case, even though brutality is master of the moment, I continue in *wanting* to think in a disinterested way and believe that *it is that which is above all important*: thought, *sympathy,* ahimsa!...I must continue to work, to think freely, to be good, to be humble. That is now a duty more necessary than ever, a sacred duty that can be accomplished.[4]

The past year has been devoted almost entirely to *Isis*. "The work does not satisfy me: without speaking about material results, which are rather discouraging, the intellectual results are also not good enough. The periodical is not yet acceptable, far from it; and my personal work has not advanced very far." George sees philosophy as a counterpart of history of science. Although in the coming year, he projects a great deal of writing, practically speaking, he must earn money. Ideally he would like to live from his writings, "but I can hardly count on that for a long time to come, because what I write can interest only a few people."

---

1. NYPL.
2. NYPL.
3. NYPL.
4. "Scholar's Wife."

George pens a will on 7 September 1914, leaving everything to Mabel. It urges her to give to the library at the University of Ghent those of his books that she does not want.[5] On 19 September Mabel writes to Méta Budry that they "*will stay* here until the last minute, to protect the house if possible, to help and protect the peasants and to help keep them *calm*, if that becomes necessary, and also because we are involved with a local ambulance." For 3 weeks they have had 5 refugees from Malines staying in the house.[6]

The German army enters Brussels on 20 August, but its main strength diverts to Mons and points east, leaving a relatively light occupying army in Flanders. The army, in fact, passes Ghent by for 3 weeks, and the front lines are porous there. George travels twice to Brussels after the Germans take the city, bringing newspapers to his Uncle Arthur. Late in September, George has the prescience to appeal in English to Robert S. Woodward, president of the Carnegie Institution of Washington. He sends Woodward the numbers of *Isis* which have appeared. His pitch is from the top: "We think indeed, that the history of science is the very heart of the whole history of civilisation, and the rational basis of all truly scientific philosophy and organisation. The past only interests us in view of the future." George plans to publish additional series: "a series devoted to the systematical study of different epochs or of the intellectual work accomplished by different races," and "a series devoted to the historical and critical study of scientific problems," such as the questions of a universal language, of bibliography, and of statistics. George had hoped to make *Isis* self-sustaining within 3 years, but the war intervened. He asks the Carnegie Institution to come to his rescue to the tune of a few thousand francs. "May I not count upon America's financial help, until Europa's wounds and more particularly those of poor little Belgium, have had time to heal up?" George is willing to settle in Washington to continue *Isis*, if the Carnegie Institution thought it desirable. He includes a specimen of his latest bibliography printed in the format of "Encyclopédie sur fiches."[7] The appeal is not unreasonable. The Carnegie Institution is in the business of supporting the research of exceptional men, and Woodward in particular is a firm believer in the moral and intellectual superiority of science over the humanities.[8]

---

5. YMS.

6. NYPL; YMS.

7. CIW, George Sarton to Robert S. Woodward, 26 September 1914.

8. Nathan Reingold, "National Science Policy in a Private Foundation: The Carnegie Institution of Washington," in Reingold, *Science American Style* (New Brunswick, N.J., 1991), pp. 190–223, on pp. 212–12; Nathan Reingold and Ida H. Reingold, *Science in America: A Documentary History, 1900–1939* (Chicago, 1981), p. 9.

Evidently, it is Mabel who crafts George's appeal to the Carnegie Institution. She persists as his collaborator. At the end of September, George and Mabel together send a proposal for a post-war armistice to Hamilton Holt, the pacifist editor of the *Independent* in New York. They take for granted that the Allies will be victorious. The peace must be generous. Maurice Maeterlinck's notion of crushing retribution is, in the Sartons' view, mistaken. French *revanchisme* in the present war derives precisely from Bismarck's terms of the Franco-Prussian War. "In spite of all we believe that hate & violence can only give birth to hate again: For peace to be lasting, it does not suffice that it can be forced upon the vanquished—but above all, these must be able to accept it without utter degradation." The victors will have to protect the rights of "all small oppressed nationalities." They shall be able to reconstitute the Polish Kingdom. They shall be able to indemnify states whose neutrality has been violated, such as Belgium. They should go further, and charge a penalty to the vanquished for their transgression against international law, to be paid not to the victim but "to found an International Fund to be used for the furtherance of World Peace and Civilisation." The peace treaty must seal the death warrant for "Military Caesarism, Imperialistic plots &...all attacks upon the liberty of peoples dreamed of by Tyrants." Europe should be left only with "nations armed for the Right," secure behind powerful fortresses on their frontier. Defensive establishments do not contradict pacifism—an allusion to the views of George's uncle, Jules Sarton, who more than 20 years previously had advocated a strong Belgian military.[9] If Belgium had erected stronger defenses, it would not have been invaded by Germany. The Sartons conclude with a concrete and urgent proposal: They ask the Carnegie Endowment, which investigated atrocities in the Balkan Wars, to investigate atrocities in the present war.[10]

According to his diary and agenda, the outbreak of war brings George's scholarship to a halt. But on 11 August he resumes several hours of bibliographic work daily, while occasionally serving night duty in the civil guard. In September, he records 82 hours of scholarship, and the pace continues until 10 October, when Wondelgem is bombarded and along with Antwerp occupied by the Germans. In the last days of the invasion, George receives a gun, whistle, and lantern and is told to guard the railway crossing at night—presumably at the new Wondelgem train station. On the day before

9. Jules Sarton, *La Défense nationale et la question sociale* (Antwerp, 1893), and *L'Armée de demain* (Brussels, 1893), appearing anonymously. Jules Sarton, *Les Grandes Manoeuvres* (Brussels, 1903) and *Défense nationale, Anvers-extensions, démolition de l'Enceinte* (Antwerp, 1904); *Phoenix*, p. 68.

10. The *Independent* rejected the proposal on 8 October 1914; the Sartons received the rejection in Cambridge, Massachusetts, on 24 February 1919. YMS.

the Germans occupy Wondelgem, the bridge over the Terneuzen Canal is destroyed. German soldiers are quartered in the Sartons' house around 17 October. "In her Anglo-Dutch jargon," Mabel scolded some of the soldiers who were stealing fruit, George recalls. "When 26 Germans—officers & men—were quartered upon us in our Wondelgem home, & we were held responsible for their safety—nothing terrible happened. Those Germans were decent people."[11] Decent, but not above suspicion. George writes later that although he surrendered his rifle at the Town Hall, he kept a revolver at home—a risky and provocative decision.[12]

George shows the first numbers of *Isis* to one of the German officers quartered in his house, a judge in civilian life. On the strength of German collaboration in the journal, the officer supports George's request to leave for the neutral Netherlands to continue publication.[13] On 9 October George has his passport stamped. On 15 October he obtains a visa for neutral Spain.[14] Although George's Italian colleague Aldo Mieli offers refuge at his home near Siena, the Sartons set out for England.[15] On 20 October the Sarton family depart for the Dutch frontier, some 15 miles distant. "We were sufficiently young not to worry over much," he recalls. George buries his notes in the garden, in a metal trunk used by Mabel's father in India.[16] They hire a peasant cart and are consequently able to take valises. George carries published numbers of *Isis* and a volume of Emerson's essays—Emerson, a close friend of Thomas Carlyle's and a harbinger of a career in greater Boston. ("I carried that volume with me everywhere and never used it.")[17] From the Netherlands they arrive in Ipswich on 22 October. On 25 October, Mabel reports to Méta Budry. "We had to lodge & feed many Germans. Food was scarce. We had hardly any more money. Fighting was near, only two miles off & all these reasons forced us to leave." George leaves Ipswich to find a post in London.[18] The Sartons become one family among as many as 100,000 families seeking refuge in Britain. Most of the refugees, George notes in retrospect, were not "rich parasites," but rather "people—rich or not—whose regular income or salary was cut off, and who could not live without it, and first of all, all the literary people and the artists."[19]

11. "Scholar's Wife"; Diary, 17 October 1915, for the retrospective date.

12. George Sarton, "The Future of Belgium," *Open Court, 29* (1915), 257–72, on p. 262.

13. "Scholar's Wife."

14. YMS.

15. E. J. Dijksterhuis, "In Memoriam George Sarton," *Centaurus, 4* (1956), 369–81, on p. 370.

16. *Phoenix*, p. 79.

17. "Scholar's Wife."

18. NYPL.

19. YMS, George commenting on a letter to the New York *Times* of 11 October 1915.

Mabel writes to George probably on 24 October 1914 from Ipswich: "I have every confidence in you and in our future. I am ready to endure a great deal, and I have much hope despite the disappointments and the struggles. We will triumph over bad luck, you will see! And *Isis* 'will live and multiply.'" On 28 October George replies from the Salisbury Hotel in London. He found her letter with the Strangways family. Arthur Henry Fox Strangways is a specialist on Indian music and Indian art, whom George had met in London in 1910.[20] George reports that it is extremely difficult to find a position. He met the chemist Sir William Ramsay, a supporter of *Isis*, who thinks that there is little chance of a university job, at least in London. If George goes to Oxford, Sir William will put him up and introduce him to the right people. Sir William presented him to the provost of University College, and they will lunch. Lectures can be arranged at University College, but they would be unpaid.[21] George has written to the *Times* without much hope, but Sir William has also introduced him to the editor of *Nature*. Fox Strangways has found him a post in the Hostile Country Censor Branch of the War Office. It pays £3 per week for a 10-hour day that finishes at 7 p.m., he notes later in his diary on 14 November 1914. Fox Strangways cannot host George, and he paid for a night in a hotel, but now George can manage on his own. Fox Strangways is far from rich. He is the music critic for the *Times*, and his income has stopped. He is 55 years old and single.[22]

From Ipswich, George writes again in English—a letter corrected by Mabel—to Robert S. Woodward at the Carnegie Institution of Washington. George indicates that he is a refugee. He hopes to obtain a university post in England through the intervention of William Ramsay, although such an appointment will be difficult to deliver. He wonders if he might give a course at American universities on the history and organization of science. He would "have to give it in French as that language is the only one I speak altogether correctly." He could come to America only if his travel were paid in advance, "for I am almost at the end of my financial ressources." He concludes: "I have nothing left but my brains & more courage & enthusiasm than ever!"[23] George writes once more to Woodward upon ar-

---

20. George Sarton, "Why *Isis?*" *Isis*, *44* (1953), 232–42, p. 235. Arthur Henry Fox Strangways, *The Music of Hindostan* (Oxford, 1914); *Ajanta Frescoes; being reproductions in colour and monochrome of frescoes in some of the caves at Ajanta after copies taken in the years 1909–1911 by Lady Herringham and her assistants; with introductory essays by various member of the India Society* (Oxford, 1915). The Fox Strangways surname was carried by the Earl of Ilchester; Arthur Henry, who was not a peer, would have been suitable company for Lady Herringham.

21. HGS, William Ramsay to George Sarton, 18 March 1912 for an early letter of support, 26 October 1914 for the unpaid lectures.

22. NYPL; YMS for the War Office.

23. CIW, George Sarton to Robert S. Woodward, 25 October 1914.

riving in London, this time in French, repeating his plea, adding that he has found a job paying £3 per week and that in January he will lecture at University College London on the history of science.[24] Woodward sends his reply to George in care of Fox Strangways. The Carnegie Institution can take care only of its own people; it "is not engaged in educational work," and so has no real contacts with the academic world; and the institution respects all the conventions of neutrality.[25] George comments when he receives Woodward's reply: "No help, holy water."[26] Woodward's supplicant, in 4 years, will become an employee.

Mabel on 29 October writes that before receiving George's letter, she imagined his arrest as an alien. George's position nets them an important income, but she knows how discouraging it is to work at an unsatisfying job. May is being taken at a school without cost along with a girl about May's age, the daughter of a French woman. Mabel continually wants to cry, she thinks from fatigue. She is vexed because she can't work to help the family situation financially. "As often happens, the duty imposed on me (simply taking care of May and living here) seems to me the most difficult....Sometimes it seems *maddening* not to have even a quiet hour to *organize my thoughts* and think about the future." On 30 October, May spends her first day in school, a kindergarten of 20 children. May is nasty to Mabel's mother, and the feeling is mutual. On First November, Mabel writes that she wants George to spend Friday, Saturday, and Sunday on his own work. She hopes that her £7 will last a long time. She proposes to stay at Ipswich for 2 weeks; May's time in school will allow Mabel to sew their clothes. Mabel thinks to go to Sophie Jackson in London; the Jacksons have a maid, and Sophie can help with May. Mabel loves George "despite my momentary unkindnesses toward you." May "has come to show me a letter just stolen from my drawer and demands a kiss from daddy! She seems to have understood that that 'is sent' by letter!" On 31 October Mabel is sad that George can't visit, not only because of money but also because she lives in a restricted area.

George, on 3 November, sends Mabel a letter for the US ambassador, asking her to translate it for him. He thinks that Mabel should indeed leave Ipswich, but not for financial reasons; he is able to send her £1 weekly. "Mother is a maniac and an egoist, who is better off alone." Their difficulties are not May's fault. "Children see extremely clearly; they are less liable to fool themselves than we are." George hopes that Mabel can be in London in January for his lectures. George closes: "We are still privileged. Don't forget it." Mabel translates and affirms that she will not stay on at Ipswich even the 2 weeks she had planned. Mabel agrees that

24. CIW, George Sarton to Robert S. Woodward, 1 November 1914.

25. CIW, Robert S. Woodward to George Sarton, 9 November 1914.

26. George Sarton to Mabel Sarton, 26 November 1914.

they are privileged, and that George's job is an enormous piece of luck. Mabel's mother adds that May attends a day school and likes it, but "it is very hard labour to keep her in order." She knows that George's "present employment is not what you would have chosen, but 'a half loaf is better than no bread.'"

The letter for translation is addressed to Wickliffe Rose, in charge of the Rockefeller Foundation's International Health Commission. George read about the Commission's work in the *Times*.[27] In his letter, George notes the Commission's inquiry into the misery resulting from the European war. George proposes to accompany Rose in Belgium. He has perfect command of both Belgian languages and he knows enough English "to speak and write if not very correctly, at least in an intelligible manner." George asks for no salary, only subsistence, but he wants his wife and child to be taken care of. Mabel continues to translate and edit George's English writings, and on occasion she takes the initiative to write on his behalf. She sends him a translation on 26 November, for example, advising him to delete one paragraph or at least place it as a postscript.

George sets down sad thoughts in his diary on 12 November 1914. "The principal accusation made against them [the Germans] is that they have made *brutality into a system*, applying science to the systematic organization of horrible tasks." The "great crime" of Germany is the violation of Belgian neutrality. "Never have I loved my little country more than now." He loves other countries, too, "but it is impossible for me to hate Germany. I am incapable of hatred. Am I missing a little spring in my heart?" He is discouraged about his prospects. The War Office work is tiring, yet interesting. He is committed to give lectures at University College in January, "and I don't yet have the courage to think about them." Despite all this, "my flesh is animated with desire, to the point of clouding my mind." He has sold 2 gold bracelets and the gold medal he received in the university competition of 1908; they netted £7.6. He still has his father's gold chain, which he thinks can fetch £2.3, and also his mother's rings. On 13 November he writes: "The immense catastrophe impinging on all of humanity now allows me to verify once more in a striking manner that external circumstances change our life and our character very little. The deep wave of our being remains almost unchanged."[28]

George writes to Mabel at the Jacksons in London on 11 November 1914 that he has spoken with the Coomaraswamys, and they are glad to host Mabel and May. They live in the country and have a daughter the same age as May. (Coomaraswamy, son of the first South Asian to be knighted, is indeed wealthy.) George anticipates lecturing in London

27. "American Relief for Non-Combattants: Rockefeller Foundation's Scheme," London *Times*, 3 November 1914.

28. NYPL.

in January and February, and perhaps in Cambridge in March, if his expenses are paid. On 13 November, he urges Mabel to ask the Army & Navy Stores (where the Sartons have on occasion shopped during visits to England) if they will cancel a portion of his outstanding debt, a matter of a few pounds, as a gesture of solidarity with a Belgian member of the society. Mabel is not doing well, she writes on 19 November. "I am so full of tears that the least disturbance brings them on, and when I begin to cry I can't regain my composure." May does not like Sophie Jackson. May is so strong that it is hard to dress her for school if she does not help. "You have no doubts about your ability as a father. I have terrible doubts about mine as a mother. I often feel absolutely sidetracked and disarmed. I have only one hope, that I can guide her better later."

George, on 20 November, reports that he uses lunch for writing. All the libraries are closed in the evening. There is no heat in his room, and the drawing room where he lives is impossible, so after dinner he takes long walks in the rain. George thinks that Mabel is worn out by taking care of May. "And the real way of a possible cure is to confide May to friends for several months." Mabel could then be with George. "*I will arrange things so that you can truly rest*." George imagines that they could work together happily in London. He thinks that Mabel should stay with May for another month to allow May to learn English better. George is concerned on 26 November: "I hope that you are not sick or that little May has not tyrannized you too much." He sends a letter for translation; it is very important, and he will have it typed in many copies. The next weekend will be the last they have together before Christmas. George and Mabel both live in London, but George sees wife and child only on weekends.

George on 27 November has come up with an idea for a London "Clearing Library," or union catalogue, for the libraries of Greater London. It would cost £2000 yearly.[29] He turns to a natural extension of his dedication to bibliography and his pre-war plans for an encyclopedia on cards. There is, of course, no money for his scheme, as pacifist Norman Angell and geologist Archibald Geike both emphasize.[30] George avows to Irénée van der Ghinst, in 1915, that leaving Wondelgem liberated him from the claims of his books. His personal library was a "thorn in his paw," and without it he is as "free as the air."[31] But George cannot live without books. He organizes his life around great libraries, just as

29. NYPL, Proposal for Clearing Library.

30. HGS, Norman Angell to George Sarton, 8 December 1914; Archibald Geike to George Sarton, 5 December 1914.

31. George Sarton to Irénée van der Ghinst, 1915, in A. M. Simon-Van der Meersch, *George Sarton's European Roots: Introduction to His Life and Times* [Exhibition organized by the Archives of the State University of Ghent and the Department of Communication and Cognition] (Ghent, 1984), p. 9.

he organizes *Isis* around its Critical Bibliographies, and on occasion he purchases books when his family is on the brink of starvation.

Mabel writes to George as Daddy. On First December, Mabel reprimands George for complaining about his room: "You must remember that I have lived with my parents in rooms compared to which your present room is quite luxurious!" Mabel wrote to her cousin Alice Cranbrook and to Dorothy Chambers, both of whom offered to take May. "I have no doubt that the best cure would be for her to have a brother or sister, but oh! my body is *so* tired I don't think I could nourish a second little life ever again." Mabel thinks "I must be a woman with an insufficiency of the 'mother' in her, or it would not tire me so much. I feel angry with myself & deeply sad too and promise myself that I will be & feel otherwise, but that is not much help." Mabel wants to confide in George, and wishes he would do the same, "but you are very proud & strong enough to be able to do without anything further...." She received 2 letters from Lausanne. Méta and Berthy invite Mabel and May to stay with them as long as necessary, even a year. Méta Budry would help with May, who could play with the other children. Mabel does not want to go without George.

On 3 December 1914, George writes that he is hurrying to send out his circular about the Clearing Library, which cost him a guinea. He reminds Mabel that she must obtain permission to travel to Clare, where she intends to stay with Lady Barker, because she is a foreigner. He spent an evening with an administrator of the Continental Edition of the *Daily Mail*. On 5 December, George reports that he sent out 90 copies of his circular without having corrected the errors, above all those of punctuation. He received so far 10 replies, wishing him well but hardly encouraging him. At lunch Fox Strangways told him that in the circular he should not have referred to his position at the War Offfice, and that the reference could be grounds for dismissal. George "*has very great need*" of a recent photograph of Mabel and May, now that they are apart. He attended a "Bohemian Concert" sponsored by the Guild of the City. It featured a transvestite singer, which shocked him but pleased the audience. He will eat in a Soho restaurant with some of his coworkers, one of whom was Maeterlinck's secretary when the *Blue Bird* ran in London. George accepts Mabel's mother's invitation for Christmas.

George worries that Britain is not recruiting enough soldiers. He thinks that married Belgians, like himself, must take the place of unmarried Englishmen. He hopes Mabel will agree that he should volunteer. He has made provision for her in case he disappears. She would have about £175 to £180 per year, plus a pension from the British or Belgian government. This would be enough for Mabel and May, "to make of this little devil a good little heart and later a good and noble woman, *that is the main thing*, and later to help her find an honest and courageous companion who may continue your care of her." He wants to serve in the British army, for then his training would be near Mabel. If May were not ill, George would ask Mabel to send her to Clare or to Mabel's cousin and in this way Mabel could join him. The war is a grand lottery, and no one knows what it has in store. He wants to spend as much time as possible with Mabel. "I would also have liked to see little May, but you are the one I need most, for I cannot get along much with this very small rat." He needs his clothes mended. How about Norman Angell's *Great Illusion: A Study of the Relation of Military Power in Nations to Their Economic and Social Advantage* (London, 1911) at 2/6 as a Christmas gift for Céline?[32] He is going to the National Gallery in the afternoon, to focus on beautiful things. In what may be a reference to Mabel's Belgian therapies, on 9 December George relates that he has reconsidered and not written to Sir William Osler about the "medical question, but rather to chemist Sir William Ramsay, whom I know better." (In Belgium he was in correspondence with both Osler and Ramsay.) Sir William telephoned a friend who will interview George. The University of London officially invited him to give his lectures, but now he thinks to decline the offer. At noon he is at the Athenaeum, in the evening at Caxton Hall with the Library Association. Referred by Victor Gustave Plarr, Librarian at the Royal College of Surgeons, George meets members—bibliographers—of the Panizzi Club, from which meeting George hopes to "attack" the British Museum, the reading room of which derived from the 19th-century labors of Sir Anthony Panizzi. Nothing comes of these initiatives. Alfred William Pollard in the Department of Printed Books at the British Museum, a founder of the Panizzi Club, reports that it is hopeless to undertake the Clearing Library.[33]

Mabel and May move to Katherine Weston Elwes Lady Barker's, in Clare, Suffolk, on 9 December 1914. Mabel spoke to Katherine, widow of a general, about George's military future, and she has written about it to a cousin, a retired artillery colonel. On the 10th, Katherine is very helpful, and May is being very good, despite an ear infection. Mabel thinks that George should consider the Belgian army, rather than the British one. She offers three reasons. First, George would be with his countrymen; second, he would be less likely sent to Egypt or another remote theater; third, there are serious difficulties in being admitted into the British army. George, on 12 December, conveys his concern about

32. Also appearing in a French edition in 1911. George meets Angell on First March 1915 in Angell's quarters at Inner Temple, London. YMS, Memoranda 1915

33. HGS, Victor Gustave Plarr to George Sarton, 7 December 1914, for the referral; Alfred William Pollard to George Sarton, 4 December 1914. The Panizzi Club's secretary is the Librarian at the Patent Office.

May's health and behavior. On 12 December, Mabel reports that May's ear is improving. Mabel saw a doctor herself, hoping for a "tonic." She signs herself, "Your Wild Bird." On 13 December George writes that he wanted to buy May a toy or a book, but he found nothing suitable, a situation complicated by his horror of entering shops. He wrote to the Belgian military authorities in Le Havre and in London, offering his services. He is a soldier without hate: "In truth, no nation is entirely innocent; all have participated more or less in preparing this enormous crime." His position in the War Office has given him an inside appreciation of the war. George is glad that the British Army will not have him: "I have seen many English officers, and it is very difficult for me to sympathize with them. A world of ideas and sentiments separate us." Notably, they are profligate in their spending, which annoys George. His letter for a Christmas Truce was not published, but he thinks that efforts are underway by the King of England and the Pope toward this end.

The Christmas Truce is clarified in a letter to the London *Times*, titled, "War, Christmas & the Gregorian Calendar." A fair copy is available in Mabel's hand, and we also have George's original text with Mabel's comments on it, but the letter is written in first-personal singular, for George's signature. The proposal is for a 3-day truce at Christmas,

> not only...a truce from fighting but also & especially a truce from hatred, anger & falsehood. The papers themselves would cease from slandering. Enemies would forget their quarrel for a time & would reflect for a moment rather upon the innumerable ties that the different civilisations have woven between themselves—& above all their thoughts would return & linger around the friends they have left far away, each in his own dear land.

The time of Christmas varies due to the use in Slavic Europe of the Julian calendar. George proposes that Russia adopt the Gregorian calendar as "an admirable way of tightening & strengthening the bonds that already unite her to her western allies."[34]

On 14 December, the Wild Bird writes that May's health is improving, "but she is often *extremely* difficult," and she strains Mabel's patience. Mabel's clothes "are all on the point of falling apart." Mabel, in a second letter on 14 December, wants to join George but cannot leave May in Clare, for the house is too drafty and cold. She could bring May to see George, but if he will leave in a few days, she asks him to come to Clare. On 15 December George has received word that the *Independent* will take his writing (it did not publish George and Mabel's letter from Belgium about an armistice); the *Monist* and the *Open Court*, as well.

34. NYPL.

One thousand words would net £1. Mabel must see George, she writes on 15 December, to mend his clothes and buy his military kit. If George is sent to France, she really wants to accompany him with May. "I have no more hatred for the Germans than you, except perhaps for the Prussian military party." Mabel writes to Méta Budry, also on 15 December, that George wants to join the Belgian army, and if he does, Mabel will join him in France for the time of his training, but she does not know if May will accompany them. When George goes on active duty, Mabel would like to stay with Méta. Mabel is frantic with worry, although she does not let George know it. "I will not be able to rest until the war is over."[35] George, on 18 December, writes that his landlady will allow Mabel to stay with him for a few days. He has withdrawn money from the bank and sends along 30 shillings. But George receives a flat rejection from the Belgian command at Le Havre.[36] The salary of a junior officer would be insufficient to allow him to live in Le Havre, and so the colonel in charge finds himself obliged to recruit men of independent means. George then writes, also without effect, to Guy Eden, proposing himself as an interpreter for the British army.

On 29 December 1914 Mabel writes from Framlingham, Suffolk, where she and May are staying with a cousin of Mabel's, Alice Dorling.[37] The Dorlings and their children are nice, but it is a very small house with few conveniences and a horrible atmosphere. There is a large open fire and a lamp, but no open window. Meals are taken in the main room, 8 or 9 people together. "A miserable yard" or the street is where the children can play, but they almost never go out. The youngest child, May's age, seems delicate and small. They snack all day on "spice," a small meat dish, and have mince-pie for dinner—no vegetables or fruit or milk. She will leave May there only so long as necessary. The people speak horribly, and Mabel does not always understand them. It is a small, old village, charming in its way. She admires the woman for having raised her family under difficult conditions. The house has a large number of small bedrooms, "each one with something naively cute in an old-fashioned way." She and May sleep in a small room with a fireplace, which has a connecting door to the parents' room. George notes about the arrangement in his diary that Mrs Dorling has "7 children and white cats."[38]

George on 30 December 1914 asks Mabel to prepare for making translations when she joins him: a letter to a Japanese friend and an article for the periodical *Open Court*. He adds in a second letter that May should not

35. NYPL.

36. Rejections dated 19 December 1914 and 3 January 1915, NYPL.

37. Mabel wrote to her mother, in care of the Dorlings, on 21 April 1911. YMS.

38. Diary fragment in NYPL.

eat meat. Mabel travels to Ipswich for New Year's and then on to London to meet George, leaving May at Framlingham. Beginning on 7 January, according to George's agenda, George and Mabel stay with her relatives Harold and Phoebe Stabler in Hammersmith West, London; Harold is a ceramist and heads the Art Department at the Sir John Cass Institute; Phoebe is a sculptor, the daughter of sculptor James McLeish.[39] Harold Stabler takes George through the Oriental ceramics of the Victoria and Albert Museum on 7 March 1915.[40] Mabel is without May for the early months of 1915.

When George's application to travel into the rising sun is rejected, his gaze turns west once more. He senses that America is waiting for his ministry, and he has reason for believing so. Several years previously, he received a letter from Pierre Boutroux, the mathematician, philosopher, and historian of science, supporting *Isis*; Boutroux, who like George sought the unity of science, wrote from Princeton University, where he was professor of higher mathematics.[41] On 15 January 1915, George asks Nicholas Murray Butler, president of Columbia University and a prominent pacifist, for a position. His journal *Isis* is tied up in Brussels, and the latest number, devoted to the scientific thought of India, cannot appear. "My whole ambition is to succeed in reconciling my studies with the necessity of earning a living for me and mine." He could also lecture on French and Belgian thought, on the exact sciences, or "fill some humbler role in the university." If nothing is possible at Columbia, perhaps he could be offered hospitality "while I searched elsewhere for some means of earning my living in the United States."[42] Columbia does eventually hire George for summer work, though notwithstanding this gesture, George will criticize Butler's view of higher learning in 1919.

Sir William Ramsay cautiously supports George's plan to seek his fortune in America. He offers to write introductions to Butler at Columbia University, to President Lowell at Harvard University, and to physicist Ira Remsen at Johns Hopkins University. But he thinks that George is "risking a good deal in going to the States. Work of the kind you wish is not easy to get there....What I fear is that you should not succeed in finding employment, & that your money should not last out. To be stranded in America must be horrible."[43] George does not heed the advice. His

39. John Adams, "Harold Stabler, R.D.I.: A Great Designer of Pottery," *Pottery Gazette and Glass Trade Review*, May 1945, p. 275; "Mrs Phoebe Stabler," London *Times*, 9 December 1955. On the Stablers in context: Tanya Harrod, *The Crafts in Britain in the 20th Century* (New Haven, 1999), pp. 71–3; Victor Arwas, *Art nouveau: From Mackintosh to Liberty* (London, 2000), p. 196.

40. YMS, Memoranda 1915.

41. HGS, Pierre Boutroux to George Sarton, 9 December 1913.

42. NYPL.

mind is fixed. With money borrowed from a wealthy Belgian industrialist in New York, Leo Hendrik Baekeland (who married the sister of George's chemistry professor Frédéric Swarts), probably arranged by David Eugene Smith, George ships for New York at the end of March.[44] Mabel and May stay behind until he has established a means of supporting them. George's colleagues in the War Office treat him to a farewell dinner at the Old Cheshire Cheese on 13 March 1915.[45] George visits May at Framlingham on 22 March. "Little May has forgotten all the French she knew and she has not yet learned English," he laments.

43. HGS, William Ramsay to George Sarton, 16 February 1915.

44. Margot Peters, *May Sarton: A Biography* (New York, 1997), p. 404.

45. Diary, 13 March 1915.

CHAPTER 9

West

The marriage of George and Mabel Sarton did not seem auspicious. George possessed few social graces and had no knowledge about how to acquire the necessities of life. Mabel, depressive and chronically ill, was innocent of finance. Their rocky union nevertheless prepared them for life during the war years. From the time of their courtship, the Sartons lived apart from each other for extended periods of time. The experience helped George and Mabel come to terms with their refugee status in England, and it allowed them to see George's departure for America as audacious rather than foolhardy.

In George's absence, Mabel writes on 30 March 1915 with news. Céline Limbosch has a daughter, 5 kilograms and 51 cm. Irénée and Marianne van der Ghinst arrived in London. They are well and cheerful. Irénée has won the cross of the Legion of Honor for capturing 5 enemy singlehandedly, for staying in Dixmude as the only doctor, and for taking machine guns from the Germans. As a German prisoner, he "narrowly escaped being shot" as he eluded his captors with 4 other men. There is no clear answer from relatives in Lincoln and Bromley. She forwards the invitation of Medora Amsden to host George in White Plains, New York (Medora's husband George is a physician at the Bloomingdale Asylum for the Insane there).[1] She wishes George well on the start of his new "epoch." Her family shall not abandon her. In a letter to Mabel of 5 April, Katherine Barker has promised to guarantee Mabel a small weekly sum until the end of the war or until George finds employment; the amount would be subscribed to by her entire family. Katherine asks whether £1 weekly is enough, or if 30/ would be better.

George writes letters as he ships to New York. On Palm Sunday, he enjoys a hot sea-water bath, and wishes that May could experience it, too. On 30 March he writes in English about the best plan for stimulating May's curiosity: "To explain her how the things that she sees around her were borne, how they made themselves or were made by clever men." May—"Petit coq"—must be taught not to act immediately, but "to look at, to see, to understand." George thinks, on First April, that while at sea he

1. George J. Makari and Robert Michels, "The Payne Whitney Clinic," *Images in Psychiatry, 154* (1997), 1751. George S. Amsden Papers, New York Weill Cornell Medical Center Archives, New York, Finding Aid.

should attempt no serious work: "One is just good to read Mark Twain stories or to hear cheap music." George saw May before he left:

> I cannot help thinking of Little May, of the *two* little Mays: the Belgian one and the Suffolk one. [There is a Maisie in Framlingham.] How unfortunate should I not have been if I had been obliged to go to the front without having seen my little girl, having seen another one instead. It would have been much more terrible to die in such conditions. But the gods forbid it! I am not going to the front; am simply travelling, and I hope to see her back, to find *her again*. But the idea that I could have died away, without having seen her again, just as if she had been *lost* to me, makes me too sad for words.

George is glad that he does not have to dress for dinner. "Poor old rat! I am not a waterrat." George writes on 3 April that the third-class passengers are treated worse than he is, presumably in second class. He asks Mabel to tell him about the mistakes he makes in his English letters. George has spent much time speaking with former prisoners on board, a German and a Hungarian. "Since I have met these prisoners, I have spoken much more German than English, but I should have been a fool if I had scorned such a valuable source of information." George sends a thousand kisses to the Big Bunny, Mabel.

Mabel writes on 6 April that she has moved to Ipswich, although she dreaded being received by her mother. The move was so exhausting that Mabel is waiting a few days before transporting May. They will finally set off together for Grantham, near Lincoln, on Monday. Her new hosts " have had sad experiences with the other Belgians: one 'aristocratic' (?) family jumped on the sofas & chairs & bullied the housekeeper." Mabel calculates that she has enough money for nearly 2 months, "so you see we have *quite* a long 'respite.'" On 8 April, Mabel is at Framlingham to collect May. In French: "Her *character* is almost unchanged, in my opinion, but she has forgotten *all* the French she knew, and I do not believe that it will come back. She has also forgotten almost completely all her past." May recognized Mabel, but she forgot Mabel's name. "She no longer knows how to say 'maman,' and that hurts me absurdly." Mabel thinks that May's intelligence and character have not been compromised, but neither have they gone forward. The Wild Bird returns to Ipswich with May. On 12 April, Mabel reports that according to a doctor, May has been suffering from malnutrition and has conjunctivitis. "She is distinctly knock-kneed & her right leg is 'bandy.'" She will probably grow out of it. "May is certainly a very tiring child to be with *all day long*. She is *noisy* & does love to talk all the time & sing loudly (she is not a bit 'subdued' like you thought!)." Mabel feels, as her mother told George, that she is "an incapable woman." Mabel has replied to Katherine Barker, saying that an allowance is not necessary at the present. George has received a telegraphic offer of employment, which Mabel sends along. "I do not know why I strongly mistrust this offer. Why not say for *whom* and for *how much*?" George must endorse a check and return it to Mabel so she can pay her mother the 10/- she owes her for her travel. Mabel has ordered the *Times Literary Supplement* and *Educational Supplement* for George, and she hopes that he receives them. On 17 May from her cousin Alice Cranbrook's at Caythorpe Court, Grantham, Mabel reports that May has the measles. She vomited a number of times on the way to Grantham, and is running a fever. Both May and Mabel are confined to a small room. She has received a telegraphic offer for renting their house from a Wondelgem notary.[2]

In her letters, Mabel is conscious of cleanliness, which relates to her sense of purity. On 18 April 1915 she has help from, in English, "a nice, quiet, clean-looking girl." Mabel writes again on the 18th. Her residence is an estate in the middle of farmland, practically denuded of trees. The soil is brown and the roads are made of white stone, filled with fossil shells, creating a dust like new silver. Sheep graze amidst fields of barley and wheat. Mabel loves it. In French: "It is so simple and big and peaceful." The distant blue skies and singing birds soften the somewhat harsh aspect of the land. Mabel has found "a new source of courage and calm this morning. I *know* that we will pass through this period having learned a great deal." On 20 April Mabel writes that her resources are disappearing faster than she had thought, but she will resist asking Katherine for the allowance. Mabel has received letters from Belgium. Céline and Raymond have a grocery business selling the produce they raise; they would be glad to put Mabel up, but they advise against it. They wonder how George has found money for his projects. Mabel has heard from her brother Hugh in Mexico. Hugh's wife has moved to Mexico City for safety; in the countryside the "bandits" are rampant. "Even the assassin [Pancho] Villa is better than the vacillating [Venustiano] Carranza," Hugh writes. Mabel thinks that it will be some time before she is strong. Mabel has not heard from George for more than three weeks.

The water rat writes in English from New York on 8 April 1915. His first impression of Broadway is frightful. He landed in a cheap boarding house recommended by a fellow passenger, a detective. He leaves the next day for Washington, to meet with members of the Belgian Scholarship Committee, among other people. George hopes to obtain a university lectureship for the fall. He will look for a job with a large newspaper until then, and in this way perfect his English. David Eugene Smith has

2. At the beginning of the war, the notary in question De Hénau and the Sartons worked together organizing a first-aid station in Wondelgem. YMS, circular letter, 16 August 1914, with the names of De Hénau and Sarton.

been very kind. Smith lives in a hotel, so cannot offer hospitality. The Belgian Scholarship Committee, an invention of Nevil Monroe Hopkins, Professor of Chemistry at George Washington University, collects money for refugee scholars and generally helps American universities profit from the influx of European talent. George will be the beneficiary, then an employee, of the Committee, which eventually extends its activity to rebuilding universities in Belgium after the war.[3]

On 11 April George writes from the Cosmos Club in Washington. He thinks about Mabel night and day. "It is impossible for me to be happy without you." But practically speaking, it is better for him to be alone at the beginning, "to be rid of the child etc." He explains that an invitation to Washington was waiting for him with David E. Smith. In Washington, he went directly to the home of Nevil Monroe Hopkins. There he met Charles Edward Munroe, the graduate dean of the university, and Admiral George W. Baird. Together they went to the house of the university president, Admiral Charles Herbert Stockton. The president made an appointment with George on Saturday to talk business. The result is that George will be appointed lecturer at a salary of $1200 per year. To this small but sufficient amount, George hopes to add $300 to $400 by writing for newspapers and magazines. "*We are safe*."

George asks Mabel to join him in Washington. He will send money for her passage and $50 for new clothes for Mabel and May. If it is not enough, Mabel can borrow from Stabler, and George will repay it.[4] He advises light clothes, for it is very warm: "I much enjoyed this night to sit naked in my rocking chair—and you know that I generally don't like to be naked." Clothes are inexpensive in Washington, but laundry makes up for it. He estimates the cost: £10 for second class on the ship, 10/- for trains, and "£3, I think for the child." Mabel can pay a portion in England and the balance upon arrival. George calculates other fees, including $6.50 for the train from New York to Washington. George doesn't know if he can afford to meet Mabel in New York; it would cost $15, or £3. She can wire the time of her arrival from sea. Mabel has lost her Belgian identification, and George is writing to the Belgian legation in London to say that she will stop by for a passport. "From a certain point of view it is rather a good thing that you have lost this paper. I don't want the people of the Belgian diplomatic service to know the circumstances of your birth. This silly thing could do me a lot of harm, and break my career. We don't deserve that." Here George recalls Mabel's illegitimate birth. Katharine Woods, sister-in-law of Nevil Monroe Hopkins, will help Mabel in all ways. "It will be so long before I have you again—both." George

3. CIW, George Sarton, "A Work of Reconstruction," 5 March 1916.

4. HGS, Harold Stabler to George Sarton, 29 April 1915, agreeing to loan Mabel what she needs.

experiments with his own version of racist American argot in describing New York for Mabel, in English: "Many things are very ugly, negrotasted, but others there are very tasteful, much more *artistic* than what we see in England."

On 16 April 1915 George, in New York, arranges to send Mabel a draft for $100 to cover her voyage to America. He is soon returning to Washington, where he will stay at the Cosmos Club. In Washington he met an astronomer from the solar observatory at Pasadena who knows his publications. The "best Washington evening paper" wrote a feature about George, introducing him "as one of the leading men of the Old World!!!!" George finds this exaggeration preposterous, but it is the American way.

George apprises Robert S. Woodward at the Carnegie Institution of Washington of his changed circumstances. He shall be lecturing on the history of science and also on French topics at George Washington University, he writes on 21 April on the letterhead of the Cosmos Club. He will be working at the Library of Congress, and also lecturing during the summer to earn more money. He would like to call on Woodward any afternoon before 5 p.m., when he lectures.[5] Woodward promptly replies that he will call on George at the Cosmos Club to arrange for an interview.[6] Woodward never follows up on his lines, but George has prepared the groundwork.

On 26 April George delivers his first lecture on Maeterlinck, partly in French and partly in English. The pupils are mostly girls, who offered him a lunch of sweet peas. He gave his maiden speech in English on Monday at the University Banquet. George finds Washington a very dry climate. "When it rains, the children don't go to school! And the climate as a rule is so dry that it is difficult to keep many years the books that have been bound in Europe." He has had his portrait taken by Washington's best photographer. George on 27 April: "I think so much of you both—I think more now of the *Mother*." George is allowing Mabel to draw on his reserve funds in a London bank, kept "for when we come back," in the event of an emergency. George thinks about May: "I am so grateful to you, for having given me this child. How happy the women are, that they can give *children*. What can we do, we men, to be compared with that? *Nothing* indeed, *nothing*." George urges Mabel to see what remains of Isaac Newton near Grantham (George alludes to Newton's father's house in the hamlet of Woolsthorpe, where he was raised), before she leaves the place.

The effect on Mabel of George's letters is to evoke resentment and recrimination. On 26 [?] April 1915, Mabel sees in George's lines a criticism

5. CIW, George Sarton to Robert S. Woodward, 21 April 1915.

6. CIW, Robert S. Woodward to George Sarton, 26 April 1915.

of her care for May; she senses George's view, in English, "that I loved her less than you." But, Mabel reminds George, he threw everything on her shoulders, without a real plan. "You seemed not to realize that a plan must be made!" George "never asked..., was I strong enough to take entire charge of her at once?" George "just thought of *her* and your own pain at what you *thought* she had become. That was not fair." Neither George nor Mabel imagined that a child could change so quickly.

> Were we not foolish, *both* of us to imagine a child could be so transformed in 3 months? You, dear Scientist who thought that environment can change little of essentials? It was real *panic* for both of us & panics are always demoralizing. We were both demoralized, I the most because I felt very doubtful of my strength to remedy things.

Mabel switches to French, commenting that George never apologized for suggesting, before he left, that each additional day of separation between mother and child "was a crime against her." It will take 6 months "to erase completely the traces, which are *superficial*, of her stay." In English: "She is *dreadfully spoilt*, has grown into rude & selfish ways, & all the nice little cleanlinesses of speech & manners seem to have been lost & slovenliness to have replaced them." Mabel is mostly concerned with May's speech: "I have made a list of 30 words she pronounces *quite* wrong. There are only *3* which she begins to pronounce properly: (yes, no, again)." Mabel continues: "Some-times her ugly speech makes me so sad & annoys me so that tears come into my eyes." Mabel wants pity:

> Be just a little sorry for *me* because I am not a good enough, "essential" enough mother, to begin my task simply & cheerfully, never counting the cost, never looking back. That is what a *really* good mother would be. But I feel very strongly indeed that I am a most *im*perfect mother & that very lack of unconscious, sublime devotion, makes the whole thing a grave trial, a greater expense of energy to me; by understanding this, you can help me, & so help her.

Still, Mabel recognizes "how strongly attached to her you are." Mabel evaluates May:

> I do not yet feel that she is a "noble" little creature, but perhaps I don't quite understand your use of the word. I think she has in her, as far as I can divine her, the makings of a reasonable, generous, fairly intelligent, affectionate woman. I shall be very proud and happy if it turns out to be so.

Then Mabel comments on George's English letter, "very ugly." She points out his stylistic infelicities. On her days: "I do not try to do things except sew when I am with her. It tires me til I feel I shall lose my reason."

On 27 April 1915 Mabel acknowledges 3 of George's letters. She can't reply at length. She has her period, and she is running a fever. The doctor thinks that Mabel should separate from May for 2 weeks and rest, and that Mabel cannot contemplate a journey for a month. Dr George Amsden has written to Mabel, inviting her to stay with him in New York and promising to meet her at the dock. Mabel has not lost her Belgian passport, but "for once in my life" she will "tell a story & say I *have* lost it." Mabel does not know where to take May, and she dreads the journey; all journeys tire her. She signs the letter, "Your wounded Wild-Bird."

The wild bird catches measles (which she thought she had as a child), we read in a letter from Alice Cranbrook at Grantham to George.[7] On 4 May, Mabel is well enough to write. She wants to send some of her Wondelgem furniture to Washington, particularly the grey, inlaid writing table, and in this way L'Art décoratif—Céline's establishment—could become known there. If Mabel obtains orders, the Limbosches can fill them by shipping through Rotterdam. Mabel can make most of her wardrobe for Washington if helped by her mother, but she will not go to Ipswich with May, whom she thinks to leave either with fellow exiles Mélanie and Victor Alter in Yorkshire or with Dorothy Chambers at Bromley.[8] On 7 May, Mabel is stronger. Katherine Barker has sent her £5 toward her clothing. The doctor will not charge her. On 10 May, Mabel writes that there is no news about the house in Wondelgem: "I love it, *even down to its faults*!!" Mabel does not like Fox Strangways: "*If he is not ill* he is to me in the future *Mr Bear of the Strange Ways*." She has written to him several times, without reply. She wonders if George told him about her illegitimacy. Mabel is uneasy about taking May across the Atlantic, given the U-boat menace. "I do not think it would deter me from coming along if we could both make up our minds to leaving May." Mabel wants

---

7. NYPL, Alice Cranbrook to George Sarton, 29 April 1915; Francis A. Hardy (the physician who treated May) to George Sarton, 30 April 1915.

8. In 1913 Mélanie Lorein married Victor (Wiktor) Alter, a Polish socialist who belonged to George Sarton's socialist student organization in Ghent and who studied engineering there in 1910. Alexander Erlich, "The Life of Victor Alter," in *Henryk Erlich and Victor Alter: Two Heroes and Martyrs for Jewish Socialism*, trans. Samuel A. Portnoy (Hoboken, N.J., 1990), pp. 43–70, on pp. 44–5, where Ghent is not mentioned. Alter returned to Poland and was exiled to Siberia for political organizing. He escaped and returned to live with his wife in Liège, where he continued his studies. NYPL, Mabel Sarton to Méta Budry, 3 May 1914, for Alter's return to Belgium. Alter later led the Algemeyner Yidisher Arbeter Bund in Lite, Poyln un Rusland (The General Union of Jewish Workers in Lithuania, Poland, and Russia), known as the Bund. He was executed on orders from Joseph Stalin in 1943. David Engel, *Facing a Holocaust: The Polish Government-in-Exile and the Jews 1943–1945* (Chapel Hill, 1993), p. 11. I am grateful to Christophe Verbruggen for information about Alter's marriage and membership in George Sarton's organization.

more details; she had not realized that he would begin lecturing right away. Mabel thinks that she could have helped George on Maeterlinck, for "I believe I know the plays better than you do." She is disturbed that George shows no relief at Mabel's report of "May's moral & mental condition" which has improved. She ends: "Ah dearest husband & lover, do not doubt for one moment that at present my heart is like a caged wild bird that beats its wings ever & ever against the bars. *I want you so*."

On 14 May, Mabel reports that the doctor will not let her travel. "He strongly advises 'getting rid' of May. Poor little daughter, I do not *want* to be rid of her. She is my greatest consolation & she needs me." The Wondelgem house has been let for 2 months for F 50 per month. She has received £5 from cousins, to help with moving to America. On 21 May, Mabel recounts a walk with May:

> Yesterday we stood & watched the brown pony who is a friend of ours, drawing the grass-cutting machine. A man leads him who is also a friend of May's (I say of *May's* because he is rather a grumpy man but he is always nice to *her*). He had a tiny, long stick in his hand & May ran when she saw it, most excitedly up to him & said "Oh Mr Beech, you're not going to hit him, are you?" in tones of bitter reproach. So Mr Beech said "Well, you see Miss, he's a bit lazy & fat so I does give him a little switch now & then." But May looked distressed & said so earnestly "I'm sure he's going to be good *now*" that Mr Beech actually laughed & threw his little switch away & May was all smiles again & came back to me saying, "They don't *like* sticks, do they mother?"

On occasion May understands goodness. About a disagreement:

> I believe yesterday she had her first dim perception *why* it is wrong. But it will want many & many a struggle of course before she understands. While she is angry it is no good explaining. But when she is good & quiet again, *some times* she likes having things explained. Other times it bores her & she very cleverly & decidedly changes the subject!

Mabel promises to go on a "pilgrimage" to see the Newton sites in Grantham. She observes that her Wondelgem passport gives her name as Mabel Eleanore Cole and she signed it E. Mabel Elwes-Cole. There is no mention of her parents. Mabel thinks that her signature does not reveal illegitimacy. "All this is of course prevarication. I know it quite well. You must tell me what you think about this *at once*."

The sinking of the "Lusitania" impels George on 10 May 1915 to advise Mabel not to sail. But he learns that she must wait for the measles to pass, in any case. "Now I do ask you: Would it not be better that you gave up for an unlimited time the idea of coming over here?" Mabel will decide what she wants to do, but "without you both, I am a very poor fellow indeed. I feel as poor as a sparrow in a snowstorm." When Mabel comes, George may be off lecturing in other states, "and of course I shall have to go wherever I am called." But in that case George will make arrangements. George is called, on 17 May 1915, to sit on the examining jury for the doctoral defense of Samuel Palkin at George Washington University. It is an experimental investigation in organic chemistry.[9] Within several months of landing in the New World, George Sarton is certifying graduate students.

On 22 May George writes that he is going to Boston for the summer months. He will try to find a bungalow for Mabel in the country near the shore. He has arranged for Mabel and May to meet his correspondent Henri Lafontaine in London; Lafontaine is a Belgian senator for the Socialist Party and bibliographer who won the Nobel Prize for Peace in 1913. George knows Lafontaine as editor of the journal *La Vie internationale*, where he published several articles. George has not yet found another position, but he hopes to find one in New England. "Be sure that I do my best; I have never struggled with more enthusiasm." George looks forward to their new house, but he doesn't know where it will be. It will be a house filled with love. "I am afraid I have not loved you enough in the first years of our marriage. I was too busy. I did not see you, neither the child, but will make amends for all that and retrieve the time lost." On 23 May, George recalls their happy days on vacation in Cromer, near Ipswich. "There is not a single detail that is not living in my memory." He urges Mabel to write a short story for *Life*. He will work on it with her. On the 25th, George thinks that the war may only be in its early years; the American Civil War lasted 5 years, he notes. George wants May to grow strong and self-reliant. On 29 May, George is leaving for Leo Baekeland's house in Yonkers, on the way to Boston. The "Old Rat" sends Mabel a hat protector.

Mabel, on 25 May, thinks that she may set off for America in a month. "May tires me, till I have to cry every time I have her rather a long time." She will need help with May for a few months after arriving. May has had "such violent passions of temper that it frightens me." Mabel continues: "I have not strength, moral or physical, to grapple with her & that makes me very worried." On 29 May, Mabel writes that her room has been fumigated (for the measles). She has had difficulty controlling May, and she provides reasons for May's bad behavior: Framlingham, and their consistent failure in making May "obedient." May "'wants' things passionately &...has a fairly strong will (we cannot know if it is *very* strong yet as it may just be obstinacy)." Mabel writes a plan of activity that will allow her to depart in about a month. It means traveling with May, leaving her for 2 weeks to visit her mother and sew clothes, go to London and purchase

9. NYPL.

essentials, and obtain her new passport. Mabel tells George that he will not have to spend much money on her.

> No, Dear *Heart*, you will have to spend lavishly out of that *heart*, & also be very clever in helping me arrange to spend the little money we have in a *truly* economical way, that means in some way to help me not to *have* to be "*very* busy" for a while after I come. Oh! Do not hurt me, & say or think I am a coward & not ready to take my share of trials & privations. *Something in me tells me now so clearly so warningly that I must be careful, that I might even die or lose my reason, if I do not guard the strength that I still have.* I feel it. I feel it deeply, keenly that I am in danger & I appeal to you to help me out.

Reflecting on Fox Strangways, Mabel senses "that it would have been better for you in some ways to have had no wife, or another." But Mabel has uniquely allowed George to give love: "*Perhaps,* the stronger, better woman would not have called out all that as well as I?" She asked her doctor about Isaac Newton. "He knows nothing & thought I was a bit mad." She has read the recent issue of the *Open Court.* She thinks that George's article, which she translated for him, is "one of the best by far." Except for the article on Socrates, the others "seem to me to lack essential qualities—they are *weak.*" Mabel does not sympathize with Paul Carus's apology for German aggression (Carus believed that Britain manipulated world affairs to force Germany into war, and he wanted the United States to side with Germany). She is especially incensed about poison gas. "I wish Eng. chemists could invent some sort that only made the men unconscious & killed them painlessly."

George's connection with the German-American Paul Carus and his journals began in 1913. George's faithful collaborator on *Isis* at Cambridge was Philip Edward Bertrand Jourdain, a mathematician of promise who suffered from Friedreich's ataxia, a hereditary and progressive paralysis. Jourdain was also the British editor for Carus's *Open Court.* In 1913, Jourdain arranged for an exchange of periodicals between Carus and George, as well as an exchange of published advertisements for each other's enterprises. Writing to Carus to confirm the exchange, George offered that *Isis* and the *Monist* have "identical aims."[10] We shall see that at the end of the war, *Isis* and the *Monist* almost merge.

George's article, "The Future of Belgium," which in his letter to Mabel of 22 May 1915 he calls "our" article, opens the May 1915 issue of the

---

10. George Sarton, "Philip E. B. Jourdain (1879–1923)," *Isis*, 5 (1923), 126–9; Laura Jourdain, "Some Reminiscences on the Moral Personality of Philip E. B. Jourdain," *Isis*, 5 (1923), 129–33. An advertisement for the *Monist* appears in the first number of *Isis*. Paul Carus papers, Special Collections Research Center, Morris Library, Southern Illinois University, Carbondale, George Sarton to Paul Carus, 1 May 1913.

Open Court, an issue devoted to Belgium; the frontispiece is Peter Paul Rubens's "Assumption of the Virgin." George begins by observing that one-eight of the population of Belgium is in exile; the people who remain are subject to deprivation and destruction. The cause is Germany's violation of Belgian neutrality. He does not want to discuss German atrocities. Rather, he emphasizes the broad social and economic reasons for the war, "a whole world of realities, arguments, sentiments, above all of instincts, where the worst is mingled with the best; a great deal of unconscious ignorance and of kindness worked upon by a few selfish and criminal intellects; rare ideals and a mass of human mire." George is sure that Belgium will emerge from the war wiser and showered in glory, while Germany will be dishonored for its treachery. He observes the tension resulting from 110,000 Belgian refugees in England, whose temperament does not fit well with English temperament. George concludes with a plea for maintaining Belgian integrity and independence after the war. "Small countries, well fortified and well armed, interposed between bigger ones—*états tampons* as they have been called—appear to me the surest factors in the European equilibrium."[11] Defensive fortresses are the key. "I believe personally that for long centuries to come it will be the peoples' duty to protect themselves by this means as it is the duty of individuals to put locks on their doors." The article repeats the contentions of George and Mabel in their letter to the *Independent* of 1914, where they insist that defensive armament does not contradict pacifism. George is in favor of heavy reparations from Germany after the war, both to Belgium and to an international fund.

George then recalls the plans for a world city designed by Hendrik Christian Andersen, aided by architect Ernest Hébrard and a phalanx of collaborators associated with Paul Otlet at Solvay's Parc Léopold in Brussels. In October 1914, Andersen published a summary of his plans in the *Independent*.[12] George, an omnivorous reader, may well have seen the summary, which would have recalled to him the third number of *Isis*, dated 6 November 1913, where he reviewed Andersen's plans. He said then: "Three centuries ago he would have been placed on the block; a century ago, he would have been confined to an asylum; but today, an intellectual élite from all corners of the world gather around him to express gratitude and sympathy." Andersen's undertaking promotes internationalism. "If

11. In his agenda for 15 January 1904, George notes the "utility of états tampons for world peace."

12. Hendrik Christian Andersen, "A World Center for Peace," *Independent*, *84* (October 1915), 152. Giuliano Gresleri and Dario Matteoni, *La Città mondiale, Andersen, Hébrard, Otlet, Le Corbusier* (Venice, 1982); Annick Brauman, "The Parc Léopold: The Home of Scientific Imagination," in *The History of Garden Design*, ed. Monique Mosser and Georges Teyssot (London, 1991), pp. 442–4.

the world center never arises out of bricks and mortar, will not at least the City of God, where original thinkers from all countries and all times find refuge, become more solid and real?"[13] In 1915, George is convinced that the world city "should be erected in Belgium, in the country sanctified by glorious wounds, and I believe that none of the other nations would protest." He is confident that Belgium will become "healthier and inspired with higher ideals." He urges America to organize a commission to investigate war crimes.[14] Andersen concurs.[15]

In modern and postmodern times, people devoting themselves to ideas face the crucial problem of patronage. It might be thought that everything turns on practicality, for in the 19th-century incarnation of Francis Bacon's New Atlantis, knowledge will cure illness and bring about a life of ease. But the 19th century also sees the rise of great museums and libraries, along with eminently useless scientific institutions like astronomical observatories and vertebrate-paleontological collections. Morally uplifting endeavor is indeed prized by captains of industry and merchant princes, as George knows from his years in Ghent. With his move to America, George is struck by the pragmatism of his new environment. He senses that it is insufficient to contend that history of science makes for better scientists, and he tries to hang his plea on whichever hook presents itself. In "The Future of Belgium," the hook is American shock at the rape of Belgium. Several years later, in an éloge of Emile Waxweiler, the hook is a new sociology. Neither hook holds firm. George's support, from the moment he steps on American shores, derives from patrons who are interested in abstract ideas. The greatest supporters of arts and sciences wish to go beyond mere utility.

In her letter to George of 29 May 1915, Mabel feels like George's "sparrow in a snowstorm." Raymond Limbosch writes to her that he has no faith in her "commercial qualities" as a furniture agent. She affirms: "I daresay he's right." George does not say anything more about his lectures. "Are they to be all literary?" She wants to know how much it costs George weekly to live. Mabel has received an invitation from Sophie Jackson, who is now a Child Welfare Medical Officer in Bradford, living at the home of a woman friend who has been appointed to head a hospital corps in Serbia; there is also an invitation to stay in Manchester. The latter

13. George Sarton, "La Conscience mondiale," *Isis, 1* (1913), 488–9. Hendrik Christian Andersen, *Creation of a World Centre of Communication* (Paris, 1913), historical part by Gabriel Leroux, architectural part by Ernest M. Hébrard and Jean Hébrard. HGS, Hendrik C. Andersen to George Sarton, 11 September 1913, sending along the introduction to *Creation* for review in *Isis*.

14. George Sarton, "The Future of Belgium," *Open Court, 29* (1915), 257–72, pp. 263, 270, 271.

15. HGS, Hendrik C. Andersen to George Sarton, 10 July 1915.

tempts her, "but I feel more & more that if I am to get my nerves steadier, I ought once more to separate from May." Mabel tentatively plans to take May to Dorothy and Arthur Chambers at Bromley . Then she will stay at Manchester and subsequently with her mother to sew her wardrobe. She calculates she will leave Ipswich for London-Liverpool on 6 July and arrive in Washington on 20 July. On 4 June Mabel makes plans to begin her final journeys in England. "I have given up on the Newton Pilgrimage. I must save every bit of my strength for my two journeys."

On 8 June, Mabel writes from Bromley in South Kent that she has received copies of the *Open Court* article, George's photograph, and the fashion book he sent. She thinks the photograph is a bit out of perspective. "I don't think any photographer will ever do any good with you!" She feels stronger. "May is very frightened I am going to run away from her." She will need a servant in America: "I really *could* not manage however simply we lived, & even if you helped me, for the first month or so." Mabel is quick to tell George that she also remembers the good times together. "Unfortunately I think pain (*heart's* pain) impresses itself more deeply upon me than joy. (Amongst my *childish* memories, the 'land-marks' are *nearly* all sad ones.)" Mabel "cannot come without May." On 11 June Mabel writes from Whitevale in Didsbury, Manchester, at the Brockbank house.[16] "I have been startled & rather upset by a letter from Katherine enclosing yours, with the suggestion of her taking May for an unlimited time." Mabel reflects: "Ah Beloved, it has been such a revelation to me, that letter of yours & if it has not at all succeeded in its set purpose, it will yet have brought much good to us all three." She appreciates George's desire to care for her, but "it is absolutely impossible for me to give up May to *any-one*, however much I like & trust them, for more than a few weeks or months. I hope it will never be 'months' again." Mabel is grateful that George wrote the letter, "except for one thing: I have a hurt, sad feeling because you thought I would consent to give up our little daughter for so long, perhaps years. May-be it is my own fault though." Mabel's mother is against her sailing to America. "May cried bitterly when I left (I had carefully explained to her that it would not be for so long as at Auntie Alice's [Dorling]), but Dorothy wrote me just a word wh. I got this morning saying she was comforted very quickly & was very good." Mabel thinks that Dorothy and Arthur Chambers have been very good friends, and the stay with Alice Cranbrook at Caythorpe helped her mend.

On 15 June, Mabel writes that the years of their marriage seem like a lifetime, "so full they have been of experiences of both joy & pain." George has taught her that "it is *so necessary* to go on learning. Sometimes in-

16. Early in the 20th century, Oliver Brockbank led the independent Ivy Cottage Church in Didsbury.

deed I feel as if these last years had been a sort of *prelude* where I was just learning to live & love, & that now my *real* life is just beginning." Mabel looks forward to a second honeymoon in the New World. She can travel directly by steamer to Boston, via Fall River, which is included in her ticket. Her passage with May is £20, and the Brockbanks in Manchester (he runs a metals firm and his relations are chemical dealers, supplying products for making poison gas) have given her an additional £20 for clothes: "a plain tailor costume, a simple evening dress, shoes, underlinen." Mabel feels strong and optimistic. Mabel's mother has made no effort to receive Mabel, as she is going to Bournemouth for a month. Mabel does not know where to do her sewing; maybe she will take a room in Bromley, if Dorothy can keep May. "I still am rather frightened of being alone on the boat with May to cope with, her angry, naughty fits are a great trial & upset me absurdly. I get then this queer 'collapsed' feeling & tremble so violently I can't even hold a glass to drink out of!" Mabel does not look forward to the prospect of saying goodbye to her mother at Bournemouth, at her mother's cousin's house. She is reading Henry Sessions Souttar's *A Surgeon in Belgium* (London, 1915) and likes it very much. She gave the Brockbanks a little volume of essays by Arthur Clutton-Brock, a Fabian socialist and an admirer of William Morris.

George, on 8 June in Boston, where Mabel thinks to join him, is "very sad." He urges Mabel not to come until September. He elaborates on 10 June. David E. Smith "has been really a splendid friend." Before George arrived, Smith wrote to many people on his behalf: "His idea was to found a new lectureship in Columbia University devoted to the History of Science, the lecturer being myself of course." Of all the people George has met in America, Smith knows his work best. Nevil Monroe Hopkins kindly arranged the lectureship at George Washington University, but George gets along better with his sister-in-law Katharine Woods. His best friends are the Baekelands, both natives of Ghent. Leo Baekeland "is a very clever man, who has a very clear sense of duty and is broadminded, but not very refined. He is neither a poet, nor an artist, but a very intelligent and honest man." He reminds George of Mme Dangotte. Céline Baekeland is a fine person who would care for Mabel. "She is a vegetarian, a teetotaler, a suffragist, a theosophist, and I don't know what more." George is hopeless for American newspaper work. He dined with a department-store owner in Boston, who proposed making history-of-science movies; the enterprising man will place George in contact with a film-maker in New York, "but for that too, I feel *hopeless*." George thinks that it will be very important for women to get the vote in New York State. He urges Mabel to do nothing about her passport, now that she has found her papers. George is unable to support both Mabel and May until September. He relates what Mabel knows, that he asked Katherine Barker to take May "for an *unlimited* time." George thinks that this is best for May, who needs "peace and steadiness." He is glad for Mabel's press clippings. George desperately needs Mabel to correct his English letters. He writes on 17 June in response to Mabel's fears of dying or going mad. She must, he repeats, separate from May: "I think that the wisest would be to part from the child for a long time—till you feel much stronger." George met with Charles William Eliot, former president of Harvard, but nothing resulted.[17] He hopes to be correspondent of the *Petit Journal* of Paris. They will pay F 100 for each of his printed letters.[18] George asks which language May speaks and whether her English is good.

George had contacted the *Petit Journal* from London on 19 January 1915,[19] and he sent the journal letters beginning with his arrival in America (his first, 8-page letter is dated 25 April).[20] Through August 1915, he writes from Washington, Yonkers, Boston, Urbana, and Chicago. He comments on everything: the entrance of Italy into the war, the report of James Bryce on war atrocities, the deficiencies of the American navy, the American commission for relief in Belgium, a new electrified dreadnought, William Jennings Bryan and world peace, the Mexican revolution and American intervention in it, and German intrigues in the United States.[21]

In his letter of 10 June 1915, George reiterates his attraction to things Oriental and his desire to try his fortune in Japan after the war ends. He dreams about training May in Japanese and Chinese and seeing her become curator in a museum. (Soon his friend Ananda Coomaraswamy obtains just such a post at the Boston Museum of Fine Arts.) "But of course, I hope that she will marry and have a lot of children, which is much better!" George has been interested in the Orient since his student years, and he pursues the possibility of moving there with his family. He writes to Nicholas Murray Butler of the Carnegie Endowment for International Peace and also to Robert S. Woodward of the Carnegie Institution, asking for a mission to East Asia to complete a general history of science.[22] Harold Stabler encourages George's plans, offering that China is "*the* place in the world."[23] In December 1915 George asks Okuma Shigenobu,

17. HGS, Charles William Eliot to George Sarton, 18 June 1915.

18. Desperation is seen in George's decision to write for the *Petit Journal*. It was well known for being a detractor of Alfred Dreyfus and also for conducting a campaign against Marie Curie in 1911 for her affair with Paul Langevin. Susan Quinn, *Marie Curie: A Life* (New York, 1995), p. 305.

19. Agenda.

20. Agenda, 25 April 1915.

21. NYPL, "Lettres au Petit Journal."

22. Diary, 6 October 1915; CIW, George Sarton to Robert S. Woodward, 6 October 1915.

23. HGS, Harold Stabler to George Sarton, 29 April 1915.

chancellor of Waseda University, about a trip to Japan.[24] He also enlists the support of the former Belgian ambassador to China, Emile de Cartier de Marchienne, for a position at the University of Peking, but de Cartier advises against the initiative: The salaries are low, and life is both costly and dangerous.[25] As late as February 1918 George considers traveling to Asia with Mabel, a plan that the former Belgian ambassador to China continues to discourage.[26] But Mabel supports George's plans to travel East. If he is offered a post that "would not earn enough for me & May to live on," he should "*take it the same*." Mabel could manage a small apartment with a loan from the Baekelands for 6 months. "I think I could become independent after that." It would be welcome if George could send along $50 per month, in that case.[27] George abandons the plan when he obtains continuing support from the Carnegie Institution of Washington. He nevertheless spends increasing amounts of time in the great Northeastern art museums studying Asian painting and sculpture.

On 25 June 1915, George writes that he has landed a summer job lecturing at the University of Illinois in Urbana. Seven weeks of work pay $400. He shall save enough for their return fare to Europe, and he will bank that amount. George has lined up a place for May in a private school for international pupils. Now George urges Mabel to leave for America in mid-August, stay with the Baekelands for a few weeks, then return with him to Washington. George is not surprised that May is naughty. "She could not be naughtier than I was myself at her age." Mabel is eager to travel as soon as possible, she writes as "Your Wild Bird" on 18 June from Manchester. Mabel's mother has given up all her plans for Bournemouth and will be at Mabel's disposal at Ipswich; May will stay on at Bromley with Dorothy Chambers.

George writes on First July from Urbana that he cannot care for Mabel and May there. George thinks that Urbana has one of the most successful state universities, growing in 25 years from 300 students to more than 5000 students. He finds the town healthy and handsome. George dined with the university's president, Edmund Janes James, to whom he was known through *Isis*. "You see, short sighted people said that I was a fool to spend so much money (and by the way, it just amounted to a few hundred dollars) for this publication. I never made a better investment." He asks Mabel not to send books, for he cannot carry them around. He recommends that Mabel bring a good watch with luminous dial when she

24. Diary, 3 December 1915.

25. HGS, Emile de Cartier to George Sarton, 23 December 1915 and 11 May 1916. DeCartier received an appointment as Belgian ambassador to the United States in 1917. Cyril Arthur Player, *Arms—and the Men* (Detroit, 1922), pp. 21–2.

26. NYPL, Mabel Sarton to George Sarton, February 1918.

27. NYPL, Mabel Sarton to George Sarton, received on 18 February 1918.

comes to America. On 7 July, George learns that Mabel will wait a month longer before coming. He gave his first lecture to hundreds of people, mostly women, who had gathered mistakenly to hear about "songs of the birds." He is fascinated by German sympathy in Illinois and by the prohibition movement. He implores Mabel to correct his letters. "*I must know english*." Initially, on 7 July, George finds the libraries disappointing; on 13 July: "There is a splendid library, very efficiently organized."

Around 7 July 1915 George outlines his plans for the future. He must survive, and then he must secure *Isis*. In the first place, he notes the American Institute in Brussels—his idea for the American contribution to Andersen's World City. Then come the study of Oriental science, editing *Isis*, a popular novel in astronomy, and a list of English history books, among whose authors are Henry Thomas Buckle, John William Draper, and William Edward Hartpole Lecky. Before concluding with "union writing," he indicates three other writing projects: "Why I became a Buddhist, A Buddhist on the war, Buddhist art."[28] George has been absorbing Buddhism for several years, largely in art but also in the writings of Moncure Daniel Conway.[29] Buddhist thought, with its emphasis on self-denial and introspection, appeals to him even more at a time of material deprivation and scholarly stasis. We also read an affinity with Paul Carus, a devotee of Buddhism, who publishes George's second article in the New World.

The second American article, in the *Monist*, is a revision of George's programmatic statement in the first number of *Isis*. At the outset, George observes that there is no good, reliable survey of history of science. The reason, he thinks, is an estrangement between humanists and scientists, who view each other with mistrust or even contempt. The organization of the article follows his text of 1913. History of science is indispensable for philosophers, for that is where philosophers' examples reside, even though the "resounding discoveries" in science of the past quarter century have discouraged philosophers from looking into the more remote past and formulating a grand synthesis; both intuitionist pragmatists and objective positivists respect science. George offers that history of science, as a discipline, originated with Auguste Comte and passed to Paul Tannery. George's own conception follows Tannery's, although George attaches "more importance to the psycho-sociological point of view." The history of science "is indeed a history of human civilization, considered from its highest point of view," and as a consequence the study of history of science must proceed not by country or specialty but only chronologically in time. George then reviews the relation of history of science to general

28. NYPL.

29. An early and enthusiastic supporter of *Isis* is the Orientalist Paul Masson-Oursel, an expert on Buddhist thought. HGS, Paul Masson-Oursel to George Sarton, 4 January 1913 and 23 January 1913.

history, to the history of technology, to the history of religion, and to the history of art. The treatment introduces new examples, for example Andrew Dickson White's *Warfare of Science with Theology*, a generous and balanced book that "had to bear the attacks of a great many fanatics."[30] George affirms once more that science is cumulative, in contrast to art, but each complements the other. He repeats his plea for the utility of history of science in the community of scientists and among teachers of science. George strengthens his earlier contention, that history of science is "an instrument of culture."

Appealing to American concern about civilizing undergraduates, George emphasizes that history of science "is a general education itself." And when discussing the connection between psychology and sociology, George modifies his French article to appeal to the spirit of social reform in America by introducing Karl Marx's laws of historical materialism, which "exerted such a deep influence on the thought of the nineteenth century." George deletes nearly all of his 1913 discussion of racial proclivities and eugenics, asking, more generally, in the direction of what is later known as social history: "To what extent does the scientific environment exert its influence upon the scientists, and *vice versa*? How do social activities evidence themselves in the domain of science? By what mental processes are the ideas of the initiators integrated in the collective thought, to become, by and by, common notions?" In a new conclusion, George calls for a "new 'humanism'" based in history of science, which "will broaden our horizon and sympathy; it will raise our intellectual and moral standards; it will deepen our comprehension of men and nature." Science is the one endeavor common to all humanity; it is international in character. A true history of science would "include, on the same level as the Western experience, the immense experience of the East." George's new humanism is a new synthesis for the modern age. It is his battle-cry over the next quarter century.

In an appendix, George summarizes an article of 1915 by Frederick E. Brasch surveying academic courses in history of science in the United States. The oldest tradition resides at Harvard University, where Theodore William Richards has been teaching history of chemistry since 1890, although Massachusetts Institute of Technology offered general history of science, by William Thompson Sedgwick and Harry Walter Tyler, beginning in 1905; David Eugene Smith has a seminar for history of mathematics at Teachers College of Columbia University; occasional lecturers include George himself and also Lawrence Joseph Henderson at Harvard. In sum, 162 courses are offered in 113 schools. George thinks that the time has come to require a course on history of science for both undergraduates and graduate students, although he suspects that there would be greater impact on the graduate

30. Andrew Dickson White, *A History of the Warfare of Science with Theology in Christendom* (New York, 1896).

level. He asks for creation of a "seminary" devoted to "the history of science as a whole."[31] It is transparent self-promotion, deriving from hunger. The tone is evident in much of George's writing over the next decade, as he struggles to keep his head above water intellectually and to keep his family together materially. It is, however, the right tone for early 20th-century America, which, as George notes to Mabel on 16 April 1915, proceeds by bluff: "Every one bluffs, but they are so big-hearted that one forgets it."[32]

A closer examination of Brasch's long survey suggests that history of science has many apologists in America. In 1906, physicist Charles Riborg Mann at the University of Chicago saw history of science as "the much-desired humanizing of science."[33] Brasch cites Wilhelm Ostwald's view that history of science provides "the most authoritative material for the study of law in the evolution of mankind." Brasch contends that history of science is a manifestation of "idealistic-historical" reaction to scientific materialism. "The final message of the history of science is to show the high plane of science—that which has given life, stability, truth and wealth—in its universal activities and its established international character as the arbiter of the future of man and of peace." These are precisely George's points, made without referring to George. Like George, Brasch (who has moved to the John Crear Library in Chicago) knows the value of bibliography, and Brasch is clear that D. E. Smith and Henderson teach the largest number of students in history of science.[34]

31. George Sarton, "The History of Science," *Monist, 26* (1916), 321–63; Frederick E. Brasch, "The Teaching of the History of Science: Its Present Status in Our Universities, Colleges and Technical Schools," *Science*, 26 November 1915, 746–60. Early teaching initiatives in history of science are recounted by Arnold Thackray, "The Pre-History of an Academic Discipline: The Study of the History of Science in the United States, 1891–1941," *Minerva, 18* (1980), 448–73.

32. Bluff is not a prerogative of Americans, for George. George Sarton, "L'Organisation scientifique du journalisme," *La Vie internationale, 5* (1913), 391–428, on p. 422, where he writes about how an international press service might expose the bluffs of irresponsible journalists.

33. Mann collaborated with Robert A. Millikan in translating Paul Drude's texbook, *Theory of Optics* (New York, 1901).

34. Both Smith and Henderson teach undergraduates. Smith informs George that he conducts a seminar from time to time when he has an advanced student, and he mentions Edward Everett Whitford's doctoral dissertation of 1912 on the Pell Equation as well as Louis Charles Karpinski's work on medieval algebras. HGS, David Eugene Smith to George Sarton, 16 December 1915. Karpinski received a doctorate in 1903 under Heinrich Weber at Strasbourg; he collaborated with Smith in 1911 on Hindu Arabic numerals. Joseph W. Dauben, "Louis Charles Karpinski," in *Writing the History of Mathematics: Its Historical Development*, ed. Dauben and Christoph J. Scriba (Basle, 2002), pp. 454–5. On Brasch, Thackray, "Pre-History," pp. 462–3.

George chooses carefully from among the historians of science active in America.[35] He has connected with Smith, he shall shortly find a patron in Henderson, and Brasch will become one of his first students.

On 13 July, George reminds Mabel that the war is just beginning, and that they shall be in America for 3 or 4 years. He advises her not to bring too many clothes: "*We have plenty of all that here.* You are not going to the Congo, but to a most civilized country." George keeps track of Mabel's and May's weight, May's height, and Mabel's periods; he asks for their weight now. George reports that he is undertaking "a great work for Belgium. It will cost a lot of pain and $2,000,000 which I have not yet succeeded to raise." George will be a new man for Mabel, and "for you it is positively a new marriage." He indicates how different: The Wondelgem man "was fat, I am rather slender. One sees my bones everywhere. He did not shave himself, and I shave myself every day at midnight. He was a home-staying bookworm, and I am a traveler."

Then George relates a dream. He was in Wondelgem, with its white-and-blue peasants' houses, each with sunflowers all around. He was in their garden. A bumblebee sounded like a flying machine, a slug seemed like a steam-roller. The flowers spoke, "with the noise of a boarding school when the classes are finished." One asked: "Where is She? She used to come here many times a day. She took great pains to give us water. She was so kind." Another said: "What is the use of being a beautiful flower, if nobody looks at you and *loves* you..." Another: "She did all the work herself: *He* did not help her a bit. Is that not a shame...." George grew sad in his dream. "Was I really guilty?" Then the old currant tree spoke: "What is all that talk about Him. Is that your business, young folks...What do you know?" The tree added: "He saved my life." George saw the sweet-peas and thousands more flowers. George wanted to send her poems by Christina Rossetti (irreligious Mabel never refers to this religious poet among the Pre-Raphaelites), but he could not find an edition.

Mabel telegraphs on 25 June that she is sailing on 10 July. But on 27 June she receives George's cable, urging her to postpone her departure. From Ipswich, she writes on 28 June, at 6:30 a.m., about her disappointment. Everything has been arranged, and many people have made commitments. But she thinks that Mr Brockbank will be able to defer the reservations, which he made especially for her. Mabel will try to correct George's letters and send him clippings from the

35. Joseph W. Dauben, "United States of America," in *Writing the History of Mathematics*, pp. 263–85. Early in the 20th century, Smith and Florian Cajori were the most prominent historians of mathematics in the United States (p. 269).

Times.[36] She has found a girl to help her with the sewing. She had "no trouble at all at the Legation & did not say I had found my passport." George, on 14 June, follows the thought about Rossetti by wondering if Christian Science could help Mabel; Mabel is skeptical about all organized religion and doubtful about the New England sect: "I do not think Christian Scientists could cure my body of its very 'material' weaknesses consequent on trying to bear 3 children, previous appendicitis, & subsequent measles!! I think all I want is a reasonable period of rest & quiet for my mind & my body, & *that is all*, except your love."[37]

Mabel writes from Ipswich on 9 July: "I feel tired & a bit depressed but I will try to cheer up, dear Old Rat!" Dorothy is very fond of May, writes the Wild Bird. Mabel, on 10 July, has an idea for a story about a *béguine*, and when she writes it up she will send it to George. Mabel again asks whether George told Fox Strangways about her birth. Then to a theme from Thomas Carlyle's *Sartor Resartus*:

> I wonder do you guess that in doing so much sewing, it is not only the wish to be suitably & as beautifully clothed as I honestly can which makes me do it (& in that is the wish & the hope to be as beautiful as possible in your eyes chiefly) but there is love for you stitched into these things & so I have a pleasure in doing them quite apart from a very real enjoyment of working with my hands.

Mabel thinks that "People are consciously or unconsciously much influenced by dress," and if Mabel's illegitimate birth might have hurt George's career, "then my outward aspect can surely also have some importance." Mabel prefers to make her own clothes. "Generally I am uncomfortable in new clothes until their newness has wore off & I can forget them." Mabel is impressed at how much George is earning at Urbana, and she will try to earn £5 over the coming month. Mabel is traveling to Bromley, then to London to meet Méta Budry, who has "dire *money and* sentimental troubles." She closes, "Bravo, dear Old Rat, so wise & so practical."

On 20 July Mabel writes from Bromley, where she has found a room, that her friends Berthe Jaeger and Juliette Mauwelaerts have invited her and May to spend the war years with them at "Aigle" in Switzerland, or

---

36. Mabel indeed serves as George's secretary for years. For example, Mabel Sarton to George Sarton, 4 April 1917, where Mabel thinks that she "ought to help you more with your correspondence." She is worried that George is "dangerously tired." Knowing George as she does, she could answer many letters for him, if he were only to indicate "the gist of the answers." In a letter of 6 April 1917, Mabel agrees to the "test" that George proposes about the correspondence question.

37. George notes Alfred Mayor's *Mary Baker Eddy et la science chrétienne* (1912) in the first Critical Bibliography of *Isis*, *1* (1913), 187, under "science et occultisme."

"to take charge of May alone while I am away." Mabel cannot accept: "I have not hesitated at all to decide I *cannot* leave her to go to you." Mabel is concerned about her health in America. She will improve "if we find some way of taking May off my hands (a way which must of course mean happiness & good *to her too*) a part of each day & she must sleep in *your* room, not mine, that means 2 bed-rooms." (In Wondelgem, we recall, Mabel had her own bed and probably her own workroom.) How shall they manage? George has given her no practical clues about their life in Washington. If they cannot live on his salary of £240, then "from the purely practical point of view, it would be best I should *not* come." Perhaps George should write to Berthe and Juliette, offering to pay the household expenses of Mabel and May. Mabel thinks that George's work might be compromised by their reunion in America. May is looking better, and "her legs are distinctly straighter." May speaks better, but still "mispronounces the vowel 'o.'" Here, as in other places, Mabel signs herself "Mother." On 20 July Mabel reports that May is attending school. There are so few faults in George's letters "that I am amazed & *very pleased*." She has not heard from Céline, and none of her embroidery or miniatures has arrived from Belgium. Méta is with her for a few days; she just missed landing a good post. On 23 July, Mabel asks George to thank Dorothy and Arthur Chambers for taking care of May. May is learning to use scissors (a counterindication for lefthandedness, which George suspected earlier). "She is distinctly clever with her fingers—the mistresses here all say the same thing." Mabel has seen a doctor. He thinks that she should join George, "as you are probably the person who best knows how to look after me & as only then shall I probably *be able to rest*."

Mabel writes on 30 July that she was very ill, she thinks from taking the strychnine prescribed by her doctor. The most recent diagnosis is Graves disease, or a defective thyroid gland, and she has been suffering from it for some time, according to the doctor. Mabel told him that Adolphe Miele had ruled out the thyroid, but the doctor disagrees. He thinks that the disease began around the time of her father's death, and the pseudo-angina-pectoris was connected with it. The doctor will give her a pelvic examination the next day and verify that her hernia has healed. At May's school, "*all* the mistresses say she is clever for her age & so *reasonable* & *reasoning*, also that she is so *precise*." May has improved her manners and speech, and she loves rough play. Mabel's doctor decides that she is "suffering from a general disturbance of the sympathetic nervous system," as well as problems with secretions of the ovaries, thyroid and suprarenal glands, along with an irritable heart; he treats her with arsenic and strychnine.

Notwithstanding the faith placed in chemistry by the Regius Professor of Medicine at Oxford Sir William Osler, scientific medicine is in its infancy in 1915. Mabel relates on 3 August that her hernia discomfort has been diagnosed as ovarian neuralgia. Her urine shows only a great loss of phosphates.[38] She is in good spirits, for she has found a girl to look after May before they leave for Ipswich and then the high seas. Méta Budry will accept a position as companion-nurse in Wales at £35. Most significantly, Mabel will meet Ambrose Heal of Heal & Sons, for a meeting provoked by Méta's recognition of Mabel's painted cradle in their shop, presumably in Tottenham Court Road in London; Mrs Heal bought it at Céline's Ghent exhibit. Mabel adds in a postscript that Heal (later, Sir Ambrose), a principal supplier of Arts & Crafts furniture for the mass market in England, wants to hire her.[39] For a designer, this connection is much like, for an academic, an invitation to lecture at a major university.

On 20 July George receives a photograph of Mabel and May. He likes Mabel's pose. "Little May looks very fat; too fat, I should say. I shall have to plane her down when she comes over here." He asks Mabel to cable the Baekelands with her arrival time. George, still in Boston, will arrange to meet her. He corrects her: "It is not fair to say that you bore *three* children; you bore *two*. However it was too much, and we shall have to be extremely careful. Try to have some good information about that, because it is impossible to get at it here."[40] On 24 July George calls up memories. He sees himself pretending to study mathematics in a beautiful garden in Assels, and a girl is gardening and picking flowers. Then he recalls a concert in Ghent where "Sheherazade" is playing. "The same girl is there in a box on the third floor." Then he sees "a small white room in an old convent. No furniture: Just a table, two chairs. Some pictures on the wall. A slave by Michelangelo on the mantelpiece. White cotton curtains. A girl is sitting on one of the chairs, and a boy is kneeling." Then a boy and a girl in a canoe on the Lys, a cliff in Cromer in Norfolk. Then a room in Dordrecht, with the Scheldt gray and brown. A little house in Ghent with a small garden, and a big house in Wondelgem, "the white gallery with a blue carpet, the library, the old trees, and there She is again, the light and the joy of the house." George thinks that they "have had a lot of happiness." He looks forward to going out with May for a good part of the day. George alerts Mabel to his heavy work schedule. First he has lectures in history of science. "I

38. In 1915 medicine is certainly inexact. Sophie Jackson, recovering from a broken jaw, writes to Mabel on 14 and 15 August 1915; Jackson, the daughter of a physician, does not think that Mabel's slightly prominent eyes indicate a thyroid condition, which had worried Mabel. NYPL.

39. Nikolaus Pevsner, *Pioneers of Modern Design, from William Morris to Walter Gropius* (London, 1974), pp. 154–5.

40. Mabel gave birth to a boy who did not survive early in 1913; in her letter of 2 July 1915, she apparently refers to a subsequent desire to conceive.

have to deliver those lectures in english, and most of my inspiration is in german." Then he must study "the American method and ideas," so that he can contribute to an institute "in Brussels after the war, where it shall be possible to get every information about American affairs." Third is his writing for the French periodical, *Petit Journal*, which has already sent him $70. Then he must also socialize. "I have at last realized that the best way to get a job is to ask for it." George is very interested in learning how the Limbosches are raising their daughter Claire. George is not sympathetic to Méta Budry's problems, for she was nasty to him.[41]

On 27 July, George urges Mabel not to be anxious about the crossing. He affirms that he mentioned nothing about Mabel's birth to Fox Strangways. "I was *very stupid* indeed to say that a matter like that would break my career." That would be true only in diplomatic circles. George has spent more on his maintenance than he had thought, but "it *pays* here to live well," for important people will be impressed that he stays in a club. George thinks that when Mabel works as an artist in America, she will earn as much as a man. In America, "people are judged by what they earn: There is something right in that in an industrial type of society." George is glad that Mabel is working on her "trousseau," but he thinks that it would be better for her to buy clothes off the rack. She may draw on his London account, and if that is not enough, then she may borrow from friends and George will repay. He leaves Urbana on 10 August. Returning by way of Chicago and Buffalo, he wants to meet Henri Lafontaine and American bibliographer Melvil Dewey; both men were active before the war in the International Institute of Bibliography at Brussels.

Mabel loves George's dream, she writes in an undated letter received by George on 6 August. "It was the sweetest Birthday Present you could send." May is "*not* too fat really." May's teacher, who is also Mabel's landlady, remembers "a terribly sad impression" George had of May when he said goodbye, but she says "what a joyful revelation she will be to him now." She is a noble but "really *ugly* person," and May likes her. "I do not think May judges people by their ugliness as so many children do." Mabel is convalescing. "I have great waves of energy come over me & the longing to do & think & feel & work ardently, & then always comes the reaction & I feel bruised & tremblingly weak. That is when I need you." Mabel now thinks that she is fated to join George, and to help him in America, notwithstanding the difficulties.

> I do not believe in people who would suppose that their husbands & lovers can triumph over or *ignore* these things, especially the material

41. May's authorized biographer Margot Peters writes that Mabel was physically attracted to Méta Budry, and that shortly after Mabel's death, May Sarton had an affair with her. Peters, *May Sarton: A Biography* (New York, 1997), pp. 10, 191.

> one. However passionately & deeply you love me, it does not dismiss the question! Nay, because I know you love me, I know too you would suffer if I seemed to want, to *need*, things we could not afford.

She returns to Heal & Sons. Ambrose Heal wants Mabel to work for him, and he does not mind long-distance designing from America. He is boyish, charming, and frank. His eldest son at age 19 just died in the war. She had tea at the Stablers; they are both busy designing, in part for Marshall & Snelgrove, who are also interested in Mabel's work.

George writes on First August that he cannot advise Mabel about the possibility of going to Switzerland. "All what I have to say is: use your common sense." Still, he is sanguine about their American future: "I think that you could have a very happy life here, and I have not fear at all, neither about you nor about the child." George knows that he will work less when Mabel and May are with him, "which does not mean that I shall earn less money, because by far the greatest part of my work now does not bring money." George calculates how much he has earned in America. He has saved about $600, out of a gross income of almost $1000. To this amount is some money in his London bank. If Mabel decides to go to Switzerland, George can send $60 per month for maintenance. He has sent $60 to Aunt Hélène, the Limbosches, and the Bounys through the Commission for Relief in Belgium. Before leaving Urbana, he meets Jakob Kunz, Assistant Professor of Physics, and on his way to New York he stops by Benton Harbor to see Paul Carus (he travels by lake steamer from Chicago to Buffalo).[42] On 30 August, George writes to Mabel, perhaps in care of the Baekelands, that he will meet her at the dock, a promise made several times before.

George's attraction to Lafontaine and Dewey relates to his continuing interest in matters bibliographical and to the more pressing matter of securing his personal future. Lafontaine is in exile because, as with George, his Belgian income ceased with the German invasion. The main part of funding for his bibliographical institute in Brussels derived from the Carnegie Endowment for International Peace, Lafontaine explains in a letter to George, and the foundation suspended its grant with the onset of war.[43] George corresponds with Lafontaine during the summer of 1915, and he persuades Lafontaine to accept the idea of hosting an American Institute near or even in the bibliographical institute at

42. Agenda, 4 and 12 August 1915.

43. HGS, Henri Lafontaine to George Sarton, fragment, n.d., where Lafontaine asks George to make discrete inquiries to determine the grounds for the suspension. Nicholas Murray Butler, "The Carnegie Endowment for International Peace," *Independent*, 27 November 1913, reprinted as no. 75 (1914) of the series *International Conciliation* by the American Association for International Conciliation, p. 9 of the reprint for the funding of Lafontaine's institute.

Brussels. The twinning, according to Lafontaine, follows from Hendrik Andersen's utopian notion of placing national buildings alongside international ones. Turning American attention to Europe would have the advantage of attenuating American domination of Latin America. Lafontaine urges George, however, to avoid referring to Andersen's world city. To attract funding for their schemes, they must appeal to practical men, who are disinclined to listen to dreamers. Lafontaine will ask for help from Theodore Marburg, the former United States Ambassador to Belgium. Lafontaine invites George to meet him and Melvil Dewey at Dewey's retreat in Lake Placid.[44] As a result of the meeting on 18 August 1915, George and Lafontaine send out a prospectus in 60 copies from Leo Baekeland's home in Yonkers.[45] It succeeds only in circulating George's name. George is continually urged by his supporters to be succinct in his schemes. His strongest ally, David Eugene Smith, emphasizes about the draft of one of George's proposals: "We have weakened the Announcement by making it too philosophical. We are aiming at interesting men of means in the project, and I know that the American business man wants things put succinctly and very straight."[46]

Lafontaine and George make common cause, even though their views of pacifism differ. George is inclined to be internationalist to a fault, while Lafontaine is more attuned to the compromises of practical politics. Lafontaine, for example, does not share George's view that Germany and Austria are victims of militarism, and he cannot imagine a Paneuropean political union.[47] If anything, George is more resolute in honoring his principles than Lafontaine is. When George declines to lecture to the Navy League because it is too jingoistic—even though he desperately needs the honorarium—he records by way of justification: "I am resolutely and at base a pacifist. But I believe that one of the conditions of peace is a solid national defense."[48]

44. HGS, Henri Lafontaine to George Sarton, 6 August 1915. Dewey, who retired from a position as librarian for the State of New York, runs a vacation lodge.

45. YMS.

46. HGS, David Eugene Smith to George Sarton, 5 February 1917.

47. HGS, Henri Lafontaine to George Sarton, 26 July 1915, for Lafontaine's comments on Alfred Hermann Fried's views, transmitted by George, about the Central Powers as victims of militarism; 13 August 1915 for skepticism about a Paneuropean political structure.

48. Diary, 16 November 1915.

Part IV

A Cultural Wilderness, 1915-1919

CHAPTER 10

Terra Nova

At the beginning of the 20th century immigrants arriving in New York saw an America that was youthful, vigorous, and enterprising while at the same time superficial, meretricious, and intemperate. Before them opened a great many ways to make a life. Among the most obscure paths, by comparison with Europe, was the one intellectuals trod. The obscurity derives from American attitudes toward products of the mind. In Europe, wealth supported innovative art and new writing; in America, wealth purchased old art and old writings. Experiments in painting and literature found appreciative European constituencies; America imported the finished products. America, having spanned a continent and having acquired a world-girdling empire, was only beginning to receive attention for its ideas; it valued European visibility, just as it resented European tutelage. These circumstances placed a heavy burden on promoters of new disciplines and styles, especially in areas broadly dominated by European creators. In this climate the Sartons labor to realize their dreams.

Mabel and May arrive in New York on 2 September 1915 and stay with the Baekelands. There Mabel unites with George. Together they charm Henri Lafontaine and his wife: "You have made us forget exile," the Nobel laureate writes to George.[1] Although exile is indeed easy to take at Baekeland's estate, the refugees must move forward. George has been offered the post of librarian at Rice Institute in Houston. Henri Lafontaine argues against taking the appointment: The abominable climate would force George to send his family away for 9 months yearly. The post of librarian "in a state that is very backward will be quite difficult for you, and if your wife wants to devote herself to applied art, she will find few openings in the region."[2] But George sees possibilities in Texas: "I have accepted with certain conditions and I wait for the answer of President Edgar Odell Lovett."[3] The conditions, possibly relating to publishing *Isis*, are not met, and George declines the offer. Family considerations do not enter into the decision, for Mabel writes

1. HGS, Henri Lafontaine to George Sarton, 8 September 1915.
2. HGS, Henri Lafontaine to George Sarton, 8 September 1915.
3. Diary, 19 September 1915. In 1914 George reviewed Emile Borel's lecture about the mathematics of molecular physics, given at Rice's inauguration in 1912. George Sarton, "La Nouvelle Physique," *Isis*, 2 (1914), 193–8, on p. 197.

on 9 September 1915, "I shall extremely regret it if really it is simply my health & May's that would prevent your accepting. *I would be ready to agree to spend the hot months away from you*, half year even, if you thought it good so, & possible." George continues to look for employment. He writes to New York newspapers and meets with the publishers of the *New International Encyclopedia* about extending their work to include his bibliographical project.[4]

George focuses on his plans for a post-war American institute in Brussels. He writes in his diary on 19 September 1915, "My friend Henri Lafontaine is helping me. He is one of the best men in tiny Belgium, and we have many interests in common." Other supporters include Charles William Eliot of Harvard and David Starr Jordan of Stanford, the latter a prominent pacifist before the war. He has also written to chemist William Henry Nichols, Mrs Russell Sage, Jerome Davis Greene of the Rockefeller Foundation, and John David Rockefeller himself.[5] The planning is troublesome. Eliot, who comments to George about the plan,[6] declines to be president of the institute, and Lafontaine tries unsuccessfully to recruit Theodore Marburg, former United States ambassador to Belgium, spending 41 hours with Marburg at Lake Placid.[7] George proposes Elihu Root.[8] Lafontaine thinks that Root will not agree to serve, especially since Root is involved with the Carnegie Endowment which has frozen payments to Lafontaine's institute in Brussels.[9] Lafontaine explains that he and George share the view that Americans "are in many respects superior to other peoples and there is no doubt that they have a more precise idea of democracy than European democrats." If America is not corrupted by prosperity after the war, it can become a model for other countries. But Americans are unaware of their task, and "authorized and progressive Europeans could forcefully help enlighten them."[10] Lafontaine seeks his fortune as a lecturer in California and withdraws from George's project; ironically, in 1919, with the help of Boston philanthropist Edward A. Filene, Lafontaine sets up a small operation in Brussels like the one

4. Diary, 21 September 1915. Daniel Coit Gilman, Harry Thurston Peck, and Frank Moore Colby, eds, *New International Encyclopedia* (New York, 1909), 20 vols. By 1915, only Colby is alive.

5. Diary, 19 September 1915.

6. HGS, Charles William Eliot to George Sarton, 13 July 1915.

7. HGS, Henri Lafontaine to George Sarton, 13 August 1915, 8 September 1915, 24 September 1915.

8. Diary, 29 September 1915.

9. HGS, Henri Lafontaine to George Sarton, 1 October 1915.

10. HGS, Henri Lafontaine to George Sarton, 24 September 1915.

imagined by George.[11] George, for his part, soon finds a position in defense of Belgian interests.

The Sartons live for nearly 2 months at a boarding house near Morningside Park in upper Manhattan for $25/week. George takes May to the aquarium and to the beach on Staten Island. He sees how hard it is to raise children.[12] New York pleases George, who works daily at the library of Columbia University.[13] His eclectic study there ranges from the sublime to the ephemeral. He reads the autobiography of Andrew Dickson White, founder of Cornell University and an active member of the peace conference in The Hague in 1899.[14] He reads William Morgan Shuster's *Strangling of Persia* (New York, 1912).[15] He begins Thomas Buckle's *History of Civilisation in England.* He finds Buckle's reasoning slack, judgments wrong, and notes irrelevant. "Buckle seems to have wanted to give his book, by these voluminous annotations, an encyclopedic character. However this is a big mistake in composition: a book must not be made like an encyclopedia. On the contrary, it must contain only what relates directly to its subject." No stranger to rambling prose, George recognizes his weakness from the start of his career.[16] And George enjoys the *Ladies' Home Journal.*[17] He is certainly no snob.

The position George lands is related to his correspondence with Robert S. Woodward of the Carnegie Institution, whom Lafontaine specifically enjoined him from contacting, and to his interest in things Asian.[18] George obtains Woodward's support to plan for an American center of information in Belgium after the war. George then asks Woodward to fund his travel to East Asia in preparation for writing a synthetic history of science, for "there is not, at the present time, a single manual on the history of science, which is of real value." Woodward is pleased at the prospect of George's history of science, for he is "profoundly interested" in the subject. About 8 years previously, he corresponded with John Theodore Merz, the

11. HGS, Henri Lafontaine to George Sarton, 8 September 1915; Henri de Man to George Sarton, 10 July 1919, about the Belgian Committee in Brussels set up by Paul Kennaday, manager of the Foreign Press Service, with the help of Edward A. Filene and Henri Lafontaine.

12. Diary, 1 and 3 October 1915, 24 and 27 September 1915.

13. Diary, 5 October 1915.

14. Diary, 19 September 1915. Andrew Dickson White, *A History of the Warfare of Science with Theology in Christendom* (New York, 1896).

15. Diary, 29 September 1915.

16. Diary, 24 & 27 September 1915.

17. Diary, 1 October 1915.

18. George thinks that the Catholic view of the world is more sane than the Protestant one. Catholics, unlike Protestants, can understand the eroticism of India. Diary 6 October 1915.

polymath author of an enduring history of science and philosophy in the 19th century, about extending the topic to achieve something like William Whewell's history of the inductive sciences; Woodward has also discussed the subject with his former colleagues at Columbia University. "But the project has, in general, appeared too large and to involve too much co-operation to be practicable of execution." Woodward has no money for George, but he will be glad to talk with him, for he has received a favorable recommendation from Leo Baekeland. George replies promptly with irritated comments about William Whewell, the man who banned Darwin's *Origin of Species* from the Library of Trinity College, Cambridge. Whewell's book "is so full of errors, that it is really better not to use it at all....What makes the value of his book, is that there is hardly any other," for Merz (whom George noted in his Critical Bibliographies) deals only with the 19th century.[19] "The history of everything has been elaborately studied," George concludes, "except of that which is the most obvious characteristic of our civilization!!"

Powerful men can be flattered by receiving personal letters about high-minded things. As Woodward emphasizes to George later: "Of the very considerable number of the correspondents of the Institution you are one of only three or four who are disposed to give attention to theoretical and idealistic considerations."[20] Idealism, certainly, distinguished George from more pragmatic correspondents; Plato's heavens over Aristotle's practical compromises. George notes in his diary at this time that the war has proved "that our civilization is only a veneer, an illusion. We are not morally adapted to the marvelous scientific and industrial progress of our time."[21] George speaks with Nevil Monroe Hopkins and Charles Edward Munroe at the Carnegie Institution in Washington. He gives a talk on history of science and is promptly named secretary of the Belgian Scholarship Committee, the organization for supporting Belgian academic exiles which George had contacted in April 1915.[22] A measure of Lafontaine's isolation in America is his complete ignorance of the Committee.[23]

George has lunch with the D. E. Smiths on 20 October, and sees the Belgian consul Pierre Mali in the evening. He packs up the family for a

19. John Theodore Merz, *A History of European Thought in the Nineteenth Century* (Edinburgh, 1896–1914); William Whewell, *History of the Inductive Sciences, from the Earliest to the Present Time*, (3 vols, London, 1837; 2 vols, New York, 1865).

20. CIW, George Sarton to Robert S. Woodward, 6 October 1915; Woodward to Sarton, 9 October 1915; Sarton to Woodward, 18 October 1915; Woodward to Sarton, 10 July 1918.

21. Diary, 18 October 1915.

22. Diary, 15 October 1915.

23. HGS, Henri Lafontaine to George Sarton, 22 December 1915.

boarding house in Washington, where they arrive on 22 October. George counts $430 to his name.[24] In his diary, George notes on 22 October 1915 that he has £10 in a London bank, and that he left 2 books with Harold Stabler: *The Making of the English Language*[25] and Henry Watson Fowler and Francis George Fowler's *The King's English* (Oxford, 1908).[26] George rarely refers to his inventory of shirts or socks, or to the stock of his pantry. No tailor's bills remain in his papers. But there is a clear trace of books. Later George tallies his American income for 1915. It comes to $1812.43. His revenue derives principally from the Belgian Scholarship Committee, although significant additions originate with Paul Carus at the *Open Court* and with his lecturing at George Washington University and the University of Illinois.[27]

The Washington stay begins badly. The boarding house is expensive and unpleasant. Mabel and May complain. May attends the Little Potomac School, paid by Mrs Hopkins. George works mornings at the Belgian Scholarship Committee in the Wilkins Building, afternoons on history of science lectures, sometimes in the Library of Congress, where he has a desk, courtesy of Herbert Putnam, the Librarian.[28] He worries about money. He worries about Mabel and May: "If they hold on, I will hold on," he paraphrases a cartoon about soldiers in the war. Mabel sees George's private remarks as a recrimination. She writes in his journal, in French, that she is not weak or selfish. She is not a civilian remaining far from the trenches. She is fighting as hard as she can. She and May have not been complaining. She is trying to find work. She will move heaven and earth to earn money.[29] And almost as soon as he is installed as its secretary, George wants to quit the Belgian Scholarship Committee to find a better-paying university position.[30] He prefers to call his own tunes. If he promotes schemes, he wants to benefit from them. In his diary on 28 October, for example, George notes that Hopkins spoke to him about a project to create a naval school on a ship, where George would teach the students. Water-rat George never mentions the project again.

On First November George calls on Robert W. Woodward, about "the world city to be located in Belgium." Through the war and into the peace, George schemes to construct a large institute in Belgium, ostensibly for

24. Diary, 20, 21, 22 October 1915.
25. Possibly, Henry Bradley, *The Making of English* (New York, 1904).
26. NYPL.
27. Agenda.
28. Diary, 16 August 1955, for Putnam's kindness. George likes the Library of Congress. Diary, 10 June 1916.
29. Diary, 23, 25, 27 October 1915.
30. Diary, 11 November 1915.

the promotion of American culture and ideals; the schemers pay him for his trouble. Here we see the institute in terms of something grander, Hendrik Christian Andersen's world city, which George analyzed in his essay on Belgium in the *Open Court*. George imagines adding history of science to the institute and in this way realize Andersen's dream. Among the supporters on George's masthead, we read Franklin Delano Roosevelt (as Assistant Secretary of the Navy, no doubt enlisted by the admirals at George Washington University) and the National Geographic Society's Gilbert Hovey Grosvenor. George asks Woodward for his name on the letterhead, too, and he adds in a postscript, "I do not give up the idea of having an American Institute created in Brussels."[31]

George begins lecturing at George Washington University at the end of October.[32] His official appointment comes through a month later. The appointment pays only $200, about the same amount as a few articles in the *Open Court*.[33] Having sought similar positions in Belgium, George is no stranger to this kind of academic exploitation. The commitment of time is about a dozen hours per week, although it gives the lecturer entree to colleagues and graduate students. Both the institution and the lecturer find advantages in the arrangement. Later in the fall semester, Otis Dow Swett, assistant professor of chemistry at George Washington University, arranges to take a minor field in history of science under George for his doctorate.[34] Swett becomes George's first student in history of science.

The Sartons spend Thanksgiving with the T. Wayland Vaughans and with Edith and Ruth Putnam. Vaughan, a product of Tulane and Harvard universities who studied music in Europe, is a sedimentologist who eventually directs the Scripps Institution of Oceanography—a boring speaker, according to George—whose daughter is friends with May. The Sartons move into an apartment at $52/month, where Mabel will have a part-time servant. On Sundays, George typically takes May to the nearby Washington Zoo. It is "marvelous," a kind of Eden in the city. But years later May recalls only that she fell asleep listening to the howling wolves.[35] May contracts chicken pox, but she behaves well at the Christmas party thrown by President Stockton of George Washington University, where she is spoiled with toys, and also at the children's dinner arranged by the Belgian Embassy.[36]

31. CIW, George Sarton to Robert S. Woodward, 2 November 1915, first and second letter.

32. Diary, 30 October 1915.

33. HGS, Charles Herbert Stockton to George Sarton, 23 November 1915.

34. Diary, 3 December 1915.

35. Diary, 26 November and 30 November 1915; 9 December 1915 for a boring talk by T. Wayland Vaughan at the Cosmos Club; *Phoenix*, p. 93 for the wolves at the zoo.

36. Diary, 9 and 25 December 1915.

Henri Lafontaine introduces George to Harriette P. Shadd, a "very friendly socialist comrade" who will help the Sartons in Washington.[37] George meets her in the fall of 1915.[38] On 17 January 1916 he dines with Mrs and Miss Shadd and with the socialist journalist Charles Edward Bellyse Russell and his wife.[39] Mabel and Harriette Shadd conceive of a business partnership to sell Mabel's designs. George will not consider it. He writes to Harriette Shadd on 18 February 1916 that while he welcomes Mabel's serving as an "art contributor & adviser," he considers "with distrust and displeasure any closer partnership between you, especially any commercial partnership." If Mabel goes ahead with the partnership, it will be against George's advice and at the risk of displeasing him. Mabel needs leisure and tranquility, which business will not provide.[40] He confides in his diary that Mabel is too weak physically for the enterprise. He "often worries that she has too many concerns, that the task of living is too hard for her." May worries him, too. "Her future already gives me great concern. She will turn out very well or very bad."[41] George concludes that he must find another apartment. "It is quite sad, especially for Mabel." He is unhappy with intellectual life in Washington. "It would be hard to find a type more representative of snob and dilettante than Nevil Monroe Hopkins."[42]

The Sartons move, the new apartment is furnished by their friends, and Mrs Vaughan takes May for a few days. Social engagements pick up. George and Mabel dine with Clinton Paul Townsend and his wife,[43] then with Hendrik Willem Van Loon at the Vaughans'. Van Loon, author of several books on the history of the Netherlands, is teaching at Cornell University, and he has undertaken to organize a lecture by George.[44] "He is a very spiritual and dilettantish boy." He has two children, Willem and Häntschel (Hansje). (Willem becomes a close friend of May Sarton's.) George lunches with Oscar Terry Crosby, the engineer and explorer. George meets with Emile de Cartier, who again strongly advises against going to China.[45] George dines with Arthur Gleason, an American cited in the report

37. HGS, Henri Lafontaine to George Sarton, 8 September 1915. Agenda, after 18 October 1915.

38. Agenda, after 30 October 1915.

39. Diary, 17 January 1916.

40. YMS.

41. Diary, 21 February 1916.

42. Diary 21 January 1916.

43. Diary, 27 & 31 January 1916. Townsend, a lawyer, joins the Board of Directors of Frederick Gardner Cottrell's Research Corporation in 1925.

44. HGS, Hendrik Van Loon to George Sarton, 15 November 1915. Cornelis A. van Minnen, *Van Loon: Popular Historian, Journalist, and FDR Confidant* (New York, 2005), pp. 41–6 ff. for the Van Loons in Washington at the beginning of the war.

45. Diary, 8 & 10 January 1916.

by James Bryce, Viscount of Dechmont, on German atrocities in Belgium.[46] Gleason relates the bravery of Irénée van der Ghinst. The Sartons dine at the Van Loons with DeWitt Clinton Poole, US ambassador to France at the outbreak of war.[47] George meets philologist Albert Joseph Carnoy of Louvain, teaching at the University of Pennsylvania. His wife is a childhood friend, Madeleine Silvercruys, who pledged to marry no one but George.[48]

George lectures. He speaks on Joan Baptista van Helmont at Washington University's College of Pharmacy.[49] On 15 January 1916, in Washington, the topic is history of science in antiquity. Arnold Carl Klebs of Nyon, Switzerland—in Washington for several years—whose publications were noted in *Isis* before the war, is in the audience. Klebs reports that Karl Sudhoff, professor of history of medicine at Leipzig, spoke about George often.[50] Klebs, a bibliophile, is vitally interested in George's Critical Bibliographies. Klebs wants to invent a method for producing bibliographies automatically: "Our crying need is a bibliography made by any average bibliographer who will run through tools & articles after some pre-conceived scheme and list the items so that the scholar can pick without having to wade through endless mud."[51] Klebs, who remains one of George's fellow-travelers for the next quarter century, is a harbinger of Eugene Garfield's Institute for Scientific Information, founded in the 1960s.[52]

The Sarton family suffers from the strain of dislocation and poverty. On 22 February 1916 Mabel writes to George that they don't understand each other as well as they could:

---

46. Great Britain, Committee on Alleged German Outrages, *Report...Appointed by His Majesty's Government and Presided over by Viscount Bryce* (London, [1915]).

47. Diary, 27 & 31 January 1916.

48. Diary, 13 February 1916. George adds that he would not have married her. In 1914 the Carnoys received hospitality from Catherine Durning Holt Dampier-Whetham in Cambridge. Dampier-Whetham to George Sarton, 8 November 1914.

49. Diary, 3 February 1916.

50. Diary, 15 January 1916.

51. HGS, Arnold C. Klebs to George Sarton, 19 April 1916.

52. For example, Eugene Garfield's inauguration of "a *total information system for the social sciences*" in 1973. Garfield, *Essays of an Information Scientist* (Philadelphia, 1977), *1*, 397–9. It shall be seen that George falls out with Klebs, but he nonetheless honors Klebs by publishing Klebs's bibliography of incunabula in volume 4 of *Osiris*, the journal companion to *Isis*, in 1938. The other monograph in volume 4 is Robert King Merton's revised doctoral dissertation on science in 17th–century England. About Klebs's text, Sarton's student I. Bernard Cohen is unambiguous: "This wholly unreadable bare-bones checklist would not have been of much interest or use to scholars in general. I, for example, have never before had occasion to refer to it during an active scholarly career spanning almost five decades, nor have I ever had occasion to make a careful examination of its contents." I. Bernard Cohen, "Introduction: The Impact of the Merton Thesis," in *Puritanism and the Rise of Modern Science: The Merton Thesis*, ed. Cohen (New Brunswick, N.J., 1990), pp. 1–111, on p. 31.

> It is not your fault if you don't quite understand me because I answered you so badly and so stupidly when you talked to me the other morning. And afterwards I knew I had only made matters worse but I was as if paralysed. I feel like a criminal against whom the circumstantial evidence is so strong that though innocent, he gives up his defense. The judge will not believe him. What is the good?
>
> Also I am very tired and your disapproval is like a dead weight on my shoulders that grows heavier each hour, that and other things together, a feeling of failure as regards May. I do not feel I am successful with her bringing up, & often I am filled with dread that she is not strong through *me* (I mean by direct inheritance.)

Mabel longs to rest, but she is restless, driven to create. It is something beyond Mabel's power to control, like George's melancholy. George "spoke of worry driving you nearly mad. I have often—no—sometimes felt a strain as if I were on the verge of madness." Mabel committed herself to the enterprise with Harriette Shadd "so far that it was extremely difficult to draw back *before* you shewed so strongly your absolute disapproval." Shadd received a letter from George, and she did not show it to Mabel. Mabel "believed that this venture could help us (& that belief & hope dies hard in spite of the blows you have dealt at it!)." The enterprise would help Mabel's emotional state, and it could help the family materially: "I did it principally I do believe in the—to you, mad & childish hope—of being of material help to you & us all!" Mabel is now committed to help Shadd in her business, showing her about color and design. Mabel thinks that she is not as capable of raising May well as of designing well. "I do not think that it is good for her to be much with me—that is a sad thing to say & I almost *pray* that it may change." The Wild Bird speaks. George opens a bank account for Mabel. She will have $100 per month for household expenses, except rent and personal items, and in addition $50 in reserve.[53] Mabel continues with Harriette Shadd, who runs a retail establishment. On 28 April 1916, Mabel reports that Shadd is away and so she has had extra work, including "much work for the Gift Shop."

George lectures at City College of New York at the end of February 1916. Upon arriving, he sees that his valise is filled with a woman's clothes. He took the wrong bag upon leaving the train. As a result, he had to invent a new talk and speak without notes, "and this is certainly the *best* talk I ever gave! Very great success." Afterwards, at the Chemists' Club as the guest of Baekeland, he meets, among others, biologist Jacques Loeb and physiological chemist Takamine Jokichi. Elwood Hendrik, the correspondent of the Greek emigré to Louisiana and Japan, Lafcadio Hearn, is his host for dinner. Baekeland is a scientific genius who understands nothing artistic or religious. "He is essentially utilitarian. What a pity that the men

53. Diary, 7 April 1916.

who govern us are not of his mold! Men of state should be like him."[54] Still in New York on 8 March, George spends an evening with Alexander Graham Bell at one of his salons. There he meets Gilbert H. Grosvenor, a relative of Bell's. Plant explorer David Fairchild speaks to him enthusiastically about the botanical gardens at Buitenzorg on Java. May accompanies George and stays with the Baekelands. George thinks that she may have diphtheria, on 27 February; he has her inoculated against the disease.[55]

Before the war, George had applied to the Paris Academy of Sciences for the Prix Binoux to support his publication of *Isis*. In 1916, the Academy gives him a portion of the award. Astronomer Guillaume Bigourdan has the task of informing George of his prize. Before George hears from the relaxed astronomer, George van Biesbroeck, his sometime colleague at the Uccle Observatory in Belgium, conveys the news. Van Biesbroeck began a 9-month appointment at the Yerkes Observatory in Williams Bay in September 1915, obtaining leave to do so from the occupying German authorities. Van Biesbroeck copies out the announcement in the *Comptes-Rendus* of the Academy.[56] George nets F 666. George requests that F 500 go to the American Commission for Relief in London, for transmission to François Bouny. The rest is for George's London bank account. This means that George has repaid F 610 of the F 1000 that Bouny lent him in 1914 at the war's onset. [57]

George finishes Giovanni Jacopo Casanova's memoirs, which he enjoys. He is the better for it, "ironically."[58] His libido is at a height in March, to judge from the entries for sex in his agenda. It is with Mabel, circumstantially. George returns to Washington and dines with Mabel at the Vaughans.[59] He gives his last talk at George Washington University. He does not like the students, who are distracted. In the middle of speaking about the spread of Arabic numerals, he dismisses the class. Afterwards, Herbert Horace Bunzell, professor of physiology, speaks to him about the country's sexual life.[60]

George's fellow exile Henri Lafontaine is unhappy with California, he writes to George. "The race for the dollar is the main preoccupation, and capitalism abuses its power with impunity." At Berkeley, President

54. Diary, 16 February 1916.

55. Diary.

56. HGS, Guillaume Bigourdan to George Sarton, 11 April 1916; George van Biesbroeck to George Sarton, received 10 March 1916; George van Biesbroeck to George Sarton, 3 March 1916. Van Biesbroeck left his wife and children in Belgium.

57. Diary, 10 and 12 March 1916.

58. Diary, 20 March 1916; presumably not all 8 volumes of the French edition of 1885.

59. Diary, 22 March 1916.

60. Diary, 22 March and 24 March 1916.

Benjamin Ide Wheeler is pro-German and a notorious conservative, and he has denied Lafontaine the opportunity to speak.[61] Lafontaine has tried to change the mind of Nicholas Murray Butler at the Carnegie Endowment about releasing funds to resume the periodical, *La Vie internationale*, but Butler will not change his policy of "hopeful abstention" from dealing with nations involved in the war. Lafontaine explained to Butler that his bibliographical center was not really pacifist—an unusual argument from the Nobel laureate in peace. Lafontaine and his wife Mathilde are inspired by a Theosophist center that he toured, and he forwards information about it.[62] Lafontaine imagines a war with the Americas on one side, perhaps allied with Great Britain, against Europe and Asia:

> All the campaign and the preparations are leading to this adventure. The capitalists who inspire and finance them have this foolish idea that they have the right to dominate the world. The profits that they continue to make from the war and the accumulation of gold in American vaults have literally aged them, and they have become victims of the same megalomania that has ravaged Germany.

He fears a silent governmental coup by the plutocrats.[63]

Like Lafontaine, George dreams, but he is also practical. He writes to university presidents for next year's job. If nothing comes up, "I will no doubt become what is called a publicist."[64] George reflects on his first year in America. It "has been for me the best education for the time, a true revelation, a considerable spur to work." He records his notion for a history of science, beginning with an image recalling Ramón Llull's medieval *Arbor scientiae*.[65] It will be:

> a history of humanity through the evolution of the sciences. Of course, the basis of the picture would be the principles of general history, but the center of interest would be the growth of the Tree of Sciences. Frequent allusions would be made to the development of social conditions, conditions of religion, literature, arts, etc.
>
> It will be truly a history of civilization from the highest point of view: A history of the conquests by Man of Nature and of Mystery.
>
> The general history forming the background of the picture will itself have an original character. I want to place a very great importance on international relations, not morbid relations (war, etc.) which historians

61. HGS, Henri Lafontaine to George Sarton, 22 December 1915.
62. HGS, Henri Lafontaine to George Sarton, 27 February 1916 and 9 April 1916.
63. HGS, Henri Lafontaine to George Sarton, 12 May 1916.
64. Diary, 27 March 1916.
65. George Sarton, *Introduction to the History of Science, volume 2: From Rabbi ben Ezra to Roger Bacon* (Baltimore, 1931), p. 907.

> are more interested in, but normal and healthy relations: intellectual and commercial exchanges between peoples. This history of international relations, especially in antiquity and the middle ages, is very interesting in itself, and has not been written. My history of science will be thus the first history of internationalism.

He will emphasize East-West relations. He is already assembling information for the first volume on pre-Hellenic science. When the book is published, he will reserve pages in *Isis* for its criticism. To help the reader, he will number the book's paragraphs. On the topic of his scholarly journal, when he resumes *Isis*, he proposes some changes: Articles will be brief and "to the point"; there will be many plates and illustrations; there will be a critical bibliography and popular articles; he will emphasize international relations; it will feature addenda and corrections of his work; there will be a section on questions and responses. "I am Belgian," he concludes, "but I am above all a *Citizen of the World*."[66]

George finishes lecturing to graduate students at George Washington University on 14 April 1916; on the 16th he arrives in Worcester, Massachusetts, for 6 talks on history of science plus one on rebuilding Belgium. Worcester, a center of industry in the 19th century, is a "sad city." He is hosted by Ralph Stayner Lillie, head of Biology at Clark University. In his diary he notes that the Lillie home has "no beauty nor imagination." It is a lesson for him to treat Mabel better and have their home provided with "sun, gaiety, and imagination."[67] He writes on 19 April to "Petit Coq": "The atmosphere of the house is a bit grey. There are few beautiful things and they themselves look so wise and good!" The Lillies summer at Woods Hole, their usual retreat.[68] George had been looking forward to the spring, and he finds himself back in winter. He has no enthusiasm for his first lecture.

Mabel writes from their Washington apartment on 14 April. Mabel reports that everyone finds Worcester gloomy. Her mother wrote about a zeppelin raid on Ipswich, where an explosion threw her to the ground. Mabel and May had Easter dinner with the Van Loons. "The table was so pretty, with a dear little grey rabbit in the middle that was for May, & coloured eggs all round, [illegible] for pudding, ice-cream eggs of all colours on nests of spun white sugar!" They rolled eggs at the White House. In the afternoon, they looked for eggs at the Zoo. The Lafontaines visited. Mabel finds Mathilde Lafontaine "the very antithesis" of Henri-Marie, "& she must surely do him

---

66. Diary, 13 April 1916.

67. Diary, 14 and 16 April 1916.

68. Ralph S. Lillie was the brother of Frank Rattray Lillie, at the time director of the Marine Biological Laboratory at Wood's Hole and chair of the Zoology Department at the University of Chicago; the brothers were born and raised in Toronto.

a good deal of harm by her tactlessness & fanatical ideas." In French: "Truly I find her a terrible woman." Mathilde Lafontaine came to her studio, where Eliza Van Loon was helping Mabel with embroidery. "Mme La Fon. behaved *shamefully*." In French: "She is a *hateful*, nasty, *stupid* woman." And the worst of it: "Then she spoke about Niggers, finally a detestable mind." Mabel finds Mathilde Lafontaine "just as the worst Germans are, with their stupid and criminal hatred." She hopes that the young German woman whom she engages to embroider does not meet her.

George, on 22 April 1916, has given 3 lectures and "gained the sympathy of my audiences." He has still heard nothing from Harvard, and he assumes that he shall lecture in the coming week. There are "*very fine people*" at Clark University, "but the city itself is gray and cold." He has visited the art museum—a nice Whistler, a doubtful Bronzino. George worries about the war. He fears German intrigue in Mexico, but he hopes that the Netherlands will endorse Woodrow Wilson's ultimatum. He dines with physicist Arthur Gordon Webster.[69] On 25 April, George wonders why Wifie (a usual salutation) does not write. He will dine with the university president, psychologist G. Stanley Hall, "a very interesting & broad-minded man." When he goes to Harvard to lecture shortly, he will dine with Lawrence J. Henderson. Henri Lafontaine, George reports, wrote "a very complimentary letter about you both." He is reading William Makepeace Thackeray's *Vanity Fair*. He encloses a check for $100 "for the next month, lest I forget it."

On 27 April 1916 George writes in French from Harvard, on a postcard, that his talk "made a very good impression. It seems increasingly probable that I will be appointed Lecturer for 2 years." He imagines that his salary will be $2100. In his diary, he notes that his lecture was attended by all the students of chemist Theodore W. Richards and the philosophical seminar; he dined with Henderson, who is Richards's brother-in-law.[70] Back in Worcester, he tours Webster's laboratory, James Edmund Ives's laboratory on Righi rays, and John W. Baird's laboratory in psychology, and he lectures on Poincaré; he visits Worcester Polytechnic Institute.[71] On the 29th, George is saddened but not surprised at Mabel's impression of Mme Lafontaine. "Do not forget however that she is the faithful wife and companion of Mr Lafontaine." Later, when George meets Henri and Mathilde Lafontaine on

69. Diary, 23 April 1916.

70. Diary, 27 April 1916.

71. Diary, 28 and 29 April 1916. In 1908 Augusto Righi, a highly-respected Italian physicist, affirmed the existence of a new kind of ray produced in cathode-ray tubes, composed of electrons bound with positive ions. The so-called magnetic rays generated scores of publications until 1918, when a special identity for Righi's rays evaporated. Helge Kragh, *Quantum Generations: A History of Physics in the Twentieth Century* (Princeton, 1999), p. 36. Ives received a PhD from Clark University in 1901.

27 May 1916, he finds them both greying and sad. "She is certainly a woman with little sense, and it is sure that she will greatly harm her husband. She has so much hate, and can't keep herself from criticizing...Poor people." But Lafontaine has spread ideas that George believes in: "pacifism, internationalism, universal bibliography, socialism. He is a great spirit, and one of the great Belgians of our time."[72]

George is back in Harvard on 30 April, hosted by Henderson, who hopes to work with Richards and Edward Clark Streeter, a rich devotee of history of medicine, to appoint George as a lecturer and share the burden of teaching history of science.[73] George lunches with the librarian Archibald Coolidge, Streeter, the specialist on Indian philology Reinhold Friedrich Alfred Hoernlé (whose work is noted in volume 1 of *Isis* and whom George had enlisted for the aborted no. 6 of the journal),[74] and William S. Ferguson in ancient history. George meets Léon Dupriez of the University of Louvain, who, on account of the long-standing admiration for his publications by Harvard's President Lowell, moved to Harvard directly after the German invasion of Belgium.[75] George is typically unimpressed by this and other Belgian professors who receive good positions in exile. In George's view, Dupriez is an old schoolteacher with a bourgeois, "narrow-minded" wife, Marie-Louise-Caroline *née* Verriest; it is a "discouraging visit."[76] Notwithstanding George's sentiment, the Dupriez family grow close to Mabel.

Within a week George's Harvard appointment comes through. He will be engaged by Harvard for 2 years, 1916–1918.[77] His salary has been arranged by Baekeland and several other friends, who collected $3400 for the purpose,[78] although in the first year he receives only $2000.[79] In retrospect George refers to Harvard as a stern mistress that deprived him of a regular position until 1940. Though a university is commonly known as an *alma*

72. Diary, 27 May 1916. Lafontaine's plan for a "Magnissima Carta" for the League of Nations in Theodore Marburg, *Development of the League of Nations Idea: Documents and Correspondence of Theodore Marburg*, ed. John H. Latané (New York, 1932), *2*, 786–7.

73. John T. Edsall, "Lawrence J. Henderson and George Sarton," *Isis, 75* (1984), 11–13.

74. *Isis, 1* (1913), 547; George Sarton, "Why *Isis?*" *Isis, 44* (1953), 232–42, on p. 242.

75. Henry Aaron Yeomans, *Abbott Lawrence Lowell, 1856–1943* (Cambridge, Mass., 1948), pp. 499–500.

76. Diary, 30 April and 1 May 1916. Dupriez was a liberal who promoted the vote for women. Léon Dupriez, *L'Organisation du suffrage universel en Belgique: Vote plural, vote obligatoire, représentation proportionnelle* (Paris, 1901).

77. CIW, George Sarton to Robert S. Woodward, 6 May 1916, announcing the appointment. The official notice: HGS, James Haughton Woods to George Sarton, 10 May 1916.

78. For the origin of the money: Hugo Collumbien, "Even memoreren: 100 Jaar geleden werd George Sarton Geboren," *Ghendtsche Tydinghen*, 1984, pp. 314–318; Roger Calcoen, "George Sarton," *Handelingen der Maatschappij voor Geschiedenis en Oudheidkunde te Gent, 34* (1980), 181–91.

79. HGS, Lawrence J. Henderson to George Sarton, 3 May 1916.

mater, literally a fostering mother, as an institution it possesses neither conscience nor morals: Human qualities reside in the university's staff. For better and worse, Lawrence Joseph Henderson is indissociably connected with George's life at Harvard. Henderson identifies George as a *locum tenens* for his own course in history of science, which takes time away from his biochemical research. Henderson sells George as good value to President Lowell, an old friend; George indeed costs Harvard very little, and the university benefits greatly from his presence.[80] He will be lodged with the philosophers, most of whom—with the conspicuous exception of Hugo Münsterberg—staunchly support the French in the world war.[81]

George returns to Washington on 2 May. He is happy to see Mabel and May. He is troubled that the Germans are trying to make use of the racial question in Belgium, setting Flemings against Walloons. He wonders if it will lead to "civil war after the war." The Germans have opened a Flemish University in Ghent—a goal of Flemish nationalists—with only 7 professors. Henri Pirenne and Paul Fredericq have been deported to Germany.[82] Anne Pelseneer was also deported, sentenced to 9 to 12 months in prison for being impolite to a German general on the tram.[83] These disquieting events are a prelude to sex on 4 sequential days, from 10 through 13 May, as George records in his diary.

Through the offices of the Carnegie Foundation, which probably contributes the money, George receives an offer of $300 for several months of summer school at the University of Illinois.[84] The university's president,

80. Yeomans, *Abbott Lawrence Lowell*, p. 541. John Parascandola, "L. J. Henderson and the Mutual Dependence of Variables: From Physical Chemistry to Pareto," in *Science at Harvard University: Historical Perspectives*, ed. Clark A. Elliott and Margaret W. Rossiter (Bethlehem, Pa., 1992), pp. 167–90, on pp. 177–8.

81. Bruce Kuklick, *The Rise of American Philosophy, Cambridge, Massachusetts, 1860–1930* (New Haven, 1977), p. 439. Kuklick provides much information about the philosophical views of many of George's colleagues.

82. Gilbert Lannoy, *La Renaissance de la communauté scientifique flamande* (M.Sc. thesis, University of Montreal, 1975), pp. 68–9; Bryce Lyon, *Henri Pirenne: A Biographical and Intellectual Study* (Ghent, 1974), p. 215. On Fredericq: Louis Fredericq, *Notes sur la famille Fredericq-Beaucarne* (Ghent, n.d.), pp. 95–9, lent to me by Suzanne Fredericq. Paul Fredericq's crime was "fomenting public discontent against the projected opening by the German administration of the University of Ghent as a Flemish university." He stated in 1916 that while he was able to lecture in Flemish, he did not want to do so. Hoogeschool te Gent, *Plechtige Opening op 21 en 24 October 1916* (Ghent, 1916), pp. 39–42, lists 12 full professors and 23 faculty with other ranks.

83. Diary, 2 May 1916. HGS, Paul Pelseneer to George Sarton, 17 December 1918, where Paul elaborates on the prison sentence of Anne.

84. Diary, 17 May 1916. HGS, Edmund Janes James to George Sarton, 15 May 1916; W. S. Miller to George Sarton, 19 May 1916: "The Carnegie Foundation has arranged with him [President James] to have you give a series of lectures–one each day–throughout our Summer Session." Miller asks George for the general topic of his lectures.

Edmund Janes James, is solicitous on George's behalf. He places George in contact with his friend Walter Libby at the Carnegie Institute of Technology in Pittsburgh, who hopes to publish an introduction to the history of science with a preface by Sir William Osler. James reports that Libby has spent the summer in England and has learned about a series in history of science to appear with the Clarendon Press at Oxford. In 1917 there will be a summer research institute in history of science at Oxford.[85] George is aware of Osler's interest in history of science, but here he learns about the accelerating plans of Osler and his collaborators Charles Singer and Dorothea Waley Singer. It takes George only 3 years to cement a firm relationship with the Oxford group.

Given George's emphasis on scientists over humanists, it is natural for him to form a connection with Cambridge before finding his way to Oxford. George's contact in Cambridge, Philip Jourdain, shares George's belief about the idealistic end of history of science: "It is possible that all the wars, cruelties and sufferings of the past will never teach mankind to place reason forever above brutal might, but it is certain that their knowledge of the world will gradually and incessantly increase."[86] In 1914 Jourdain applauds George's scheme for a British "Clearing Library"; the English editor for Paul Carus's *Open Court*, Joudain encourages George to write about *Isis* and also about Belgium, offering £1 per thousand words.[87] In April 1916, Jourdain reviews Aldo Mieli's *Scuole ionica* for the British philosophical periodical, *Mind*. There Jourdain signals George's pioneering efforts (Mieli's book is dedicated to George):

> The man who did most to found the new school [of history of science] was the gifted Paul Tannery, and the ambitious work of organising and co-ordinating historical work on science in general was carried on with great enthusiasm and ability by M. George Sarton, who published his excellent review *Isis* near Ghent for rather more than a year until the atrocious devastation of Belgium took place. M. Sarton, like many other literary men of Belgium, was forced to leave his work and his country. Prof. Mieli was one of those who worked with M. Sarton to make *Isis* worthy of its founder's ideals.[88]

Throughout the war, Jourdain, in declining health, accumulates reviews for George. In August 1916 he proposes a triumvirate to run *Isis*, when

85. HGS, Edmund Janes James to George Sarton, 28 June 1916. Libby is author of *An Introduction to the History of Science* (Boston, 1917).

86. George Sarton, "Philip E. B. Jourdain (1879–1919)," *Isis*, 5 (1923), 126–9.

87. HGS, Philip E. B. Jourdain to George Sarton, 8 December 1914 and 22 December 1914.

88. Philip E. B. Jourdain, "Aldo Mieli, *La Scuole ionica*," *Mind*, no. 98 (April 1916), pp. 261–5, on p. 262.

it reappears: Under George's direction, he could take the mathematical and logical part of the journal; Aldo Mieli would handle the chemical and physical part, for Jourdain sees George and Mieli as rising stars in history of science.[89] Further discussion of the relationship stalls with Jourdain's disability and premature death in 1919.

On 18 May 1916 George travels from Washington to Cornell University. George Lincoln Burr, librarian and professor of medieval history there, wrote to him several days before, suggesting that he contact historian John Franklin Jameson at the Carnegie Institution about setting up an independent institute for history of science; Burr also relates the American appeal for the release of Pirenne and Fredericq from detention.[90] George notes in his diary on 19 May that Mrs Baekeland heard his talk at Cornell. He spoke for a long time with Andrew Dickson White. "That a great institution like this one could develop in less than 50 years, nothing shows better not only the material power of this country, but more especially its moral power, its idealism." In several centuries, Ithaca will be a "pilgrimage" site, like Bologna, Oxford, and Leiden. Americans will come here to venerate "the cradle of the American *renaissance*." George returns to Washington and elaborates plans for an institute of history of science. After his Lowell Lectures in the fall, he will write an article about it. He will contact David Eugene Smith, Streeter, Henderson, Richards, Klebs, Louis Charles Karpinski, Florian Cajori, Herbert C. Hoover, A. D. White, Frederick E. Brasch, and university presidents.[91] He does not record that virtually no science professors attended his talk at Cornell.[92]

In a typewritten text of an article appearing in March 1917, composed after June 1916, George outlines "an *Institute* dedicated to the History of Science and Civilization." He emphasizes that "science is the very bedrock of our modern society, it is 'man's true universal language,' and it is the strongest force that makes for the social unity of our civilization." Immanuel Kant proposed that "the object of universal history" was "the growth of a universal community pursuing a common purpose." More than a century later, that purpose is seen as "the creation and diffusion

89. HGS, Philip E. B. Jourdain to George Sarton, 29 July 1915 and 12 August 1916.

90. HGS, George Lincoln Burr to George Sarton, 15 May 1916. Peter Novick, *That Noble Dream: The "Objectivity Question" and the American Historical Profession* (Cambridge, 1988), p. 114 for the petition to release Fredericq and Pirenne. Burr assisted Andrew Dickson White on White's *Warfare of Science with Theology in Christendom.* Henry Guerlac, "Sartoniana and Forward," *Lychnos*, 1986, pp. 1–27, on pp. 17–18.

91. Diary, 25 May 1916.

92. HGS, Jacob Schurman to George Sarton, 27 October 1917, citing a dean at Cornell: "And I am not sure that we could interest the professors and students in the various sciences of physics, chemistry, biology, etc., in Professor Sarton's lectures; unless I am mistaken, there was hardly a representative of the teaching staffs of these departments present at his lecture before."

of beauty and knowledge." Science alone is universal and progressive. History of science has as a goal "the genesis and the development of scientific facts and ideas." All the sciences must be studied together, a point emphasized by Auguste Comte, who understood that the history of science "is the logical basis of any philosophy of science and indeed of any philosophy." George desires the institute to be engaged primarily in research and only secondarily in the training of teachers. His model is the astronomical observatory, where each associate has a task "according to a general plan and in proportion to his own abilities." This kind of institution is just the thing "to raise the intellectual level of the nation," and it "would make it easier to graft on the young and strong American tree all that was fine and great in the earlier Eastern and Western civilizations." If his plan succeeds, it would be "the cradle of a New Humanism."

George's notion of the material conditions of the institute includes a library, offices for the fellows, "iconographical collections," and photographical facilities. He indicates institutes as models for what he proposes: Karl Sudhoff's institute for the history of medicine in Leipzig, and the late Karl Lamprecht's historical institute, also in Leipzig. Assets that could be acquired by the institute include George's own library in Wondelgem, Arnold Klebs's library in Switzerland, and David Eugene Smith's library in New York. The institute could be supported by a society, "in the same manner as geographic or astronomic institutes." He gives as supporters a number of prominent American scientists and historians, and he provides 9 historians of science and medicine who approve of the plan: Smith, Klebs, historian of biology William Albert Locy at Northwestern University, historian of Asian technology Berthold Laufer at the Field Museum of Natural History in Chicago, medievalist and historian of medicine James Joseph Walsh at Fordham University, William T. Sedgwick and Harry Walter Tyler of Massachusetts Institute of Technology, Edward Clark Streeter, and medical bibliographer Fielding Hudson Garrison in Washington.[93] Sarton included, we see here a minyon for the discipline of history of science in the United States. Within the decade, a number of these players will rally to George's cause.

The Sartons' social engagements split between the academic and the artistic world. George and Mabel dine at the Stocktons and also at the Bush-Browns, who are artists. Lydia Bush-Brown, the daughter, is a designer of dresses and boxes, and within a year she arranges for Mabel to exhibit her work.[94] George does not neglect May, taking her to see

93. NYPL.

94. Diary, 6 and 7 June 1916. "Exhibit of Industrial Art," Washington *Post*, 7 March 1917, p. 7. Mabel and Bush-Brown exhibited in a collection of house furnishings and interior decoration at the Arts Club of Washington.

Buffalo Bill.[95] He and May go to New York. In a letter of 12 June, Mabel speaks of her loneliness in Washington while George will be lecturing at the University of Illinois and while May is at the Baekelands, where, Céline Baekeland writes on 13 June, "May is perfect."[96] Mabel: "I feel as if I were a poor battered bird that has been whirled in a storm till it is dizzy." (Possibly it is an affliction with a physical basis, since Mabel has incipient cataracts in both eyes.[97]) She has calculated her budget, and it is more encouraging than she had thought. She may have enough to pay the doctor. She must buy some underclothes, "especially if I go to Mrs Van Loon. I would really be a little shy of her seeing how I manage at present." Mabel has much design and decoration to do. "Miss Shadd is of course most anxious to have her things done. She has no idea *at all* that it is often the having to *invent & design* in a hurry *that is so tiring*." Mabel wonders if the "Peace People" are working to avoid war with Mexico, given the American intervention in the revolution there. On 14 June, Mabel reports that Eliza Van Loon has asked her to embroider a dress for an upcoming trip to Boston, where her husband has a Harvard reunion. The Van Loons leave for Ithaca about 14 July. They have invited Mabel to travel with them to New York and then Ithaca "when I leave these rooms." Mabel would like to stay with the Van Loons all summer, although she will "go with May somewhere for a fortnight" in July. Mabel apologizes for having "provoked the crisis & then somehow failed to help you meet it."

The next day, Mabel writes that she knows how hard George works to support the family:

> I am sometimes painfully conscious of the efforts you make, your weariness & of the fact that being married forces you to give up other, dearer work. Don't forget that you yourself are conscious of this & that years ago you had determined *not* to allow wife or child to intervene. I know you have changed, but I cannot altogether forget.

Few men "have as definite & passionate a desire" to fulfill their dream, as George does.

> If I revolt at your sacrifice, however willingly you make it, it is because I know & feel all this. It is because after all, in my own imperfect & often blundering way, I love you dearly & long for you to fulfill yourself. I long for it & feel it more sharply because I too have dreams & passionately wish to realize them. Dreams less important to the world than yours, but probably as important & dear to *me* as yours to you.

95. Diary, 30 May 1916.

96. NYPL, Céline Baekeland to Mabel Sarton, dated 13 June 1916.

97. Diary, 3 June 1916.

George writes from the Baekelands on 12 June, apologizing. "I am so sad dear Wifie, that I have not been a better & brighter husband these last days." Mabel writes on 17 June to George in Urbana. Hendrik Van Loon cannot go to Europe because the British Embassy will not give him papers due to his German mother-in-law, German servants, "& certain of his writings on the neutrality of Holland. France would have welcomed him. [French diplomat and literary historian Jean Jules] Jusserand is charming to him." Mabel is uncertain that her time in Ithaca with the Van Loons will be entirely cordial.[98] She finished Margaret Gillespie's blouse. A long conversation reveals that Gillespie is an unhappy pessimist. The important thing, Mabel concludes, is to have an ideal and to live for it.

Notwithstanding George's displeasure at the prospect, Mabel shall shortly form a partnership with Margaret Gillespie—by 1919 known as the Studio of Belgian Embroidery—to sell her designs and embroideries.[99] Over the next 15 years the enterprise does produce revenue, at times providing a crucial supplement for household expenses, but as George foresees, Mabel allows herself to be exploited—her partner (or, rather, her employer) generally being in her debt. The studio gives Mabel a vital creative outlet, as a successor to the Dangotte enterprise in Belgium, until it fails during the Great Depression.[100] Margaret Gillespie recognizes Mabel's *feu sacré* and talent in just the way that Lawrence J. Henderson recognizes George's energy and commitment. The studio is Mabel's *Isis*. Just as George lectures widely on history of science, so Mabel travels periodically in connection with her designing. And just as in Belgium Mabel received international notice as a designer before George did as a scholar, so in America does she exercise her creative talents before George is able either to do much publishing or even to resume *Isis*.

On 14 June 1916, George discusses his institute with David Eugene Smith. Smith thinks that Archer Milton Huntington, who endowed a building for the Hispanic Museum, the Geographical Society, and the Numismatic Society, would help. Smith would give his books to the institute, as would George and George Arthur Plimpton, who chairs Ginn

98. Van Minnen, *Van Loon*, p. 53, for Van Loon as persona non grata in the United Kingdom. Van Minnen, *Van Loon*, pp. 60–2, for the instability of the Van Loon marriage, leading to a separation in 1918.

99. NYPL, Mabel Sarton to George Sarton, received 14 June 1919, on the Studio's letterhead designed by Mabel; the address: 1517 H Street, NE, Room 15, Washington, D.C. Mabel Sarton to George Sarton, 14 March 1924, on different stationery with the same address; at the top, Mabel's design and the legend, "Margaret Gillespie, Proprietor," without giving a name for the enterprise.

100. NYPL, Mabel Sarton to George Sarton, 3 February 1933: "If I have not spoken much of my friends here it is perhaps that I am very depressed about them. Financially they are both in dire difficulties. Mrs Gillespie seems to me on the edge of nervous collapse. Margaret is not so bad but–Well. I don't want to write about it."

and Company.[101] Smith raises issues with George's scheme. First, George should drop the word "civilization" from his plans: "It is not an institute of civilization, but one of science *as related to*, or *as a part of* civilization." Second, George must dissociate his name from the plan, otherwise a reader will find the proposal self-serving. Smith will correct George's prospectus. When George has copies available, and when he sends one to Huntington, Smith will contact George Frederick Kunz, a vice-president at Tiffany's and a member of the Board of Directors of the Research Corporation, founded to support science in 1912 by inventor Frederick Gardner Cottrell and Smithsonian Secretary Charles Doolittle Walcott.[102]

George is in Buffalo on 15 June, and he finds it sad and dirty. He watches the parade of a German athletic club. He confides that he has been offered positions in Switzerland and Italy where he could continue *Isis*, but he declined because he wants to stay in the United States and work for the reconstruction of Belgium and the creation of an institute for the history of science.[103] George writes to Mabel on 17 June from Buffalo. He thinks that May "will be good, so much the more because she gets everything that she wants" with the Baekelands. George is impressed with Niagra Falls, but "I feel that I am now unable to entirely enjoy nature without you." Nature is unlike a museum or library, where George "must be alone." George hopes to "put my Lowell lectures into shape" during a week with Mabel and May in the country. George thinks about his coming anniversary: "I did not understand quite well what it all meant, when I married you; but I understand it every year better. If I had not married you five years ago, well beloved, I would marry you *now*." He sends her $110 ($10 for his bookcase and her incidentals), and he will send more in August from Urbana.

During the early years in America, George focuses first on history of science, and then on learning American psychology and language.[104] He views the United States through the prism of European experience. On Memorial Day 1916, George observes that the ideals of reconciling the Civil War "have not penetrated into all hearts, above all in the South."[105] He does not like Chicago, he notes in his diary. He concludes that the best preparation for peace is anticipation of war. "There is more social justice in Germany than elsewhere; it is natural that patriotism—i.e., devotion to

101. Diary, 14 June 1916. The connection: David Eugene Smith, *Rara arithmetica: A Catalogue of the Arithmetics Written before the Year MDCI with a Description of Those in the Library of George Arthur Plimpton, of New York* (Boston, 1908). Plimpton is sometime president of the Board of Trustees of Amherst College.

102. HGS, David Eugene Smith to George Sarton, 13 July 1916.

103. Diary, 15 and 16 June 1916.

104. Diary, 18 February 1916.

105. Diary, 30 May 1916.

the state—is more intense there." He continues: "If the entire movement for military preparation—so popular now in the United States, ends in more social justice, an immense progress will be achieved. If not, it will be in vain, or rather it will only produce evil."[106]

Mabel writes to Urbana from Washington on 18 June 1916 that she misses May "but I miss you a great deal more." She feels that they "hardly ever talked to each other," and the thought makes her sad. She knows that George's work away from home brings him into contact with interesting people and ideas, but he does not "even attempt to tell me about these people & things." A point of contention between them is George's belief "that the bringing up of May should fill the rest of my life." George confirms his own loneliness upon receiving Mabel's letter. The students at Urbana are less interested in his work than those at Clark and Cornell. "But never mind: I do my best." George is grateful to the Van Loons for their help, but "I do not sympathize with them. To my taste they are a little bit too superficial and too showy." Furthermore, George is hurt by their pro-German and anti-British feelings. George thinks that he is right to be married. "It is a man's duty to be married & to have children." George does "not at all object to your art work; on the contrary, I love to see you do it. But I resented your going into business, because you are not businesslike, and not strong enough to do it with the necessary regularity and efficiency." He wants Mabel to work for a large enterprise in decoration. He has heard, as Mabel has heard, about the movement to bring Fredericq and Pirenne to the United States. Harvard would be their destination. He urges Mabel to tell Margaret Gillespie

> that life indeed has no sense unless one gives a sense to it by doing some disinteres[te]d work: a work of science, of art, of love. Not everybody is an artist or a scientist, but everybody can be kind, helpful, cheerful. Everybody can do a little bit more than is necessary for the mere business p[oin]t of view, and do that little bit, *by love*: That is what saves the whole thing.

On 19 June Mabel pledges her love and assistance "to fight the hard times, & to stand the test of happiness together too." She wants to spend a few weeks with George by the sea, perhaps Ogunquit, before they move to Cambridge.

George appreciates Mabel's letters. In his diary on 21 June: "I received such a good letter from Mabel, and another from little May. Mabel hurts as much as I from our separation. May the gods give us many years of intimacy and love. There is nothing better in the world." George, on 23

106. Diary, 18 June 1916.

June, will try to speak more openly with Mabel. He has met a lecturer of the American Association for International Conciliation. The man makes a poor impression. He is "a pacifist to the extreme, a *bigot*." The man "is unable to understand the heroism of all those young men, who are ready to give their life for an idea. He refuses to acknowledge all that is *great* & *noble* in that. Such bigot pacifists do not make for peace."

On 24 June George records in his diary an evening with Edmund Janes James, president of the University of Illinois. "He has a rather pessimistic view, which I attribute in part to his German sympathies. It was the same with the *anti-dreyfusards*." The Republican Party offered James a gubernatorial candidacy in Illinois, but the university trustees would not allow him to resign. On 26 June, George reports on a talk scheduled by Ian Bernard Stoughton Holborn of Oxford on Greek life; there were only a handful in the audience, and the speaker did not show due to illness. George mentions receiving a letter from Yuen Ren Chao, a Harvard student who wants to take his course in the fall. Chao studied mathematics and philosophy at Cornell University and in 1914 was one of the organizers there of the influential journal *K'o Hseuh*, or *Science*, published in Shanghai, and with it the first scientific association in modern China, the Science Society.[107] Chao has already taken Henderson's course in history of science, which consisted mainly in reading through John Theodore Merz's history of science in the 19th century, and he wants to make history of science one of his fields for the doctorate.[108] Chao will indeed attend George's course at Harvard over the coming years, and he will become George's first assistant.

Mabel, on 22 June 1916, tells about meeting with George van Biesbroeck, whose wife's family experienced German atrocities in a village near Liège:

> We talked of Romain Rolland whose writings he knows from the *Journal de quinze* [*Cahiers de la quinzaine*], but I realized keenly how *impossible* it will be to forgive these things for a generation, & as he too said, "They have been *too many*, too systematic," & it is as if the ones whom R. Rolland wishes us to think are the race, *they*, the good ones, are the *exceptions*.

Mabel concludes: "I cannot divide the sin from the sinners. I hate both." On 22 June Mabel returns to the issue of sacrifice in marriage. "I feel as if

107. Yuen Ren Chao, *Life with Chaos: The Autobiography of a Chinese Family, 2: Yuen Ren Chao's Autobiography, First 30 Years, 1892–1921* (Ithaca, 1975), pp. 79, 84. Zuoyue Wang, "Saving China through Science: The Science Society of China, Scientific Nationalism, and Civil Society in Republican China," *Osiris, 17* (2002), 291–322, on p. 304. In Pinyin, Chao's name is given as Zhao Yuanren.

108. HGS, Yuen Ren Chao to George Sarton, 30 June 1916.

when the man or woman sacrifices his or her *health*, that that comes near to being a very real sacrifice, tho' I think the best women are often absolutely unconscious of it. And women have sometimes to sacrifice much more than their liberty." She continues:

> Don't other women watch their husbands' weariness day by day & suffer through it & long to help? The truth I believe is I am too proud (with what is not the best pride) to accept willingly your sacrifices. Just to see you once as you were then, once as you were when you wrote to me in Wondelgem just after the 2nd baby was born, makes it impossible. I feel that is not right. I fight the feeling down, but it will not die. I believe I can only accept absolutely that you provide for me & have to pay for anything that seems extravagance to you, when I have proved to myself & you that I can earn enough to do it myself. I would like to be several years wholly independent financially. I fear it will hurt you to know this—yet it is something that is in me—& I would rather tell it you quietly now when my heart is full of love & not an atom of resentment.

George replies on 25 June. He begins with the thought that happiness starts with the recognition of one's limitations. "Where there is love, there cannot be any sacrifice." George feels "that you do not accept your destiny, as you should accept it: You think too much of what you have not, you should think more of what you have." He asks Mabel what she thinks of Poet Laureate Robert Seymour Bridges's *Soul of Man* (He means *The Spirit of Man* [London and New York, 1916]), a collection of French and English poetry. "I think it is a great & noble book."[109] On 25 June, Mabel describes her design for the new studio in Washington. The next day she relates the "heavy cloud that never lifts," dominating her life since they left Belgium. Mabel is too shy to talk about the problem, which we might identify as clinical depression. "Besides talking it over is useless. One has just got to bear it. One help to bearing it is certainly to do satisfying work." Mabel tells George that "every night I am troubled by thoughts concerning the ultimate fate of Germany." Mabel finds a chasm between reports among George's papers and reports from George van Biesbroeck, on the one hand, and a book of Roman Rolland's (probably *Au dessus de la mêlée* [Paris, 1915], which George read and liked),[110] on the other hand. The Sartons' naïve pacifism in the pre-war years evolves. George allows pacificists to prepare for war; Mabel wrestles with the notion of war-guilt.

In a second letter of the 26th, Mabel responds to some of George's concerns. "We are both very foolish but we will grow less so I do believe." She

109. On 31 July 1916, George wanted to send *The Spirit of Man* to Mabel for her birthday, but he settles instead on a guide book for birds.

110. Diary, 13 April 1916.

confides: "I too wanted passionately to come to you those last nights & stretched my arms to you for hours just as you have once told me, but you were filled with angry pain & revolt, & I was frightened of you, *not too proud*." She talks about bearing problems in silence. "After all *why* is it good to bear one's own troubles? *Chiefly so as not to burden others,* but if by so doing we hurt the others, let them see our suffering but refuse their help, *that* is not right at all." George should not keep things to himself.

> For your nature, I believe it to be a *better victory* over yourself, to quietly share your wishes and troubles.
>
> And a very real danger of such theories & *systems* is: to be proud of possessing & exercising them, to be self-conscious of one's own superior attitude. *That* may lead anyone to unkindness & lack of generosity while they are all the time feeling quite satisfied with their superiority.
>
> This sounds very severe. I am not applying it to you, who have in you a wonderful & beautiful quality of humility that leavens all your *real* self. But I have felt in you a superficial self, clothed in these habits of mind that you were educated in, that troubles me.
>
> Who am I to criticise you? Just a companion, full of failings & wrong pride & contradictory instincts myself. *Help me*, too, for my heart is proud too & a reproach from you enters deep.

On 28 June, Mabel disputes George's contention that marriage has harmed her health, although "I felt that you might have made a bigger effort to avoid the birth of a 2nd child coming so quickly, that it would have been better to think of it & take radical, sure measures however distasteful & difficult." Mabel declines to accept her ill-health or the fate of having children, for, she asks George (recapitulating an exchange of letters with Raymond Limbosch several years earlier),[111] "the whole question is to decide if it *is* one's destiny." Rather, "my fate is to fight my destiny." She does not understand George, who lives "on God-like heights." She "*was* ready to lay down my life when May came, but I was not joyous over it, how could I be even if only to think of leaving you alone *and the child if it lived*?" She reflects: "Surely we have not to give that unborn child its best chance? It's not just you and I and *our* joy and love that's involved. No, that little child of ours that died, it was *not* fate, *it was our selfishness*, & not our love." Nor does Mabel think that married life is the only place where love may be expressed fully. And "when Death comes, I hope I will face him very quietly and humbly, but I will fight him with every ounce of strength in me first." On 29 June 1916 Mabel writes from Washington to George, in Urbana, recalling their garden in Wondelgem. She asks George not to mind her letter of yesterday. She will try to submit to things she cannot change. "Nothing really matters if we love each other."

111. NYPL, Raymond Limbosch to Mabel Sarton, 18 June 1913.

George, on 29 June, thanks Mabel for her "loving and wise letters." He asks her to help him put together his Lowell lectures on science and civilization at the time of Leonardo da Vinci. "It is this year that I shall have to fight the decisive battle and establish my reputation. It will cost me an enormous amount of work, but I must and *will* succeed." Mabel's assistance for Leonardo "will be invaluable." The last of six lectures "will be devoted to the explanation of my studies of what I call the New Humanism as opposed to the Humanism of the Renaissance. Will you help me for that?....I shall be able with your help to make a good and beautiful piece of work."

George has referred to Leonardo on occasion since 1907, when he read Gabriel Séailles's biography of the Renaissance man.[112] More recently, his friend and potential patron Arnold C. Klebs is delivering popular talks and publishing papers on Leonardo.[113] Now, with Mabel's support, he will focus on Leonardo as an exemplar of the art of science. But what George really wants from Mabel, in the way of assistance, is not her knowledge of Renaissance painting but rather her ear for colloquial English and her hand for correspondence. In Wondelgem, Mabel helped George place *Isis* on its feet; in England, she revised and transcribed his letters in English; Mabel is a coauthor of George's first English publication in 1915, a fact George acknowledges to her. Mabel continues in this capacity for a number of years in America.[114] She writes letters of appeal on behalf of George's institutional schemes.[115] Her support figures explicitly in George's eventual appointment with the Carnegie Institution of Washington. William Sedgwick, recommending George to the Carnegie Institution's Robert Simpson Woodward, emphasizes that George "has also an English wife who heartily seconds his endeavours." Sedgwick, who serves on the advisory council of the Shady Hill School where May is a pupil, knows first-hand about the Sarton family.[116]

---

112. Gabriel Séailles, *Léonard de Vinci* (Paris, 1907). Diary, 15 December 1951.

113. HGS, Arnold C. Klebs to George Sarton, 4 April and 19 April 1916.

114. For example, George writes on 4 July 1916 to Mabel in New York, asking her to correct a manuscript for him.

115. HGS, James Harvey Robinson to George Sarton, 23 November 1916, thanking George for the letter Mabel wrote on his behalf for "coordinating the contributions that are being made to the history of science"; George Lincoln Burr to George Sarton, 8 September 1917, apologizing for not answering his letter "from Mme Sarton writing on your behalf." Burr sees no way to realize George's plan.

116. CIW, William Sedgwick to Robert S. Woodward, 24 April 1918. Margot Peters, *May Sarton: A Biography* (New York, 1997), p. 404. On Sedgwick, Arnold Thackray, "The Pre-History of an Academic Discipline: The Study of the History of Science in the United States, 1891–1941," *Minerva, 18* (1980), 448–73, p. 454. In 1917 the Shady Hill School moved from the Hockings' house at 16 Quincy Street to part of the Charles Eliot Norton estate, Shady Hill Square. The school, its staff, and the families of the students figure closely in the Sartons' life at Cambridge.

George wonders, in his letter of 29 June, about Germany: "Is it possible to ever forgive them?" George will publish an advertisement for the Bryce Report in *Isis*,

> But we must also appeal to a better Germany: It is impossible that well informed Germans can approve of that. However few they may be, I shall collaborate with them, for the glory of Old Germany. What is disheartening in a book like the Report Bryce is that it proves how thin the skin of civilization is. It is perhaps thinner in Germany than in England or here, but I know that it is nowhere very thick. We are misled by the orderly appearances of our society: In our modern civilization, social control & repression have become so efficient that it is more & more difficult to discriminate between the men who behave well—by necessity, by *fear*—and those who behave well from a sense of duty.

On First July, George thinks that the expense of Mabel's going to Ithaca to visit the Van Loons (about which she has misgivings) does not justify the trip. He thinks that Mabel could spend a few weeks in Ogunquit before joining him in Boston, where he arrives on 16 August. On 7 July, George reflects on Mabel's description of their garden at Wondelgem: "I was a bit disappointed that you had not a word of sympathy for my old redberry tree and my thistle. I am perhaps the single soul in the world who tends over them. How will they all look when we go back?" George wants May to pay more attention when she writes to him. "You must learn her to write better, & not send the letter if it is not properly written."

Mabel arrived in Yonkers on 2 July, she writes on 3 July. "May has tired me with her continual talking since early this morning, yet you know I am deeply happy to have her again!" She reports an epidemic of infantile paralysis, nearly all in Brooklyn, with 300 cases and 50 deaths. On 5 July Mabel reports, in French, on May's new words: "Put it in the right 'persition' mummy, and immediately (pronounced mediately!)." On 8 July George describes his lunch with President James and the art historian Stoughton Holborn. He urges Mabel not to take May into New York because of the epidemic. Mabel, on 6 July, will begin on a manuscript of George's, although she has "a good many orders of dresses, etc., & now some embroidery for Mrs Baekeland." Céline Baekeland spoke about her hard life, and Mabel "suddenly realized how wonderfully kind and dear and generous you are compared to many really superior men in their way, and I felt very humble and thankful." On 8 July, she hopes her corrections of his letter are understandable. She notes that he writes to her as "Darling" for the first time, although George has used it previously for May. Mabel writes in French on 9 July to "Cher Vieux Rat." On 11 July, Mabel is glad to help George with his Lowell lectures. "I did think of your thistle! & nearly added a p.s. to my letter about him." In another letter on 11 July, Mabel describes a temper tantrum of May's: "She was stamping and grimacing with rage like a little savage. I told her when she had got quiet again that she had looked like a little savage, & that very much impressed her."

Lecturing during the summer at Urbana is an inconstant affair. George reports in his diary on 5 July that he turned up to give his lecture on mathematics in the Renaissance and faced no students at all. Four students turned up on 4 August for his last lecture, on Paracelsus and Vesalius. He is also an inconstant husband, recording sex twice at Urbana, on 24 and 25 July 1916.

On 12 July, George looks forward to meeting Mabel and May in Ogunquit. In French: "It will be our first long time by the sea since the happy days in Cromer." On 16 July 1916, George comments on Mabel's trip to the movies in New York: "The movies are the greatest problem of our time, my greatest source of anxiety. What will come out of it? When a whole generation will have been fed upon these shadows? There is an enormous amount of cleverness in them, unlimited imagination wantonly spent, no art, no beauty." After seeing the movies every week for 20 years, a girl will have destroyed herself: "What of her brains, of her heart? Will she have any soul left? Hang it all." Yet he does use cinema for relaxation, and Mabel accompanies him. He writes to Mabel nearly a decade later, on 29 March 1925, that when he is too tired to work he goes to the movies, "and of course this makes the matter worse. Yet it rests me in one way; it is like going to an old nurse who tells silly stories."

Mabel, around 19 July, thanks George for thinking to introduce her in Boston to a purchaser for her designs, "but I have a great deal still on order. I have about 20 to 22 orders for embroidered dresses etc!" On 25 July, Mabel has delayed her trip to Boston. Harriette Shadd was to have met them in New York and taken them to Boston by boat, but Mabel wants more certain information about Maine before leaving, and she is concerned about the possibility of quarantine from the epidemic. Mabel spent the day in New York on business with Shadd. She showed embroideries to Medora Amsden, who showed them to friends, and Mabel has more orders. The next day Mabel decides to leave at once for Maine with May, because quarantine restrictions may be imposed shortly. Mabel writes several days later from Ogunquit, where she has found an expensive room in a hotel. "This is a *lovely* place, in all sorts of ways."

George, in Urbana, revisits his notion of writing a book about philosophy of science. On 30 June he writes in his diary that the book would provide a succinct treatment, "for each science," of principles, deductions, and facts. He might add historical notes, but he thinks that it would be better to keep the history for a second, independent book. He makes little headway on the project. George, on 29 July, is depressed and "very anxious to know the circumstances of your journey & arrival in Ogunquit." It is very hot in Urbana. "Think that I am here alone in hell, working as much as I can, which is very little, discouraged every evening because so little has been done, notwithstanding my pains; tired, ill, maddened by the heat." So George has been reading: *A Connecticut Yankee in King Arthur's Court, Tom Jones, Tristram Shandy, Madame Bovary* in French and English, Gustave Flaubert's novel about North Africa *Salammbô* ("a wonderful book"), Rousseau's memoirs, "a saddening book: *cher et grand Jean Jacques*."

George notes in his diary on 30 July 1916 that Emile Waxweiler has died, in London, the third member of the *Isis* advisory board, after Karl Lamprecht and William Ramsay, to disappear. His éloge of Waxweiler appears in *The Nation* on 8 February 1917. George emphasizes that Waxweiler, trained as an engineer at Ghent, became an expert in social statistics and read widely in biology. He put all this erudition into his journal *Archives sociologiques* and his book, *Esquisse d'une sociologie* (Brussels, 1906), and into his administration of the Institut de sociologie of the Solvay Institute in Brussels. Waxweiler's is a functional approach to sociology, where fluctuations take precedence over forms. George is silent about Waxweiler's socialist beliefs. He concludes with the hope that Waxweiler's spirit "will continue to dwell in that hospitable house—one of the places in Belgium most dear to me," and that after the war his institute will be reconstituted according to the wishes of Ernest Solvay and confided to "the right man."[117] Speaking about the dead, George would be the first to agree, should serve the living.

On First August 1916 Mabel hopes that Harriette Shadd will lend her camera when George comes to Ogunquit. On 3 August, Mabel and May are at the home of Dr & Mrs Spalding in Portland, Maine. May "is very brown, with pink cheeks & such shiny clear eyes." George travels to Ogunquit by lake steamer and train. The train makes him ill, the heat aggravating his "habitual indisposition [diarrhea]," he records in his diary on 6 August. The next day he notes that Montreal reminds him "of a city in the French provinces—same qualities, same faults: dirt, vulgarity. Perhaps not more vulgar than in a British milieu, but the vulgarity is more apparent." He much prefers the "Anglo-Saxon" environment of Toronto. He is struck by how French Canadians, externally American, have kept "their prejudices and their French faults." He thinks that the persistence is hereditary, "due to the race," as well as cultural, due to religion. This unappealing line of thought, present before the war, relates to George's flirtation with eugenics, which he elaborated in *Isis*.

George likes the rocky shores of Ogunquit, and the entire family goes swimming every day. They visit art galleries, above all the one run by Lucy May Stanton, who had arrived in Ogunquit several months previously, the first of 10 consecutive summers she spends there. Stanton, an

117. George Sarton, "Emile Waxweiler (1867–1916)," *Nation, 104* (1917), 168–9.

artist from Atlanta, Georgia, is "a lovely person who produces very beautiful miniatures," as well as oil portraits and seascapes.[118] Lucy Stanton, who like Mabel had started out painting miniature portraits on ivory, will be a close friend of the Sartons over the coming years.[119] George receives a letter from Frederick Edward Brasch, associated with the library at Stanford, who will be in Cambridge during the fall and wants to take George's course.[120] Brasch will become a mainstay of George's organizational enterprise. George suffers from headache and toothache. He is troubled by H. G. Wells's *Research Magnificent* (London, 1915), about a single-minded savant, because the hero is a "maniac," and the book will provide readers with "an excuse for their lack of idealism." He loves Robert Louis Stevenson's *South Seas*.[121] Stevenson's book *A Child's Garden of Verses* is the only non-scientific volume that he carried with him throughout the summer.[122] George condemns May for being taken in by a guest in their hotel, a perfumed woman with bleached hair. "It worries me greatly....It will be difficult later to protect her from herself." George is worked on by a Boston dentist who summers at Ogunquit; there is no charge for the quick fix, which does not last.[123]

118. Diary, 8–19 August 1916.

119. Erica E. Hirshler, Janet L. Comey, and Ellen E. Roberts, *A Studio of Her Own: Women Artists in Boston, 1870–1940* (Boston, 2001), p. 197, for a short biography; Betty Alice Fowler and Andrew Ladis, *The Art of Lucy May Stanton* (Athens, Ga., 2002), for a full treatment.

120. Diary, 23 August 1916.

121. No doubt Robert Louis Stevenson, *In the South Seas: A Footnote to History* (New York, 1901), and other editions, suggested to him previously by President Stockton of George Washington University. Diary, 25 and 27 December 1915.

122. Diary, 25 August 1916.

123. Diary, 27 August 1916.

CHAPTER 11

Truth

"In September, we will have to start a new, real life," George wrote to Mabel from Urbana, Illinois. On 28 August 1916 he arrives in Boston to lay the foundation for it. His toothache begins again; he finds a new dentist. He walks by evangelist Billy Sunday's church and visits the Boston Museum of Fine Arts.[1] He writes to Mabel from the YMCA in Boston on 30 August that he has been pounding the streets of Cambridge and Boston. Northern Cambridge seems cold compared to southern Washington, "as austere as a 'begijnhof.' I think you will come to like it." Boston is, in the words of one of George's correspondents, "the American Ghent."[2] Mabel and May join him, and they lodge with Edward C. Streeter and his family.[3] George and Mabel are visited by Bolton Carrington, Head of Prints at the Museum of Fine Arts.[4]

It is significant that even before finding an apartment, the Sartons connect with the Museum of Fine Arts. Museums are a continual source of inspiration for both George and Mabel, and the Boston Museum of Fine Arts becomes a favorite haunt for the rest of their life. George writes in his diary on 27 January 1917 that he is beginning to know the Oriental section of the museum. "I will owe much to this museum in Boston. It has given me a new education." The connection is cemented in 1917 when George's friend Ananda Coomaraswamy establishes himself in Boston as a conscientious-objector refugee from England, after having visited America in 1916 with the music tour of his wife. During the tour, Coomaraswamy came into contact with Harvard's artist Denman Waldo Ross, who purchased Coomaraswamy's collection of Indian art for the museum and then had Coomaraswamy appointed curator. For the next 30 years, Coomaraswamy assembles South Asian art and publishes an enormous corpus on philosophy, philology, and aesthetics.[5] George often

1. Diary, 28–30 August 1916.
2. HGA, Gertrude Richardson Brigham to George Sarton, 9 May 1916.
3. Diary, 31 August 1916.
4. Diary, 1 September 1916.
5. Many of Coomaraswamy's papers are held by the Princeton University Archives, and their finding aid elaborates on biographical matters. On Coomaraswamy and Denman Ross, Walter Muir Whitehill, *Museum of Fine Arts Boston: A Centennial History* (Cambridge, Mass., 1970), *1*, 137–41, 362–71; Roger Lipsey, *Coomaraswamy: His Life and Work* (Princeton, 1977), pp. 123–6 for the move to Boston.

consults with Coomaraswamy, and it is likely that Coomaraswamy introduces him to Denman Ross, who becomes George's close friend.

The Sartons settle in Cambridge at 10 Avon Street—3 rooms, with kitchen, bath, and porch, furnished with a loan of $300 from Baekeland.[6] The Streeters had taken care of May while the Sartons sought an apartment, and after nearly 2 weeks the family is reunited.[7] The Sartons engage a West Indian servant girl, who impresses George more than their black servants in Washington.[8] Initially Henderson denies George a room in the library, but George is able to get one by applying directly to the authorities.[9] On 28 September George occupies Widener Library 573, "a real cell." There is a long table, bookshelves, 2 chairs and a table beneath the light, and a big window facing north. He must stand on a chair to gaze out. "I like that—no noise, no view to distract me. I am surrounded only by white walls, my books, and a bit of blue sky."[10]

As George has known, during the first years of the war Harvard is sponsoring distinguished scholars from devastated Europe, men who could be engaged for a contract limited in both extent and remuneration. George continually bristles when he meets senior Belgian scholars who receive such appointments, for he imagines that he, better than anyone, knows the value of a Harvard berth. Harvard has staff more cultured and accomplished than staff elsewhere, although George perceives that "for many reasons the number of real scholars in the Faculty is rather small."[11] More than the staff, whom he finds agreeable, there are unparalleled resources for George's career in the new Widener Library.

Widener—the great vessel for 20th-century scholarship—floated on an equally great maritime catastrophe. Among those going down with the "Titanic" in 1912 was Harry Elkins Widener of the Harvard class of 1907, one of the heirs of the Philadelphia fortune (considered to be as high as $75 million) of his grandfathers Peter A. B. Widener and William L. Elkins. Harry was an enthusiastic bibliophile, and he had willed his collection to Harvard. The Widener family (who lost both Harry and his father George Dunton Widener on the "Titanic," Harry's mother Eleanor Elkins Widener having found her way to a lifeboat) thought immediately of endowing a memorial building to house the books. Plans dramatically escalated to embrace millions of Harvard's volumes. The building, in the hands of the Widener and Elkins families' favorite architect Horace

6. Diary, 10–12 September 1916.

7. Diary, 21 September 1916.

8. Diary, 23 September 1916.

9. HGS, Lawrence J. Henderson to George Sarton, 22 May 1916, for the refusal.

10. Diary, 28 September and 3 October 1916.

11. CIW, George Sarton to Robert S. Woodward, 25 August 1918.

Trumbauer, became an imposing neoclassical monolith. The Harry Elkins Widener Memorial Library was dedicated on Commencement Day, 24 June 1915, when Harvard gratified Mrs Widener by granting honorary degrees both to Trumbauer and to her future, second husband, physician Alexander Hamilton Rice.[12] George Sarton had spent the preceding weeks in Boston, knocking on doors and acting as a Belgian propagandist.[13]

Architect Trumbauer and his associates created an Olympian hall. What made Widener more than a warehouse for the writings of 'dead white guys' was the inspired labor of the director, Archibald Cary Coolidge. In the years around 1910, Coolidge was much troubled by Harvard's antiquated, lacunary, and entirely inadequate cataloguing system, which was coming apart under an avalanche of acquisitions. Beginning in 1910 he reorganized the cataloguing work, having Harvard's cards follow the ongoing classification undertaken by a succession of men at the Library of Congress.[14] Coolidge constituted, in the years before Widener opened, a union catalogue for titles squirreled away in numerous departmental and special collections. Over the quinquennium ending in June 1915, his staff completed and filed more than 2 million new cards—extraordinary industry apparently exceeding the assiduous cataloguing at the Library of Congress.[15] Coolidge also arranged, through his own private fortune, to have cards printed and sold for titles not in the Library of Congress or at the John Crerar Library in Chicago, whose cards (intercalated with newly clipped or rewritten cards from Harvard's unique accessions) formed the basis of the system. Coolidge's printed cards were his effort to market Harvard's collection and help pay for the massive cataloguing effort.[16]

12. William Bentinck-Smith, *Building a Great Library: The Coolidge Years at Harvard* (Cambridge, Mass., 1976), especially pp. 54–103. In 1915, the firm of Trumbauer, Zantzinger, and Borie was also chosen to design the Philadelphia Museum of Art, the entrance of which bears a strong resemblance to Widener Library. Steven Conn, *Museums and American Intellectual Life, 1876–1926* (Chicago, 1998), p. 222.

13. YMS, George Sarton, "Lettres au *Petit Journal*," diary page where George gives the datelines for letters sent to the Paris publication, *Le Petit Journal*. George is in Boston at least from 10 June through 22 June.

14. Francis Miksa, *The Development of Classification at the Library of Congress* (Urbana, 1984) [*University of Illinois, Graduate School of Library and Information Science, Occasional Papers, 164*], pp. 16–40. The decision to follow Washington's lead was to some extend a homecoming, since in the first decade of the 20th century the Library of Congress decided to borrow heavily from the flexible classification system devised by the sometime Harvard librarian Charles Ammi Cutter in the 1870s and 1880s.

15. Miksa, *Development of Classification*, pp. 39, 43, according to which the Library of Congress catalogued or recatalogued 1,169,200 volumes in 1903–10, and 1,059,300 volumes in 1911–21.

16. Bentinck-Smith, *Great Library*, pp. 37–49.

We have seen that George Sarton is an inveterate list-maker. We have also seen that his plan for an "encyclopedia on cards" on the eve of the First World War found an extension during the dark months in London, when he wanted to establish a union catalogue for British libraries. Even when he is safely ensconced at Harvard, financed by the Carnegie Institution of Washington, his close supporter Edward Clark Streeter sees his fundamental contribution as bibliographical: "Quite a terrier for work—a bibliog. whot *Isis*, whot?"[17] And with his later appointment at the Carnegie Institution, George makes sure that he stays in his eyrie at Widener.

Once there, no force can dislodge him. It is neither love of nor loyalty to Harvard that keeps him on, for he knows that the university considers him as an independent luminary displaced on an eccentric. It is Widener. From his youth he loves books. "A brain is best compared to a library," he notes in his diary in 1903. Late in life he writes: "I do not love Harvard so much, a hard stepmother, but I love the library, and I am grateful to all its officers, from the top Dr [Keyes DeWitt] Metcalf, to the girls who replace books on the shelves."[18] In notes for an autobiography he emphasizes his gratitude "to H[arvard] C[ollege] L[ibrary] & to every employee of it, up & down."[19] Or again: "H. U. would have allowed me to starve."[20] Even at the end of his life, George finds it difficult to host visiting lecturers from abroad: "As I have a good deal of influence abroad, foreigners think that I have some influence in Harvard. They are sadly mistaken, for I have none and my credit is zero."[21] In George's attitude toward Harvard, we see his European sensibility, following which the center of academia lies with professorial institutes—the university itself being merely a swarm of clerks and janitors orbiting a library. George's rooms in Widener and his threadbare budget become his institute. And when he addresses the world from his modest quarters, eventually funded by the Carnegie Institution and lodged by usufruct, George Sarton radiates authority. George imagines that he commands Widener, but in the end the library masters him. Widener provides George with more than sobering counsel. It domesticates him. By his tireless glossing and conflation, George becomes Widener's firmest apologist. It is typical that in an unpublished outline for reminiscences, written in 1944, George devotes one of 11 chapters to Harvard's library.[22]

---

17. BFG, Edward Clark Streeter to Fielding Garrison, 9 April 1920.
18. May Sarton, *A World of Light* (New York, 1976), p. 37
19. YMS. George Sarton, notes for an autobiography.
20. YMS. George Sarton, notes for an autobiography.
21. Diary, 24 September 1951.
22. YMS. George Sarton, "Reminiscences," 17 September 1944.

On 29 September 1916 George attends a meeting of the philosophy faculty. He speaks with Hugo Münsterberg, a man with an enormous head. Others at the meeting: Henry Maurice Sheffer, Ralph Barton Perry, Edwin Bissel Holt, Reinhold Friedrich Alfred Hoernlé, William Ernest Hocking, and Herbert Sidney Langfeld. Although when George publishes *Isis* once more Sheffer becomes a regular reviewer for it, George's new circle does not displace his existing one, especially given a lack of sympathy for pacifism on the part of Perry and Hocking.[23] At just this time, George emphasizes that Arnold C. Klebs is actively seeking help for his institute, notably with Simon Flexner, director of the Rockefeller Institute for Medical Research, and George F. Kunz.[24]

George delivers his first lecture on the evolution of the idea of number to 4 students: Yuen Ren Chao, Frederick Edward Brasch, Ralph Monroe Eaton, and James Lockhart Mursell.[25] Modeling his career on independent writers like Carlyle, Maeterlinck, and Romain Rolland, not primarily on university professors, George expresses little interest in guiding students in research. In the course of his career, he directs the doctoral dissertation of only 2 students. Nevertheless, from the beginning at Harvard he teaches talented people. Chao goes on to a career at the University of California at Berkeley, Brasch finds a position at the Department of Terrestrial Magnetism of the Carnegie Institution of Washington and becomes a loyal supporter of *Isis*, Eaton becomes an authority on Descartes, and Mursell becomes a music educator.

We have a record of George's lectures on history of science at Harvard, compiled by Yuen Ren Chao, who unaccountably recalls that he was "the only student in the class," which met in Widener Library.[26] On 5 October, George begins by emphasizing the indeterminate origins of science, which cannot be known from the records of ancient civilizations, from the cultures of primitive peoples, or from children. He moves from numbers to astronomy. By 22 January 1917 he is lecturing on the Renaissance of the 13th century. The new year is devoted

---

23. Bruce Kuklick, *The Rise of American Philosophy, Cambridge, Massachusetts, 1860–1930* (New Haven, 1977), pp. 442–4.

24. Diary, 29 September 1916; HGS, David Eugene Smith to George Sarton, 30 September 1916.

25. Diary, 5 October 1916.

26. Yuen Ren Chao to W. James King, 22 April 1964. This letter and the meticulous record are found on note cards kept in the Niels Bohr Library of the American Institute of Physics in College Park, Maryland. HGS, Yuen Ren Chao to George Sarton, 1 June 1938, where Chao reports the destruction of all his books in Nanking and the survival of his notes, including those for George's course. Zhu Kezhen, studying for a PhD in meteorology, also took Sarton's courses; Zhu attracted Chao to Harvard. Zuoyue Wang, "Saving China through Science: The Science Society of China, Scientific Nationalism, and Civil Society in Republican China," *Osiris*, *17* (2002), p. 304.

to mechanics, electricity and magnetism, optics, heat, and sound. In September 1917, Chao follows the second part of George's course, devoted to mechanics.

Chao's notes provide insight into how George views his field of study. In view of George's alleged antipathy to technology, it is relevant to extract from his opening lecture of 27 September 1917:

> It is in science alone that we can properly speak of *progress*. Literary men ask what good? Well, material development is necessary for moral dev., or we should be living in our caves. Hydraulic engineering, e.g., was necessary for developm. of Babylonia. Communication, irrigation are necessary. It has never been, & ought to be done, to write a history of civilization, centering about the history of science. Usually, a history is written with kings & wars and a few pages of hist. of science in the appendix....
>
> Democracy also depends for its possibility upon technical knowledge, of press, of communication, etc.

In the very opposite of an abstract approach, George lectures on 29 September 1917 about the history of tools, and the first example is the prosaic handsaw. George is sometimes criticized for concluding his *opus magnum* in the 14th century, but he moves quickly in his course, ending the first part of it with Newton. The second part of the course begins with the Enlightenment, and in it George lavishes attention on topics that sometimes escape historians of science today: musical harmony, the steam engine, photography, the phonograph, and flying machines. George's is a history focusing on experimental results. He introduces Chao to Willem Kamerlingh Onnes's liquified helium, and to Ernest Rutherford's work in radioactivity. The "new theory of matter," formulated by Hendrik Antoon Lorentz, Henri Poincaré, and Joseph Larmor, features a positive nucleus surrounded by a ring of electrons. George concludes his course in May 1918: "In 1904, Max Planck offered his quantum theory. This made energy atomic. Both this and the second law of thermodynamics can thus be treated by the statistical method." It marks the first time that the history of civilization is told from handsaws to quantum statistics.

Sarton's respect for technology continues after his course. He writes in May 1918:

> In my book on XIXth century physics, I intend to show the continuous interactions between pure science and technology. I will give a short history of the main applications of physics (for ex. history of the dynamo, steam turbine, telephone, telegraph...)....I must avoid being lopsided, the more so that it will be much easier for me to prove the value of pure science, & that I will do it more effectively, if I show that I fully

> understand the importance of its applications & that I am comprehensive and generous.[27]

The book never appears, but technology figures in *Isis* over the next 30 years. Utility is George's watchword. He writes also in 1918:

> We must not forget that if the people of the Old World seemed often to show more interest in disinterested investigations, this interest in most cases is far from being of the right kind. On account of the prejudices fostered by the ruling classes, scientists and scholars are encouraged and honored—*not* because the people believe their work to be of special value and perceive its remote usefulness—but quite on the contrary because their work is believed to be *useless*. A noble man does not work, or if he work, it is understood that his work must be useless, that it cannot possibly pay. Now this attitude is perhaps worse than mere indifference fostered in the universities—on the same footing as the most pregnant investigations. Of course, if I were convinced that any kind of knowledge were useless, I would say, stop it. That is the root of my antagonism to so much knowledge which is simply *word* knowledge, that is: false knowledge, scholasticism, criminal waste.[28]

In these quotations, George plays to the sensibilities of his employer, but it is clear that by this time he views science as interventionist rather than speculative.

Harvard receives good value from George's lectures, and his colleagues in philosophy are friendly to him. On 6 October, George and Mabel attend an evening at Hocking's home along with faculty and graduate students. He hears George Herbert Palmer speak, the last of the three founding philosophy department members (the others were William James and Josiah Royce). George speaks briefly with Münsterberg, Richard Clarke Cabot, and James Jackson Putnam, but he and Mabel leave early because their servant cannot stay late with May. (Several months later, George is shocked by Hugo Münsterberg's sudden death during a lecture at Radcliffe College; they had been corresponding, and Münsterberg had sent books to George.[29]) The service for the reception is arranged by Agnes Hocking, Ralph Barton Perry's wife Rachel (sister of art historian Bernard Berenson), and Mabel.[30] Chemist Theodore Richards, Nobel laureate and director of the Wolcott Gibbs Laboratory, calls upon the Sartons on 8 October 1916. On 13 October George dines with William T. Sedgwick and Harry W. Tyler from Massachusetts Institute of Technology, who will

27. CIW, George Sarton to Robert S. Woodward, 30 May 1918.

28. CIW, George Sarton to Robert S. Woodward, 17 July 1918.

29. Diary, 16 December 1916.

30. Diary, 6 October 1916.

shortly publish a textbook on history of science; they promise to help him with his plans for an institute. On 14 October George dines with astronomer George Ellery Hale, his wife and daughter. George notes that he and Hale became engrossed in conversation, ignoring the women; Hale forgot to eat. Hale thinks that the institute should be placed under the auspices of the National Academy of Sciences. D. E. Smith and Arnold C. Klebs conclude that the institute would cost $2.5 million.[31]

According to his diary, on 18 October 1916, George declines an offer of a position from Arthur Dehon Little in Boston to manage the library of his successful chemical-research company. This is the fifth offer that he has declined: Rice Institute in Houston and 2 Midwestern universities had "insufficient" libraries, and a post at the Belgian legation would give him no time for scholarship. He is happy in Cambridge. On 5 November he and Mabel light a wood fire in their apartment, the first fire they have enjoyed since Wondelgem.[32] The couple are intimate, George records in his agenda on 18 November: "Afternoon lost."

Beginning in September 1916, George works to translate documents into French for the League to Enforce Peace, netting him $100.[33] It is probable that George received the assignment through Henderson's recommendation to Harvard's President Lowell, chair of the League's executive committee.[34] The League to Enforce Peace derived from concern among prominent New Yorkers, among them Hamilton Holt of the *Independent* and Theodore Marburg, about the First World War.[35] Under the presidency of William Howard Taft, it aimed to persuade US authorities to establish an international organization for guaranteeing peace; it declined to associate with pacificist organizations, and it became a vocal supporter of the League of Nations. On 4 November 1916, George notes a letter from Theodore Marburg, inviting him to write the introduction to the French translation of all the documents of the League to Enforce Peace. George accepts. "It is a great opportunity for me." He travels to New York to meet with Marburg, "very friendly, kind, not very smart." George hardly sleeps on the train, and he is exhausted for the meeting. He lunches with Marburg and Jarousse de Sillac, secretary of the French committee planning for a League of Nations. "Essentially a sad day."[36] George elaborates in a letter to Marburg that the league's plans will not

31. All from Diary.

32. Diary.

33. Diary, 14 September 1916.

34. Henry Aaron Yeomans, *Abbott Lawrence Lowell, 1856–1943* (Cambridge, Mass., 1948), pp. 430–43.

35. Ruhl J. Bartlett, *The League to Enforce Peace* (Chapel Hill, 1944).

36. Diary, 19 November 1916.

bear fruit unless Taft goes to Europe, for Taft "is the only man who would be listened to not only by diplomats but also by kings."[37] From George's short time as an astronomer in Uccle and his somewhat longer term as secretary of the Belgian Scholarship Committee, we know that he is temperamentally unsuited to be a hired hand. Even when his very survival is in question, he will not march to the tune that others call.

Theodore Marburg urges that George's introduction be completed quickly. The "handy little volume" appears late in February 1917 under the title *Ligue pour imposer la paix.*[38] George's preface, dated December 1916, emphasizes that the league seeks in no way to intervene in the ongoing war. Rather, the goal is to establish a political organization that may operate, after the war, to reduce the risks of armed conflict. The idea of the league is for nations to submit their conflicts to arbitration before undertaking war. Nations that do not consent to this procedure would be subject to league sanction, both economic and military. The volume includes extracts from speeches at a meeting held in Independence Hall in 1915 and one at Washington in 1916. George explains the league's view of political history, which is also substantially his own: "Every time that peace has been established and maintained, it has been possible only through the preponderant force of actors more numerous than their adversaries and acting in concert for the common good." Among the speeches he extracts is one by A. Lawrence Lowell.[39]

In Fall 1916, George is working a number of angles. Arnold C. Klebs, in Washington, acts as a contact with the bourgeoisie who will be required to finance George's institute for the history of science. Klebs and D. E. Smith come up with a yearly budget of $125,000, including a salary of $10,000 for the director; the endowment required is about $2 million.[40] Sociologist Hugo Münsterberg and Asian ethnologist Berthold Laufer, according to Klebs, support the project.[41] Associating Münsterberg with the scheme would be the kiss of death, however, for George realizes that "the insertion of Münsterberg's name during his lifetime would have brought about the resignation of the greater number of endorsers."[42] Despite

37. George Sarton to Theodore Marburg, 21 November 1916, in *Development of the League of Nations Idea: Documents and Correspondence of Theodore Marburg*, ed. John H. Latané (New York, 1932), *1*, 204.

38. Theodore Marburg to George Sarton, 7 November 1916, in *Development of the League of Nations Idea, 1*, 191; Theodore Marburg to Jarousse de Sillac, 1 March 1917, in *Development, 1*, 275.

39. *Ligue pour imposer la paix, William Howard Taft, Président*, trans. and ed. George Sarton (New York, 1917), p. 9.

40. HGS, Arnold C. Klebs to George Sarton, 29 September and 5 October 1916.

41. HGS, Arnold C. Klebs to George Sarton, 15 December 1916.

42. HGS, Arnold C. Klebs to George Sarton, 21 January 1917, paraphrasing George.

anointment by William James, Münsterberg's industrial psychology finds few American admirers, and his pro-German sympathy alienates him in Harvard circles.

Shortly before his death, Münsterberg elaborates his views in correspondence with George. Münsterberg observes that nationalistic pride is an unfortunate consequence of the war, and that Germany is certainly not different from other countries in this regard. Moreover, he contends that "Germany had the right to fear that Belgium's neutrality could not be conserved under any circumstances and that if Germany did not march through Belgium France would do it immediately." France should have watched as Germany rolled to its borders by the English Channel: "If Belgium had not been in a tacit alliance with France against Germany it would have been perfectly justified in confining its actions to a vehement protest against Germany's actions and otherwise yielding to it as Luxembourg did, instead of offering the armed resistance which had to lead to the horrible fight." Münsterberg rejects outrage against German atrocities in Belgium, observing that the Russians did worse in East Prussia. Then he moves to George's plans for an institute for the history of science. America is not the place for it—Belgium is—even though Münsterberg shares Karl Lamprecht's notion that synthesis of the kind George seeks is the work of individuals, not groups.[43] Münsterberg observes that the great synthetical works in history of science will proceed by the force of a great personality.

Arnold Klebs writes in October that William Welch is enthusiastically in favor of the plan for an institute devoted to history of science, will arrange for George to speak at Johns Hopkins University about his project, and recommends astronomer George Ellery Hale as a potential supporter; Hale has persuaded Henry Edwards Huntington to contribute to a building for the National Academy of Sciences.[44] Chicago astrophysicist Edwin Brant Frost advises George to associate his institute with an existing library or academy.[45] George naturally imagines a place in the proposed National Academy of Sciences building in Washington, a notion supported by D. E. Smith but discouraged by Frost's colleague Hale: "Do not mention the National Academy plan—it appeals to me, but there are no funds available and I don't know what other members would think of it."[46] Simon Flexner, initially sympathetic, quickly decides that his own Rockefeller Institute in New York and his involvement with the National Research Council preclude his supporting George's endeavor. In the words of Klebs: "We must

43. HGS, Hugo Münsterberg to George Sarton, 28 November 1916.

44. HGS, Arnold C. Klebs to George Sarton, 12 October 1916.

45. HGS, Edwin B. Frost to George Sarton, 7 November 1916.

46. HGS, David Eugene Smith to George Sarton, 17 October 1916, where he mentions that he will meet with Hale on the 18th; George Ellery Hale to George Sarton, 17 October 1916.

realize that we have the Rockefeller & Carnegie officials if not against at least not with us. We are their competitors in a way."[47] Advice was not only conflicting but also contentious. Thomas Jefferson Jackson See, an idiosyncratic naval astronomer at Mare Island, California, is lavish with praise: "You could not possibly engage in a more useful work, or one more needed above all in America, where the progress of pure Truth is so largely lost in the grosser things which are saleable." Pure truth, for See, relates to his own mechanical theory of gravitation, which he offers to George as a case-study of scientific creativity.[48] D. E. Smith, as he promotes George's plans for an institute, arranges for George to receive a contract to teach at Columbia University in the summer of 1917, one course in history of science and one in history of mathematics; Smith offers $500 for the assignments.[49] George has found 66 supporters for his institute, but among the philosophers at Harvard only Hocking has promised to help.[50]

During his début at Harvard, in addition to placing his ducks in order over the past three thousand years, George plunges ahead with more specialized inquiries, notably the Lowell Institute lectures on "Science and Civilization in the Time of Leonardo da Vinci." His contact there is the curator, historian of science William T. Sedgwick (Harvard's President Lowell, a descendant of the Institute's founder, is the trustee). George's first talk is mediocre:

> I speak rather correctly and rather quickly, but my diction is bad. It is bad in French; how would it be good in English? I am not a speaker. I do not like this trade. It shocks me to decry, to speak in a loud voice, more exactly, in a large room to be able to scream. This is what I think essentially, what I sense. I hope to God that I soon may no longer have to bear this burden, that I may return to work in my own corner.[51]

He feels cold during his second lecture, but the talk goes well.[52] Christmas 1916: Mabel is sick, May is difficult, there are too many gifts, and George

47. HGS, Arnold C. Klebs to George Sarton, 17 October 1916.

48. HGS, Thomas Jefferson Jackson See to George Sarton, 7 November 1916. On See's campaign against Einstein's relativity: Jeffrey Crelinsten, *The Reception of Einstein's General Theory of Relativity among American Astronomers, 1910–1930* (diss., University of Montreal, 1981), pp. 102–8, 273–86; Ronald C. Tobey, *The American Ideology of National Science, 1919–1930* (Pittsburgh, 1971), pp. 107–9. See criticized Einstein's general relativity in December: T. J. J. See, "Einstein's Theory of Gravitation," *Observatory, 39* (1916), 511–12.

49. HGS, David Eugene Smith to George Sarton, 17 October 1916; Diary, 25 November 1916.

50. Diary, 25–30 November 1916.

51. Diary, 6 December 1916.

52. Diary, 23 December 1916.

thinks about Belgium.[53] The Sartons enjoy dinner on New Year's Day 1917 with Lawrence Joseph and Edith J. Henderson and then visit the Dupriez and the Charles Homer Haskins families.[54] And George balances the books.

George and Mabel kept track of centimes in Belgium, and for the rest of their life they are conscious of pennies. In his married life at Wondelgem, George resembled some sides of the private scholar Peter Kien in Elias Canetti's novel *Die Blendung*—an emotionally-arrested man surrounded by books and close to the end of his inherited money. From the beginning George is unrealistic about the cost of a life, and his point of repose is a tight-fisted miserliness, the sobriquet *vieux rat* used with pride in his correspondence with Mabel. The attitude creates much friction later with his daughter May, who always spends more than she has, and it leads to ridicule from Henry Guerlac, his younger rival at Cornell University in the 1940s.[55] But George grew up in an industrial society lacking in social services; he knows about and fears a life of destitution. Mabel shares his sensibility, though she comes to resent his patronizing management of the family income. During their married life, the Sartons generally send monthly support to relatives and friends, even when they themselves have barely enough for food. So it is with some satisfaction that, for 1916, George calculates his net worth as $300, a figure that takes account of the value of his furniture but does not include his property in Belgium, a Brazilian bond in deposit in London, and debts of nearly $250, in addition to a dentist's bill and $50 remaining for the *Encyclopaedia Britannica*.[56] The calculation is made notwithstanding David Eugene Smith's loan of $200 late in October and early in November.[57]

George is off to New York from 16–21 January 1917 to line up support for his institute, while Mabel and May are at Brick Lodge in Auburndale, Massachusetts. He is met by the Smiths, who have arranged the meetings for his visit.[58] He visits the Egyptian section of the Metropolitan Museum of Art. In the afternoon he meets medievalist Henry Osborn Taylor and he dines with Baekeland's son. He meets George F. Kunz at Tiffany's. Kunz is an expert on precious stones, but in George's opinion he is otherwise not well educated. He sees James Matthew Barrie's *A Kiss for Cinderella* featuring actress Maude Adams, whom he adores. He dines

53. Diary, 25 December 1916.

54. Diary, 1 January 1917.

55. Henry Guerlac, "Sartoniana and Forward," *Lychnos*, 1986, pp. 1–27, notably p. 23.

56. Diary, 30 December 1916.

57. HGS, David Eugene Smith to George Sarton, 26 October 1916.

58. HGS, David Eugene Smith to George Sarton, 3 January and 9 January 1917, for Smith as impresario.

at the Smiths with the Fogg Museum's assistant director Paul Joseph Sachs, his wife Meta, and a physicist from Barnard College, Margaret E. Malsby.[59] He visits Archer M. Huntington, "perhaps the only American millionaire who understands me." He meets the materialist biologist at the Rockefeller Institute Jacques Loeb, an original member of the committee for *Isis*, "clever and sarcastic." Loeb has "vitriolic opinions. Any trace of sentimentality or mysticism for him is madness." George has dinner at the Columbia Club, meeting historians James Harvey Robinson and James Thomson Shotwell, along with psychologist James McKeen Cattell and chemist Frederick Barry.[60] He spends an evening with George A. Plimpton, arranged previously by D. E. Smith.[61] He dines with the legendary Brussels architect Victor *baron* Horta, like George a Belgian refugee in America since 1915 who lectures at universities.[62] In George's view, Horta is an unimpressive man. "I hardly think that he will be able to succeed here." Horta, in fact, obtains an appointment at George Washington University.[63] George returns to Cambridge in time to give a lecture on Roger Bacon.[64] The result of George's trip to New York is a prospectus for his institute, to be prepared and apparently financed by D. E. Smith.[65]

Mabel writes from Brick Lodge, on 18 January 1917, misdating the year 1916. She is resting there with May, but in separate rooms. Mabel, too, is attracted to Buddhism. "It is quite true, I am very far from being a Buddhist, yet of all religions of which I know anything at all, it is the one which appeals both to my heart & mind the most. Perhaps after many more imperfect lives, after realizing the emptiness & pain that are at the bottom of every earthly happiness, perhaps *then* I shall set out towards the Great Peace." She is glad that George has enlisted Huntington and the Columbia people, and she thinks that the institute is moving forward. She wonders if it will be in America. "Nothing in me foretells that I shall live my life here, possibly because I do not wish it so!" Mabel adds in another letter, "A sort of wave of agitation still comes over me sometimes, the sudden thought of

59. Or possibly the educator Julius Sachs.

60. James Harvey Robinson, in the History Department, is notable for having supported the doctoral dissertations of Lynn Thorndike, Dorothy Stimson, and Martha Ornstein, all authors of significant books in history of science. Nathan Reingold, "National Science Policy in a Private Foundation: The Carnegie Institution of Washington," in Reingold, *Science American Style* (New Brunswick, N.J., 1991), pp. 190–223, on pp. 366–7.

61. HGS, David Eugene Smith to George Sarton, 2 November and 18 November 1916

62. Françoise Aubry and Jos Vandenbreeden, eds, *Horta: Van Art Nouveau tot Modernisme* (Ghent/Bruges, 1996), p. 227.

63. HGS, Charles Herbert Stockton to George Sarton, 19 August 1917.

64. Diary, 16–21 January 1917.

65. HGS, David Eugene Smith to George Sarton, 23 and 24 January 1917.

our financial troubles (at least of how inadequately *I* have met them) or of arrears of work & then my heart beats wildly for a few minutes, but I can drive the feeling away." She signs it: "Always your Wild Bird."

On 3 February 1917 George notes in his diary that the United States has broken diplomatic relations with Germany. "May the other neutral nations of Europe and America act in this way and show their condemnation. The poor German people, led by fools into the pit." The event threatens not only the peace but also his plans for an institute devoted to the history of science. David Eugene Smith writes on 5 February: "I think you are quite right in saying that the recent international complication has made it wise not to push things too rapidly." Still, Smith thinks that the plans hang on Arnold Klebs, who is respected by scientists as a scholar and whose "acquaintance with men of means is such that I have always felt that the money would come through him, if at all."[66] But Klebs refuses to sign a draft text written by George.[67] Klebs is in favor of brevity and practicality—"subordination of ideas to practical exigencies"—while George is prolix and philosophical; Klebs is sympathetic to Germany, while George is a victim of Germany.[68] Klebs resents George's meddling and his patronizing tone. After all, who is George Sarton? "Your remarkable interference with Dr Smith has shown me how little trust you were willing to bestow in others, who are far above you in actual accomplishment."[69] Klebs is cross with George because of George's advocacy of war against Germany, which he sees as a blow against internationalism: "I was particularly attracted to you because you pleaded for internationalism in science, a plea very rarely made by Frenchmen & Belgians."[70] Smith tries to make peace between the men, suggesting that George pen his own personal views, to be included with the prospectus signed by his committee. "If three good, honest men cannot brush away misunderstandings, then we should all of us turn over the founding of an Institute to somebody else."[71] George is not as conciliatory as Smith would like, to Smith's chagrin, but the damage is repaired within several months.[72] Klebs concludes that the task "offers greater difficulties of execution than was at first thought."[73]

---

66. HGS, David Eugene Smith to George Sarton, 5 February 1917.
67. HGS, Arnold C. Klebs to George Sarton, 31 January 1917.
68. HGS, Arnold C. Klebs to George Sarton, 5 February 1917.
69. HGS, Arnold C. Klebs to George Sarton, 9 February 1917.
70. HGS, Arnold C. Klebs to George Sarton, 12 February 1917.
71. HGS, David Eugene Smith to George Sarton, 9 February 1917.
72. HGS, David Eugene Smith to George Sarton, 12 February 1917; Arnold C. Klebs to George Sarton, 10 April 1917.
73. HGS, Arnold C. Klebs to George Sarton, 12 March 1917.

Two months later, on 3 April 1917, George writes to Mabel: "Darling, I spent the whole morning writing letters. This is the 12th! War is declared, and Mr Wilson's speech of yesterday will remain as an imperishable monument of American idealism." He hopes that Mabel, beginning the third trimester of pregnancy, is resting at Auburndale. He encloses 2 checks, but "I could not pay your check entirely now." George is alone with May in Cambridge. On 4 April, Mabel comments about George's correspondence with D. E. Smith and Klebs. It "brings back the feeling that has long been lurking in my mind....that I might & ought to help you more with your correspondence." She feels "just now that you are in special need of help—more than I am—Your head is dangerously tired—I am worried about you." Mabel observes: "Having read & heard you speak of your ideas, knowing your habits of thought to a certain extent at least, I could I believe, with very little help, answer many letters for you." George could set out her task before lunch, and she would accomplish the task in the afternoon. George must not forget to eat apples and water her plants. On 5 April George feels lonely in the apartment. He had dinner with Edith Henderson, "who said that May was very good." (Evidently May is staying with the Hendersons.) He reluctantly agrees to a trial period when Mabel helps with his correspondence. On 6 April, Good Friday, Mabel has received $14 from Margaret Gillespie, who has on order work that will earn Mabel $50 more. On 7 April, Mabel thanks George for the checks and thinks that Wilson's declaration of war "is splendid & most heartening for the future."

War might hasten the millennial world city in a free Belgium. But war gives and takes away. Several weeks before America's entry into the war, George's proposal for an independent institute devoted to the history of science, rewritten and elaborated, appears in *Science*. It is a clear plea for money, and George concludes with a long list of his supporters. Commentators offer their opinions. Edward Sapir, the Head of the Division of Anthropology in the Geological Survey of Canada, urges George to consider including an ethnologist in his scheme to "look after the primitive stages of the scientific quest."[74] George's supporter Francis Sydney Marvin writes that the institute has "the cordial good wishes of all kindred spirits in the Old World and especially of England and of Positivists."[75] Philip Jourdain, from Cambridge, imagines combining the institute with plans that he and Eugenio Rignano are formulating for a "kind of scientific union between the allied countries"—essentially Henri Lafontaine's establishment before the war.[76] With American mobilization,

74. HGS, Edward Sapir to George Sarton, 28 March 1917.

75. Francis Sydney Marvin, "An Institute for the History of Science and Civilisation," *Positivist Review, 294* (1917), 138–40.

76. HGS, Philip Jourdain to George Sarton, 5 May 1917.

however, all bets are off. As David Eugene Smith writes, "this is no time to start new ventures in this country."[77] In a second article in *Science*, George mentions his "ill-timed" proposal.[78] But in addition to America's late entry into the war, George faces the problem of intellectual renascences—which by their nature are necrophilic—and in particular humanistic rebirths. His New Humanism, notwithstanding an appeal to pre-war American sentiment in favor of "humanizing" science, has the dubious distinction of an association with Neohumanism, the rededication to classics that dominated the 19th-century schools.[79]

On 24 April 1917, Mabel reports that Mary Bouton, wife of Harvard mathematician Charles Leonard Bouton, invites her to use her garden across the street from the Sartons in her absence; the Boutons become close friends. On 23 April, George writes from Baltimore. President Frank Johnson Goodnow called at breakfast and took him to Johns Hopkins University; Goodnow was sent to China by the US when the emperor was dethroned. On 25 April, George writes that he has received the letters forwarded by Mabel. "One was from Mr Herbert Hoover, cordially endorsing" his plans; in 1914 Hoover directed the Commission for Relief in Belgium, which distributed substantial aid to civilians. In a second letter on the 25th, George writes, "My lectures thus far have been very successful." George is "glad that [actress Alla] Nazimova's blue jerkin has been mailed. She will have at last something to put on." On 26 April, Mabel writes from 10 Avon Street, Cambridge: "Oh Daddy! I am jealous of those other mothers because they have children whom you are always so enthusiastic over. Never, *never* have you shewn or told me you thought *our* little girl was even ordinarily sweet & cute or had an interesting little face!" Mabel knows "it is a sort of principle with you. You are *so* afraid I may be too proud of her!" Mabel imagines that she had pains over the night in her ovaries. "Little son is wildly excited each morning. Truly it is as if he woke up then & was

77. HGS, David Eugene Smith to George Sarton, 21 May 1917.

78. George Sarton, "An Institute for the History of Science and Civilization," *Science, 45* (1917), 284–6, *46* (1917), 399–402. George notes in 1924 that when the United States entered the war, "the idea of creating such an institute became preposterous and the whole thing fell flat." George Sarton, "An Institute for the History of Science," *Isis, 6* (1924), 408–10, which reproduces plans announced by Friedrich Dannemann for an institute in Munich. George's first article of 1917 found a sympathetic reader in Kuwaki Ayao, a founder of the history of science in Japan. Setsuko Tanaka, "Kuwaki Turns to History of Science," *Historia Scientiarum, 2*, no. 2 (1992), 85–108, on p. 96.

79. Lewis Pyenson, *Neohumanism and the Persistence of Pure Mathematics in Wilhelmian Germany* (Philadelphia, 1983), pp. 5, 109–24. Friedrich Immanuel Niethammer's invocation of a new Humanism in 1808 was popularized as Neuhumanismus by Friedrich Paulsen, the 19th-century historian of education. Arnold Thackray, "The Pre-History of an Academic Discipline: The Study of the History of Science in the United States, 1891–1941," *Minerva, 18* (1980), 448–73, p. 455 for pre-war sentiment.

tired of his cramped position!" This is the first mention of her pregnancy. It inconveniences George, who records sex in Detroit on 31 March 1917.[80]

George is in New York during July and August 1917, teaching at Columbia University. Mabel is in Cambridge, waiting the birth of a child. Mabel writes in French, uncharacteristically, on 5 July. "I have a large and heavy heart to feel so alone, and these 6 or 7 weeks seem terribly long." But she will be brave. "The little one is doing his evening exercises. He seems to want to remind me that, after all, I am not alone." On 6 July, in English, she reads Bertrand Russell's

> wonderful and beautiful book. I feel an almost personal sense of gratitude to him for having voiced many things so clearly & so generously, amongst others, the changing position of women & their attitude to marriage & child-bearing; but also many other subjects. I really got quite excited about it last night. It entirely took me out of myself. I felt as if morally & mentally I had taken deep breaths of pure air from great spaces, after having been choked in stuffy shut-in rooms of thought.[81]

She has had time for her correspondence "& been carefully through your article which shall be posted before this even."[82] Her maid "is very thoughtful & kind & we shall get on nicely I hope." She concludes: "I still get a lump in my throat when I see certain eloquent signs of yours & May's absence."

After depositing May with the Baekelands, on 9 July George describes his room at Columbia University to Mabel: "It is noisy: noise from Amsterdam Avenue, & noise from the inside; prisonlike, ugly, & generally uncomfortable. The rooms in the YMCA are smaller but better kept & on the whole more pleasant." May travelled well to New York. "May was very good & sweet indeed. You know, she is really a darling." He has dinner with Mr and Mrs Edouard Potjes. "He is a well known pianist & composer. Mrs Potjes [J. Potjes-de Keghel] has known you & Céline & Margaret Sorel at the Institut de Kerchove & sends her regards. They are perfect specimens of Flemish 'bourgeois' but I had pleasure in talking with them." He walked for 2 hours along the Hudson with May. "Dr Baekeland bored us to death & I began reading [H. G. Wells's] *Mr Britling Sees It Through* [London, 1916], although Mr Britling also bores me." Hendrik Van Loon came for dinner. He has been waiting to obtain permission to travel to England. "I have

80. Agenda.

81. Bertrand Russell, "Marriage and the Population Question," in *Principles of Social Reconstruction* (London, 1916), pp. 168–96. In 1916 Russell was dismissed from Trinity College for antiwar activity, also discussed in this book.

82. Possibly the second part of George Sarton, "An Institute for the History of Science and Civilization," *Science, 45* (1917), 284–6, *46* (1917), 399–402, or (less likely) "The Teaching of the History of Science," *Scientific Monthly,* 7 (1918), 193–211.

been thinking almost all the time of you, darling wife," and he thanks her for the corrected manuscript, which has arrived. "I am very glad that you enjoyed reading Bertrand Russell's book & felt in sympathy with him." George has not yet read the chapter on women and marriage. On 10 July, George writes that he has "some good students in my course on the history of mathematics, almost all girls (teachers in different colleges)." He dined with philosopher Horace Meyer Kallen at the Harvard Club. Kallen, a former student of George Santayana's at Harvard and an instructor of philosophy and psychology at the University of Wisconsin, is also teaching summer school at Columbia University and supports George's idea for an institute for history of science. Kallen was one of George's contacts when he arrived in America, according to his agenda.[83]

Mabel is friendly with Horace Kallen's sister Deborah.[84] Mabel writes on 7 July that Mrs Kallen and her sister, no doubt meaning Deborah Kallen although possibly another sister Frances Kallen, have called on her, also Mrs Withington. Dr Withington thinks "that the places where Infantile Paralysis was most rampant last year will be immune this!"[85] Dr M. W. Richardson "who talked to a Mothers' Club & who lost 2 children of it in 1908" thinks that it is a germ carried by rats (on 8 July Mabel addresses "Cher Vieux Rat," in French). On 13 July, George explains that there are 4 students taking his course in history of science (2 women and 2 men) and 5 in history of mathematics. "The 2nd class is a good one: they ask questions, they are alive." If their child is a girl, George thinks Annette would be a good name: Annette Léonie, Annette Franklin, Annette Mabel.Mabel has decided on the sex. She reports on 9 July that her "little son stretches and kicks most violently. Soon, soon he will get out of prison, I am sure." She has received 2 packets of *The Belgian Deportations* and will distribute them.[86] On 10 July, Mabel addresses "Pauvre Vieux Rat," advising him to get a better room

83. Milton R. Konvitz, "Horace Meyer Kallen (1882–1974): In Praise of Hyphenation and Orchestration," in *The Legacy of Horace M. Kallen*, ed. Konvitz (Rutherford, N.J., 1987), pp. 15–35. The connection may have been through Kallen's friend Norman Angell. Kallen resigned from the University of Wisconsin in 1918, when the university would not allow the pacifist Angell to speak; in 1919 Kallen, a prominent Zionist, became a founding professor at the New School for Social Research in New York. The undated reference to "Abe Kallen" in his Agenda for 1915; CAJ, Horace Meyer Kallen to George Sarton, 12 May 1917. Louis Menand, *The Metaphysical Club* (New York, 2001), p. 394 for Kallen's promoting pluralism as "a formula for a kind of noninvidious segregation" in America.

84. CAJ, George Sarton to Horace Kallen, 14 May 1917.

85. Probably Edward Theodore Withington, a persistent and accomplished historian of medicine beginning with the publication of his survey, *Medical History from the Earliest Times: A Popular History of the Healing Art* (London, 1894).

86. Arnold Joseph Toynbee, *The Belgian Deportations*, with a statement by Viscount Bryce (London, 1917).

and rent an armchair, unless Columbia can supply one. On 11 July, Mabel has almost finished the cradle. "Do you remember helping me nail in the white muslin lining in *your* cradle, on the very day May was born??" Mabel thinks "somehow Alfred Hugo will not arrive until about the 20th, but he is active, extremely so!" On 12 July, her doctor has advised Mabel to discontinue taking her pills, as she thinks they give her heart palpitations. George asks on 14 July if Mabel goes to the hospital that day, "or has our little Alfred not yet come to a decision?" At the Baekelands', May behaves just as George did, when he visited his grandmother. "Everybody said: 'What a good and brave little boy.'" George reports on a meeting at Baekeland's house: Fabri of Nancy (probably, Charles Fabry of Marseilles) and Henri Abraham of Paris, "two of the greatest French physicists"; Nobel laureate chemist François Auguste Victor Grignard of Nancy; and an Italian astronomer. "The French scientists knew my work, which just, I noticed, made a great impression on Dr B." George is traveling with May, Céline Baekeland, and Nina Baekeland to the Adirondacks—Upstate New York, but not back to Mabel in Cambridge.

On 16 July Mabel writes to "O Dearest, Wisest of Old Rats!": "I do believe that even if things were hard or difficult, I would come through alright & the baby be healthy & strong." On 17 July, without salutation, "I have a certain proud satisfaction that those scientists shewed Dr B. they knew your work." She had dinner with the Hendersons. They walked her back. Lawrence J. Henderson "was so pleased & struck with our little apartment. I did not know he had never been in it!" She is glad that May will attend camp in the Adirondacks. On 18 July, George writes that he has been given a rocking chair. "I have not moved to another room. That would have obliged me to repack all my things & I would rather die." Mabel writes on 18 July that she is now due to give birth between 25 and 31 July. "Mrs Henderson will be allowed to see me *in your place* & will be told everything." On 19 July, Mabel reports that she thanked all 7 women who sent her a cradle. On 20 July, Mabel has nearly finished a dress for her neighbor, feminist Elizabeth McCracken; it will bring in $5, which Mabel wants to send to her mother. She asks George to write to her mother after the child is born. "Last time & when May was born I managed to write to her often, almost each day, but at the hospital they won't allow any such thing." She closes: "Little son kicks & stretches *dreadfully* till I want to cry, but I try to laugh instead." On 22 July, Mabel relates that Deborah Kallen will send out birth notices. "I have planned some beautiful little things for Little Son & have designed and begun a little bonnet. It is the only really *beautiful* thing I shall have made for him & it has made me happy all day long to be doing it. I want it to be beautiful enough to be worn & then *kept* for him when he grows up & has a baby of his own!!"

On 22 July, George writes that he returned from an outing with Teacher's College colleagues, which exhausted him: "Many hours in an automobile had made me very nervous & it was extremely hot." The excursion took him to the Palisades on the New Jersey side of the Hudson, where the group caught copperhead snakes. "The company did not please me much. Most were of the schoolmaster type, most uncongenial to us." He lunched with a potential patron, Mrs Cope, at Forest Hill, "a kind of garden city built like an old Surrey village. It will be very beautiful in fifty years hence, if it be not replaced before by an old Dutch or a Japanese village." Mrs Cope now understands what George wants to do, but he doesn't know if she will help him. Mabel writes on 23 July that she has trouble sleeping because of the heat and humidity. "Surely, surely little son will arrive to-morrow or next day." Mabel regrets that George's Palisades companions were uncongenial, but "Do you know I have worried so about *snakes*, not for you, but for May in the woods!" On 24 July, Mabel sets out plant boxes. An awning over the porch on their apartment, which Mabel had installed, will provide "great pleasure & benefit...& so the expense will be justified." Her maid asked if Mabel would have a nurse for the first weeks after she comes home from the hospital, "but I told her no." Mabel on 25 July, had dinner at the McCrackens. "She is interesting & knows lots of interesting people, knew & liked Münsterberg."

George, on 25 July, reports that he dined with Kallen at the Faculty Club and met Norman Angell, returning from a peace conference at Stockholm. "I had a rather unpleasant discussion with him. Have no use whatever for his 'negotiated peace.'" He read Sir Edmund Backhouse and John Otway Percy Bland's *Annals and Memoirs of the Court of Peking* (Boston, 1914), "the most depressing book I have read in years." He also read Thomas Babington Macaulay on the first governor-general of British India, Warren Hastings.[87] He urges Mabel not to eat too much ice cream and to use the shower, which he fixed for her. "I am very anxious to see the little bonnet that you [are] making for Alfred." George, on 27 July, writes that he will visit the Baekelands the next day, "as I need a rest, and am most anxious to see our little May, & to hear from her own lips what she saw in the forest." He imagines the screened porch as a room for the new child. "It will be a very good place for him to sleep in day and night & it will be very convenient to you." He is reading more of the *Arabian Nights* in the French translation of Joseph Charles Mardrus, "far superior to [Sir Richard Francis] Burton's translation." Mabel, on 28 July, is tired of waiting: "Sometimes I look at myself & wonder, I don't mean with *fear*, simply with puzzlement! *How* can he find a way out through such a narrow door?? I think he will need a lot of pushing!"

87. Hastings, impeached by Parliament and exonerated, is the subject of an essay by Macaulay.

On 28 July George writes a sombre letter to Mabel:

> Sweet & good wifie,
>
> I am extremely unhappy because I am anxious about you & the little one, & also—and not less—because I am not at peace with myself. I cannot be happy when I do not work, and I feel that I do not work enough. It is true I am delivering some good—some very good lectures—especially in my course in the history of mathematics. But I do not do any serious work outside of that, & I have never felt more lazy & dissipated. I have been reading—I do not say studying—a lot of Arabic & Chinese literature, but without enthusiasm & my soul is slumbering. It is true that I do not read a line without a purpose, a single purpose, & that these desultory readings contain much fresh material for my own thinking, but my soul is slumbering, and I am cold.
>
> I am also unhappy because I have just realized—too late—that I should have left yesterday night to visit you in Cambridge. I might have been there in time to welcome our little son & cheer you up. But maybe you will not be delivered until your own birthday.
>
> I do not call you "beloved," although you are the beloved; neither dare I kiss you because I feel in disgrace.
>
> Your husband
> George

Mabel replies on 29 July:

> I am a little tired of waiting & of the heat & of our separation, that is all. Don't pay any attention if I seem depressed. There is *no* real reason to be so on "our" account. Quite the contrary. So let your heart take courage.
>
> But oh! How lovely it would have been if you had come. Still I would rather it was at *the* time & not before! & though I long for it dreadfully I do not feel we ought to give way to our longing because of the expense. Do you know it helps me so just to know *you wanted to come*!

Mabel tries to cheer him up: "Just put your hand in mine & feel that I am not only the Beloved, but the Comrade too, so near, so near to you. Don't you feel that? I feel it so for you! You are sometimes a hero to me, but you are sometimes just a dear, faulty human being, to whom I dare confess my weaknesses & be helped & comforted, as no 'always-hero' could do!"

On 30 July, Mabel writes that Edith Henderson is leaving town, but Lawrence J. Henderson "is staying on for a few days anyhow, *partly* on my account, as he feels some-one should be here who *knows you* & could take the responsibility of a decision for instance if the doctor wished one." Caroline Eliza Hall, wife of physics professor Edwin Herbert Hall, will write each day with a report. On 31 July, Mabel reports that the temperature is 103 degrees in the shade. She has received propaganda pamphlets from Henryk Arctowski, the Polish meteorologist who had been employed at the Uccle Observatory in Belgium just before George signed on there, and a compilation by Jean Massart, *La Press clandestine dans la Belgique occupée* (Paris, 1917). Mabel asks when George gets paid. She will have to dip into her reserve of $50 if she goes into the hospital; George gave her $20 less the past month. She signs herself, "the Wild Bird on her nest = (our nest)." George on 31 July, visits May in Yonkers. May received the doll Mabel sent. Céline Baekeland is attentive toward May, but she lacks Mabel's touch. "*You are really a splendid, a unique mother.*" In the Adirondacks, "May did not see any animal, except mosquitoes which were extremely abundant."

On First August, George sends Mabel a check for $107. "I know how thrifty you are, & to be sure, I do not fail to appreciate it. You would deserve to receive much more money, both for housekeeping & for your own private needs, and it is a cause of deep & continuing sorrow to me to be unable to give you more. And I do not see the dawn of better days." George wishes Mabel a happy birthday, for 3 August. He has left her 2 presents in his desk drawer. One is a thermos bottle. The other is a little book for telephone numbers. On 2 August, Mabel writes about poverty:

> Dear Husband & Comrade, *don't* be sad not to give me more money. Don't worry over it. I don't "deserve" that you should struggle nor work more than you do, for I often feel distressed that you have *to do so much. You do your utmost*, no one can do more. I wish we could earn more or save some to send to our people. That is really what weighs on my mind.

Mabel is glad that she can contribute to the household expenses because she is troubled when George directs his energies away from what he considers most vital, his scholarship. On 3 August, Mabel is pleased with her birthday gifts and George's thoughtfulness. "I did see the thermos bottle one day & wondered what the queer thing was, but I did not connect it with myself at all!" Céline Baekeland sent her a silk peignoir, valued at $12. "I just feel it wicked to have or use such a useless, luxurous thing, & I felt a hypocrite in writing to thank her." Mabel took a long drive with Marie Dupriez to Concord. She switches to French. It was in the automobile of the wife of Henry James, with whom she was staying, and the car had an English chauffeur.[88] Dupriez gave Mabel a check for $50 to send to relatives or friends in Belgium. If they add in $5, which Mabel will earn from finishing a dress for Miss McCracken very soon, it would mean F 100 for Aunt Hélène, Uncle Arthur, and Céline Limbosch.

George visits Cambridge on Sunday, 5 August. On 7 August (misdated 8 August) Mabel writes that soon they will know if they have a son or

88. Henry James, lawyer son of William James, was a major force in Harvard administration, becoming in 1922 one of the Overseers of the College.

daughter. Her doctor has decided to induce labor, for the baby is now 8 days overdue. Mabel wishes that the procedure had been scheduled for Sunday, when George was there. On the morning of 8 August, Mabel reports that she took herself to the hospital. Her room is not attractive, but she has glimpses of the river. She hopes that George will receive a wire that a son is born "before you leave for your class this afternoon." George cables on 8 August his love for Mabel and his son, who has been born.

On 10 August, George receives Caroline Hall's letter, where Mabel and Alfred "are both doing well." He will visit Yonkers to see the Baekelands and May. Mabel writes on 10 August that she is nursing Alfred every 3 hours. On 11 August, George relates that Hall reports "that Dr Corbert considers that you are both doing very well." He has sent a check for $100 [he writes $ 00] to the Commission for Relief in Belgium for Aunt Hélène, Uncle Arthur, and the Limbosches, in the name of Alfred Hugh Sarton, Esq. Mabel, on 12 August, writes that she has an abundance of milk, and Alfred is nursing well. On 13 August, Alfred's face is changing for the better, "but he is not pretty like May was." Mabel will contribute milk for other infants by using a breast pump. Alfred lost 9 oz since birth, but he has regained one ounce. On 14 August, Mabel writes that Alfred cannot keep down his milk, nor have his bowels moved. Mabel thinks that he looks thin. She is fine. One "inside stitch had to be made," and it causes discomfort when she receives an enema.

George, on 13 August, reports that he spent a whole day with May. "I had never spent as much time with her before." May "is so intimate with Mrs B. that she does not like to call her 'Mrs B.,' & having no other name to give to her, she gives none."[89] On 15 August Mabel reports that Alfred has violent indigestion, but that the doctors find "*nothing* organically wrong." Céline Baekeland is bringing May to see her on Saturday, on their way to Annisquam. On 16 August, Mabel writes that Alfred has been seriously ill. The doctors fear that there is "an obstruction of the opening of the stomach into the intestines," but they think that it is probably "a *spasmatic contraction*," and if there is an obstruction, it is only a partial one. Mabel is skeptical of the explanations. "I was worried about his not being well." She resents the way she has been treated: "*If only they told the truth here!* But I don't trust them." The baby screams in pain. He has lost one pound since birth. On Friday evening, 17 August, Mabel reports that they still cannot say if the obstruction is structural or spasmodic.

Alfred dies, possibly before George arrives on Saturday evening or Sunday morning. The baby's ashes are taken to Annisquam by Mabel and George, probably in the company of Céline Baekeland, with the thought

89. Compare *Phoenix*, pp. 87–9, where May knows Céline Baekeland as "Bonbon."

of scattering them over the water. When the couple arrive at the sea, they face low tide, with crabs scuttling "in the black ooze."[90]

In the second week of September, the Sartons reunite at Annisquam with May, who had remained with the Baekelands during the summer.[91] Then Mabel writes to George on 22 September 1917 "to clear way the little cloud that seems growing up between us." George thinks that Mabel does not understand how things big and small are weighing on him. "Probably it is true," Mabel admits. But George does not know Mabel's deepest pain. "The worst pain *can* never be told, is not meant to be told, I think." Yet they can help each other. She makes the point also in French.

Mabel travels to Washington in October 1917, exploring business possibilities and staying with Margaret Gillespie. George reports on 17 October that "there is a kind of epidemic in the school. I will not let May go there." He sends regards to Harriette Shadd, the Gillespies, and the Vaughans. May dictates a letter to Mabel on 19 October. She is going to play with Harry. She got her rubbers and a "frange cut." Their maid Marian was not in on Friday. "Daddy made the supper." George adds that he will lecture in St Louis, Madison, and Chicago during the first week in December. Vassar College has asked him to lecture on the Ellen H. Richards Memorial Foundation funds, which will publish the talk as a pamphlet. On 21 October, George writes that the maid is following Mabel's instructions, but she served a chicken, "than which I have seldom eaten aught more tough and stringy. Nay, the Devil himself could not be tougher." He will also lecture at the University of Toronto, "the first time that I would address a British audience"—3 lectures for which he receives $150 plus lodging along with an introduction to James Playfair McMurrich, Professor of Anatomy, a Canadian who took a doctorate at Johns Hopkins University and taught at US institutions.[92] On Sunday George and May visited the Laskis. (British political economist Harold Laski, like George teaching at Harvard since 1916, resigned later under pressure from Harvard's President Lowell for defending a strike of Boston police.[93]) On 22 October George writes: "Yes, yesterday was a gray day. A Sunday without you is not a Sunday, wifie."

Late in 1917 and early in 1918, when the Sartons are apart, their letters record affection. On 3 December, Mabel writes to George that Sunday was "a long, weary day." First she plunged into the slums to meet a woman who had worked for her before. She waited on a filthy landing for 90 minutes, but the woman did not turn up. Then to the Manhattan sugar

90. May Sarton, *A World of Light* (New York, 1976), p. 62.

91. YMS, Mabel Sarton to May Sarton, 3 September 1917.

92. HGS, Robert Falconer to George Sarton, 27 October 1917. On McMurrich, Thackray, "Pre-History of an Academic Discipline," on p. 451.

93. Menand, *Metaphysical Club*, pp. 428–9.

company, where she bought a good supply at a premium, 30% beyond the usual price. She walked with May to the Dupriez's house to consult about dresses that Mabel was making. Then to more dressmaking and caring for May. She wants to know, in French, exactly when George returns. "Come back as soon as possible, my dear man." On 18 January 1918 Mabel writes to "Olde Ratte" to thank him for flowers, the "fairy-like bush of pink-ness! Do you remember that they were the fashion in Ghent the last years before we left? Madeleine [van Thorenburg] loved them & often had them, & I loved them too." May lunched with a friend, and Mabel went to a "hen-party." "May misses you in many ways just as I do. She has said several times 'it's lonely without Daddy' & 'I don't like it when Daddy's away.'"

In February 1918 Mabel writes to "Mon cher Vieux Rat" from Gillespie & Gillespie's Shop of Gifts, 1203 F Street Northwest, in Washington. She is staying with Margaret Gillespie. On 18 February George receives Mabel's condemnation in French of evangelist Billy Sunday, whom she heard. George writes on 28 February that their small apartment has a "terrible, almost uncanny, quietness" in Mabel's absence. George offers thoughts on Mabel's Washington business. "All that you are doing now is simply experimental & preparatory work, upon which basis it will perhaps be possible to build something more important, as soon as the war is over, especially if you can secure Céline's entire & exclusive collaboration." During the war, the prospects "do not seem to be very brilliant," but he thinks that the ground can be prepared for success.

Both Mabel and George are concerned about what George will do after his contract ends at Harvard. There is the possibility of working for Emile de Cartier, who earlier expressed interest in George's project for an institute in Belgium, although in his letter to Mabel of 28 February George comments that he already declined one offer from the Belgian diplomat.[94] On 4 March, George informs Mabel that he has written to William Ernest Hocking about a position in the Department of Philosophy at Harvard; at the time, May is attending Shady Hill Cooperative Open-Air School, run by Hocking's wife Agnes.[95] Dupriez called to say adieu upon his departure on a California lecture tour. "I feel he is living in another world. So is likely de Cartier. On account of that I have very little hope in him."[96]

94. HGS, Emile de Cartier to George Sarton, 4 January 1918, offering George the position of paid secretary of the committee for the Fonds du Roi Albert, a Belgian propaganda initiative in the United States. De Cartier emphasizes that George can carry out the work while he is at Harvard.

95. *Phoenix*, pp. 101–19; Margot Peters, *May Sarton: A Biography* (New York, 1997), p. 404.

96. HGS, Emile de Cartier to George Sarton, 24 January 1918, where de Cartier expresses interest in George's idea for an American Institute in Brussels.

George is bucked up by speaking at dinner with Harvard linguist Leo Wiener, who arrived in New Orleans from Europe 40 years previously. "I could but admire his pluck & perseverance."

Another letter of 4 March 1918 from George reveals that he has "more students than I ever had—17: 9 taking one course, and 8 the other one." Even in the best of times and at the finest institutions these are slim numbers. Parnassus, in war-time, still requires crowds to absorb the teachings of its sages, and Harvard's generosity, such as it was, is fast running out. In a letter received by George on 5 March, Mabel finds his letters to Hocking interesting, but she still hopes that he may find a post with the Belgian government. George reports on 8 March that Alfred Hoernlé, Dean of the Department of Philosophy, will nominate George to President Lowell, but Harvard has a hiring freeze; the next day, the president turns down Hoernlé's request.[97] George emphasizes to Mabel on 9 March 1918: "Hence we shall not live any more in Cambridge, unless I decide to begin a different career, for inst. as a literary man (with Baekeland's assistance). That implies that I must give up publishing my book on Leonardo, & all the work which I have done in that direction is so to say lost." George will travel to Washington for lunch with de Cartier and Mabel, but he expects nothing to come of it. "I feel rather discouraged, dear wife." George persists and writes to David Eugene Smith, who replies that there is nothing for him at Columbia, but: "I admire tremendously your grit and perseverance and optimistic spirit, and I feel sure that things will turn out all right in one way or another."[98]

In a letter received by George on 9 March 1918, Mabel has a jaundiced view of Harvard: "Some instinct keeps telling me to warn you against landing yourself at Harvard because I feel & believe that de Cartier will want you. He certainly gave Dr Baekeland the impression that he would like to use your abilities & your sympathies for & knowledge of the best American ideas." Mabel really thinks that it is important for George to travel to Washington to see de Cartier in person. She dines with the Vaughans and has lunch with Harriette Shadd. Mabel has in fact telephoned the Belgian Embassy, spoken with de Cartier, and arranged a lunch with him on 14 March, to which George is also invited.

Clear-sighted Mabel knows the formidable obstacles to an affiliation with the university where George eventually becomes a fixture. There are 2 kinds of professor: those with an independent income, and those without one. Given the charitable origin of his salary in 1916, George is in the latter category, and the administrators at Harvard know it.

---

97. HGS, Alfred Hoernlé to George Sarton, 7 March 1918 and 9 March 1918. Following the rejection, Hoernlé writes to Ralph Barton Perry, who has moved to Berkeley, on George's behalf.

98. HGS, David Eugene Smith to George Sarton, 8 April 1918.

Furthermore, a pattern of professorial penury had been set by Charles William Eliot, the Unitarian scientist who brought Harvard to world prominence and who retired as president of the university in 1909. Eliot required both asceticism and devotion of his professors, and he was disinclined to feather their nest.[99] Eliot's successor, lawyer Abbott Lawrence Lowell II, brother of astronomer Percival Lowell who predicted the location of the planet Pluto and of poet Amy Lowell who wrote about lilacs, disciplined undergraduates, attracted substantial endowments, and dramatically expanded Harvard's offerings. Lowell was no special friend of research, and George's pleas for universities to be directed by sensitive scientists would have left Lowell cold. Yet, we shall see, Lowell could be bought.

George has several problems at Harvard. One problem is a penchant for alienating colleagues—his second article in America is a stirring rejection of American pragmatism, made into a philosophy at Harvard by William James, although by 1918 the Harvard philosophers come in many shapes and hues.[100] Another problem is scholarly invisibility. Lowell continues Eliot's policy of hiring only the very best staff, and others among George's talented philosophy colleagues—Hoernlé and Sheffer—managed to stay on only by divine grace.[101] George is 33 years old with no major book to his name, the editor of a journal that has not appeared for 4 years. He pecks away at the Renaissance. His gaze extends from antiquity to the 20th century, and his energies are spent on politics as well as teaching. Is this, indeed, an auspicious beginning for a scholarly career?

At the end of his life, in notes for a biography of his wife, George writes: "Loafing during early years in W[ondelgem] & C[ambridge], (not to mention the War Office). I had not yet completely found myself, my way was not yet clear & my methods were as yet undetermined. Economic insecurity had much to do with that."[102] On Easter Sunday 1918, in desperation, George writes once more to Robert S. Woodward, notwithstanding Woodward's hostility to his efforts at founding an institute for the history of science. As Arnold C. Klebs observed in 1916: "Woodward whom I know is unprogressive. He has made no satisfactory success of his Institution & he is temperamentally inclined to view others attempt with a vague

99. Henry James, *Charles W. Eliot, President of Harvard University, 1869–1909* (Boston, 1930), *2*, 81–2.

100. George Sarton, "The History of Science," *Monist, 26* (1916), 321–63; Bruce Kuklick, *A History of Philosophy in America, 1720–2000* (Oxford, 2001), pp. 129–78. George and his friend Andries MacLeod discussed William James's pragmatism in the spring of 1912; MacLeod: "I do not like this way of thinking!" HGS, Andries MacLeod to George Sarton, 11 April 1912.

101. Kuklick, *Rise of American Philosophy*, p. 411 for Hoernlé, p. 458 for Sheffer.

102. YMS.

suspicion and perhaps a little jealousy."[103] By the spring of 1918 George had learned how to use the new scholarly resources of the New World to recover the old science of the Old World. This understanding, he pleads to Woodward, is his unique gift.

103. HGS, Arnold C. Klebs to George Sarton, 30 September 1916, although on 4 November 1916 Klebs is quick to agree with George that the *Index Medicus*, a bibliography funded by the Carnegie Institution, "alone is sufficient for my own appreciation."

104. CIW, George Sarton to Robert S. Woodward, 4 July 1919.

CHAPTER 12

Faustian Bargain

George Sarton's Easter Rising of 1918 has significant implications for his life and for his discipline in America. In his letter to Robert S. Woodward, George cries that his lectureship at Harvard is ending and he has no prospects elsewhere. He has lectured conscientiously at 20 universities and given a number of courses, which delayed his own publications. But disaster looms: "If I be not reappointed or rescued in some other way, I must give up these studies. I cannot wait—not even a fortnight." He continues that he must "*stop that work or starve*." Why cannot America support just one person "to devote himself to the study of the positive foundations of human progress?" He needs only a salary: "All that I need is to be kept alive and safe."[1]

Robert S. Woodward is an engineer from the University of Michigan, who rose through the Coast and Geodetic Survey to a chair of mathematics at Columbia University before becoming the second head of the Carnegie Institution in 1905. From the beginning, his institution had a bureau, later a department, of historical research, created following a proposal by Frederick Jackson Turner "for a historical school for graduate students in Washington." Supported by the American Historical Association, direction of the bureau fell to J. Franklin Jameson, founding editor of the Association's journal. Jameson had been professor at Brown University and the University of Chicago. His mandate at the Carnegie Institution was to help place historians in positions ("a national clearing house"), to publish documents, to collect material from European archives, and to edit the *American Historical Review*. Jameson was a workmanlike historian, and the Carnegie Institution published his diplomatic documents in an enormous and tedious series—a pale, New World reflection of great European projects, notably the *Monumenta Germaniae Historica* that secured the profession of historian at German universities. Facts and their establishment were everything for Jameson and his patron Woodward. By 1918 Jameson focused his attention on establishing the United States National

1. CIW, George Sarton to Robert S. Woodward, Easter Sunday, 1918.

Archives.[2] But Jameson by no means monopolized Woodward's attention in matters historical. During the first decades of the 20th century, the Carnegie Institution supported a number of works relating to history of science.[3] George's encyclopedic and bibliographical vision of history of science confirmed Woodward's sentiments.

Woodward writes that he will consider appointing George as a researcher, essentially as a substitute for John Theodore Merz. "Please tell me what you would like most to do in case the Institution could find it practicable to furnish you a salary whereon you could support yourself and your family while engaged in your proper work. Would it be practicable for you to prepare a monograph on the history of the evolution of what might be called the scientific method?" George answers at length: a study of Leonardo's scientific manuscripts; a book about teaching the history of science, drawn from his previous publications. He thinks that the teaching of history of science must be entrusted to people who are primarily scientists, not philosophers. He prefers not to write on scientific method, for he is not yet ready to do so and would regret the result, "just as Mr Merz perhaps regrets—at least I hope so—the immaturity of so many parts of his *History of European Thought*." In an appendix, George emphasizes that he proposes to continue writing simultaneously about physics in the 19th century and also science and civilization in antiquity. He closes:

> I am now already 33 years of age, and although I have garnered considerably, I have produced comparatively little. I have only 30 or 40 years left for my work, at best. This is what makes me so nervous and anxious. My anxiety is less due to financial difficulties—however acute they may be—than to my fear that I may be prevented from doing the work—which I know—I am able to do.

Woodward inquires with Harvard and with William Thompson Sedgwick at MIT about George. The answer is favorable. Harvard reports: "He is

2. Victor Gondos, Jr, *J. Franklin Jameson and the Birth of the National Archives* (Philadelphia, 1981). In 1929 George becomes a member of Jameson's old Department of Historical Research, renamed the Division of Historical Research by Woodward's successor John Campbell Merriam. George Sarton, "The History of Science in the Carnegie Institution," *Osiris*, 9 (1950), 624–38, on p. 625.

3. Leonard Eugene Dickson, *History of the Theory of Numbers* (Washington, D.C., 1919–23); Victor Selden Clark, *History of Manufactures in the United States* (Washington, D.C., 1916–28); Edward Ball Knobel, *Ulugh Beg's Catalogue of Stars* (Washington, D.C., 1917); Balthasar Henry Meyer and Caroline E. MacGill, et al., *History of Transportation in the United States before 1860* (Washington, D.C., 1917); Christian Heinrich Friedrich Peters and Edward Ball Knobel, *Ptolemy's Catalogue of Stars* (Washington, D.C., 1915); James Byrnie Shaw, *Synopsis of Linear Associative Algebra: A Report on Its Natural Development and Results Reached up to the Present Time* (Washington, D.C., 1907). Sarton, "The History of Science in the Carnegie Institution," pp. 634–8

well equipped to proceed with the scholarly and scientific plan of work which he has mapped out for himself." Harry Walter Tyler and Sedgwick also support Sarton, Sedgwick emphasizing George's passion for research—he is "the most promising man in the world today, to undertake this greatly needed endeavour." George lines up testimonial on his own, notably from Andrew Dickson White at Cornell. Woodward dithers, asking if George would consider a 2-year contract. George agrees. He underscores that on Leonardo, he wants "to write a book which will be as authoritative and as final as any book can be, and which will be also a *beautiful* book." As to salary, George received $2000 yearly from Harvard, "certainly insufficient," and he estimates that this was matched by outside lecturing. His monthly expenses in 1916 and 1917 were about $265. He affirms nevertheless: "It would be unfair, however, to say that we suffered privations. Our sufferings were more due to anxiety than to actual need." He asks for a salary of between $250 and $300 monthly. He wonders whether his writings become the property of the institution. He again insists on the value of beautiful books, a concession to Mabel's talents as a book-designer. Woodward makes favorable noises and relates that the Carnegie Institution issues the books of its employees and keeps the copyright.[4]

The matter is resolved early in May 1918. Woodward signs George on as a Research Associate for 2 years, with an annual salary of $3600.[5] George quickly decides that he must remain at Harvard. It is a calculated gamble. He writes many years afterwards, in notes for an autobiography, that it was the "only method for promoting scholarship," for Harvard "would not have permitted me to live" due to "petty administration, immense waste due to adminis. pedantry, pettiness." He continues that Woodward "saved my life, the life of Mabel, May and *Isis*, ...saved my life's work." Harvard "would have allowed me to starve & even if it did give me a pittance it would not have paid for the public. of my introd. (4000 p.). Very grateful." Again: "The h. of sci. was a luxury for H. in an age of science but the hist. of education was not, the hist. of philosophy was not, the hist. of music was not, the hist. of art was not.... Let us laugh together. It is better to laugh than to cry." He summarizes: "My soul life a powerful gyroscope: A hundred deans & provosts pulling together could not have

4. CIW, Robert S. Woodward to George Sarton, 3 and 8 April 1918; Sarton to Woodward, 10 April 1918 (for Merz and the appendix); Roger Pierce to Woodward, 15 April 1918; H. W. Tyler to William T. Sedgwick, 25 April 1918; Sedgwick to Woodward, 24 April 1918; Woodward to Sarton, 22 April 1918; Sarton to Woodward, 28 April 1918; Woodward to Sarton, 1 May 1918. HGS, Andrew Dickson White to George Sarton, 3 April 1918, who writes from his sick-bed, and 12 April 1918, where White is glad "that the prospects of continued work in your field are so bright. Please find enclosed, Dr Woodward's letter."

5. CIW, George Sarton to Robert S. Woodward, 14 May 1918; Woodward to Sarton, 5 June 1918.

swerved me from my path."[6] We shall see that George's internal gyroscope is not powerful enough to prevent him from leaving Cambridge in the summer of 1919.

George writes to his new employer on 23 June 1918, thanking him for the Carnegie Institution yearbooks. Then he elaborates his "plans for the humanization of science," which "will sooner or later entail a deep educational revolution." He quotes from his recent publication on the matter, a manifesto for the New Humanism.[7] George has talked about the new humanism for years, sometimes in the context of an almost medieval expression of faith, but he writes the first long version of his call in French, and he publishes it in the Italian journal *Scientia*, which he promoted in the years before the war as a key outlet for history of science.[8] It is a call to unite the separate cultures of science and humanities. Before the war the debate about the merits of scientific or humanistic education took on an academic hue. George believes that the war has brought the matter to a head. The war has shown that science merits much more attention than it has previously received, and this necessarily requires, in education, a reduction in the time given to the humanities. But scientists currently "retail information and technical familiarity," while humanists "are the true educators." The only solution to avoiding a precipitous decline in the overall quality of education is for "instruction in the sciences [to] be *humanized*."

Humanists emphasize the importance of history and the disinterested nature of their inquiry. In fact, however, science is richer in aesthetic pleasure than any other field. By teaching science in a general way, by emphasizing large principles and grand lines of development, this human side of science can be conveyed. Here is a way to apply the sense of beauty that animates scientists, which is identical with the humanists' sense of beauty. Furthermore, if one seeks evidence for the progress of humanity, that which humanists loudly proclaim, then scientists are the best situated to provide it: "Kings and statesmen succeed each other, wars occur, but the conquests of science remain: They are the common heritage of all people." The unifying thread shall be the history of science. Appropriate education for the present age consists of instruction in a material language and one of the major spoken languages; the study of science, in-

6. YMS. In his obituary of Woodward, George repeats that Woodward "saved my lifeand my life's work." George Sarton, "Robert Simpson Woodward (1849–1924)," *Isis*, 7 (1925), 112–14.

7. CIW, George Sarton to Robert S. Woodward, 23 June 1918.

8. Diary, 9 February 1917 for the medieval-humanist juxtaposition. George records a talk at Clark University in Worcester. The president and the students dressed in academic robes. They sang a hymn before his talk, and the school song after his talk. "It impressed me! I spoke about the new humanism."

cluding its history; and the history of human progress. Students should be encouraged to learn non-European languages. Who shall direct this new educational program? It shall be a new class of scholars, educated in centers like an institute for the history of science and civilization, as George has described in *Science*. The time has come to end the mandarin domination of the humanists and to supersede the hyperspecialization of the scientists. The New Humanism points to the future.[9]

The text is widely appreciated. Physicist Henry Crew at Northwestern University, at whose home George stayed when he lectured there in 1917, reports favorably about the *nouvel humanisme:* "I am in sympathy with every word."[10] Emile Masson sees the text following directly from *Isis* and from George's early novellas, where George consistently evokes the unity of the physical and spiritual world. Masson thinks that George takes Ernest Hello's call to heart: "The unity of God must be established." It is a unity that has been destroyed by Descartes's separation of matter and spirit and eroded by years of specialist studies.[11] Masson is struck by the similarity between George's program for education and his own ideal program "in a quite different form, actually imaginary, which I have just expressed several months ago in a kind of novel called *Utopie des Iles bienheureuses dans le Pacifique en 1980*."[12]

George receives a letter from Mabel on 27 May 1918 where she confirms that she will correct proofs of his. May asks when he is returning. "I do hope you are feeling a little quieter & better in every way." She adds: "I stayed while May & other children went through a little sort of exam in reading. She knows more than I thought but is extremely diffident, is not *sure* enough of her knowledge." The proofs in question are probably George's first publication from his new post with the Carnegie Institution of Washington, titled (ironically, in view of his employment not as a teacher but as a researcher) "The Teaching of the History of Science." It leads the September 1918 issue of the *Scientific Monthly*. George begins by taking on the philosophers, who have claimed to teach history of science but who "were so deeply ignorant of the most elementary facts of science, so unaware of their real significance, so innocent of the true spirit of science." Just as "no amount of culinary legerdemain can make up for the lack of chicken" in a chicken pie—George has first-hand experience

9. George Sarton, "Le Nouvel Humanisme," *Scientia, 23* (1918), 161–175.

10. HGS. Henry Crew to George Sarton, 23 October 1917, 26 December 1917, and 8 May 1918.

11. Emile Masson, "George Sarton et le 'nouvel humanisme,'" *Les Humbles*, no. 10 (February 1919), 7–19, on p. 14.

12. HGS, Emile Masson to George Sarton, 27 November 1918. Masson, *Utopie des Iles bienheureuses dans le Pacifique en l'an 1980* (Paris, 1921), going through 4 editions in that year.

eating on the edge—"no amount of philosophic legerdemain can make up" for the absence of science in a course on history of science.

Launching into his imperialist vision for having history of science form the core of a history of civilization, George emphasizes that the development of science is as complex as the development of a life-form, and that the philosophers have not done justice to the topic. A case in point is the ghost haunting the Carnegie Institution, John Theodore Merz:

> His *History of European Thought in the Nineteenth Century*, enjoying as it does a kind of monopoly, is unanimously praised, especially by those who would make us believe that they have read it. This book certainly conceals a considerable amount of material, but it is so prolix and discursive that its rich substance has to be almost entirely redigested to be of any great service.

(Similar criticism shall be leveled against George's own *Introduction to the History of Science*, when it finally appears, by succeeding generations of scholars.) George's answer is for history of science courses to be "experimental, concrete, matter of fact." History should concentrate on "human achievement in the realms of knowledge, beauty, and justice." That is "the real thing; the rest is merely anecdotic." Students should be presented with creator heroes. But the conveyer must be someone who knows science and explains it with the use of instruments and visual aids. Experiments are essential for introducing students "to the handicraft part, the most living part, of science"—as George knows well from the tedium of his own experiments in chemistry. No history of science has yet been written from the point of view of the experiment, although he thinks that Ernst Mach and David Eugene Smith have advanced in the right direction. George proposes an ideal lecture hall in the form of a science museum. Historians of science should become the great coordinators and synthesizers of science, seeking unity in disciplinary division. In an appendix, George lists books and supporters, including a number of the scholars on the cover of *Isis*.

George concludes that educational leaders cannot come uniquely from the tribe of humanists, who "see only with one eye." He goes further. "In truth, a literary [university] president is as hopeless as a blind admiral." Historians of science, under the banner of the New Humanism, will provide the right balance: "Idealism alone is blind and powerless; knowledge without compensation is brutal. We need the golden combination of both."[13] Who can dispute the Aristotelian golden mean? In appealing to the wisdom of scientists, George raises his employer, Robert

---

13. George Sarton, "The Teaching of the History of Science," *Scientific Monthly*, 7 (1918), 193–211, Merz on p. 195, blind admiral on p. 210.

S. Woodward, to the ranks of the presidents of Columbia, Harvard, and Yale—all literary men.

One must know how to flatter a superior. George comes perilously close to losing Woodward's affection. When the article appears, Woodward is clear: "What you say there is, I think, profoundly true, but perhaps we should hesitate to state the truth in any such phrases as 'a literary president is as hopeless as a blind admiral.'"[14] George also risks losing supporters by continually attacking Merz. Mathematician George Birkhoff at Harvard, a sympathetic although powerless voice among the faculty there, while agreeing to help George organize his institute for the history of science, nevertheless hopes that it will direct some attention to history of science in general, in the line of Merz—and indeed in the line of George's early programmatic statements in *Isis*.[15] And philosopher Alfred Hoernlé finds George's fulminations about philosophers "needlessly sweeping and aggressive and expressed in a language too violent, considering that the realization of your aims cannot but be hampered by your arousing their opposition."[16] Hoernlé, indeed, is sympathetic toward George's nemesis John Theodore Merz, whose history of 19th-century science "is equally distinguished by the range and accuracy of its learning, the lucidity of its style, and the sanity of its judgments."[17] George has a knack for alienating precisely the people who appreciate him best. Even David Eugene Smith, his firmest supporter, doubts the wisdom of "calling your antagonists names," as George does in this article.[18]

At the end of June Mabel is in Washington. Mabel found George's note in her diary, which she opened on 28 June 1918 to write about "the breaking up of our tiny home." She is sad to read his note. She "so often" wrote in her diary "when I was sad or weak, much oftener than when I was happy, so it is not a *fair* picture." She continues: "I do love you, & I feel your love around me like strong arms. I *know* you love me in many ways, in happy, *comforting* ways. So this is a new beginning once more, *on we go*, together, & with our little May."

During the summer of 1918 George, with the wolves at bay, is free to defend his interests. He writes to Woodward to ask if the Carnegie Institution will announce his work, because the Italians are publishing their own 400th-anniversary volume on Leonardo. Woodward wants nothing to do with the Italians, but he will entertain publishing George's book, when the manuscript is finished. George needs the good will of Italian colleagues,

14. CIW, Robert S. Woodward to George Sarton, 18 September 1918.

15. HGS, George Birkhoff to George Sarton, received 17 January 1917.

16. HGS, R. F. A. Hoernlé to George Sarton, 22 October 1918.

17. R. F. A. Hoernlé, "The Philosophical Work of Dr John Theodore Merz," *Proceedings of the University of Durham Philosophical Society*, 6 (1922–23), 233–90, on p. 239.

18. HGS, David Eugene Smith to George Sarton, 18 September 1918.

however, and it is necessary to inform them of his plans, so that people on each side of the Atlantic may work fruitfully toward a common goal.[19]

Woodward asks his tame savant about the environment of Italy that led to Leonardo and Galileo. He reports that he asked the same question of physicist Hendrik Antoon Lorentz in Leiden, who was however less sanguine about the coming years, notably because of German belligerence. George answers at length. There is no explaining the appearance of genius. "I am a deep believer in eugenics, but here again we have to deal with averages, and genius is the abnormal." There is a practical lesson for the head of the Carnegie Institution: He must "improve the organization of the whole...and...select the right men." Woodward disagrees. In his view, genius is not pathological; in fact, the very word genius "has become offensive to me." George elaborates at length, and Woodward agrees with him. George knows the value of epistolary exchange. He asks Woodward to intervene with the Harvard library staff to secure his room in Widener Library, which George imagines might be repossessed, and Woodward is glad to do so.[20] Then he expands on Woodward's notion of talent. The important social point is to shift the mean in a population by "both eugenic and educational methods." There must be an aristocracy of talent:

> Education must be based on duty & reverence. Unfortunately, many people have such mistaken notions on what they call "democracy"—that they think all the time of their rights, little of their duties, & believe that they do not owe reverence to anybody. "Why! Are they not just as good as any other man?" Yes, politically speaking; *not* otherwise. Without the sense of duty & reverence, democracy is bound to fail. We must inculcate reverence; & who are more worthy of it but these men of "genius," who lead mankind—these pioneers of progress?
>
> I fully understand your impatience with words such as "genius" because of the confusion of ideas which they entail. Yet you must see the good side of it—that is, hero worship. I think that a moderate amount of it is distinctly beneficial, almost necessary, at least for younger people.

George had heroes in his youth—an echo of Carlyle—and the idolatry did him no harm. But he does not condone "the loathsome counterfeit of it too often exhibited by *snobbish* & crazy women." George is in favor of meritocracy, neither Bolshevism nor hereditary aristocracy. In George's view, the main task of an administrator lies in "selecting the right men

19. CIW, George Sarton to Robert S. Woodward, 22 July 1918; Woodward to Sarton, 27 July 1918; Sarton to Woodward, 31 July 1918.

20. Arnold Thackray and Robert K. Merton, "On Discipline Building: The Paradoxes of George Sarton," *Isis*, *63* (1972), 473–95, p. 487, conclude that Woodward's intervention prevented George from being expelled from Widener. The correspondence does not allow one to establish cause and effect, although George is clearly petrified at the thought of banishment.

for the existing jobs. This requires a vast amount of wisdom & intuition. It is not a science, but an art." George continues that people make their own opportunities. He compliments Woodward for having appointed him, and he pats himself on the back. He conveys his youth, "a motherless child brought up in utter loneliness," and he offers that he will never be either wealthy or "'a man of the hour,'—nor do I at all wish to ever become one."[21] In his correspondence with Woodward, George also alludes to related notions popularized by Henry Adams—notably a "*degradation* of intellectual energy."[22]

George is also free to dwell on matters of historical interpretation. He becomes firmly persuaded about presenting scientific ideas in the context of their time, and he realizes the unreliability of retrospective accounts by scientists. He conveys these realizations to Henry Crew. Crew is appreciative of George's observation about "the impossibility of 'a final history of unclear ideas.'" George wrote to Crew about the "last straw" that breaks the camel's back of old notions. Crew:

> Until I read your letter I had not realized that it was really that "last straw" of which you speak—the one that broke the camel's back, for which I had been looking; in other words, for the fellow who gave the modern quantitative touch to the idea of mass. I am much impressed by what you say about the discoverer himself being unable to say just how much of his own idea is original and how much inherited.[23]

Mabel writes on 21 August from the Baekelands' that "May looks very well indeed, but speaks so loudly & so badly it is astonishing." On 22 August, George hopes that May has not turned out to be "a little snob." She is too loud. "I often wonder whether she will have any order or not. I am thinking of using her later (temporarily of course) as a secretary; it would be a very good training for her. But if she have no order, it would be out of the question." On 23 August Mabel is again concerned about May's

21. CIW, Robert S. Woodward to George Sarton, 3 August 1918; Sarton to Woodward, 11 August 1918; Woodward to Sarton, 15 August 1918; Woodward to Sarton, 27 August 1918; Sarton to Woodward, 25 August (long quotation) and 5 October 1918. The discussion of genius concerns correspondence between Woodward and Lester Frank Ward, geologist at Brown University and first president of the American Sociological Society; Woodward compared George with Ward and recommended that George read Ward's collected essays, *Glimpses of the Cosmos* (New York, 1913–18). George Sarton, "Emily Palmer Cape, *Lester F. Ward*," *Isis*, 7 (1925), 244. Ward is seen by his colleagues as the founder of sociology in America. Samuel Chugerman, *Lester F. Ward: The American Aristotle* (Durham, 1939). Ward appears frequently in Herman Schwendinger and Julia R. Schwendinger, *The Sociologists of the Chair: A Radical Analysis of the Formative Years of North American Sociology (1883–1922)* (New York, 1974).

22. CIW, George Sarton to Robert S. Woodward, 7 July 1918, 25 August 1918.

23. HGS, Henry Crew to George Sarton, 14 July 1918.

behavior. In French: "She speaks *horribly* badly." On 24 August, Mabel does not think that May has become a snob. She offers that May will never become a secretary satisfactory for George. She writes in French so that her thoughts remain confidential: "Sometimes you leave my letters on your desk." Mabel on 24 August, in French: "The little one is however annoying. I will be *so* happy to be back in the nest." George, on 25 August, hopes that they may put Alfred out of their mind and look forward, in 2 or 3 years, to a sibling for May. "I had not forgotten this anniversary, the birth of little Alfred, although I had done my best to forget it. I really think we should try to forget it, especially at a time when mothers are losing not just born babies but big boys, big men, whom they have learned to love for many & many a year." On 26 August, George reports that Mary Bouton has lost her 9-month-old son.

The Sartons vacation in Ogunquit early in September. George's article in the *Scientific Monthly* appears. Woodward likes George's call for public administrators to be drawn from the ranks of scientists, but he thinks that George is a bit strident. Good works are needed. "The policy I have pursued...is that of letting the 'heathen rage' and working as diligently as practicable in the production of results that will stand the test of time."[24] The word practicable reappears in Woodward's correspondence. In his mind, activity must produce conclusions in a reasonable length of time.

Will the work on Leonardo, that which George has promised to the Carnegie Institution, materialize? George decides, in October, that he does not know enough medicine to comment on the anatomical parts of Leonardo's manuscripts, so he subcontracts them to Playfair McMurrich, whom Woodward pays $500 for his trouble.[25] The McMurrich collaboration drags on for years, although the Canadian anatomist does produce a book for the Carnegie Institution.[26] To execute the Leonardo project with the least amount of trouble, George enlists a second collaborator, art historian Osvald Sirén, sometime lecturer at Yale and also professor at the University of Stockholm, author of books on Leonardo and on Giotto, to produce a study on Leonardo's art. Around 1920 Sirén progressively turns toward Oriental art. To satisfy his engagement with the Carnegie Institution, Sirén revises his English book on Leonardo and publishes it in French.[27]

24. CIW, Robert S. Woodward to George Sarton, 18 September 1918.

25. CIW, George Sarton to Robert S. Woodward, 30 October and 4 November 1918; Woodward to Sarton, 2 November and 8 November 1918, with appendix.

26. J. Playfair McMurrich, *Leonardo da Vinci the Anatomist (1452–1519)* (Baltimore, 1930), with a preface by George Sarton.

27. CIW, George Sarton to Robert S. Woodward, 4 February 1919; HGS, Osvald Sirén to George Sarton, 2 February 1919. Sirén, *Leonardo da Vincis studier rörande flygproblemet* (Stockholm, 1910), a brochure of 22 pages; Sirén, *Leonardo da Vinci, the Artist and the Man* (New Haven, 1916), produced with the help of William Rankin, revised as *Léonard de Vinci, l'artiste et l'homme*, trans. Jean Buhot (Brussels, 1928).

The First World War ended on the eleventh hour of the eleventh month of 1918—in Europe. That meant 5:00 a.m. in Cambridge, Massachusetts, where "the tooting of the factory whistles and the ringing of bells" wakes George Sarton. He promptly writes to his employer. He hopes that Bolshevism will be extinguished in Germany, although he wonders if the next conflict will be between American liberals and the Bolsheviks. George signals his desire to visit Belgium in 1919, to settle his affairs there. He assures Woodward of his love for America. "The *moral atmosphere* is so much purer than in the old world, especially the Latin countries." George has taken steps to share *Isis* with Charles Singer of Oxford, Sir William Osler's protégé, and this will allow George more time for his labor with the Carnegie Institution, notably his work on Leonardo (both of which he places before his family in order of priority). He argues against residing in Washington, as Woodward has suggested. The summer heat is oppressive to him, a European, and the libraries are not as serviceable as Harvard's. If he were to return to Washington, he would need a reference library of 2000 to 3000 volumes, supplemented by his personal collection.[28] We shall see that George does move to Washington early in 1920. The experience is short and unhappy.

George's connection with Sir William Osler goes back to the beginning of *Isis*. Osler, the Canadian phenomenon who propelled McGill University, the University of Pennsylvania, and Johns Hopkins University to medical heights, in 1904 became Regius Professor of Medicine at Oxford. There he elaborated his passion for history of medicine and science, in 1912 founding the medical history section of the Royal Society of Medicine.[29] George noted the new society in *Isis* in 1913, and he sent Osler a reprint of his éloge of Henri Poincaré, where he emphasizes that "the secret of the future" would be the mathematical elaboration of biology.[30] Osler found a collaborator in Charles Joseph Singer, a young physician from Oxford who had begun to publish in medical history. In 1914 Osler arranged for Singer to obtain the Phillip Walker Studentship in Pathology at Oxford, which Singer accepted with the understanding that he and his collaborator wife, Dorothea Waley Singer, could devote their time to historical topics. Osler prevailed on the staff of the Bodleian Library (with whom he had been working closely for a decade as a devoted bibliophile) to organize a history of science room in a bay of the Radcliffe Camera; the Singers donated £500 to the project (more than the Sartons' annual income in 1914) and took charge of it. In November 1914, Osler crowed

28. CIW, George Sarton to Robert S. Woodward, 11 November 1918.

29. Clelia Pighetti, "William Osler tra scienza e utopia," *Physis, 30* (1993), 79–103, for a useful survey by a distinguished historian of science.

30. George Sarton, "Une Nouvelle Société d'histoire de la médecine," *Isis, 1* (1913), 249–50; "Henri Poincaré (1854–1912)," *Ciel et terre, 34* (1913), 1–11, inscribed reprint in MO.

to his American colleague Walter Libby at Pittsburgh: "We have already got six workers at various subjects" in the Camera.[31]

Osler was proud of his *pépinière.* He defended his project to Sir John Young Walker MacAlister at the Royal Society of Medicine in London:

> It is not a good way to pull up the turnips to watch their growth, and we cannot make medical historians in a couple of years. I think if your friend will look over the material presented to the [historical] section, while perhaps it does not indicate much research, I do not think there is much that could be called folk-lore or gossip. What is wanted in this country is not dilettante students like myself and some others but real scholars and your friend will be interested to know that some of these are at work on serious medical research. [Sir William Turner] Thistleton Dyer, for instance, is doing for us at the Press a splendid piece of work in Greek botanical terms, etc.;[32] [Edward Theodore] Withington is making a special study for us in Galen and Hippocrates for revision of the medical terms in Liddell & Scott.[33] We have formed at the Bodleian a separate little department for the study of the history of science and medicine at which at present five are working, one making a comprehensive study of the English medical manuscripts; Singer is doing a very interesting unknown MS of Manfredi, another is collating and will edit an unknown MS of Maimonides;[34] another is working at the Spanish medical MSS in the Library; while we have one of the Belgian professors at work on some interesting iatromathematical

31. E. Ashworth Underwood, "Charles Singer (1876–1960)," *Medical History, 4* (1960), 353–8 for a detailed portrait; MO, Acc. 417/115, Charles Singer to William Osler, 1914, for the endowment; Osler to Walter Libby, 30 November 1914. Osler elaborates on the arrangements in his introduction to the first volume of Charles Singer's *Studies in the History and Method of Science* (Oxford, 1917), pp. v–viii. Also: Geoffrey Cantor, "Charles Singer and the Early Years of the British Society for the History of Science," *British Journal for the History of Science, 30* (1997), 5–23, on p. 9.

32. William Turner Thiselton-Dyer, "On Some Ancient Plant-Names," *Journal of Philology, 34* (1918), 290–312.

33. Henry Liddell and Robert Scott, *A Greek-English Lexicon*, 8th edition (Oxford, 1897); George Sarton, review of the revised, 1925 edition for the importance of the work for history of science: *Isis, 8* (1926), 200–2. Edward Theodore Withington, "Some Greek Medical Terms with Reference to St Luke and 'Liddell and Scott,'" *Proceedings of the Royal Society of Medicine, Section of the History of Medicine, 13* (1919–20), 122–32. Edward Theodore Withington wrote on Galen, Plato, and immortality for the festschrift, *Contributions to Medical and Biological Research, dedicated to Sir William Osler, in honour of his seventieth birthday* (New York, 1919). In 1927 Withington published an English translation of the surgical works of Hippocrates with the Loeb Classical Library (Harvard University Press).

34. In the first volume of Singer's series *Studies in the History and Method of Science*, Singer published on Geronimo Manfredi's *Anathomia* (translated by A. Mildred Westland), and Reuben Levy wrote on the *Tractatus de Causis et Judiciis Morborum*, attributed to Maimonides.

> literature.[35] Of course, this is the sort of stuff we need, and I think it is quite possible that we may gradually get associated with the history section a group of scholars capable of doing spade work. If your friend wants a job in the historical branches, send him along. The harvest is plenteous but the labourers are few. Thank you all the same for his criticism, but if he looks over the papers he will come to my view that it is a bit bilious.[36]

In the penultimate sentence, Osler repeats Scripture (Luke 10:2, Matthew 9:37); the most illustrious physician in the English-speaking world did see himself as a prophet of medicine and its history.

Osler's nursery, to judge from the publications that emanated from it, was a respectable seminar. The letter reflects the support by Sir William and Lady Osler for 16 professors fleeing the University of Louvain. Dorothea Singer was the researcher occupied with organizing the English medieval manuscripts. When Charles Singer went on active military duty around 1915, his wife reigned supreme in the Radcliffe Camera, at Osler's insistence. As Osler wrote to Dorothea when Charles enlisted: "If he goes off you must come here and run the Science Room and do the proofs and lay wires and act generally as medico-historical hormone." She also figured in Osler's ambitious plans for a summer historical school: "We shall hold a Seminar on the History of Science every summer. We can get from 15–20 students from America. You shall be the Secretary and 'me and the Col' can keep the men answered with kindergarten work."[37] The Singers fairly bristle with enterprise. They want to photograph medical manuscripts and deposit the product in America; they were 40,000 cards into a catalogue of all Anglo-Saxon botanical texts. But, like the British expeditionary forces and their sometime officer Charles Singer, Osler's plans fail to gain ground. Into October 1918, Osler was unable to appoint Singer either as Lecturer in History of Biology or as Director of Studies in History of Science at the Bodleian.[38]

35. Georges Dreyer and Ernest William Ainley Walker wrote an article on iatromathematics–medical astrology—for Osler's 70th-birthday festschrift, *Contributions to Medical and Biological Research*. In 1907 Dreyer, a Dane, became the first Professor of Pathology in the Sir William Dunn School of Pathology at Oxford.

36. MO, Acc. 417/115, William Osler to J. Y. W. MacAlister, 17 November 1914. MacAlister, at the Royal Society of Medicine, was then Secretary of the Library Association. HGS, Victor Gustave Plarr to George Sarton, 7 December 1914.

37. MO, Acc. 417/115. William Osler to Arnold C. Klebs, 4 November 1914 for the number of Belgian professors; Osler to Mrs H. B. Chapin, 6 November 1914 for the financing. Osler to Dorothea Singer [1914?] and 1914. Hormones were used with abandon in metaphor as well as in therapy.

38. Lewis Pyenson, "What Is the Good of History of Science?" *History of Science*, 27 (1989), 353–89, on pp. 372–3, for an assessment.

George Sarton knew about Osler's operation. He writes to Osler in 1917 to express his condolence at the death of Osler's son Revere. Early in 1918 he sends along thoughts on creating an institute for the history of science. Later in 1918 he proposes a scheme for developing the history of science in association with London's great libraries. In his article on teaching the history of science in the *Scientific Monthly*, he congratulates the "splendid movement" in Oxford undertaken by the Singers and Osler.[39] Sarton sends reprints to Osler, and Osler replies, alerting Sarton to the Radcliffe Camera and the key rôle of the Singers, as well as to plans for a summer school.[40] It is a plum, and George is ready to pick it. George contacts Singer and signs him on as coeditor of *Isis*. The new arrangement is announced in a letter appearing in *Science* in February 1919: Singer will handle medieval and medical topics, George the rest; the series that Singer already began in 1917, *Studies in the History and Method of Science*, will remain for longer articles.[41] With no substantial publications to his name and with an inadequate salary and title, George connected first with the best university in America, then with the most well-endowed private research institution in the world, and now with Britain's most venerable scholarly institution and with one of history's most respected healers.

His star is rising, and not only in symbolic capital. George calculates that in 1918 he earned $3852.01, considerably more than $2750 in 1917.[42] But the Old Rat, who contemplates the renewed publication of *Isis*, is still worried about finances. Late in 1918 George writes a note to "Dear Wifie," wondering "whether we would not be more happy without servant, I doing part of the work." He would agree to take on unspecified chores. Mabel replies:

> I have too clearly understood that the house keeping is *my job*. If I choose to struggle to earn enough to have a servant: allright. But I must not expect you to pay for one & I must not & will not expect you to help me with *my* job. 2ndly you are a dear to want to help but you can't do it well. You "do the bath" but you don't get it clean. You can't really give sufficient effort somehow & to really help keep the house clean & tidy would cost an effort, more than you imagine.

Mabel writes to George on Christmas 1918, "Daddy—dear," in French. She apologizes for making George wait for her and being distracted before he left. "I am sad for so many things this evening! It seems that I

39. Sarton, "Teaching of the History of Science," p. 211.

40. HGS, William Osler to George Sarton, 15 January 1918, Osler to Sarton (at the Carnegie Institution), 5th of the month and 10th of the month; also in MO. The Poincaré reprint in MO.

41. George Sarton, "The Publication of *Isis*," *Science*, *49* (1919), 170–1.

42. NYPL.

have never had a Christmas Eve so sad." In Belgium they were close to Céline, and they prepared gifts for the children. What Christmas could they anticipate? And so much fear for the future. Mabel writes to George in Yonkers on 6 January that she took May to Sunday School. In French: "I do not find it very effective, perhaps it is better than nothing. It will help teach her how she should conduct herself in a church, but that's not very much!" Her lessons at school are going very well. May played nearly all day with her Flexible Flyer sled. "The snow is dry & firm & tho the wind is very cold it is really glorious weather." Mabel apologizes for being irritable. "I will try hard to be a better 'help-mate' when you come back. It *should* not be hard when I love you so." She signs herself "Your Wild Bird." On 7 January, Mabel highly recommends "The Russian" in the latest *Atlantic Monthly*, transcriptions of comments by Russian soldiers at the front in 1915–17.[43] On 8 January 1919 she asks if George is well. "Only in the way of all foolish wives & mothers, I get flashes of panic through my mind like a painful electric shock. I suddenly wonder are you *ill* or have you had an accident?"

George is fine, he writes on 8 January 1919. He dined with the Smiths in New York. "Mrs [Fannie *née* Taylor] Smith talked a great deal of you & May." He spent most of Monday at the Metropolitan Museum of Art. It is his favorite place. "There are so many treasures there, such an abundance of beauty, a splendor of such multiplicity, that I cannot help having a feeling of revelry, & that my soul gets very tired at the end of it." He called on John Williams Lieb, Vice-President of the New York Edison Company, and they talked about Leonardo, a hobby of Lieb's. Lieb will drive George to New Rochelle, where he lives, and then return him to the New York Harvard Club. George does not think that Mabel lacked patience with him. "But much as is necessary for our happiness that we know & share each other *serious* troubles, it is our duty to keep for ourselves the little ones, especially those which are due to our own negligence. I find these little annoyances to be more difficult to bear with than the real difficulties of life, & they tire me most when I cannot do anything to straighten them up." On 9 January, George spent the morning at the Public Library. Then to the museum, and then tea at the home of Henry Osborn Taylor, then dining with the Klebses and seeing Molière's *L'Avare*. Mabel writes on 20 January from Cambridge that "May is good, but so rough & *noisy*. I do want to teach her to be gentler & not to shout & 'squeak.'"

Around the beginning of January 1919, George writes to Mabel that he has read all her letters, and he thinks that she is entirely right. In particular, he should have been more attentive "to avoid your becoming pregnant." Their years of married life have given George great happiness.

43. Mme Fedorchenko, "The Russian," trans. Thomas Whittemore, *Atlantic Monthly, 123* (1919), 1–12.

Before she takes any extreme step, she should ask whether she would be "more happy without me." To this, George answers: "At least you know *me*, imperfect as I am." From 1919, in another note of George's, "not to be opened by anyone but May," George confesses that he read the whole of Mabel's diary, and it makes him realize how selfish and brutish he is sometimes. If he is not worse than he could be, it is due to Mabel. "She might have had a worse husband and I could not have found anywhere in the world a better, a more lovable wife. She is not simply a wife to me, not also a mother—it is really she who has educated me—and a friend, my best friend." Mabel uses her diary in the manner of Jane Welsh Carlyle, to record her pain about her husband's inattention. Unlike Jane Carlyle, Mabel considers leaving George. Unlike Thomas Carlyle, George reads the diary of his wife while she lives.

George's appointment at the Carnegie Institution suits his temperament in one way: Woodward wants detailed plans and projections of George's work, which George is happy to supply. George writes to Woodward on 23 January 1919 with a forecast. He will publish *Isis* and the *Studies* series with Charles Singer; he will work toward a collaborative history of science and civilization; he will write a history of physics in the 19th century; he will produce a history of the dawn of science; he will establish a photographic archive of manuscripts and rare books, beginning with a selection from Singer's recent inventory of scientific manuscripts in England before 1500; he will produce an encyclopedia about science in China; he would plan for a large exhibit dealing with the history of science, pointing to the need for a museum; he would prepare a catalogue of scientific instruments, promoting an archaeology of science. The realization of these plans "would be secured" if the Carnegie Institution brought Singer to Washington, which would then rival Leipzig as a center for the discipline.[44]

Sir William Osler is not consulted about George's plan to recruit Charles Singer, but the Regius Professor of Medicine at Oxford welcomes the joint editing of *Isis*. He writes to Singer:

> So glad to see by his reprint that you are joining Sarton on the *Isis*. If it appears in England, I suppose it will save the multiplication of a journal. Get both your energetic minds on the question of a summer school here next year—two weeks of lectures and demonstrations and an additional four for the research students. I would see that the Bodley gave every encouragement, and we could use my rooms at the Museum for overflow work and demonstrations.

Singer indeed begins the collaboration by contributing bibliographical entries and notes to George. Osler sings the praises of George's New

---

44. CIW, George Sarton to Robert S. Woodward, 23 January 1919.

Humanism in his presidential address to the Classical Association. Sarton responds to his new confederates in Oxford by delivering an enthusiastic review in *Science* of the first volume of Singer's *Studies*, where *Isis* figures in Osler's introduction.[45] The new cooperative arrangement appears on the title page of *Isis*, no. 8, dated autumn 1920.

Another possibility for publishing *Isis* presents itself to George early in 1919. Philip Jourdain writes from Cambridge, England, that he shall be named in Paul Carus's will as editor of the *Monist* and the *Open Court*. When this comes to pass, Jourdain would like to coordinate publication with *Isis*; his journals would take over philosophy and logic of science, while he would handle mathematical physics for George's journal.[46] He recommends a younger colleague endorsed by Joseph John Thomson, Archibald E. Heath, for work on modern physics: "We all three agree, I think, that the history of science must be rescued from the ordinary *a priori* philosopher; and feel acutely that the philosophy of the future must be founded on an intimate study of the conceptions and methods of science."[47] But George prefers to work in the glow of Sir William Osler's Oxford.

Mabel writes on 12 February 1919, from Margaret Gillespie's in Washington, that she is preparing work for 3 or 4 people, and it tires her. George, on 13 February in Cambridge, has heard from Raymond and Céline Limbosch, and he encloses their letters. He is finishing his article on Leonardo, which has been read by the wife of his philosopher colleague, Alfred Hoernlé: "I have already given the first part to Mrs Hoernlé for correction, & hope to be able to give the whole of it to morrow midday to Miss Moldenhauer for typewriting."[48] He will take a rest on the week-end. "May & I are getting on together very well, as usual. There is nothing to report about her. She is a good little girl, & expects so little from me that I am able to give her every day something more. And to get more

45. MO, William Osler to Charles Singer, 31 March 1919. George Sarton, "Eighth Critical Bibliography," *Isis, 3* (1920), 316–17 for the new arrangement. William Osler, "The Old Humanities and the New Science," in *Selected Writings of Sir William Osler* (Oxford, 1951), pp. 8–33, on p. 29; Osler cites both the French and the English version of George's new doctrine. George Sarton, "The History of Science," *Science, 49* (1919), 497.

46. HGS, Philip Jourdain to George Sarton, 26 March 1919.

47. HGS, Philip Jourdain to George Sarton, 27 April 1919, extracted in George Sarton, "Philip E. B. Jourdain (1879–1919)," *Isis, 5* (1923), 126–9, on p. 128.

48. George Sarton records that his manuscripts were corrected by Agnes Winifred Hoernlé (*née* Tucker). In 1923, the Hoernlés returned to South Africa. Alfred became a professor of philosophy at the University of Witwatersrand and Winifred became a distinguished anthropologist. Alfred was not opposed to apartheid. Martin Legassick, "Race, Industrialization and Social Change in South Africa: The Case of R. F. A. Hoernlé," *African Affairs, 75* (1976), 224–39.

than one expects, what else is happiness?" On 15 February George relates that he mailed his "Message of Leonardo." He is "glad that I am through with that, but fear that *Scribners* will not love it, because it is much longer than I had said—7000 words instead of 5000! This morning I felt like lounging and got up only at 7:30! Instead of 7:15, although there was no little imp to wake me up. May stayed overnight at the Boutons." He continues: "May is very good indeed, but of course to keep her in good order, prepare her supper, supervise her washing is nagging, & reading to her takes a considerable time: But I love to do it, the more so that I feel it is a real vacation for you to be for once free from these exacting household duties." George studied the League of Nations Covenant. "So my faith that that war would irresistibly lead to it is justified! But what must all the old Harvard Tories think of it? It is a great privilege to be living at a time like this: The world has suddenly made a gigantic step forwards; it has become far smaller & far greater."

George is concerned about support for the League of Nations. "*The need is great and urgent*," he writes to the *Nation* on 29 December 1918. "A wind of justice & generosity is blowing across the world, but great as this sacred enthusiasm may be, it would soon die down if each of us does not keep the sacred flame burning in himself, & it would then be of little avail against the solid strength of obdurate traditions & related interests." He appeals to people in the name of the newly conceived League of Nations, an appeal that should be seen in the same light as the Red Cross appeal. He urges the *Nation* to start a fund for the League of Nations or the League to Enforce Peace, calling it the Golden Gross.[49] His enthusiasm then wanes. He writes to Theodore Marburg: "I am simply disgusted at the lack of imagination and of wisdom of the French." He is most distressed by "their lack of appreciation of the idealism" underlying Wilson's initiative.[50]

Mabel writes from Washington on 14 February that she is glad that George forwarded letters from Belgium, but she can't reply. "It is impossible to write 'real letters' during the week. I am here at the Studio all day long & work all the time until I am too tired to go on. I enjoy it you know & I feel so much better these 2 days." On 15 February Mabel adds: "I have designed & had executed two really lovely things & several 'average' things. Everyone works splendidly to help me carry them out. It is wonderful to have such co-operation." She signs "Mumsie." George, on 17 February, is proposing to the Fogg Museum at Harvard that they celebrate the 400th anniversary of Leonardo's death. George would speak on his personality, then someone else (in the event, George Harold Edgell) on his art. He trans-

49. YMS.

50. George Sarton to Theodore Marburg, 16 February 1919, in *Development of the League of Nations Idea: Documents and Correspondence of Theodore Marburg*, ed. John H. Latané (New York, 1932), 2, 614.

mits a letter from Robert Bridges at *Scribner's*, who will pay George more than £4 per page for his article on Leonardo, which stretches to about 10 pages. He will bank the amount for their trip to Belgium. "May be I shall not be obliged to borrow any money from the Baekelands....That I have been able to write this article alone (Mrs Hoernlé made but few changes—tenses of verbs etc.) makes me feel very proud. If somebody had told me five years ago that I would be paid 1200 Fr for writing 10 p. in English, I would certainly not have believed!" On 18 February, George dined at the Club as Henderson's guest with geneticist Louis Blaringhem (one of the earliest French disciples of geneticist Hugo de Vries), the historian Roger Bigelow Merriman, the physiologist Walter Bradford Cannon—"he was in Brussels in December & says it was by far the gayest city he saw!"—chemist Theodore W. Richards, and mathematician Edwin Bidwell Wilson. On 19 February Mabel thanks George for writing. A letter from him affects her "just as it did before we were married. I always have a sort of happiness that is almost fearful, almost a *pain*, when I read them."

George's "Leonardo,"—his maiden publication on the topic about which the Carnegie Institution expects a substantial volume—appears in *Scribner's* later in 1919. George focuses on themes that relate him to the virtuoso. He emphasizes that Leonardo was taken from his mother at an early age, and that his father had 4 wives serially: "A motherless, brotherless, lonely childhood, we cannot lay too much stress on this; it accounts for much." Leonardo served many masters, some of whom failed to appreciate his genius. He was a lonely man, who died before his time. "Only those who have known suffering and anxiety can fully understand the drama and the beauty of this life." Then to Leonardo's observations of nature and his mechanical contrivances, including flying machines. He practiced inductive philosophy 150 years before Francis Bacon. His manuscripts provide a unique window into the creative genius who applied the "spirit of craftsmanship and experiment to the quest for truth, its sudden extension from the realm of beauty to the realm of science." Elaborating 5800 pages of Leonardo's manuscripts is George's task. He concludes by placing a message in Leonardo's mouth: "To know is to love...My life was one long struggle with nature, to unravel her secrets and tame her wild forces to the purpose of man....A literary education is no education. All the classics of the past cannot make men....The study of nature is the substance of education—the rest is only the ornament. Study it with your brains and with your hands." Then George speaks: "We must try to reconcile idealism and knowledge, science and art, truth and beauty....In the last analysis, that is what Leonardo tells us, and it is also the message of the New Humanism."[51]

---

51. George Sarton, "The Message of Leonardo," *Scribner's Magazine, 65* (1919), 531–40. The editor at *Scribner's*, Robert Bridges, pays George $200 for the article. HGS, Robert Bridges to George Sarton, 15 February 1919.

The Italians pose a thorny problem for George, for they control Leonardo's language and nationality. Antonio Favaro looks west and sees a camel drinking from Carnegie's inexhaustible well. He asks George for a copy of Charles Singer's *Studies*, which George sends along with words to the effect that "America is Italy's best friend." George detects "nervousness" in Favaro's request.[52] He has no choice but to publish a description of his project in the Italian periodical devoted to Leonardo studies. His text, appearing in May 1919, is a short advertisement. He explains his unique appointment as a researcher in history of science at the Carnegie Institution. Leonardo is one of his ongoing projects, the other one being a history of 19th-century physics. He chooses Leonardo because it is the best way to humanize science. His book, a collaboration with Playfair McMurrich, will present Leonardo in the glory of his time and place. George will use all the primary and secondary material available. Pierre Duhem, who disparaged George's endeavor in 1912, receives stern rebuke. Duhem, in his writings on Leonardo, is unpersuasive about the debt Leonardo owes to medieval thought; Duhem's reasoning is forced; Duhem's style is adversarial, repetitive, and fatiguing; and Duhem's method is to collect fragmentary essays published over a long period of time. George, after publishing his book, wants to publish all Leonardo's writings, organized chronologically by subject matter. He concludes with the hope that Italians will join him in his struggle for the unity and beauty of science.[53] Favaro is overwhelmed: "Your plan is really colossal: You dispose indeed of such a means of which we have no idea."[54]

Not all George's commentaries find their way into press. On 24 March 1919, George withdraws a letter, set up in type for James McKeen Cattell's *School and Society*, in which he criticizes the president of Yale University, Arthur Twining Hadley, for suggesting that professors be paid less and given the opportunity of earning as much as 60% of their salary through "outside work," notably during summer vacation. George, the exploited and unremunerated lecturer at Harvard, is uncompromising: "I have read many crude statements on the subject of university administration, but I must confess that I have never read anything quite so crude and brutal." He is astonished that Hadley does not appreciate that "genuine research work and moneymaking are irreconcilable pursuits." It is reprehensible that a university would "take advantage of the disinterestedness of its servants." George emphasizes that the summer is when professors dedicate themselves to research and uninterrupted thought. George shudders to think if Josiah Willard Gibbs had been at Yale under Hadley.

52. HGS, Antonio Favaro to George Sarton, 9 April 1919, George's comments in pencil.

53. George Sarton, "Une Encyclopédie léonardesque," *Raccolta vinciana: Nel quarto centenario dalla morte di Leonardo da Vinci*, no. 10 (May 1919), 235–42.

54. HGS, Antonio Favaro to George Sarton, 9 April 1919 and 5 August 1919.

> Try to imagine the good president suggesting to a Gibbs to eke out his salary by writing essays or spending his summer on the Chautauqua circuit. But that would scarcely happen, since Dr Hadley explicitly proposes to select as instructors only men who are able to supplement their salaries by outside work. He will not be embarrassed by men of real genius; his ideal apparently is to have a staff of newspapermen and magazine writers. I must say that I rather prefer Dr [Nicholas Murray] Butler's idea [at Columbia University] of having a staff of millionaires.

George understands that universities are in difficult shape financially, but he does not think "that it is possible to save the universities by killing the teachers." If Hadley's notion comes to pass, universities will sink into senescence. George proposes that universities scale back their offerings, and in this way continue to support the professors who remain.

James Cattell, in a letter to George of 9 April 1919, affirms that "Dr Hadley deserves what you say." Encouraged, George sends the withdrawn galleys to Robert S. Woodward, his boss, who is shocked by George's boldness. Woodward counsels silence: "Pass such amazing outbursts as those of Presidents Butler and Hadley in dignified silence," he writes to George on 15 April. Woodward acknowledges George's point, that universities are spending much on buildings and little on staff. He thinks this situation will change. It is true that universities are in trouble; the future lies with private research institutions, like the Carnegie Institution: "While there are indications that the higher forms of learning must be more or less eclipsed in our colleges and universities in the near future, I think there are at work causes and influences which are very promising for learning under other auspices than those of educational establishments." Science, Woodward observes, is a recent development in civilization. "We should bear in mind that we are still on trial and that nothing will be taken for granted in our favor yet awhile. We must demonstrate."[55] George is admonished to stick to his last.

George does continue to cobble for the bourgeoisie. He writes to Woodward in April 1919:

> There certainly is in this country a *latent* interest in the history of science not simply among scientists but also among business men. I have already spoken to you of Herbert Hoover, J. W. Lieb & Clemens Herschel. Recently, I have got into touch with a 4th business man: Mr. J[ohn]. A[lexander]. Mathews, President of the Halcomb Steel Co., Syracuse, who has a collection of early books on *metallurgy*! If my position has no other advantage than that of bringing this latent interest to the surface,

55. YMS, draft and Robert S. Woodward to George Sarton, 15 April 1919.

> transforming these hobbies into useful work, & coordinating these desultory efforts, it would already be worth while.[56]

Spoken like a true social democrat, whose great fear is the displacement of rich patrons. In May, George's salary rises to $400 monthly.[57]

On 11 June 1919 Mabel writes from the Gillespies' establishment in Washington that she is "so very happy to see some of the lovely things which I had designed *on paper* in Cambridge." She is writing to May at Beverley, Massachusetts. In her correspondence from this time, Mabel calls May variously the sprite, Little Brownie, or My Little Squirrel.[58] George on 12 June reports that, after depositing May at North Station for the trip to Beverley, he dined at the faculty club with medievalist Henry Osborn Taylor.

> Tuesday night a young girl came with a bunch of roses for you. I told her that you had left, which seemed to chagrin her. After a little while, she asked me whether I would accept them. I said: I certainly would. Thereupon she gave them to me and I thanked her. But I forgot to ask who she was.
>
> I was glad to see May's report. I want you to mail it to Mrs Baekeland, after which I shall likely mail it to Mother. She will thus see that May is not hopeless & it will help her to bear up with her.

On 13 June Mabel has not heard from George and wonders if he is ill or vexed with her, "Your drooping winded Wild Bird." George, on 14 June, is printing the seventh Critical Bibliography, "which bores me immensely." On 15 June, "I have not been happy a single week, a single day, nay a single hour every since you left. Therefore, I do hope that you will come back as soon as possible. The house is dreadful without you, I *hate* it, and even the club annoys me when you are not home." Mabel, on 15 or 16 June, writes to George:

> Will you understand if I tell you that out of the to me terrifying & deeply hurting discussion we had the other day, out of that *dark hole* has come up this flower of love, the realization that we only nearly quarrelled *because* we are still only just entering the real kingdom of Love. We are only just beginning to really know & understand each other, our needs & strengths &

---

56. CIW, George Sarton to Robert S. Woodward, 9 April 1919. Clemens Herschel, who developed the Venturi flow-meter, had an active interest in history of science, beginning with his essay, *Frontinus and his II Books on the Water Supply of the City of Rome, A.D. 97, a lecture delivered before the engineering students of Cornell University, February 2nd, 1894* (Ithaca, 1894).

57. CIW, Robert S. Woodward to George Sarton, 7 May 1919.

58. NYPL, Mabel Sarton to Marthe Patyn, 30 May 1919, commenting on May's climbing trees.

> weaknesses. There is a long, long, beautiful, adventurous, mysterious road to travel before we find the heart of our kingdom. *Aren't you glad?* I am!

These lines, George writes on 16 June, have "made me another man, as when the sun reappears after a long darkness." But Mabel, on 17 June, writes about George's "pathetic note saying you have been so unhappy without me." She cannot change her plans to return immediately. She is gratified that she means so much to him. The exchange has become their *pas de deux*: George self-centered and remorseful, Mabel enterprising and forgiving.

Behind their words is the anticipation of returning to England and Belgium. George has obtained leave from the Carnegie Institution for a semester in Europe, notably to work up his Italian in Florence. The family will see what remains of their property in Wondelgem. Mabel will spend time with her mother and with the Limbosches. May will be reintroduced to French. They wonder if post-war circumstances shall allow them to pick up their European life, or if they are captives of the New World. Notwithstanding much inconvenience, the family has adapted to America. The Sartons have tasted the experiences of their rich friends and colleagues, for whom they are tame intellectuals. Mabel has established herself as a designer who appeals to the bourgeoisie. May attends élite private schools and summers in the country. George has climbed beyond penury to join the new class of 20th-century scholars employed by a private foundation.

On 24 June 1919 George records thoughts upon packing up home and office. He is sad to leave Cambridge. The family visits Denman Ross, who "fully knows the value of beauty & of truth." George is fond of Ross, whose home is "a museum & a sanctuary."[59] George takes May to the Baekelands for 3 weeks, leaving Mabel to close up the apartment. They have made no plans to continue residing in Cambridge, but George hesitates over relocating in Belgium. He writes to his patron in Washington: "I am now more of an Anglo Saxon than the Anglo Saxons, and more of an American than the Americans."[60]

---

59. Diary for quotation. George notes Ross's book, *A Theory of Pure Design* (Boston, 1907), in his review, "The Principle of Symmetry and Its Application to Science and Art," *Isis*, *4* (1921), 32–8, on p. 37.

60. CIW, George Sarton to Robert S. Woodward, 4 July 1919.

Part V

Dilemmas of Creativity, 1919-1925

CHAPTER 13

Europe

Historians see the First World War as a watershed, ending the long 19th century. During the war, science became an instrument of slaughter and oppression, and religious faith evaporated in the trenches. After the war, Old Europe, reciting the mantra of national self-determination, took new forms, whether fascist, communist, or capitalist. The war occupied a large part of George and Mabel Sarton's thoughts, and the deprivation they experienced in America was easier to bear when contrasted to the deprivation of German-occupied Belgium. From the Sartons' impressions of post-war Europe comes a sense of dislocation. Unable to describe the change, they nevertheless feel out of place. They retain many of their ante-bellum sensibilities while recognizing that the world of their youth and young adulthood has disappeared. They understand that their place is in America.

In 1915, the Sartons steamed into the setting sun; in 1919 they travel east. On the crossing, the Sartons have relatively inexpensive cabins on the promenade deck. George uses the time profitably. He completes his essay, "Secret History," which he shall have typed upon landfall in England and then sent to Robert Bridges for *Scribner's*. May shares a berth with another girl several years older than she, and it is a good arrangement. "Mabel is also enjoying the trip, but does not rest—because she will never be able to rest." George takes a daily hot salt-water bath. He enjoys reading about the America he does not know, *Huckleberry Finn*; Mark Twain "was a wise and kind soul." He finds William Makepeace Thackeray's *Virginians* interesting but less likeable.[1]

"Secret History" is George's professional apology, and it is an affirmation of his pre-war credo. Until now, he writes, history has been told by "men of letters who mistook the obvious activities of the rulers and of the masses for the really progressive but less evident achievements of mankind." These traditional historians produce "nothing but a superior kind of gossip." The important laborers in history are the few "trying to unravel the secrets of nature, to get a little nearer to the truth, to give us a little more beauty, to adjust social relations with a little more equity." George speaks like a true social democrat. Progress is making haste slowly. His progressive minds are like Bernard Le Bouvier Fontenelle's bees,

1. Diary, 23 July 1919.

undeterred from their endeavor by natural or human calamity. They "die happy if they can but leave behind them a little more truth, beauty, or justice." Here, in English, are the sentiments of George's early novellas. It is not enough to know, George continues, one must be clear how one knows—knowledge tempered by humanity. And so, scientists must cooperate with craftsmen and historians. Notably, scientists will cure humanists of lying. George is not against the glories of art and literature—he calls for renewed interest in the ideals of classical Greece. But he insists: "The powerful men who believe that they are 'running the whole show' are in fact nothing but stewards whose main function is to enable the scientists and artists to do their work and so to accomplish the destiny of the whole race." Presidents of Harvard and the Carnegie Institution, hear this. George concludes with a plea to recognize the secret aristocracy of the intellect: "The advance of civilization is due neither to the people nor to their ruling and dominating class of whatever kind. It is due to the untiring efforts of a small, unobtrusive, and powerless aristocracy of scientists and artists."[2] Grand events like the recent war with its millions of corpses are disasters best forgotten.

In London for a week, George meets Arthur Henry Fox Strangways and Harold Stabler. On 30 July 1919 Mabel writes upon arriving in Ipswich. Her mother, she relates in French, "rented us a room in a miserable little hotel." George visits Ipswich for a few hours, and then the whole family go to Antwerp. Raymond Limbosch meets them. They dine with Uncle Jules and his daughter Marguerite. The next day the Sartons go on to Brussels where Céline takes them to her home, the Pignon Rouge in Rhode St-Genèse. The Thiery family is also there. George begins working at the Solvay Institute on 8 August. To celebrate, he makes love: "Mabel may have been pregnant tonight," he records in his agenda. In Ghent, they meet Aunt Hélène, the Willems, and the Pelseneers. Paul Pelseneer has become Perpetual Secretary of the Academy of Sciences in Brussels. The Sartons are shocked at the state of their house in Wondelgem: "Cruel and harrowing visit—everything has been soiled & desecrated—everything of all that was so dear to us—is now disgusting."[3] If on arriving in Europe they had thoughts of staying in Belgium, Wondelgem persuades them otherwise.

George rents a room in Ixelles, Brussels, to live near the Solvay Institute while avoiding the Limbosches. He reflects in his diary, on 12 September, about his good fortune at leaving Wondelgem. "The Germans were the only ones who could drive me out, & so they did. I am almost grateful to them for it." His Ixelles room is near the monument to Charles de Coster, which represents the character of Coster's whom he and Mabel love, Thyl Ulenspiegel. "I go round the Place Ste Croix—a typical old Flemish

2. George Sarton, "Secret History," *Scribner's Magazine*, 67 (1919), 187–92.

3. Diary, 12 August 1919.

square, where the odour of cheese, salted meat and fish is almost always present." He has finished Henri de Man's *The Remaking of a Mind: A Soldier's Thoughts on War and Reconstruction* (New York, 1919), the best book about the war, he thinks; De Man sent it to him earlier that year.[4] He takes Mabel to the Monnaie theatre to see Jules Massenet's *Jongleur de Notre Dame* and a ballet based on "Sheherazade." The Massenet is better than he anticipated; the Rimsky-Korsakov adaptation is "disgusting...art prostituted to our lowest self." It ruins the music for him.[5]

George writes to Mabel from the Institut de sociologie in Brussels on 8 September 1919 with social obligations and plans for their time together in Belgium, from 11 to 15 September. "I feel that I am so overburdened with work that I shall not be able to spend many more days in Wondelgem." Their house survived well until the time of the departure of the German officers, on 3 November 1918. Then it was looted by parties unknown, who left most of the books untouched.[6] The missing books are all mathematical ones.[7] On 19 September, George reports to Mabel that the Van Slabekes arrived to buy their house. They accepted George's terms: F 25,000 in cash. On 23 September, George writes that Céline has forgotten to indicate a value for their linoleum. "I would rather ask for much less, than for more."[8] The day he sells the house, he buys the collection of scientific portraits assembled by the late Professor Charles van Bambeke, who had been George's physician when he was a child, for F 2250.[9] He signs the agreement to send no. 6 of *Isis* to press. "One more link with the past restored."[10]

---

4. HGS, Henri de Man to Leo Hendrik Baekeland, 26 June 1919. But De Man "shocked" Baekeland, when they met, by his socialist views. NYPL, Mabel Sarton to George Sarton, received 23 August 1918. De Man met Baekeland, an old friend of De Man's uncle, when he visited the United States in 1918; Baekeland introduced De Man to New York scientists and inventors, just as he had introduced George to them. Hendrik de Man, *Gegen den Strom* (Stuttgart, 1953), p. 142.

5. Diary, 12 September 1919.

6. In "The Adventures of a Scholar's Wife," George refutes the claim by Siegmund Günther in Sudhoff's *Mitteilungen, 19* (1920), 110, that his house had been destroyed not by Germans but by British aviators. "The Wondelgem house had never been bombed either by the English or by anybody else." *Phoenix*, p. 80, claims two bomb holes in the roof.

7. "Scholar's Wife."

8. The Sartons paid Léopoldine Dangotte F 400 for the linoleum. HGS. Léopoldine Dangotte to George Sarton, 18 November 1911.

9. Charles van Bambeke, Professor of Botany at Ghent, taught among others Julius MacLeod. Paul van Oye, *Hugo de Vries, Julilus MacLeod en Edward Verschaffelt, Vriendschap en Wederkerige Invloed* (Brussels, 1961) [Koninklijke Vlaamse Academie voor Wetenschappen, Letteren en Schjone Kunsten van België, Klasse der Wetenschappen, Mededelingen, 23, No. 9], p. 7. Diary, 9 May 1954, for Van Bambeke's inconstant surveillance of George's health in his youth. *LM II, s.v.* Van Bambeke.

10. Diary, 30 September 1919.

And so it is. Number 6 completes, after an interruption of more than 5 years, volume 2 of *Isis*. George notes in his preface that he has been a Wandering Jew, pursuing his work "on occasion under difficult circumstances, in many widely-separated places." His library at Wondelgem was "pillaged" by the German invaders; he writes from Brussels, where his American notes and books are inaccessible. He had planned for no. 6 to be devoted exclusively to science and civilization in India. But during the intervening time, a number of the contributions have appeared elsewhere. By continuing *Isis*, George seeks to "defend the ideal of the New Humanism, that is, the reconciliation between science and life, and between idealism and common sense and experience." His review is an expression of faith in human progress. He wants it to be a living and sympathetic undertaking, and "because of that, notwithstanding the objectivity of its contents, it retains a personal character."[11]

George continues his thoughts in the lead article, "War and Civilization." He holds that war is a "temporary regression..., a moral failure" on the road to truth and justice. The recent war laid injustice bare, and for this reason good may come of it, in the way that it has led to advances in applied sciences, notably aviation. The task of the disinterested thinker has never been greater. Without mentioning it, George rejects the post-war Allied boycott of German science and learning, orchestrated by the new International Research Council in Brussels during the summer of 1919.[12] George, unlike his colleague Eugenio Rignano in Italy but like his friend Philip Jourdain in Cambridge, clearly sides with the neutral countries, who were also against any sanctions directed at learned Germans. He

11. George Sarton, "Avant-propos," *Isis, 2* (1919), 313–14. In an appendix, he reprints a letter appearing in the New York *Evening Post* of 22 February 1919, where he explains why *Isis* will publish only in English and French. The choice is in no way a reprisal against Germany, but rather a practical affirmation of faith in the "younger civilizations of the world." George offers that the war, having given rise to the League of Nations, may help internationalize the world in the way that the American Civil War unified the United States. On 20 February 1919, George writes to Mabel that the New York *Evening Post* "has devoted a rather foolish editorial to *Isis*, but I am glad of it because this gave me a chance of sending them a good answer. I have just written it, at the cost of a headache." George reaches the decision when he is in England in 1914, notwithstanding the reservations of his collaborator Philip Jourdain, who thinks that the restriction will impair the international character of the journal and will alienate the large number of German-language readers in America. HGS, Philip Jourdain to George Sarton, 22 December 1914. In fact *Isis* publishes many German papers in the middle 1920s. Issue no. 18, sent to press in October 1923, contains 38 pages in German by three German and Austrian scholars, including the inaugural lecture by Friedrich Dannemann at the University of Bonn, "Wissenschaft als Einheit," *Isis, 6* (1924), 395–9.

12. Brigitte Schroeder-Gudehus, *Les Scientifiques et la paix: La Communauté scientifique internationale au cours des années 20* (Montreal, 1978), pp. 101–30.

believes that nationalism and internationalism are each necessary and desirable, for "the cosmopolitan spirit is destructive of what is best in both the national and the international ideals." In *Isis*, he will avoid "the controversies of political history." His undertaking will help provide charity and humanity to the "conceited scientists and inventors" who have used science for "sinister purposes." *Isis* will remain an internationalist journal, "open to the expression of truth irrespectively of its source," the organ of the New Humanism.[13]

Academic innovators in America do well to announce their message in several words: Frank Aydelotte's "Honors Program" at Swarthmore College, Mortimer Jerome Adler's "Great Books" at Chicago, and Clark Kerr's "Multiversity" in California are signal innovations of this kind. So, too, George's New Humanism. Slogans hide contradictions. In George's case, he wants idealistic scientists to seize education from literary and legal minds. The central obstacle to his plan is the materialistic nature of American society, where ideas—scientific or literary—count for little. French geneticist Louis Blaringhem, in a letter to George, sees the matter acutely. Blaringhem agrees with George that the United States is the best place for instituting the New Humanism, while in the next breath he is struck by the "mad rush" in America for acquiring material possessions.[14] John Bagnell Bury is more pessimistic. He thinks that every university should have a chair in history of science. "But it will take some time to convince the democracies of its importance. They are beginning to recognize that theoretical progress is a necessary preliminary of practical progress, but it will be more difficult to make people see that the history of man's intellectual advance is more than interesting antiquarianism."[15] Nor is George's analysis universally applicable. Antonio Favaro, a senior Italian historian of science, finds that literary minds pose no danger for historians of science; in Italy, the scientists have eliminated university positions in the field.[16]

On 2 October 1919 George and Mabel spend the evening with Raymond and Céline at the home of the painter Georges-M. Balthus, a friend of the

13. George Sarton, "War and Civilization," *Isis*, 2 (1919), 315–21. HGS, Philip Jourdain to George Sarton, 4 September 1917, for the intellectual exclusion of German scholars.

14. HGS, Louis Blaringhem to George Sarton, 20 August 1919.

15. HGS, John Bagnell Bury to George Sarton, 16 July 1921.

16. HGS, Antonio Favaro to George Sarton, 2 April 1918. In 1910, universities suppressed courses taught by adjuncts, and history of mathematics accordingly disappeared from the curriculum. Umberto Bottazzini, "Italy," in *Writing the History of Mathematics: Its Historical Development*, ed. Joseph W. Dauben and Christoph J. Scriba (Basle, 2002), pp. 61–95, on p. 88. Paola Govoni, *Che cos'è la storia della scienza* (Rome, 2004), pp. 44–5, for the refusal of the education minister Giovanni Gentile, a philosopher, to create a university chair of history of science during the early years of the Fascist regime.

Limbosches.[17] George finds him "a very clever & interesting man, of considerable talent, but too much of a dilettante to suit me. He is not a man of faith." He likes Balthus's wife, the daughter of the German sculptor Adolf von Hildebrand. Then a moment of crisis:

> We left at 11. I had more or less understood that Mabel would come home with me (to my room in Ixelles) & I was wishing very hard for it, but she chose instead to return to Rhode with Raymond, May being alone in her room.
>
> This made me suffer terribly, but I am a fool to attach so much importance to such a thing. O Lord! When shall I be more indifferent....! This weakens me dreadfully.[18]

Throughout George's life runs the dialectic between indifference and passion, between the distance and isolation of individual scholarly work and the desire for solidarity with others. Did a month in Brussels, away from Mabel and May, really speed his career? Would one book notice in *Isis*, more or less, have made a difference? Would another essay have resulted in a call to a Regius professorship at Oxford or a post at the Collège de France? Until the close of his life, when he gave himself over to synthetic narrative, George Sarton is perpetually driven to accumulate small notions and accomplish punctual tasks. His legacy, the discipline that he nurtured, pays less attention to his assembled facts than to the social structures he animated.

George sets off, on a leisurely voyage south, to spend several months in Florence. He writes in his diary on 10 October that he is enchanted by the art of Basle and also by the Basle Museum's ethnographic collections, "excellent material for various articles on primitive technology for *Isis* (for ex. comparison of ploughs, looms, boats...)." Sixteen years earlier, in July 1903, George thought that the museum was "the most beautiful part of Switzerland." He is unable to see the skeleton prepared by Vesalius, but he acquires a portrait of Erasmus. He sees the sun set on Lorraine, and Alsace appears in the moonlight. "That is now again French soil. What do the people there really think of it?"[19] He visits Arnold C. Klebs in Nyon, staying at a hotel because Klebs's "palatial residence" on the lake, with its magnificent library of the history

17. Balthus illustrated a short publication of 10 quatrains of Raymond's, *Le Livre des merveilles*, published by Céline's firm (which I have not seen), and he wrote about Raymond's furniture designing (with illustrations) in the *Anthologie Limbosch* [*La Nervie*, *11*, no. 2 (1930], "Le poète et l'art décoratif," pp. 50–1. He is evidently not the painter known as Balthus, born in 1908, who was the protégé of Rainer Maria Rilke. George's and Raymond's Balthus was born in St Trond.

18. Diary, 3 October 1919.

19. Diary, 10 October 1919 and July 1903.

of medicine, is not yet ready.[20] Then he takes the train to Italy. On 15 October, George writes from Milan, which he finds "unmoded and antiquated." He has breakfast in Lausanne with Méta Budry. "She has not changed; she is the same lovable, but irresponsible, irrepressible and unweighable creature." And on 16 October he arrives in Florence.

He finds a room in the top floor of the old Lamporecchi palace (to reach it he must climb 95 steps) and collects his mail at the Biblioteca Laurenziana. Then he loses himself in Florentine art, visiting the Uffizi, San Domenico, and Fiesole, which "gave me my first intimate contact with Etruscan civilization."[21] On 21 October Aldo Mieli travels from Rome to visit George. Mieli, a distinguished historian of science, has been an enthusiastic contributor to *Isis*. George comments in his diary: "He has much the sturdy appearance of a peasant, & seems to be a simple and kind hearted man. I like him."[22] Mieli introduces George to Antonio Garbasso and the instruments of the Accademia del cimento, including those of Galileo and Torricelli.[23] Here, in the space of weeks, we see again George's enthusiasm for the history of medicine and the history of scientific instruments—topics that later commentators have claimed he disdained.[24]

On 13 October, Mabel writes to George in Florence that she will cross over to England. On 14 October, she has not yet left Rhode and is still handling finances involved with moving their household. By this time the Sartons have clearly decided to return to Washington, the source of both their incomes, in January 1920. Washington lacks the intellectual sparkle of Cambridge, but both Mabel and George imagine that it will sustain their ambitions. Notwithstanding the range of emotion in their marriage, the couple are optimistic and enterprising about making a life together.

Mabel writes on 21 October, from Ipswich, that her mother is instructing her about dispersing her belongings after her death. May has stayed in Belgium. On 26 October, Mabel goes through photographs and family treasures with her mother:

> I myself am living also in the past, only it is just the little intimate part of my own child-hood. Mother as you know is going through all her possessions, & as she takes out & lovingly handles the things, she tells long tales about them & I often very much enjoy it, though many I have heard often before. There are some delightful photos of Hugh. He must

20. Diary, 11 October 1919.

21. Diary, 17 and 21 October 1919.

22. Diary, 21 October 1919.

23. Diary, 23 October 1919.

24. Lewis Pyenson, "What Is the Good of History of Science?" *History of Science*, 27 (1989), 353–89, on p. 388 for the assertion and its correction.

> have been a quite fascinating little child, & somehow one feels very sad in looking at such an intensely happy, sunny little face to see it gradually closed over, till it gets an almost tragic look. Yet the last photo of all (the one we have too) where he is so like Mother, has something I very much admire & love in it: All the more since I begin to realize how much he has suffered through his wife, for though it has a sad, strained look, there is something tender and firm which is lacking in the others. I so wish we could meet him again, really get to know each other I mean.

On 28 October, in French: "I have thought about you, about me, about May. I believe that I am very happy. I need to be a bit removed from you both to see it completely. When I am too near, my heart is a bit dazzled, I believe, a bit blinded." On First November, "Mother & I have got on splendidly." Mabel, before going to sleep on First November, writes George a love letter. She writes about the intensity of her feeling for him, paraphrasing Isaac Newton. "Words when I try to tell you seem as mere tiny shells cast up upon the shores of a great sea." She continues:

> Sometimes my love comes murmuring shyly to you, & yet so longs for you to bend & listen! Sometimes my love comes singing to you, singing deep songs of happiness that throb & quiver till my heart near breaks with longing. Sometimes my love comes weeping to you because my body fails & lags & will make no more music. Ah Love my Love, bend you heart to me & listen & be tender still, for I do love you more than words can tell.

On 28 October, George describes his typical day in Florence. In the morning he studies Italian and writes letters in Italian. In the afternoon, he works in a library or museum. His Italian teacher, Albertina Zapparoli, is the daughter of a librarian.

> I am really happy to meet with her, because she gives me more than I had hoped. It is true that she does not give me real lessons and does not correct me as much as she should, but in return she teaches me to understand better Italian thought and the Italian soul. She is passionately Italian and we have long discussions. I still speak rather poorly (the verbs are very difficult to learn) but in sum I can say everything I want, and if my interlocutors are generous, I can discuss all questions and have my point of view understood.

George thinks that with 3 months in Italy instead of one month, he would really learn the language. He hopes that in several years, Mabel and he can spend time in Rome.

On First November George reports on a visit to San Miniato, in Florence. The smell of roses and chrysanthemums recalls his sombre youth. He must still correct proof (the seventh Critical Bibliography for *Isis*, he observes in his diary for 4 November). "Nothing brings you to reality like printer's proofs. It's like a cold shower." He has made some imprudent purchases of old books: Antonio Zarotto's edition of Juvenal's *Satires* from 1474, an edition of Sallust from 1506, Aristotle's *Zoology* in Latin from 1506; he is tempted by Erhardus Ratdolt's Euclid from 1482. He also bought a book for Mabel. In all, he acquired about 50 works. Instead of taking a daily bath at F 3, he drinks a half bottle of wine at F 2. He plans for the end of his European tour. On his way back to Belgium from Florence, George will stop by Bologna to see Federico Enriques and Padua to see Antonio Favaro. In Paris he will visit Romain Rolland. When he goes to England about 20 December, he will probably stay with the Singers rather than with Sir William Osler, who is "too great a figure," he writes in French, to stay with comfortably. He asks Mabel to file a claim for damages to their Wondelgem home on the part of the Germans. No. 6 of *Isis* has appeared. He "wants *terribly*" to have a real home. On 3 November, George is overwhelmed by Florence. He is not working much. "I do not learn words, but I am acquiring Italian thought and sensibility."

Fiesole, on the hills above Florence, attracts him greatly. The plains of Tuscany appear in soft beauty. "The faint but fresh green of the cypress brings a note of eternal youth to this automnal decor." The Convent of San Marco impresses George, not the least for the presence of "two immense souls" who fill it with "eternal clarity, that of Fra Angelico and that of Savonarola."[25] Mania would have been more accurate than clarity, but George transforms the experience into a credo, which he publishes at the beginning of the third volume of *Isis* as "The Faith of a Humanist."

The credo begins as George ascends to Fiesole, asking on a "cold and dull" day whether his labor was worth while. Then he sees the Franciscan monastery, "a magnificent assertion of virtue and charity against triumphant barbarity." After many centuries of humanity, all "power and wealth had vanished." What remained was "the amount of truth, of beauty, of justice which they had conquered,—pure gold, eternal joy extracted from the chaos." Fiesole inspires George to rededicate himself to his mission, and to seek truth, beauty, and justice, to identify "in the works of science of the past, that which is not and cannot be superseded." A true humanist must contemplate the past "from the summit of modern science, with the whole of human experience at his disposal and with a heart full of hope." A true humanist explores the past to improve the present, "combining the historical and the scientific spirit." Then to his own noble struggle: "This is a considerable task and we may not succeed in accomplishing it, but it certainly is worth while trying. Some of us at least must needs do it and they should dedicate themselves to it in the same spirit that mediæval

25. Diary, 4 November 1919.

craftsmen dedicated themselves body and soul, to their art."[26] It is an adaptation of the apocryphal motto of William of Orange, recovered from the pages of George's first novella.

Back in Rhode on 9 November, Mabel writes that she completed the paperwork for selling their home in Wondelgem. She feels "like you, *terribly homeless*." In French: "I do not have a place *of my own* in the world." On 10 November, on letterhead from "L'Art décoratif C. Dangotte," she writes that she is working for Céline. On 16 November, Mabel reports that Véra Bouny's François wants to return; his friend Hélène has had enough. Mabel urges George to write to Véra at Mons. It is another marriage in trouble.

George, on 6 November, misses Mabel, yet he is

> really unhappy with the thought that I must soon leave Florence, because I really feel that if I could stay here for 2 more months, I would master Italian, while the knowledge that I have of it now is evidently quite superficial. On the other hand, I will be glad to see you again. It seems like an eternity since we parted. Happily I am thoroughly occupied, for otherwise your absence would be hard to bear. Patience, my love, my love. Patience. Soon we will see each other again, and then a bit later we will have a new home.

On 9 November, George elaborates:

> I am now almost in the same deplorable state as May. I am trying hard to speak Italian, & that means that I think in Italian as much as I can. But the clearest result thus far, is that I seem to have lost the ability to speak French or English, without having yet gained any considerable knowledge of Italian. It is a great pity that I cannot stay here a little longer, because I am convinced that after another 2 or 3 months, I could really speak Italian, while now I can just stammer, or as the people here say, my tongue would be untied. But it is better not to think of it, because I could not stay here any longer, chiefly, because I could not stay longer away from you. I am longing for you, & have already been tempted to leave everything and hasten back to Brussels.

Soon they will begin in a new home in Washington. "*I love you*, Mabel, in so many ways that I can make every day new discoveries of *how* I love you, and so deeply that I sometimes feel that my love could never be satiated." On 17 November, he confirms their plans to abandon Europe. The Institut des hautes études in Brussels

> offered me to become Director of their Scientific Section, and then to give six lectures on Leonardo (for F 300). I have declined both propositions,

26. George Sarton, "The Faith of a Humanist," *Isis, 3* (1920), 3–6.

> but said that I would *give* them one lecture. I will give another lecture at the Royal society of medicine in London (date not yet fixed) and I fear I shall have to give another one before the Oxford University or one of the Colleges.

The mantle of a visiting American scholar rests easily on George's shoulders.

On 17 November, Mabel urges George to remember her to Romain Rolland in Paris: "I too count him a friend & have been made richer & happier through his books." On 20 November, she wants to make a "best coat" for May out of an old dressing gown of hers. On 21 November: "I have not only had my white evening dress made for 'great occasions' but a plain black evening dress to wear with my good old lace collar. I ought to manage well in this way." She is "so *very* vexed" that George must leave Italy before he has completely "conquered her. I am so sorry that I would willingly wait another 2 months for you. That is saying more than you probably guess."

George writes in his diary on 20 November about leaving Italy, and about how his Italian teacher, Albertina Zapparoli, has helped him:

> She has not only guided my first efforts to learn her beautiful language, she has done infinitely more: She has helped me to discover the Italian soul. She is truly a lovely young girl with a heart of gold, a cultured mind, and she has an exquisite sense of it....I am really happy to know her and we will stay friends. I will never forget that she introduced me to her country; she is the one who showed me Italian sensibility.

Because of her (and also because of bad weather) he extended his visit to Florence, putting off trips to other northern Italian cities. This rapsody compares in intensity with very few of George's commentaries on women. After the apotheosis of Miss Zapparoli, George indicates his desire to complete the "spiritual conquest of Italy" by returning with Mabel and May. He then lists the booty that he has acquired: 22 books published before 1550, including 8 incunabula. Then, the next day, he spends F 500 on lace for his wife ("the biggest gift that I have ever given to Mabel").[27] It is surely wergild. George notes sexual intercourse in his agenda on 3 and 13 November, twice on the latter date. If Zapparoli is the object of his passion, there is no sign of it in their later correspondence. For Christmas, she sends George a copy of the *Avventure di Pinocchio*, to be shared with Mabel and May.[28] In the draft of a letter to Zapparoli from America, George comments only that his time in Florence was the first time that he relaxed in 5 years.[29]

27. Diary, 21 November 1919.

28. HGS, Albertina Zapparoli to George Sarton, 20 December 1919.

29. HGS, George Sarton to Albertina Zapparoli, 12 April 1920, draft.

George writes a last letter from Florence on 22 November:

> Mieli came to see me again (it is the third time he comes from Rome to see me) and we had some good chats together. He dined with me tonight. As you know he is my best ally in Italy, & we shall continue to work together. Although I was very pleased to see him, I regretted not to be able to go to Fiesole again this afternoon as I had planned.

George, indeed, has been firmly anchored in Florence, declining even to visit Antonio Favaro in Padua.[30] On 25 November, George writes from Milan that the editor of *Scientia*, biologist Eugenio Rignano, has organized a soirée for him. Rignano "explained me his most interesting socialistic ideas," published in 1901 in his first major book.[31] In his letter to Mabel, George writes that his piece "War & Civilization" has been translated into French and published in *La Feuille* of Geneva.[32] Then to business:

> Yesterday, I opened a few business letters of yours to forward them without the envelopes which were too bulky, and was distressed to see that Stearns are making a draft against you for some $50, a draft which your bankers will not be able to honor. How did you leave America having debts unpaid? That is rather grave. Why did you not say me anything? When you are in trouble, I should be the first to know it. Why did you not trust me? I am bound to know of it anyway, and I would prefer to know it directly *from yourself, at the beginning.*

George arrives in Brussels on 29 November. Earlier that day, in Paris, he saw the Leonardo paintings in the Louvre and spoke with Romain Rolland: "He is very discouraged and certainly the peace we had dreamed of is not likely to be realized before long." Romain Rolland does not like Woodrow Wilson. George in English: "No man brought greater hope to the world yet leaving it worse than it was. I explained that Wilson's mission could but be one of compromise."[33] The next day George descends on Rhode. Mabel and May are not there, the Limbosches explain, they are either in Ghent or at Eecloo. George returns to a rented room in Brussels. Mabel arrives on 2 December. In

---

30. HGS, Antonio Favaro to George Sarton, 25 January 1920.

31. Diary, 26 November 1919. George Sarton, "Eugenio Rignano (1870–1930)," *Isis, 15* (1931), 158–62, for George's discussions with Rignano on 24–26 November 1919.

32. It is a partial translation by his friend Emile Masson, *La Feuille*, 14 December 1919. HGS, Emile Masson to George Sarton, 16 November 1919: "I confess that without your authorization I translated your superb article from *Isis*, titled 'War and Civilisation,' and that this translation should appear tomorrow in *La Feuille*, printed in Geneva."

33. Diary, 29 November 1919.

French: "She had not received the card where I indicated my arrival on Sunday. All is well."[34]

Mabel writes from the Limbosch household on 8 December, after returning from England. She finds May in fine fettle. In French: "May speaks French always *without hesitation*, but unfortunately always with an awful accent." On the ambience of Brussels and London, in French:

> It is sad weather here, fog, rain, cold. A penetrating humidity, large drops falling from the trees, a carpet of wet leaves on the roads. And yet it is often so beautiful! London, also I find it so beautiful, so austere, so mysterious. I love it a bit as a man must always love a woman he was in love with and whom he always finds beautiful.

George affirms in his diary on 11 December that he will write a book in English on the New Humanism. He wants it to be comparable to one of his persistent inspirations, Maeterlinck's *Sagesse et la destinée* (Paris, 1898). He visits his Aunt Hélène, not in good health, who gives Mabel and him a piece of oak furniture, an Oriental carpet, her family baptismal towel, and the portrait of a family ancestor, the *seigneur* Joseph Duvet de la Tour. George is not on good terms with the Limbosches, camping as he does in Brussels, but he nevertheless lends Raymond F 10,000 for 8 months.[35]

George leaves Brussels on 20 December. He picks up Mabel and May, and they travel to Ipswich via Antwerp. He will see the Singers in Oxford and spend several days at Windsor with the Leonardo manuscripts. Then the family will return to America.[36] George is happy with how May travels: "I was very pleased with May. She is really a very good & very sensible little person. She has a head and a heart. I am proud of her." Mabel's mother gave them "a great many treasures" brought back from her and her husband's travels, including a set of chessmen and the family Bible. At the Singers' George meets ethnologist Charles Gabriel Seligmann and chemist Frederick Soddy. Then to Oxford University Press. George is impressed: "I am more & more tempted to have, if possible, the Oxford University Press publish *Isis*, Singer becoming a joint editor." George calls on Arabella Burton Buckley Fisher, who was secretary to Charles Lyell and who published *A Short History of Natural Science* (5th edition, London, 1894). She knows *Isis* to be the organ of Oxford undergraduates, and she does not understand why its editor is located in America.[37]

34. Diary, 1 and 2 December 1919.

35. Diary, 15 December 1919.

36. Diary, 19/20 December 1919.

37. Diary and Agenda, 29 December 1919; Diary, 25 February 1953.

On 30 December George writes from Bournemouth, at "another conference with the Singers." In London, he has "promised to attend a meeting of The Positivists & to say a few words." He learned from the *Times* of Sir William Osler's death. "I was just going to call upon him, not to see him but to leave my card. Yesterday noon, we had telephoned to inquire about his health, & they said he was better." George in fact speaks to a "small but attentive and regardful" audience in the Positivists' Church of Humanity on New Year's Eve,[38] but the main outcome of his stay with the Singers is a precise understanding of the new arrangement for *Isis*.

> Spent the last day of yesteryear with the Singers at Bournemouth. Had a long walk & talk with Charles Singer & reached the following conclusions: From number 8 (iii/2) on *Isis* will be edited by me with the assistance of Singer, and from vol. iii on Singer will edit his *Studies* with my assistance. That is we will jointly edit both collections, but I remain the master of *Isis* & Singer of the *Studies*.
>
> It is likely that from vol. iv on *Isis* will be published by the Oxford University press. Singer & I will also prepare the plan of a History of Science in ten or twelve volumes (something like the Cambridge Modern History). Our plans should be ready for the end of the year, & we would then take the first step to gather the necessary collaboration; about the middle of 1921 we might make our first public announcement. The publication might be begun in 1924/5 & should be completed, including index volume by 1930.
>
> This is a very great undertaking to which I have been thinking for many years, but without Singer I would have postponed it indefinitely, and I am not yet certain that we can do it now; I am afraid that we shall not be able to collect enough contributors. At any rate it is worth while making the plans.

He has other plans. He must set *Isis* on course as the instrument of the New Humanism. He must continue his Leonardo studies. He should prepare the sections on the 15th and 16th centuries for the new History of Science series, and also write the sections on the history of physics. He will defer learning Chinese until he has remastered Greek and Latin. "It is quite clear that my hands are full!"[39]

George writes from London to Mabel in Ipswich on 11 January 1920, with plans for their voyage to America. Mabel, on 12 January, thanks him

38. George Sarton, "Auguste Comte, Historian of Science, with a Short Digression on Clotilde de Vaux and Harriet Taylor," *Osiris, 10* (1952), 328–57, on p. 357. George is introduced by his supporter Francis Sydney Marvin, after a dinner at the National Liberal Club.

39. Diary 9 January 1920.

for having "planned it most thoughtfully for me & it is wisely thought out." And, also in January, she apologizes:

> My Dearest Dear, I have been reading in your "Saints of Islam"[40] & my heart is filled with humility & love & shame for my meanness & crossness to you, & when you come in I will not find time or the courage, or opportunity to see you alone to tell you until the tenderness & sorrow have sunk back again too deep into my consciousness to *say* them. And yet I want you to know that I always suffer when I am like that. I judge myself clearly & more harshly than ever your kind heart would let *you* judge me. I say all the things to myself that a sterner, more selfrighteous man would say to me. Be sure of it! It may help you to forgive & forget. And know that your gentleness is the best cure for my faults. I love you for it more than you know. *It casts out the horrid little devil that gets into me!*

They return to America on 19 January. George has shipped 41 cases of books to Washington.[41] George reads Tacitus and Marcus Aurelius on the crossing (in 1908 he had presented Marcus Aurelius to his student association Reiner Leven) and "also Maria Bochkareva's deeply moving autobiography."[42] On the passage he is transfixed by wars old and new.

40. Perhaps, Claud Field, *Mystics and Saints of Islam* (London, 1910). I have seen no manuscript by George with this title.

41. Agenda, 19 January 1920.

42. Maria Bochkareva, *Yashka: My Life as Peasant, Officer and Exile* (New York, 1919); Diary, 28 January 1920.

CHAPTER 14

Permanent Exile

Although the Sartons kept up contact with Washington, late in 1919 they had forgotten that early in the 20th century, for a large city on the east coast of the United States, the national capital was possibly the flattest and most superficial of places. Someone with the wealth and power of Henry Adams could find intellectual satisfaction in a city of functionaries and political transients, but returning from the ferment of Europe, George and Mabel Sarton quickly see it as a desert.

When they arrive in Washington nothing seems suitable. In his diary on 31 January 1920, George is disappointed that he will not be able to have an office at the Carnegie Institution, as he had hoped. He is also discouraged about finding an apartment and a servant. Until they settle in, May is staying at the home of geologist Thomas Wayland Vaughan. "This is a poor beginning." George is "utterly disappointed & discouraged" on 3 February because the Carnegie Institution will not give him a research budget.[1] "Long talk with Woodward. No financial help possible."[2] He recognizes what he should have realized earlier, that Cambridge is where he must relocate. "I am dreaming of going back to Cambridge; this seems to me now to be [the] only place in this country where there is a sufficient depth & a sufficient density of culture to make life bearable."[3]

George writes to Lawrence J. Henderson at Harvard, who replies that George can be appointed lecturer at no salary, giving "a few lectures in the course of the year in payment for the use of a room in the library."[4] George now sees his adopted land in a new light: "I must admit that America this time thoroughly disappoints me. Much is extremely crude, & there is a frightful amount of pharisaism." But he wonders if his disappointment derives from post-war discouragement generally.[5] Henderson delivers before the month is out, and George resolves to leave Washington at the beginning of June. May and Mabel will spend

1. Diary, 3 February 1920.
2. Agenda, 3 February 1920.
3. Diary, 3 February 1920.
4. HGS, Lawrence J. Henderson to George Sarton, 9 February 1920.
5. Diary, 15 February 1920.
6. Diary, 29 February 1920; Agenda, 27 February 1920.

the summer in Maine or New Hampshire, and they will settle a new home in the fall.[6] Harvard will exploit George in this way for the next 20 years. He recalls late in life:

> When I arrived in Cambridge (for the second time) in 1920, President Lowell who was a practical man and a "great liberal" did not pay me anything for teaching (I was then lecturing more than I ever did) for it was obviously a waste of money to pay somebody having as much "feu sacré" as I had. I myself was too deeply immersed in my work to think of payment.[7]

George explains his flight from Washington in a letter to Antonio Favaro. In addition to difficulties with continuing *Isis*, "the intellectual atmosphere of Washington does not please me: Life is too political, too bureaucratic, with too much wealth and too little culture." He will move to Cambridge, "the most civilized city in America."[8] Mabel shares George's thoughts. On 7 March 1920 she writes to Eliza Van Loon (whom in her correspondence she sometimes addresses as Evanel, for E. van L.):

> People at the Institution have so cooled off about George's work, tho' he is of course still attached to it & will be (don't say so please), that we all got fearfully depressed....*and* we are going back to our dear old quiet Cambridge, where at least we have a few friends who are neither snobs nor politicians & where we can live decently if poorly (on...what should really be ample means but wh. is almost poverty in these days & absolute beggary in Washington!)

She reports on what they found in Wondelgem: a deal of their furniture, some linen sheets and towels, silver and curtains. "All our small treasures—china, glass, pewter, etc. etc. were stolen or smashed (with a few miraculous exceptions) & the house was in a most ghastly mess." She and George "are dreaming of a new little home over here...if the profiteers will let us get one!!" Mabel is for the most part glad to leave Washington,

> *but it is a blow to my work*. Mrs Gillespie sold the old-fashioned shop she had had—& her mother before her—for years & years, & now has opened one on Connecticut Avenue, a sweet little modern one, where she sells *only* my embroidered things! Open one month & doing such a roaring trade we can't keep her supplied. I really believe I am going to make some money—& we need it.

7. Diary, 14 September 1947.

8. HGS, George Sarton to Antonio Favaro, 16 March 1920, draft in Italian.

George has received 2 notes from Hendrik Van Loon about the New Humanism, Mabel continues. "He sounds sympathetic but I am skeptical."[9]

George works through the spring. *Scribner's Magazine* accepts an article of his on Herbert Spencer, which appears in June.[10] It is a meditation on the possibility of synthetic or encyclopedic knowledge, that which he seeks through history of science. The present times, he begins, are "a period of ebb, when the centrifugal forces far exceed the centripetal ones, when man is more conscious of his own precious self than of the rest of humanity, more concerned about his material than his religious needs, more prone to assert himself than to own any deep solidarity with his fellow men." Everything seems to be falling apart. "A great many men and women seem to have no higher ideal in life than to be 'different' and 'exclusive.' Our artists are determined to be original at any price. As to the scientists they shut themselves more and more out of the world and live like clams in their little shells." George identifies post-war pessimism and skepticism. To "emerge from the present anarchy," to overcome relentless specialization, will require renewed attention to synthesis and unity. He approaches the theme by considering Spencer's search for synthesis.

George recalls that Spencer was an autodidact who, after having worked as sub-editor of the *Economist*, retired with a small inheritance to write. "Such is always a very hazardous decision, never more, however, than in the case of a man who is less a writer than a thinker, a slow and hard thinker, whose ability to express himself is constantly inhibited by the fear of error." Spencer "knows exactly what he must do,...calmly estimates the immensity of the undertaking and the frailty of the means,... knows that his decision practically involves the surrender of his liberty for the rest of his days and makes of him a slave to his ideal." George emphasizes Spencer's shortcomings: his insistence upon Lamarckian evolution instead of natural selection, his emphasis on freedom over laws, and his dogmatic principles of nature, but George lauds Spencer's dedication: Despite his modest means, he invested more than £3000 in the publication of his descriptive sociology. Spencer's celibacy, "his sexual indifference," is turned into the virtue of dedication to his goal, and his dogmatism is almost a requirement of "a genuine encyclopaedic mind." We read here George's projection of his own conceits on Spencer's record, even to the extent of emphasizing Spencer's ignorance of history. On the centenary of Spencer's birth, George urges attention to "the example of moral courage and of faith which he gave us. He helped us to understand the nature and the desirability of synthetic science, to realize its possibil-

9. NYPL. The Gillespies' Washington studio remains separate from their new retail establishment.

10. Diary, 25 March 1920.

ity and to keep alive the need and the love of it."[11] The world is invited to view George's undertaking in history of science—encyclopedic, systematic, and self-financed—as an extension of Spencer's vision.

George prepares no. 8 of *Isis*, where as a lead article he elaborates, in line with his treatment of Spencer, the notion of synthetic bibliography.[12] In the article, he proposes limiting the number of printing plants to curb the expansion of scientific literature, and he deplores hyperspecialized studies of trivial topics. The proliferation of scientific literature gave rise to "an immense and ruinous publication," the *International Catalogue of Scientific Literature*, which issued annual publications until 1916. George calls for critical or "synthetic" bibliographies—like his own in *Isis*—where wheat is separated from chaff. It is one way to "surely if slowly thwart bad and immature publication."[13] George urges professors to restrain the publications of their students. Less is more. The argument leads a reader to imagine that George's meagre record of publication is a virtue.

George and Mabel attend a performance of *Hamlet*, his first Shakespearean play.[14] He ponders Guglielmo Ferrero's *Fra i due mondi* (Milan, 1913), comparing the Old World and the New World. He finds that he often agrees with Ferrero, the son-in-law and collaborator of criminologist Cesare Lombroso, whom he had admired in Ghent;[15] he thinks that an essay on Ferrero would make a good introduction to volume 4 of *Isis*, wherever his journal is to appear.[16] George is careful to accept outside work only on his own terms. *Collier's*, advised by President William Allan Neilson of Smith College, offers him $1000 for a short book about history of science, to be delivered in half a year: "It did not take me long to dispose of this foolish proposition. I believe more firmly than ever that any such inconsiderate publication is wicked. He who writes for any other motive than the internal impulse to give to the world some new parcel of truth, of beauty, of justice—is a dishonest man. I am not at all interested in that sort of book making."[17] The very words of a scholar who declined to attend the talks of the popularizer Albert-Louis du Chastain in Ghent.

George, on 3 May, describes May's eighth birthday. May gives him her savings, about $2, and asks him to buy Mabel a fountain pen which her

11. George Sarton, "Herbert Spencer," *Scribner's Magazine, 67* (1920), 695–702; reprinted as an editorial for the ninth number of *Isis, 4* (1920), 375–90, with notes added. Here pp. 375–6, 378, 381, 383, 386 of *Isis*.
12. Diary, 25 March 1920.
13. George Sarton, "Synthetic Bibliography, with Special Reference to the History of Science," *Isis, 3* (1920), 159–70, on pp. 162, 164.
14. Diary, 25 March 1920.
15. Diary, undated, 1903.
16. Diary, 27 April 1920.
17. Diary, 29 April 1920.

mother wanted. After school, May's friends arrive, and they all go for a long drive in Virginia. They return to the boarding house where the Sartons take their meals (they live in rooms nearby) for cake and ice-cream. Margaret Gillespie helps organize the party. George describes it as a happy time, but in light of the family's recent experiences in Europe, the party seems forced and sad.[18] "Melancholy," he notes in his agenda on 9 May.

On 11 May 1920 George writes to Mabel from Boston, where he is house hunting. What pleases him most is the greeting of their old postman, who read about his reappointment in the paper. He dines with Louise Inches Seton in whose 17th-century house at Medfield—where she spends the summer—the Sartons are frequent visitors.[19] He calls on Alice Martha Streeter, *née* Chase, and dines with Henderson. And in his agenda he records sex on the 10th and 11th. George returns to Washington and lunches with the inventor and health crusader Emile Berliner: "Complex personality: Agnostic, yet believing in the church as I do. Zionist, Buddhist."[20] He summarizes 4 months in Washington. He was able to survey German scholarship appearing during the war, work on *Isis*, and make friends. The Carnegie Institution's Woodwards "were very kind to us three & invited us to their home. (All the other scientific members of the Institution simply ignored our presence in Washington.)" He had good relations with Fielding Garrison, "a cultivated and lovable man, the only one in Washington having a genuine & deep interest in my work." He likes the family of May's best friend, Margaret Wayland, and he is trying to turn Mabel away from business: "I am trying to encourage her artistic activities & to discourage her business activities, for she will always be as poor a business woman as she is great in the field of art & soul." Mostly, George is struck by his modest circumstances: "Our poverty is very clear to me—more than ever—and we will have to retrench more than before."[21] George intervenes with Margaret Gillespie, writing her "a sharp letter" and then having a "disagreeable talk" with her.[22] They are so poor in Washington that they cannot live together as a family in a real apartment, but George persists in discouraging Mabel from enterprise that might supplement the family income.

Using his connection with the late Philip Jourdain, George offers himself as an adviser to Paul Carus at the *Open Court*. He writes on 4 June

18. Diary, 3 May 1920.

19. Louise Inches, née Pomeroy, wife of Charles E. Inches, is immortalized in a portrait by John Singer Sargent of 1887. The Sartons' friend is her daughter, the wife of Henry Seton.

20. Diary, 21 May 1920.

21. Diary, 7 June 1920.

22. Diary, 23 and 28 May 1920.

1920 to propose two titles for Carus's series on philosophy of science, which Jourdain founded. One title is Louis Rougier's book on relativity and quanta, "written with characteristic French clearness." The second book is Roberto Marcolongo's study of the 3-body problem. George cannot be a translator, but he is glad to write a preface. If the books sell, George is glad to advise about additional titles, "if you made it worth while to me." For advice about the two books, George suggests physicist Robert Andrews Millikan and mathematician George Birkhoff.[23] George admires Nobel-laureate Millikan for an emphasis on "moral and spiritual values," and on the vital role of scientists in improving the material qualities of life, while he ignores Millikan's antipathy to Einstein's relativity.[24] Birkhoff, an impractical mathematician, George knows from Harvard.

George installs himself in Henderson's room at Widener Library, Henderson taking leave for a semester. He unpacks the books brought back from Belgium (he estimates that one-tenth of them were destroyed or defaced). "We are both very blithe to be in Cambridge again. We feel that we are considered as members of the Harvard family & we are surrounded by friends. Besides the New England moral atmosphere is very congenial to me." He records a meeting with Henry Fairfield Osborn, head of the American Museum of Natural History, "a very cultivated & broadminded man."[25] The words are characteristic. Cultivation, in George's vocabulary, means an interest in history of science; broadmindedness indicates a sympathy for George's plans. In this case, Osborn expresses interest and sympathy for George's notions about a museum for history of science and civilization.[26]

The Sartons live through the summer of 1920 on the generosity of their Cambridge friends. May spends weeks as the guest of Louise Inches in Medfield and the Boutons in Kearsarge, New Hampshire.[27] Mabel meets George's first cousin Claire van Halmé in New York and brings her to Cambridge. Van Halmé is a pianist from Brussels seeking a career in the

---

23. Paul Carus Papers, Special Collections Research Center, Morris Library, Southern Illinois University, Carbondale, Ill., George Sarton to Paul Carus, 4 June 1920. Louis Rougier, *La Matière et l'énergie selon la théorie de la relativité et la théorie des quanta* (Paris, 1921); Roberto Marcolongo, *Il Problema dei tre corpi da Newton (1686) ai nostri giorni* (Milan, 1919).

24. Robert A. Millikan, "The Most Important Thing in the World," *Isis, 4* (1921–22), 324, reprint from *Science*. Ronald C. Tobey, *The American Ideology of National Science, 1919–1930* (Pittsburgh, 1971), pp. 146–7, p. 179 for Millikan on science and religion and for Millikan on morality, p. 107 on relativity.

25. Diary, 23 June 1920.

26. In 1924 George proposes a system for relating exhibits to postcards or brochures in Osborn's house-organ. George Sarton, "The Museum of To-morrow," *Natural History, 24* (1924), 710–12.

27. Diary, 6 July 1920.

United States.[28] In July and August, the Sartons, along with Van Halmé, occupy the house of physicist Edwin Herbert Hall while the Halls are in California.[29] George uses the summer to revive his Greek. In August, they find lodgings at 24 Agassiz Street: "We will share a house recently bought by Dr La Flamme, a dentist, and which they are now remodelling. He and his wife will live on the first floor, we on the second and third." For George's birthday, Mabel executes an *ex-libris* of the Egyptian goddess Isis, and May gives him 3 silk ties. George feels at home: "The people we know are gentle and cultivated; the whole atmosphere of this place is quiet and refined. I love Cambridge & hope to spend here the rest of my life."[30] Early in September, the Sartons leave for a week at Pemaquid Point, Maine, occupying the cottage of Isabel Eaton. Nearby are Duncan Black Macdonald and his wife; Macdonald, a distinguished Arabist at the Theological Seminary in Hartford, becomes George's first teacher of Arabic.[31] George occupies his time reading again Thackeray's *Virginian*.[32]

George completes the ninth number of *Isis* in September 1920. There George comments bitterly on the choice of Geneva over Brussels for the seat of the League of Nations. He reiterates his vision of a Belgian world city, "the intellectual capital" of the League, just as Geneva would be its political and administrative center and The Hague its judicial center. The choice of Brussels is apposite because Brussels is the seat of the Union of International Associations, founded in 1910 to organize world congresses every 3 years and the publisher of the periodical *La*

28. Agenda, 14 July 1920. The Sartons keep in contact with the cousin and consider helping her financially. Mabel writes in French from the Martha Washington Hotel in Washington, in 1921, that Claire van Halmé is "filled with energy and will and health." Claire will succeed, Mabel believes, even though her letters of introduction are social ones, without a direct connection to the musical world. Claire van Halmé married Henry Murphy.

29. Diary, 19 July 1920.

30. Diary, 31 August 1920. Possibly Frank Laflamme, Chief of the US Army Dental Corps.

31. As George relates in *Introduction to the History of Science, volume 1: From Homer to Omar Khayyam* (Washington, D.C., 1927), p. 45, and "Why *Isis*?" *Isis, 44* (1953), 232–42, on p. 237. George has Macdonald vet his comments on publishing Arabic words in the Roman alphabet. George Sarton, "Note on the Transliteration of Arabic," *Isis, 6* (1924), 410–12, on p. 412. Macdonald remarks about his long collaboration with George: "There is no part of my life as a scholar on which I look back with more satisfaction than this." William Douglas Mackenzie, "Duncan Black Macdonald, Scholar, Teacher, and Author," in *The Macdonald Presentation Volume*, ed. Robbins Wolcott Barstow (Princeton, 1933), pp. 1–9, on p. 6. Thomas F. Glick, trans. Mercè Viladrich, *George Sarton i la història de la ciència a Espanya* (Barcelona, 1990), p. 19. Macdonald is one of the 5 scholars analyzed in depth by Jean-Jacques Waardenburg in his remarkable book, *L'Islam dans le miroir de l'Occident: Comment quelques orientalistes occidentaux se sont penchés sur l'Islam et se sont formé une image de cette religion* (Paris, 1962).

32. Diary, 1–9 September 1920.

Vie internationale, where George placed two articles in 1913. George recommends the Union's plans for an international university. He reiterates his internationalist credo. *Isis* will report on the undertakings of the Union and its international scientific congresses, "but only on those which are genuinely international, not on those from which the representatives of German, Austrian and Russian science are systematically excluded."[33]

Throughout the war and into the peace, George maintains the political rôle that he forged in Belgium early in his marriage, that of Fabian-style internationalist. Europe honored the rôle, but America rejects it. George's political capital, which he announced in the piece for the *Open Court* that he and Mabel wrote in England, lands him a position as a propagandist for American intervention on the side of Belgium. He tries to enlarge the rôle by advocating Belgium as a neutral, industrial center separating the Great Powers. The vision finds few takers, and it excludes George from participating in setting the terms of the peace. His post-war writings increasingly emphasize the isolation of the greatest scientists and scholars, whom no force can turn from the pursuit of their calling. He writes to justify his poverty and loneliness. He loses nothing by condemning the American boycott of Central European and Eastern European science.

The Sartons live in furnished rooms until 4 October, when their apartment is ready to receive them.[34] George reads Havelock Ellis's *Nineteenth Century: An Utopian Retrospect* (London, 1901); Ellis "is one of the most original and fearless thinkers in England today."[35] He begins Edward Gibbon's *Decline and Fall of the Roman Empire*, while Mabel supervises moving their furniture into the new apartment.[36] Financial notes indicate that late in September Mabel is furnishing their apartment with a white enamel bed and other items costing $271.[37] The Belgian pieces have been "greatly damaged" in the transatlantic voyage.[38] "Furniture unpacked and found in a wretched state—a real catastrophe," George notes in his

33. George Sarton, "The International Organization of Science," *Isis*, 3 (1920/21), 420–1.

34. Agenda, 4 October 1920.

35. Ellis published on Leonardo in *The Nation*, a text that Arnold C. Klebs sent to George with high praise. Klebs, whose perceptions often differ from George's, is disappointed that Ellis "barely mentions" Sigmund Freud's study of Leonardo, a study that George roundly condemns. HGS, Arnold C. Klebs to George Sarton, 6 July 1919; George Sarton, "Sigmund Freud, *Leonardo da Vinci: A Psycho-Sexual Study of an Infantile Reminiscence*," *Isis*, 3 (1920), 104. George's attraction to Ellis may also relate to Ellis's studies of sexual activity. Lesley A. Hall, "Sexual Cultures in Britain: Some Persisting Themes," in *Sexual Cultures in Europe: National Histories*, ed. Franz X. Eder, Lesley A. Hall, and Gert Hekma (Manchester, 1999), pp. 29–52, on p. 37.

36. Diary, 27 September 1920.

37. YMS.

38. Diary, 1 October 1920.

agenda.[39] George plunges into his Belgian notes, which he thinks will provide the bulk of his projected *Introduction to the History of Science*; soon he will realize the limitations of his early research.[40] The spectre of poverty looks over the family: "We are again very hard up." He confides in his agenda: "Unpleasant discussion with Mabel on money matters. Not her fault, neither mine, lack of money."[41] George writes to 14 colleges and universities, proposing himself as a speaker. William Allan Neilson at Smith College replies favorably. He travels to Northampton on 19 and 20 November 1920 and gives his first special lecture in English since his visit to Montreal early in 1919.[42]

Late in November Mabel and George dine at the home of Theodore and Miriam Stuart Thayer Richards, along with their daughter Grace and her fiancé James Bryant Conant—Harvard's future president—"assistant professor of organic chemistry, a quiet young man who is far more intelligent than he looks." George and Mabel like the Richards family—"a perfect day," George records after one such dinner. They have Thanksgiving with Louise Inches in Medfield, where Mabel and May were house guests in November.[43] In Mabel's absence, George's agenda records sex a number of times. For example, he delivers May to Medfield on 7 November. He dines alone at the Faculty Club, then sex on the 8th and 10th. But George and Mabel are nevertheless intimate in December.[44]

Early in 1921, George is devoting every Monday to the pre-Ming paintings at the Boston Museum of Fine Arts. On 31 January, he attends a lecture by Michael I. Rostovtzeff, located at the University of Wisconsin since leaving Petrograd, an event worthy of special note in his diary. On 17 February he acquires at his own expense a secretary, Miss Gilpatrick. George and Mabel invite a student of Rostovtzeff's, Robert Blake, to dinner, and George picks his brains about Asian art.[45] George's libido is strong. On 23 February his diary records sex twice, with the observation, "Casanova."

---

39. Agenda, 27 September 1920. F. O. Bailey Antiquarians, *A Fine Auction of the Estate of May Sarton, Sunday, May 19, 1996* (Portland, Me., 1996), p. 19, for a photograph of the Sartons' 18th-century bahut in walnut and oak.

40. George Sarton, "Introduction to the History and Philosophy of Science (Preliminary Note)," *Isis, 4* (1921), 23–31, p. 30: "Unfortunately many of these notes [from Belgium] were written when I was still very immature and are, in their present form, of little value."

41. Agenda, 26 October 1920.

42. Diary, 1 and 19–20 November 1920; Agenda, 26 October 1920.

43. Diary, 25 November 1920; Diary, 5 November 1922, for "perfect day."

44. Agenda, 8, 10, and 11 December 1920.

45. Diary, 1 February 1921.

On 20 March 1921 he writes to Mabel to remember their anniversary:

> I can see ourselves in a little boat on the Lys, my first and last attempt of that kind. I wish we could do that again! And I remember too that before we were engaged I once told you that the accomplishment of my work would always be for me a primary consideration, that is, would be more essential in my eyes than earning money. I never dreamed then that we would be tried as we have been, nor that you would help to earn what we needed for this earthly life. I am deeply grateful to you for what you are doing, but when I see you tired and unwell, I am sometimes terribly ashamed & sorry, & I wish I could free you entirely pretty soon.

George sends Mabel 17 roses.

> The first rose stands for a day when you came to see me in my father's house & we exchanged our first kiss. The second recalls an afternoon at Assels, you worked in the garden & I studied mathematics. The third, a symphony concert in Ghent. The fourth, a little concert and tea at the house of Cecile Cauterman. The fifth, a room in the old convent of the Domenicans in Ghent, and our engagement. The sixth, a brief stay in Eecloo, when you spent much time in bed & I paid you brief visits. The seventh, a visit to London, when we stayed with the Enkes in the same hotel. The eighth, the cathedral of Norfolk. The ninth, happy days in Cromer! The tenth, our little house near the Coupure. The eleventh, our own house in Wondelgem. The twelfth, the garden in Wondelgem. The thirteenth, May's birth (a lucky number). The fourteenth, Ogunquit. The fifteenth, Luz. The sixteenth, Tardets. The seventeenth, St Joan de Luz.

A visit to the last three places located in southwestern France against the Pyrénées, by their position in the list taking place after the Sartons travel to America, would have occurred in late summer or early autumn of 1919, the only time Mabel and George are in Europe together between 1915 and 1921. During this time, however, George's diary and agenda are silent about the Basque country, and there are no letters concerning it. The Sarton family does travel together to Saint-Jean-de-Luz and Tardets in the summer of 1925, recalled by May Sarton as "after my mother and I had spent a hard winter in Belgium."[46] The unmistakable date of the letter is handwritten in English. George may refer here to a projected visit during their next European trip, but it is also possible that the reference marks a pre-war vacation; George's diaries are incomplete, and his agendas unavailable, from the early years of marriage.

---

46. May Sarton, *A World of Light: Portraits and Celebrations* (New York, 1976), p. 19. St-Jean-de-Luz is the place where Louis XIV married in 1660. George Sarton, "Notes with regard to my Journey in Europe 1925," YMS. Luz, Tardets, and St-Jean-de-Luz are indicated for late August and early September. George Sarton, "Le Sentiment du passé," *Isis, 8* (1926), 391–402, dated Luz, 12 August 1925.

Mabel writes on 15 March 1921 from Cambridge. She hopes that his lecture went well (according to George's diary, at the Women's City Club, New York), and that the Baekelands are well. In a second letter of the 15th, Mabel thanks George for his lines from Yonkers. May is helping her do the dishes. She hopes that he enjoys himself "with his Chinese friends at the Metro[politan Museum of Art]," and she signs herself, "Your Busy Bird." On 16 March, Mabel reports on having gone to the Arts & Crafts dinner as the guest of Francesca Warren. She has finished a dress and has tried to buy silk wholesale. Mabel begins teaching at the Buckingham School in April 1921.[47]

George, early in May, is "very depressed yet worked courageously" and records sex—a usual sequence.[48] George has, however, avoided the cultural pessimism of Oswald Spengler, whose *Decline of the West* takes Germany by storm late in 1920. His colleague at Heidelberg, Julius Ruska, informs him about it: "quite remarkable, with scholarly overlay, but also a grotesque [barockes] and contradictory book."[49] Ruska, a leading authority on Islamic science, thinks that scholarship is possible only in America. Western Europe, in the grips of economic uncertainty, hatred, and Bolshevism, is in its twilight years. Ruska nevertheless shares George's admiration for Hermann Alexander *Graf* von Keyserling's *Reisetagebuch eines Philosophen* (Darmstadt, 1919), with its message of spiritual regeneration, over Spengler's "often fanciful assertions."[50]

In the midst of the depressing year, George considers the sad life of mathematician Evariste Galois, who died in a duel at age 20 early in the 19th century. "When one sees how terribly fast this ardent soul, this wretched and tormented heart were consumed one can but think of the beautiful meteoric showers of a summer night." George focuses on Galois as an inattentive student, interested in nothing other than mathematics. Galois tried and failed to enter the Ecole polytechnique. He published in mathematics and sent a manuscript to Augustin Cauchy for the Academy of Sciences, which was misplaced. His father committed suicide. Unable to study at the Republican citadel of engineering, the Ecole polytechnique, he gained entry to the Ecole normale, the nursery of schoolteachers. More publications and another manuscript lost at the Academy of Sciences, this time by Jean-Baptiste Joseph Fourier. He joins a radical society in the aftermath of the Revolution of 1830, and for his pains he is dismissed from the Ecole normale. The Academy of Sciences rejects a new manuscript, and Galois plunges headlong into radical politics, lead-

47. Diary, 4 April 1921.

48. Diary, 2 and 3 May 1921.

49. HGS, Julius Ruska to George Sarton, 7 November 1920, answered 14 January 1921. Ruska's review of Spengler in *Isis, 5* (1923), 176–81.

50. HGS, Julius Ruska to George Sarton, 31 January 1921.

ing him to arrest and imprisonment. He dies in a duel over a woman, but not before writing a summary of his mathematical ideas. George introduces an observation of President Lowell's: "Talent is that which is in a man's power, genius is that in whose power man is." But George thinks that both talent and genius are inborn. George loves Galois, an unappealing boy, "because of his adversity." The lesson to be extracted: Be kind and tolerant of wayward youth.[51] The life of the martyr counts for everything in this popular piece; Galois's mathematics is nowhere to be seen. Galois is one of George's secular saints—the people of learning who interest him most. He notes near the end of his life: "I am deeply interested in saints because they are rebels against material comfort and money, defenders of spiritual life. They are the heroes of the human conscience."[52]

George receives his teaching assignments at the end of each semester. He writes on 25 May 1921 that Professor Clifford Moore told him that "one course on the History of Science was considered sufficient for next year." George proposes to offer a history of astronomy in 1922. It is designed for students having taken a year of natural sciences, and it emphasizes the relations between astronomy and other sciences, along with the "human side of this development and its influence upon the progress of civilization."[53]

In June 1921, George and Mabel are frequently intimate, and George thinks that Mabel may be pregnant.[54] The Sartons observe their tenth anniversary without celebration, "for we could not afford it."[55] On 30 June, George observes in his diary that he has reached the second century in his *Introduction* (he began writing in January).[56] He plans to write, in French, his first article on Chinese painting; it will finally appear in 1923. He thinks that after the 15th century, he will move to study each science separately, mathematics and astronomy first. He looks forward to working on an "Archaeology of Science," based on European collections of ancient instruments, among many other projects.[57]

On 6 July, when May is in Greensboro, Vermont, with the Hockings and he is alone with Mabel in Cambridge, George writes to Raymond Limbosch. He hopes that Raymond's automobile, about which May recounted nothing but trouble in her memoirs, is giving him good service.[58]

51. George Sarton, "Evariste Galois," *Scientific Monthly, 13* (1921), 363–75, on pp. 373, 375.

52. Diary, 28 March 1953.

53. NYPL.

54. Diary, 10, 14, and 15 June 1921.

55. Diary, 22 June 1921.

56. Diary, 12 January 1921.

57. Diary, 30 June 1921.

58. *Phoenix*, p. 124, for the automobile.

George has been silent because he resents Raymond's continual reproaches. George affirms that he works extremely hard and he continually tries to improve himself. He expects no reward. "My activity is determined in a very large measure by my love for Mabel & May, and when they are honored because of me, I am wildly happy." George reproaches no one, and he does not want to be reproached, *"especially in a general and abstract way."* He doesn't appreciate being treated like a child before a schoolmaster. "Friendship implies *absolute* confidence."[59] George feels "tired and empty" on 15 July, but not too tired.[60] Mabel leaves for Medfield on 21 July; George records sex on 22 July.[61]

Raymond Limbosch writes to Mabel on 22 July 1921 about a painful letter that he received from George, who is perfectly honest and open, but also extraordinarily maladroit. He has grown remote from George over the past years, holding on to a vision of their friendship from earlier times. "It is enough for not estranging us; it is too much for us to remain friends." Raymond sends the same sentiment to George directly, on the same day. Raymond regrets having pointed out George's spiritual disequilibrium, "which I am not the only one to see and to deplore." The break is not repaired, even years later when George pays for the education of one of the Limbosch children. Raymond writes to Mabel on 2 February 1939: "All exchanges of ideas between us are impossible. It's a painful thing, but after so many years of testing this impossibility without believing in it and without giving into it, today I must really give in and take things as they are." Raymond explains more on 19 July 1939: "Whether rightly or wrongly, I have taken the history of science as 'the chronology of neutral opinions.'"[62]

Mabel shows the letter to George and then explains that although she did not read it completely, Raymond is just in his comments. Her letter to George, alternating between English and French, makes it clear that her tears, when they talked about the letter, had nothing to do with Raymond's letter or George's reaction to it, but rather to a melancholy state of mind. She thinks that she knows what set Raymond off. It was George's comment that, Mabel writes in English, *"life was too short for you to defend yourself to your friends* or to discuss their criticisms." Mabel finds herself caught between husband and friends. In English:

> I have always clung to the hope that you & they would eventually understand each other better. You have persistently shewn them one side

59. NYPL.

60. Agenda, 15 July 1921.

61. Agenda, 21 and 22 July 1921.

62. NYPL, Raymond Limbosch to Mabel Sarton, 2 February 1939, for the financial support.

> of your character, & if they have seemed ungenerous to you, you most surely have seemed so to them. You have persistently suspected me of vile treachery to you through them (surely you must suffer worse than I have for it.) All that is so painful that I stuff it away deep in my heart, but someday it must all come out into the light & be burned pure again.

In her correspondence with others, Mabel will not write about George's feelings or doings again, except in a cursory way.

> How sickeningly foolish the whole thing is, as if the little misunderstandings & disappointments *really mattered* between people who love each other. They are just like little clouds in the sky, even bigger troubles are just like passing storms which only make the fair days seem fairer. I feel a huge impatience with both you & them. You are spoiling something precious, infinitely dear to me, & that would have meant, I do believe, added riches & happiness for you & them, spoiling it year by year with a persistent stupidity, & with a cruel passion that is unworthy of us all. And it seems to me as if it depended on *such a little thing* for that misunderstanding to end—a little more faith in your power to love, on their part; a little more patience & generosity on *your* part when they claim me & dream perhaps foolish dreams of my going oftener to them (each time they dare to wish I might come, you fling out at them as they were criminals trying to seduce me away from you!)

Mabel thinks that there is some justice in Raymond's observation that George rejects friendship because of his work.

> You have the intellectual sympathy of a few people like the Richards & Henderson & D. E. Smith & Miss Stanton, but *they must make no real calls on your time & energy* just as you will make none on *theirs—you cannot afford it.* You very clearly realize this tho' perhaps it may shock you to see it in words: you do not give nor ask intimacy of any-one: *not-even-of-me.*

George has doors in life that he has never opened. Mabel wants to walk through them with him. Possibly he has gone through doors into places where Mabel cannot go. She hopes so. "I too, have those which are sacred to *me alone.*" Mabel recalls that George told her that Raymond and Céline were no longer his friends before their marriage.

Mabel saw her friend Méta bearing children one after the other, too quickly. "It seems to me that I would not dare to marry, being a man, without first knowing how to control the arrival of children." Raymond and Céline criticized George for allowing Mabel to become pregnant so often. At their first apartment in Cambridge, Mabel writes in French, "you spoke to me at length about how you suffered because of my sexual frigidity."

Mabel thought that it would be better for her to leave him or for him to find another wife more to his liking. "It was like a black hole" opening before her. She then said that if he wanted, she would leave, but she "*completely forgot about May.*" George saw only that Mabel contested for her "liberty," but Mabel warned him that she would fight for the child. In any event, Mabel must reply to Raymond. It is he "who finds you superbly selfish."[63]

The crisis passes. Mabel leaves with the McDougalls—Anne *née* Hickmore and William—on 3 August to stay at their home in Intervale, New Hampshire, while May stays with the Hockings in Greensboro, Vermont. George remains in Cambridge, dining on 4 August with Lothrop Stoddard, who reads from his book, *The New World of Islam* (New York, 1921).[64] Mabel writes to George on 9 August 1921:

> Please dear Husband, don't ever speak again, nor *think* of yourself as a "poor man" in relation to me. I am so happy as we are! I truly don't want you to try to earn more. I *hate* you to feel you must. I would be happiest of all if I might take that entirely on my shoulders. I mean, the earning of just a little surplus to pay off what we owe, & create a little reserve fund for holidays & sickness. I would rather hunt about for a smaller apartment & face the bother of moving etc. if we could really save in that way, than have you worried over money & cramped in your expenses for books you want & so on.
>
> I cannot manage for food as well as some women would, I believe, but I am ready to economize in other ways, by a less expensive house & also by teaching at May's school which should save us at least part of her schooling fees. Oh Dear, my Dearest, we shall manage & we are rich & happy beyond measure I sometimes think. I am not good at telling you how happy you make me, but sometimes I offer it with my kisses. Don't they tell you?

Mabel's kisses do not speak to George. In his agenda he records sex on 5 and 6 August. On the 7th: "Lunched at the Club. Long aimless walk in Boston. Evil thoughts." Sex on 10 August and twice on the 12th. On the 18th: "Loneliness."[65]

May arrived at Intervale a day early from Greensboro, Mabel writes on 12 August 1921, and no one was at the station to meet her. So May asked a boy to carry her bag and direct her to the McDougall house: "The first thing I knew was hearing someone say 'see who's here' & there was a little 'nut-brown maid' oh so brown! With shiney eyes & rather a quivering face standing before me. Poor childie, she had really been very disappointed & a little frightened."

---

63. YMS, Mabel Sarton to George Sarton, 1921.

64. Diary, 4 August 1921; Agenda, 3 August 1921.

65. Agenda.

On 13 August, Mabel reports that William McDougall has read a chapter of the book he is writing on the psychology of laughter. The choice is significant. McDougall is the first psychologist to see laughter as an instinct, evolving to express an innate quality of sympathy.[66] McDougall arrived at Harvard from England in the fall of 1920 to occupy the chair held by William James and then Hugo Münsterberg. He is a firm advocate of eugenics and of the primacy of inherited characteristics, and he champions the instinctual pursuit of indistinct goals. McDougall dislikes behaviorism and materialism, and like William James, he is attracted to spiritism as a topic worthy of serious, experimental inquiry. (Mabel, on 9 August 1921, writes to George about a psychic at Intervale who produced a message from William James, Jr, to William McDougall.) From his arrival in America, McDougall remains aloof from most of his colleagues, who conspire against him, and in 1927 he moves to Duke University, where as chair of psychology he actively promotes the research of Joseph Banks Rhine and Louisa Weckesser Rhine in parapsychology.[67] The Sartons and the McDougalls find much in common as cultured Europeans who are marginalized at Harvard, although George notes to Mabel on 15 August 1921, "I hardly know them."

Money continues to rule the Sartons' life. George, on 14 August, does not see how they can all spend a month at Ogunquit, for it would cost $200. He reminds Mabel that they cannot spend much on home improvements until he has paid his debts. But he does want to buy an armchair for Christmas. "To think that for the last years we have not had a single chair which enabled me to take you on my lap is shocking. *It is wicked.*" About May's unannounced arrival at Intervale: "I am very glad that May behaved with so much sense when she arrived. Although one would not do it on purpose it is sure that every such experience is a positive gain." He reminds Mabel that when he is alone in the house, "I hardly hear any human speech. It is a rare day when I speak as much as twenty words." He lives with the written word. On 15 August, the proofs of *Isis*, volume 4, have arrived. He sent typewritten copy, but the printers made a terrible job of it, "the worst I have ever seen." He will go to the movies in the evening to forget about it. "Oh, when shall I be able to publish *Isis* in America?"

George has tried to interest American publishers in *Isis* (according to his diary, the University of Chicago Press rejects it early in January

66. William McDougall, "The Theory of Laughter," *Nature, 67* (1903), 318–19; "A New Theory of Laughter," *Psyche, 2* (1922), 198–300. We have seen that one of George's first publications was about Henri Bergson's notion of laughter. George Sarton, "La Signification du comique," *Flandre libérale*, 30 November 1903.

67. Seymour H. Mauskopf and Michael R. McVaugh, *The Elusive Science: Origins of Experimental Psychical Research* (Baltimore, 1980), pp. 59–70.

1921), but he will continue with Belgian printers for the next 20 years. He would have been in worse shape had *Isis* been published in Germany, but what about the plans with Oxford University Press? Was that not part of the thought in linking up with Charles Singer? The English connection would have worked, so long as Sir William Osler lived. Singer negotiated Oxford through Osler's levitating powers, however, and with Osler's death the enterprise in the Radcliffe Camera, which had been conducted with displaced scholars, fell down to earth; the new Bodley's Librarian, Arthur E. Cowley, "bundled him out." In 1920 Charles Singer was appointed University Lecturer in the History of the Biological Sciences, but without a place to roost he lost interest in the post. Singer obtained a lectureship in the history of medicine at University College, London and, although his inaugural lecture at London is an Oslerian-Sartonian synthesis, he abandoned both *Isis* and in his own series.[68] George severs his editorial partnership with Singer in May 1921.[69] The snake in Eden was Robert Theodore Gunther, zoologist and tutor at Magdalen College, unusually well-connected in the university and diplomatic world by birth and marriage, whom Osler had encouraged to restore and exhibit antique scientific instruments kept in the Old Ashmolean Museum. Osler, as president of the Classical Association, inaugurated Gunther's exhibit in May 1919. From there, Gunther set out on a career in the history of scientific instruments and in 1934 became Reader in the History of Science at Oxford.[70]

On 17 August 1921, Mabel reports that William McDougall asked about *Isis* and spoke about Charles Singer.

> They do not like him evidently tho' they said little. Especially they objected to the way he had behaved when a sort of testimonial was got up to Sir Wil. Osler, who was a good friend of theirs. (I will tell you about it later.) Evidently they do not look upon him as *dependable*, but rather as a self-seeking man, wanting honors & not over-generous. This sadly confirms your own conclusions.

The judgment seems excessive. Singer's donating £500 to the Bodleian Library not generous? Singer's desiring a real appointment ambitious?

---

68. Charles Singer, *Greek Science and Modern Science: A Comparison and a Contrast* ([London], 1920); George publishes Singer's note on history of science as a proposed BA subject at University College, London, with Singer lecturing on history of medicine in the 1919–20 session. C[harles]S[inger], "Courses on the History of Science at University College, London," *Isis, 3* (1920–21), 419–20.

69. Diary, 9 May 1921.

70. A. V. Simcock, "A Dodo in the Ark," in *Robert T. Gunther and the Old Ashmolean*, ed. Simcock (Oxford, 1985), pp. 43–92, on p. 68, Simcock reporting Gunther's description of Singer's fate.

On or about 17 August Mabel describes the New Hampshire countryside:

> I could not sleep & I got up & watched the dawn creeping up behind the mountains. There is an almost full moon & it hung in a wind-swept greenie blue space above the mountains, with banks of cloud that turned orange & pink & purple as the sun rose. The wind sank as if the daylight had sapped it of its evil power & I went back to bed & slept a deep sleep till eight o'clock this morning! The air is clear like a glass that has been washed & polished; the mountains are so near one can count the trees on them. It is a blessed, still, day & I wish more than I can tell you that I could share it with you.

A report like this one is insufficient for George, who needs constant reassurance from his wife and daughter. On 17 August he begs for a letter from May: "I am sad that May has not written to me for such a long time. When I was a little boy I used to write a card to my father every day: I do not ask for so much, but could she not write me a short card or letter once in a while?" On 19 August, "No word from either of you neither yesterday nor today, and that when I needed it so urgently. My cold has left me all in a tumble. *I am desolate*. I am going to the sea." Mabel replies on the same day: "Your this morning's letter came as a painful shock to me. I scarcely know what to say to you except to repeat that *I love you* & my thoughts have been hovering round you all these days tho' evidently you have not felt them." George, on 20 August, declines to join wife and daughter; his cold prevents travel. He looks forward to Mabel's and May's return; he has not seen May for nearly 2 months. Mabel and May return to Cambridge on 23 August.[71] On 27 August, George and Mabel see "Enter Madame" with Gilda Varesi.[72]

Early in September 1921 the Sarton family vacations at Ogunquit. George swims "almost every day, & sometimes Mabel came with me."[73] At Ogunquit, George projects his activity for the coming 15 years. He hopes to complete an introduction to the history and philosophy of science by September 1923. Then come introductions to the archaeology and the iconography of science, and an introduction to the history of art; a geography of human genius, a history of science in the 15th century, Leonardo da Vinci, and 19th-century physics.[74]

Earlier in 1921, George and Mabel became reacquainted with the painter Lucy Stanton, who is established in Boston as a teacher of art

71. Agenda.

72. Agenda.

73. Diary, 24 September 1921.

74. NYPL.

history and art at Dana Hall in Wellesley, Miss Choate's School, and Milton Academy.[75] George writes in his diary: "She is so kind and so sound, and her quiet energy is amazing. I like her." They dine with her, and on 12 May 1921 she invites them to meet her friend Dr William Lorenzo and other guests.[76] George records tea and dinner with Stanton on 15 September, and he notes her address and telephone number in his agenda—a rare entry. Mabel leaves for Malboro on 18 September. In his agenda George indicates sex on 21, 23, and 28 September. In his diary on 24 September: Lucy Stanton, "who is really a dear friend" introduces him to Percival Chubb, leader of the St Louis Ethical [Culture] Society, "himself an old friend of Havelock Ellis."[77] Mabel leaves for Washington on 2 October at 7:30 p.m. George records sex. On 30 October he dines with Lucy Stanton and walks her back to Boston.[78]

From Washington on 9 November 1921, Mabel sends George a detailed description of her activity:

> I did not write yesterday, I think. The days go so strangely, I get one confused with the other. I am getting on with my work. I have been through nearly all the old colour schemes which needed altering, choosing the new silks & wools, writing a little label of instructions & fastening the whole bunch together with a rubber band. (There are often as many as 8 cards of the *same colour* scheme introduced into *different* designs.) I feel greatly relieved to have done this. Many, many of them were really *bad* through replacing. At night I do drawing—so far no new designs!—all completing of old sets & re-arranging them in various ways. Tomorrow I may begin some new work. I am tired tonight. I have worked almost feverishly & I am rather worried over the future business arrangements which I am determined to have clearly put upon paper. M[argaret]. E. G[illespie]. told me today she had once been to a lawyer (when you wrote that letter!) to ask if you had the right should such an idea come into *your* head to simply come & take all the designs & colour charts away, as she had no written agreement. He said neither you nor I could do this, because by accepting the 10% on wholesale sales, that in itself was a tacit agreement, which could only cease by mutual agreement again. It is true that she has *not* gone on paying me the 10%, so surely *now*, I could withdraw my work past & future, if I so chose. (I do not wish or intend to do so.) But I do want to know *what I am bound to.* I would like to talk it over with you but I know it bothers you so I had better not. Indeed I ought not to write this much. (Whatever you do, *don't* write to Margaret again, not even the nicest letter in the world!)

75. Betty Alice Fowler, "An Art in Living," in Betty Alice Fowler and Andrew Ladis, *The Art of Lucy May Stanton* (Athens, Ga., 2002), pp. 16–28, on p. 22.

76. Diary, 15 May 1921.

77. Diary, 24 September 1921.

78. Agenda.

A Mr Irving, of Scots extraction, is interested in transforming the business into a corporation if he can raise money for it. Mabel writes on 14 November that she needs a week more in Washington. That will make 3 weeks' absence. On 16 November, Mabel is working all day but hopes to attend a lecture by Sylvanus Griswold Morley, an associate of the Carnegie Institution who is an authority on Mayan civilization; at the end of the 1920s, George will find himself in an administrative structure along with Morley.

At the beginning of the booming 1920s, it appeared as if Mabel would finally achieve success as a designer. She writes on a Saturday night late in 1921 from her Washington studio that Mrs Gillespie has sold $300 of goods to actress Billie Burke. For the month of November 1921, Mabel stays at the Waylands on 16th St in Washington, and then moves in with Miss Florence Stiles, who works at the Carnegie Institution and knows George. Margaret Gillespie welcomed her so warmly that she thinks all difficulties can be overcome. "The work-rooms are so orderly & systematised that it is a real inspiration. There is a tremendous lot to do. Almost every colour scheme needs re-doing because original materials are no longer available, & many, many new drawings & schemes are needed." Mabel continues: "Margaret will pay half the expenses of my trip here, & I am to have a *salary.* She has borrowed about $3000 & the business will become a corporation." In another letter from this trip:

> There is a really frightening amount of work to do for me. Much of it I *can*not do quickly. For instance, going over old colour schemes, suppressing the wools & silks which through replacing are no longer good colourings, & finding new harmonies. The whole business is on a much larger footing. I am pretty sure we shall make a little money, but am almost as sure that the artistic standards are being lowered, in spite of the fight I put up to defend them, & in spite of a real desire on M.E.G.'s part to keep them up as high as is *commercially possible*!

But it is precisely the commercial side that fails. Mabel writes to Eliza Van Loon (who has divorced her husband Hendrik) on 22 December 1921:

> I am—the same as ever. Generally feel horribly tired. Yet I don't do really much, but I have been dreadfully worried all year to know how to meet my monthly bills. I have earned scarcely anything from Washington—that business is still hanging fire—lack of capital. They owe me $1600! But much good it does me. Still I shall get it some-way, some-day. [79]

George regularly asks the Carnegie Institution for a subsidy to support *Isis*, and he is always turned down. It would not have been unreasonable

79. NYPL.

for the Institution to publish *Isis*. Certainly the journal compares favorably with J. Franklin Jameson's diplomatic documents and also with the *American Historical Review*, both produced under the Institution's watchful eye. As Mabel has learned, one must obtain clear terms going into a position. George waits too long before making a pitch, and when he does the watch has changed. Robert S. Woodward's successor John Campbell Merriam, a conservationist, eugenicist, and paleontologist from California, does borrow George's notion that scientists should lead American society (without referring to George), but he has little sympathy for art and literature as complements to science. He is more interested in saving redwood trees and saving white America than in saving old manuscripts. Merriam talks with Playfair McMurrich and convenes a meeting to look into support for *Isis*; Merriam's close confidant is Charles Homer Haskins at Harvard, an early supporter of George's plans for a history-of-science institute.[80] The matter drags on. George makes a case to Merriam on First April 1922. His losses over the past 3 years have averaged at $494.71 annually. "The expense is relatively small, but it ruins me."[81] The wonder is that Merriam, having seen the content of *Isis*, continues to support George at all.[82]

George summarizes the summer of 1922, "the best summer of our married life..., and I the best summer of my whole life." They stay in Lucy Stanton's Ogunquit studio while she is in California—Mabel and May 3 months there and George 2 months.[83] "We have enjoyed our vacation in Ogunquit very much," George writes to Stanton in August.

> Perhaps you do not realize yourself what it meant to us: this has been the longest vacation of my life (excluding the first four years); it has been the first real vacation of our married life, one which I would have been utterly unable to offer my family without you.
>
> Then how can I thank you? It can't be done. Yet I will try by and by.[84]

The Sartons also spend a week at Stonehurst, Intervale, in the house of Mrs Daniel (Helen *née* Bigelow) Merriman—the place where Lord Bryce

80. HGS, John Campbell Merriam to George Sarton, 5 April 1922 and 8 May 1922; Charles Homer Haskins to George Sarton, 14 November 1916. Tobey, *The American Ideology of National Science*, pp. 172–5 for Merriam's confidence, in 1920, that scientists are better than humanists as civilizers.

81. HGS, George Sarton to John Charles Merriam, 1 April 1922.

82. George notes in 1940 that the president of the Carnegie Institution, Vannevar Bush, "declared to me that my work of the last thirty years was 'irrelevant', and had my salary considerably reduced." HGS, George Sarton to Max Meyerhof, 11 March 1940.

83. Fowler and Ladis, *The Art of Lucy May Stanton*, p. 81 for the Sartons in Stanton's studio.

84. ALS, George Sarton to Lucy May Stanton, 29 August 1922.

summered when he was British ambassador to Washington.[85] They travel together to Ogunquit on 15 June. George returns to Cambridge without Mabel and May on 18 June. He records sex in his agenda on 23 and 25 June and on 3 July. On 8 July George rejoins Mabel and May. George's breech of faith does not prevent him from remembering Mabel. On 21 June George writes to Mabel in Ogunquit about the anniversary of their honeymoon in Holland. In a second letter, he offers to send a money order if the check he left cannot be cashed. Mabel writes on 22 June from Ogunquit: "We have been wonderfully happy together. I think we have had immeasurably more real happiness than most married people *seem* to have."

On 24 June Mabel describes the tragedy of a double drowning, witnessed by the children, including May, who "were deeply indignant with two women who apparently had behaved in a senseless & callous way. Today is the first really bright day & death always seems by contrast so stark & pitiless. They were quite young." Mabel went in the water for the first time, "and was surprised to feel a sort of 'tugging' in the water. I was not happy to have May in altho' there were a few people bathing but I realized how easily her little head could disappear & never be missed till *too* late. Surely people should be warned better here & a guard be keeping a watch." The waters are indeed treacherous, for the previous summer George "almost drowned with Mabel nr. Lobster Point."[86] These events are recounted more than 20 years later by Mabel in an unpublished short story, "Ordeal by Water," in which a man does not find the strength or desire to help his wife, who cries for help as she struggles to keep her head above water.[87] On 30 June, Mabel writes that she "can scarcely swim at all & it is a great shock to me to 'get in' but afterwards I am so happy to be in all that lovely movement & a different element. I feel as if I had been plunged into a lovely new existence where I regain my youth!" Then to finances, which George controls: "I gave the check to Miss Nowell. You were thoughtful to add the 2nd week. She appreciated it." On 3 July, she forwards a card from Lucy Stanton, to whom she is sending off a dress. Stanton begins a pastel portrait of George in September.[88]

While Mabel is laboring on *couture* in Maine, George writes an essay on Ernest Renan. He contrasts gentle Ogunquit to the rugged seacoast of Brittany where Renan grew up. Renan, preparing to become a Sulpician, was able to set out on his own when his sister gave him

85. Diary, 16 September 1922. Her son is the Harvard historian Roger Merriman. YMS, Mabel Sarton to May Sarton, 29 July [1939], commenting on the death of Helen (Effie) Merriman: "She understood & loved Daddy as no one else here did."

86. Agenda, 10 September 1921.

87. NYPL.

88. Agenda, 1 September 1922. Perhaps the pastel of George in ALS.

the savings she earned as a governess in Poland, but he always spoke highly about his training at the seminary. He lived in Paris with the young Marcellin Berthelot, later to become a distinguished chemist and author of works on medieval Arabic chemistry. The friends weathered the Revolution of 1848, and Berthelot's influence is apparent in Renan's book, appearing much later, *L'Avenir de la science*, which was a call to apply scientific methods for solving social and political problems. Then a book (his doctoral dissertation) on the philosophy of Averroes, the medieval philosopher. According to George, Renan lived in harmony in Paris with his bride, his sister, and his mother (a very difficult *ménage*, by other accounts), while he earned a pittance in the manuscripts section of the Bibliothèque nationale. Refused the chair of Hebrew at the Collège de France in 1857, he was sent to the Near East on an archaeological expedition, accompanied by sister and wife, the latter "acting as his manager, his secretary, and his beloved confidant." His wife died of malaria, and Renan fell seriously ill, but he completed a draft of his famous and controversial *Life of Jesus*. Then, George writes, Renan's "heroic years were over," and his article ends with Renan's memoirs of his youth (where Renan contends that he was destined to fail as a seminarist). That is where the interest lies, for generally "the fate of any great writer, scientist, or artist has been largely determined before he has reached thirty."[89] George recounts here the themes that he reads into his own life: religiosity, a hard childhood and the early loss of a parent, close experience with social radicalism, early programmatic writing, a doctoral study of a scientist, and a devoted wife who acts as a sometime assistant. But Renan lived for 30 years more after George loses interest in him. George is silent about the controversy surrounding Renan's eventual appointment at the Collège de France (over charges of atheism), his oscillation between democracy and authoritarianism, and his desperate search to connect science with the Christian faith.

By October 1922, George realizes that he cannot hope to write his *History of Asiatic Art* until he has completed his *Introduction to the History of Science*.[90] He would like to edit, nevertheless, a counterpart to *Isis* dealing with Asian art.[91] He finds time to visit the Italian Episcopal church in Boston, near the North Church in Salem Street. George, the lover of churches, attends services: "The minister is sincere and kind, and speaks well." The church is his opiate: "I love this church because it is so humble and poor. Wealth destroys churches as it destroys men."[92]

89. George Sarton, "Ernest Renan," *Nineteenth Century, 92* (1922), 953–61.

90. Diary, 1 October 1922.

91. George Sarton, "Why Isis?" *Isis, 44* (1953), 232–42, on p. 236.

92. Diary, 15 October 1922.

On 23 November, George records his family's plans for the coming year. Mabel and May will spend the autumn and winter in Europe without George, who will live in a student dormitory "and save enough money to pay all my debts before I leave America." He hopes to complete most of his *Introduction* (up to 1500) by the time he joins wife and daughter, in 1924.[93] But he is perpetually unrealistic about his scholarly projects, and none more so than the *opus magnum* that becomes his albatross.

George carries the burden of *Isis* as he labors to set down the record of pre-modern science. Early in 1922 he tries to raise subscriptions for his journal. He enlists Véra Bouny-Tordeur's relative, A. Tordeur, to mail 5000 circulars printed by M. Weissenbruch to scientists and physicians drawn from the academic register *Athena*, then 5000 more by using another well-known register, *Minerva*.[94] The effort and the expense bring little result. The Belgian printer does not help matters. The beginning of hyperinflation, early in 1923, brings a resolution:

> A letter, received from Weissenbruch a few days ago announces that their prices must be raised. The increase upon pre-war prices was hitherto only 250%; it will now be 475%. Even more, because they introduce many extra charges. Does that increase really correspond to an increase in cost, or to a depreciation of money or do they blackmail me? I believe it is a combination of both. At any rate it shows that it is high time to part company with those printers who have always treated me ill since the war is over. Vol. 5 will be certainly the last volume printed by them, and I am not sure that I will be able to publish it entirely at least in Belgium, for if money begins to drop faster, the cost may reach almost any price (in francs).

The inflation is the result of bad political decisions:

> I am afraid that the wicked and unwise policy of France will cause France and Belgium to follow Germany on the downward path. France cannot destroy Germany without destroying herself.
>
> This is a terrible shock to me. It made me ill. But I will pursue my own course and so my appointed task.[95]

George indeed falls ill, "vexed beyond measure by the increasing cost of publishing *Isis* and the lack of support..., about the slowness of growth of my *Introduction*." He emphasizes in his diary on 9 February: "Since my childhood, never have I been handicapped by disease to the same extent."

---

93. Diary, 23 November 1922.

94. HGS, A. Tordeur to George Sarton, 14 February, 26 March, and 16 July 1922.

95. Diary, 29 January 1923.

The Carnegie Institution declines to review its $500 grant to George "for secretarial and clerical assistance."[96] On 28 January 1923 George writes to Mabel in Washington, where she has contracted tonsillitis. He spent the whole morning redrafting his appeal to Merriam. "I have not yet recovered from the last shock and I feel very melancholy. And even while I was writing at my desk I got very cold. I am taking May to luncheon in Boston and then to see *Robin Hood*." Mabel, on 27 January, writes from Washington in French about Merriam's refusal: "I believe despite everything that *Isis will survive this crisis*." George, on 29 January, reports that he didn't like *Robin Hood* as much as *The Three Musketeers*, "the most perfect film I ever saw." On 31 January, George has read the *Life* of George Henry Borrow, the chronicler of the gypsies.[97] "Strange indeed." On 2 February George thinks that the recent rebuff "will help me to pull myself together & start anew differently." He grieves that he has "mismanaged to such a degree the affairs of our household. Well, it is not yet too late to steer in another direction." He and May play dominoes after dinner and go to bed at 8. He cautions that Mabel's going to see Merriam "would do more harm than good." Even if she sees Mrs Woodward, she should not mention *Isis*.

On 3 February, George reports that Louise Inches is organizing repetitions of Yuen Ren Chao's lecture on Chinese music, given in January. In 1916, Chao was his student at Harvard, and from his undergraduate days at Cornell, Chao had been interested in Western music.[98] "May will write today. I did not ask her to write during the week, for she is obviously very tired when she comes home from school. That Dramatic Club of hers is, I believe, a nuisance." (The nuisance becomes May's first *métier*.) He continues about *Isis*:

> My mind is quietening down. After all, all that happens to me is just punishment for not minding my own business. If I had been content with pursuing my own investigations, nothing would have happened. But I began foolishly to worry about what other people did. After all if for most people, Wells & Van Loon are the acme of history—I should worry! I am preparing a new and humbler plan of life. After all, it is not too late for a new beginning, and as I wrote to Merriam, "A mighty beautiful ship has been sunk, but the captain is still alive."

Mabel's indisposition in Washington renders her unfit for travel, she writes on First February. In French: "I urgently want to return, and yet

96. HGS, George Sarton to John Charles Merriam, 4 January 1923.

97. Possibly, Herbert George Jenkins, *The Life of George Borrow* (New York, 1912).

98. Agenda, 23 January 1923. Yuen Ren Chao, *Life with Chaos: The Autobiography of a Chinese Family, 2: Yuen Ren Chao's Autobiography, First 30 Years, 1892–1921* (Ithaca, 1975), p. 87 for music at Harvard.

M[argaret] has paid for my trip!" On 2 February Mabel is anxious about George's and May's health. She urges him to move a comforter or blanket into the living room or study downstairs, and also move May's bed there. On 3 February, Mabel is vexed to be idle, and she doesn't want the Gillespies to wait on her. Then to George's work:

> My heart is deeply troubled about *Isis*. I fear from Merriam's letter that even after your appeal, nothing will be done. And yet, I do not, *can* not agree that the publication must stop. I feel indeed as if *Isis* were our second child, the child of your brain: if we had a second child, we would manage somehow to feed & clothe & educate him. We must do the same for *Isis*. I will do anything you can think of, anything I myself can devise to help you (which will not endanger our welfare in any other way). I mean, live in a smaller house, sell some of our furniture, take a little more teaching. I am sure we could manage. I would be so proud one day if you told me I had helped *Isis* to live & so she became not only the child of your brain but of my faith in you.

George writes on 5 February that Mabel should return as soon as she is able. "We have to meet many payments soon: income tax, mother (sending money to her from Belgium is now entirely out of the question, the £ is worth more than 80 francs!), dentist. But do not worry about all that now dearie. My main worry now is thy throat. We must avoid further complications."

Merriam, at the Carnegie Institution, is intransigent. He is sorry that *Isis* will fold, although maybe it will rise "from the ashes."[99] George, on 9 February 1923, reflects on Merriam's refusal to support *Isis*. "I am still very depressed. In fact, these are the darkest days of my life; or rather the longest dark period of it, and I have an unreasonable feeling that I will never get out of it." On that very day, Mabel wants to come home but thinks that she should honor her commitment to the enterprise. "I am not only doing it for Margaret but because I still hope & believe I shall get some financial return from the business, someday. I know that they intend to make things right for me, whenever they have the money. Margaret has asked my forgiveness for allowing the debt to grow as it did, recognizing that it was wrong." On 8 February 1923, Mabel relates that the Gillespies are in a liquidity crisis. They cannot pay their rent, and so they cannot pay her. George looks to Harvard for salvation. On 9 March his diary reveals that he has written to Henderson, "my best friend at Harvard," asking for a salary from Harvard, so that he can apply all of it to *Isis* and solve his financial crisis. Spots appear before his eyes from time to time, making it impossible to read.

George certainly needs and relies on Lawrence Joseph Henderson, who has been lecturing on history of science since the early years of the

---

99. HGS, John Campbell Merriam to George Sarton, 6 February 1923.

century and who needs a pliant and authoritative subordinate. Over time, George wearies of the relationship. He writes about Henderson in his diary on 13 November 1938:

> He does not know the history of science and has never made a serious effort to know it, yet he thinks that in a way he knows it better than if he had studied it! He tries to "boss" everything in the H.U., in particular the "history of science" & no progress will be made in that field in H.U. as long as he is in charge.
>
> I do not know any man, more exactly any man of science, who is more impatient of criticism, and more ready to criticize and berate anybody around him; more sensitive to any difference of opinion, and yet less thoughtful of the feelings of others. His attitude to criticism was well illustrated when a reviewer ([physiologist] Yandell Henderson of Yale?) spoke ill of one of his works. He was raving mad & tried to "get" at the reviewer in a roundabout way, just like a politician who tries to punish an adversary....
>
> Henderson's conceit & other peculiarities do not really matter but he is blighting the normal development of the history of science in H.U., and that is very sad.
>
> I must not allow this cynicism to curdle the milk of my good nature. Then he would really hurt me. As long as I can passively resist and remain cheerful and kind he will do his worst in vain. Poor old man!

Not everyone is pleased to live as a servant, which is George's position for more than 20 years at Harvard. In matters of organizing instruction, Henderson simply ignores George, who recalls that Henderson "was instrumental in obtaining" 2 younger scholars to replace him—Georgio de Santillana and Willy Hartner—"and probably took the initiative" in recruiting them.[100] A lesser man would have been driven to inactivity or, worse, to an early grave.

History of science at the universities best designed to sustain the discipline—Harvard and Oxford—is written in the lives of overbearing, well-placed biological scientists with thin scholarly accomplishments. Henderson finds a counterpart at Oxford in Robert T. Gunther, who would have used Charles Joseph Singer, beginning in 1920, just in the way that Henderson used George Sarton. The pattern extends to other universities with enormous resources for scholarship, and it, more than any other circumstance, accounts for the attenuation of George's calling. Emile Waxweiler identified the pattern when he counseled George in 1913:

> About the future, the atmosphere of the universities is such that to present a strong dossier to the authorities [pour s'affirmer], it is almost necessary to choose a field corresponding to something that is taught.

100. Diary, 5 November 1948.

> To speak concretely, I think that you would achieve your legitimate ambitions more quickly as a mathematician, a physicist, or a chemist, as someone competent in only one part of science, rather than identifying yourself with a field that is not yet well defined.[101]

The irony is that George's New Humanism calls for scientists to control the destiny of the field. Here is, perhaps, an example of the law of unanticipated consequences, formulated later by Sarton's young collaborator Robert King Merton.[102]

At the time that George is suffering from uncertainty and depression, an uplifting preface appears in volume 5 of *Isis*. Written at Intervale during the happy summer of 1922, it is titled "Knowledge and Charity." In the text George wants to show that history of science is about both head and heart, both works and the people who accomplish them. Acquiring knowledge is a moral duty, George emphasizes, and the people who do the most to advance knowledge are a small vanguard of humanity. These are the people best suited to guide us, not politicians and generals who "do not know the world into which they are leading us." Conventional humanists, he names Matthew Arnold, equate scientific knowledge with crass materialism. George thinks that the problem concerns utility: "If I study anything for the sake of utility, this may increase my usefulness and better my position; this will not enlighten me." Science certainly "enables us to extract from nature for the benefit of all men, wealth untold and treasures beyond the wildest dreams of avarice." He affirms, "it would be mean to cultivate science only because it pays to do so, but it would be incredibly foolish to believe that it can have no spiritual value because its material value is so immense." He warms to the main point: "Science is the search of truth and, insofar as our quest has been successful, it is truth itself.... the foundation of everything, the axis of our life, the substance of our reason, the key of our fate." The revealing of truth proceeds without reversal: "Whenever a particle of truth was finally established, it was established forever; the advance might be slow but it was certain. The history of science is nothing but the history of this protracted struggle between credulity and research, of this long series of partial deceptions and progressive discoveries."

Knowledge, obtained in this way, enlightens human kindness, "it reduces the appalling danger of blind and misdirected generosity." In a phrase that seems to evoke Jesus, George offers that the noble sentiment love is the intermediary of enlightenment. The proof of this contention is in the obligation of truth seekers to give it, "liberally, to all who may

101. HGS, Emile Waxweiler to George Sarton, 17 December 1913.

102. Robert K. Merton, *The Sociology of Science: An Episodic Memoir* (Carbondale, Ill., 1979), p. 110; Arnold Thackray, "The Pre-History of an Academic Discipline: The Study of the History of Science in the United States, 1891–1941," *Minerva, 18* (1980), 448–73, p. 469 for the dependence of history of science on academic scientists.

want it." Kindness, then, as Beethoven remarked, is the surest sign of nobility, and the visible sign of kindness is charity. George's long prolegomena leads to a focus on the human circumstances of scientific discovery—including struggle, egoism, and perfidy. Revealing that side of the "Human Comedy" provides scientists with "a proper sense of intellectual values." And in pursuit of that goal, historians should strive to be brief, identifying themselves with "one great work or one single ideal." All these points relate to the New Humanism, which will complete against the Old Humanism. George concludes:

> The historian of science has in him the stuff to make a complete humanist, but he will fail altogether to be one, unless his knowledge be tempered by charity and seasoned with tolerance and humor, unless he be prepared to consider every scientific achievement almost in the same spirit as a work of art, unless he have trained himself to take the whole of life into account.[103]

Truly, *tout est dans tout*. George's logic breaks at a number of points. Alfred Hoernlé, for example, criticizes George for holding that truth is "co-extensive with, and limited to, science and the use of scientific methods," a view that does not take account of the limitations of scientific methods.[104] Heinrich Wieleitner, an authority in Augsburg on the history of exact sciences, admires George's idealism but does not share it. Wieleitner is sad that Europe will be overtaken "in frightful struggles" by America. He adds: "the worst is yet to come."[105]

Art, apparent in "Knowledge and Charity," runs like a red thread through George's life. His friend Emile Masson emphasizes that in his youth (the men have corresponded since 1905) George was captivated by Jan and Hubert van Eyck's triptych, "The Adoration of the Mystic Lamb," displayed in the cathedral of St Baafs in Ghent, and he especially appreciated Jan van Eyck's "Madonna of the Inn's Hall," showing the Virgin reading a book, which he took as an allegory of science.[106] The promoter of a discipline of history of science saw art history as the reverse of his medal. Mabel the artist was his complement. The struggle of artists offers him secular saints. He hears Feodor Chaliapin on 15 April 1923 and remarks in his diary that the singer "worked as an apprentice to a shoemaker in the same street where Maxim Gorky was toiling in an underground baker's shop."

---

103. George Sarton, "Knowledge and Charity," *Isis, 5* (1923), 5–19.

104. HGS, Alfred Hoernlé to George Sarton, 12 June 1923.

105. HGS, Heinrich Wieleitner to George Sarton, 22 June 1923, upon receipt of *Isis, 5,* no. 1.

106. Emile Masson, "George Sarton et le 'nouvel humanisme,'" *Les Humbles*, no. 10 (February 1919), 7–19, on p. 19. Masson refers to the "Madonna from the Inn's Hall," 1433, presently in the National Gallery, Melbourne, Australia. The periodical is a "revue de Jeunes." HGS, Emile Masson to George Sarton, 10 January 1919.

George selected his authorities for art carefully. He cites William De Morgan's "Result of An Experiment" (published anonymously in 1909) in his diary on 1 March 1923:

> It is the best thing on Earth—that incessant struggle....Art is more important than you think. But it must be earnest, grim life-earnestness that has no tincture of gain in it or love of earth-fame, only the strength of one's arm, & the whole power of one's being is to be given to it; and to look neither to the right nor to the left, but go straight on doing the best that is in one.

The source of the quotation is revealing. William De Morgan and his artist wife Evelyn, inspired by the Pre-Raphaelites, undertook experiments to collect automatic writing, allegedly the communications of angels and spirits. George flirted with spiritism in his youth, and he comes back to it through Winifred and Alfred Hoernlé, Daniel Comstock, and William McDougall at Harvard.[107] In middle age, George is still a youthful bohemian in temperament, spending more than he has on *Isis*—his writer's canvas. Did he sense how much he resembled Puccini's Rodolfo? The Sartons never comment on *La Bohème*, although W. Somerset Maugham's portrayal of Paul Gauguin moves them both deeply.

George publishes the first results of his study of Asian art in the Parisian journal, *Gazette des beaux-arts*. It and succeeding communications are intended to prepare the ground for a complete and synthetic history, he begins in French. In his view, the unity of Asian art derives from Buddhist inspiration, and as an authority he cites Arthur Henry Fox Strangways's publication on the Ajanta frescoes in India. He begins with a Chinese fresco from the 6th century acquired in 1918 by the Metropolitan Museum of Art, which is in the Ajanta tradition. He then considers the northern T'ang school of art, also in China, from

107. The intellectual milieu familiar to George Sarton in Ghent before 1914 radiated spiritism, temperance, vegetarianism, and socialism, a congeries found generally in Western Europe. Christophe Verbruggen, *Een Sociale Geschiedenis van het schrijverschap tijdens de Belgische belle époque* (diss., University of Ghent, 2006); Jan Romein, *The Watershed of Two Eras: Europe in 1900*, trans. Arnold Pomerans (Middletown, Conn., 1978), pp. 494–500. At the time that George meets the Hoernlés, they have just translated a work by sometime Theosophist and champion of the occult Rudolf Steiner, *The Philosophy of Freedom* (London, 1916). George talks with Daniel Frost Comstock about ways to investigate psychical phenomena. Diary, 6 July 1920. Comstock was a physicist at Massachusetts Institute of Technology until 1917, when he devoted all his effort to an independent laboratory that in 1922 developed "Technicolor." Stanley Goldberg, *Understanding Relativity: Origin and Impact of a Scientific Revolution* (Boston, 1984), p. 250. George comments on spiritism as an object of scientific study in his review of E. E. Fournier d'Albe's life of William Crookes, in *Isis, 6* (1924), 549–53. Spiritism also attracted Marie and Pierre Curie. Susan Quinn, *Marie Curie: A Life* (New York, 1995), pp. 206–9.

secondary sources, as well as paintings attributed to Li Chao-Tao, one acquired by the Boston Museum of Fine Arts in 1920 and the other acquired by the Metropolitan Museum of Art in 1918. He observes that the perspective emerged in Europe some 8 centuries after it is manifest in China. Finally, he considers the 9th-century Japanese "Hokké Mandala," a representation of Buddhist paradise acquired by the Boston Museum in 1911. His publication is in the nature of notes on recent American acquisitions of Asian masterpieces. It signals American authority in the field of Asian art, and it announces George as a privileged commentator on the topic.[108] George elaborates his view of art several years later in the *Yale Review*. There is nothing new under the sun, he offers. "Genuine originality, as distinguished from snobbish eccentricity, brings one back invariably to tradition." The proposition holds for 20th-century art as well as for art of previous times, and George is strongly attracted to Chinese art because of its appeal to tradition. He emphasizes that a principal task for art historians is to understand East Asian painting. In so doing, it is important to understand the nature of interpretive copying. The end of such knowledge will be an affirmation of "the unity of Asiatic art, and finally of the essential harmony of the human mind." The history of science is the best path to understanding Western civilization, and "the history of art is the master key to the heart of Asia."[109]

George realizes that he must become an American citizen before he can sail to Europe, because as a resident alien he would have to pay much more in taxes before going abroad, so Mabel and his plans for the coming year have foundered. Mabel still wants to go to Europe by herself for several months; George does not know how to afford it. George is "still very much alone here, working in the wilderness." Visitors buck him up: Johannes Eduard Friedrich Strohl, zoologist from Zurich who was an early subscriber to *Isis*; town planner Patrick Geddes, "one of the most extraordinary fellows I have ever seen." He takes most pleasure from meeting the director of the geological survey of Alsace and Lorraine, Emmanuel de Margerie, who had been a member of the section of historical and descriptive geography in the Ministry of Public Works in Paris: He "is very genuinely interested in the history of science and he is a true humanist."[110]

108. George Sarton, "Matériaux pour l'histoire de l'art asiatique (première série)," *Gazette des beaux-arts, 8* (1923), 1–17.

109. George Sarton, "Art as an Approach to Asia," *Yale Review, 15* (1926), 540–52, on pp. 540, 551–2.

110. Diary, 30 May 1923. Geddes may also have appealed to George because of his writings on sex. Lesley A. Hall, "The English Have Hot-Water Bottles," in *Sexual Knowledge, Sexual Science: The History of Attitudes to Sexuality*, ed. Roy Porter and Mikuláš Teich (Cambridge, 1994), pp. 350–66, on p. 359. George reviews de Margerie's survey of world cartography in 1914, a review appearing in 1919. George Sarton, "Emmanuel de Margerie, *La Carte géologique du monde*," *Isis, 2* (1919), 464.

In a long article, George reviews de Margerie's recent edition of a translation of Eduard Suess's magisterial history of geology, titled *La Face de la Terre*. George divides the history of geology into 3 periods: the heroic age of Abraham Gottlob Werner and James Hutton, extending up to 1820; the age of the first syntheses of Charles Lyell and Alexander von Humboldt, up to about 1875; and the age of Suess and his school. George addresses theory, but he sees the concrete evolution of geology in the elaboration of geological maps—the prospector's synthetic bibliography. Suess is the great synthesizer, who, in his notion of receding oceans, folding landmasses, and an ancient equatorial continent called Gondwana, culminated the classical period of geology. George concludes that Suess, an adopted Austrian, died before he could see the world, in which he was an international citizen, destroyed by war.[111] George signals the work of a classical geologist as an example of fruitful synthesis at just the time that the notion of continental drift, introduced by Alfred Wegener in 1912 and elaborated over the succeeding quarter century, explicitly challenged many of Suess's conclusions. George's admiration of Suess does not preclude his interest in Wegener. He writes in 1923: "Wegener's theories are so revolutionary that the strong opposition to them, at least among geologists, is not surprising. They can not yet be considered proved. But they are very interesting, very plausible, and deserve to be carefully studied."[112]

On 29 June 1923, George writes to Mabel, who is in Washington before she sails to Europe for 2 months alone, from New York:[113]

> I never thought that you did not work enough. I wish, on the contrary, that you did not have to work so much & chiefly that you did not have to do any commercial work, for which you are clearly unfitted. I was chagrined by your having to go to Washington before sailing, and the more so because I could not help it nor help you.

He is reading Knut Hamsun's *Growth of the Soil*, "a great book, one of the greatest I have read for a very long time." Then to the cats:

> The three cats give me considerable trouble or rather worry. The present combination is utterly unfair to them, not to speak of me. They are on my mind all the time, from 6:30 a.m. when I finally get up & let the mother out (after a struggle of half an hour or so between my tireness & my wish to help them) until midnight. The affectionateness of the

111. George Sarton, "La Synthèse géologique de 1775 à 1918," *Isis*, 2 (1919), 357–94. The work in German: Eduard Suess, *Antlitz der Erde* (Prague, 1883–1909).

112. George Sarton, "Alfred Wegener, *Die Entstehung der Kontinente und Ozeane*," *Isis*, 5 (1923), 289.

113. Agenda, 26 June 1923.

> big cat is pitiful. She needs petting & eating....Not that she is underfed but she was used to many meals. As to the kittens, they need playmates. They want mom & more play, & I have so little time to give them. They deserved a better fate than to be imprisoned in a big apartment without friends. It seemed this morning that their mother was already fighting them off. What then? The kitten will need food many times a day; they will need some nursing; may-be they will need some protection against their wild mother.
>
> That is bad. And the consciousness of all the worryness, the constant worryness it entails, are the main cause of my inability to work. I will never get out of it.

And the next day, in his farewell to Mabel: "The cats seem to have come to their senses & to have taken a cooler view of the situation. At least when I come home, they spare me those reproachful glances of their funny little eyes and that 'where-have-you-been-so-long?' attitude which got terribly on my nerves."

George reports on 8 July that he and May have "spent a delightful week together," and she is off to Duxbury, to stay by the sea with the family of Louis Joseph Alexandre Mercier, professor of French at Harvard.[114] In the evening he is reading Tolstoy's *War and Peace*, in French. On 15 July, George is weekending with May at the Merciers'. He and May celebrated Bastille Day at the home of Samuel M. Waxman, professor of French and Spanish at Boston University, with whom he shares an interest in medieval Spain. George will spend 3 days "preparing the tables of my notes devoted to the period extending from the ix. cent. B.C. to the xiiith, more than two thousand years." From Cambridge on 18 July he sends early birthday greetings to Mabel in Ipswich:

> As I was reviewing in my mind our wedded years, it occurred to me that I had not managed so badly as far as housing and feeding was concerned, but that I had not done as much as I should in order to clothe you in a seemly manner. Therefore I am sending you a cheque for five hundred francs for you to spend in Belgium to buy some lingerie. It will not carry you very far, alas! I had much liefer given you the things, than the money, but this would not have been practical. And I am always practical, first and last.

And on the cats: "Happily I was able to give one to Margaret Stearns, to be brought up in the country. The mother cat went hunting for her kitten all around the house. *You will never again get me to do all these things.* It is heartbreaking." But George is a heartbreaker. He records sex in his agenda on 17 and 24 July.

---

114. Agenda, 7 July 1923.

Mabel writes on board the "Suffren" on 10 July, sad at leaving George and May for 2 months. On 14 July 1923 in Ipswich, she has been thinking about George. In French: "I have a strange memory, I know. Yet it holds precise details (sounds, colors, smells, everything) sometimes allowing me to relive an instant or an hour of the past with an unparalleled intensity and a precision." George fetches May from Duxbury on 29 July.[115] George writes on 31 July that he is "working tolerably well, in a leisurely way, without enthusiasm or special pleasure. I kill time working as others do, playing bridge or smoking opium." On the cats, he has accepted caring for them "as a part of the general scheme," and he regrets that he wrote so much about them to Mabel's mother. He keeps the cats outside most of the time. The danger for the kitten outweighs its delight in escaping "a second floor city flat." He has shipped May off to Rockport where she will stay with Elisabeth Coit and Edith King until the middle of August, there to take part in their production of the 12th-century romance, *Aucassin and Nicolete.*[116] The kitten "is quite charming and certainly deserved a better fate. His favorite sport is to climb on me as if I were a mountain. This he will do ten or twenty times over, indefinitely. I am generally the first to get tired of it. But I dare not show any impatience, for his mother is admirably patient even under the greatest provocation."

Mabel writes from Ipswich on 18 July that she has received an unpleasant letter from Raymond Limbosch. "I am a bit afraid of seeing him again." Céline, on the other hand, has become "less categorical in her judgments." She reminds George about packing for May, having her hair cut at Filene's or Jordan Marsh, and a visit to the dentist. May has braces. "I am writing to him on this subject, explaining that we cannot continue, for the most essential part is done. But I believe that he should look at the device still on the lower teeth and perhaps place something above that he will adjust only one time in three months." On 19 July, Mabel writes that she has found a lodging for herself and her mother in the countryside near Ipswich, at Trimley St Martin's. On 24 July:

> This is a most peaceful spot & mother loves it. We are the only strangers in the house & are treated as friends indeed. There are only country sounds around us. The air is lovely. (We are only 3½ miles from the sea.) Yesterday was stormy & sea-gulls came scudding along the grey sky. I helped pick currants for the market yesterday afternoon before the storm. May would love it: There are all sorts of baby animals—pigs, ducks, chickens, kittens, etc.

115. Agenda, 29 July 1923.

116. Identified by Margaret Peters as a summer camp. Peters, *May Sarton: A Biography* (New York, 1997), p. 30. The dramatic production in Diary, 13 January 1924. From the camp, May goes to stay with the Arkins in Plymouth.

Mabel receives George's long disquisitions on the cats and is hurt by what he writes. She knows that something is not right. She is "swept by a storm of resentment & pain." George's

> care for the happiness of the cats has entirely outweighed your care of mine. You have persistently robbed me of the little peace of mind I had assembled, *& I did need it.* It was hard enough to come against your wishes & to have to accept the natural consequences but you have by your lack of intelligence & energy, added quite unnecessary & unnatural consequences & have not spared me in a single letter, a jot or tithe of them. Oh! I know you will say it is because you suffered *so much* for the poor cats. I understand & imagine it only too vividly & was at first awfully sorry for you & them & bitterly reproaching myself. I am *still sorry* for you & them, *very*, but I am also for the first time in my life, a little ashamed of you & feel that you are so at the mercy of circumstances it is time you learned to re-act & *find a way out* & if not, then, "to grin & bear it." You say I shall never get you to do these heartbreaking things again. No, *you will leave them to me to do.* (Did I ask you to drown the kittens? No, I did it myself & said as little as possible about it.) Did I ask you to help train our Mother pussie to be mean? You would not even empty the earth-pan once for her! Yet you say you enjoy her so much, & agree it is right May child have a pet-baby.

It is a good thing, in Mabel's view, for George to experience the cats by himself. She also resents his suggestion that he is alone and miserable while she is enjoying herself in Europe. Part of the problem is George's choice of words.

> Do you not sometimes feel surprised at the strange trivial words that come from your lips, when down in your heart you are feeling something quite different? Perhaps even as you write about the cats you felt it was not fair? Perhaps even you were a little ashamed that you had found no way out of the difficulties? Perhaps your resentment to me was mixed with some gratitude for taking off your shoulders these sad or disagreeable duties when I am with you?

Then Mabel offers practical solutions to George's feline woes.

Mabel continues her criticism of George, notably his phrase in a letter to her mother that he is "not a good provider." Mabel is trying to dissuade her mother of the notion. First Wondelgem, which she chose with George. "You could not have provided me with a happier little corner of the earth for May to be born in (& later *Isis*)." Then the war. "You went to America & most bravely sought to earn our livings by doing *your own work*. I could not love you as I do, if my love were not mixed with admiration for it & faith in your work." She continues: "I wonder if you realize that in some ways I am deeply grateful to Fate for *not* making you *too good* a provider....If you were, I would have less excuse, less incentive to do work of my own, & I believe I need the incentive. At least it would be an immense & joyful sense of pride to me could I someday really become a *co-provider*!" Her only reproach is that George tended to treat her as "the civilian lurking in safety, while you fought my battles!" Mabel is "deeply incomprehensibly happy & grateful that you have never seemed to doubt, till now, that I infinitely preferred the thing you gave me to riches bought at the expense of your honour." George is a true scholar, displaced from his home, "having to make a new start at a time that scholarship was a drag on the market." Until now Mabel has left the planning of their travels and their budget to George, but she wants to share in these tasks. "We will plan *together*." She senses that the summer budget was unrealistic. "I also realize the tremendous strain you have been under because of *Isis*. Your mind was saturated with that anxiety & perhaps never will we pass such a year again." But Mabel is just as unrealistic about finances as George is. On 2 August she asks George to send $50. She needs it to land once more in England, on her way home from Belgium, but she will not spend it. "I had wanted not to ask you for any money at all, even as a loan. If by chance you are short, Mrs Inches will lend it to me & you must ask her. I promised to ask her if necessary, yet I prefer to come to you for I already owe her some for this journey."

George reports on 5 August that the mother cat is weaning Tut. "My main principle in the education of cats is not to impose my personality upon theirs. I realize that their personality is as precious and important to them as mine is to me. *That is sacred.*" The principle allows the mother cat to devour a pigeon in front of the house. He sends a check for 2083 Belgian francs, the equivalent of $100. Mabel might prefer $70, with the remaining $30 deposited in her bank account. If the franc falls, the money in Europe might not be enough for the return voyage; if the franc rises, Mabel would gain. With the Dawes Plan in force and the withdrawal of the Belgians from the Ruhr, George thinks that the franc will rise. Given the discussion of cats and cash, does Mabel suspect that George is unfaithful? George records sex in his agenda on 7, 8, 14, and 21 August.

Mabel, on 16 August 1923, reports a strange vision in London. "I saw all those men's faces & saw hundreds of other imaginary ones behind them, realized how wonderful it was that I had found you from among them all: It is a miracle indeed, for I cannot imagine another man I could have loved so well, who would have appealed to me in so many different ways, & *who could have loved me*." Mabel knows her faults. She believes that "there is a sort of 'frame' to one's character, or soul which is unchangeable, but within those boundaries one can change tremendously & the fault can be so controlled that it becomes almost, if not quite, a virtue." On 24 August, Mabel reports that Raymond Limbosch thinks that his writings are mediocre. His family keeps him from killing himself. She saw Madeleine van Thorenburg at Assels. Madeleine carefully cared for Mabel's plants, some of them brought by Mabel from England 20 years previously.

George spends the first weeks of September 1923 with May at the Merriman house, in Intervale, working every morning. He completes his second essay on New Humanism, which will appear in the first number of volume 6 of *Isis*.[117] At Intervale, he meets Simon Flexner and his wife, whom D. E. Smith warned George not to approach with his scheme for a history-of-science institute and who counseled J. Franklin Jameson about *Isis*.[118] The Sarton family reunites in Cambridge on 15 September.[119] George, on 23 September 1923, sends Raymond Limbosch comments on Raymond's new collection of verse, *Vers et versets*. George is upset with the text, which seems to be an attack against the unity and beauty of nature, as revealed by science and art; George's New Humanism depends on that unity and beauty. George continues in a second letter of 4 November: "To damn the whole because of the details shocks me greatly, because the harmonies revealed by science and art are *infinitely* deeper than the small dissonances that disturb our tiny personalities." George affirms: "I believe in progress. My studies prove its existence in the past; I postulate its existence in the future. This belief implies the existence of imperfection." Struggle against the imperfection is what leads to "all the great intellectual and social reforms." To rebel against the entire world, however, makes no sense to George. "Sterile resentment" is the only outcome of such a childish attitude.

On 6 January 1924 Mabel describes her penury to Eliza Van Loon. "We both try so hard to work off old debts & meet current expenses, but it is up-hill work."[120] *Isis* is the heaviest financial burden, because of the failure of an appeal for subscribers.[121] Then 20 years of experience as a mobilizer of kindred spirits produce a useful solution. For nearly a decade, George had imagined an institute with a society attached to it; in the words of his supporter Arnold C. Klebs:

> I should like to see an Institute such as you plan it established first and then only the society. The opposite plan is apt to bring about a dissipation of forces. It has been my experience that societies are often joined

117. Diary, 13 January 1924.

118. Agenda, 3 September 1923. HGS, David Eugene Smith to George Sarton, 27 November 1916; CIW, J. Franklin Jameson to John Campbell Merriam, 18 October 1922.

119. Agenda.

120. NYPL.

121. George Sarton, "The Publication of *Isis* (An Urgent Appeal to American Scholars)," *Isis, 6 (1924)*, 35–9.

> by men who have not the slightest intention to work & only seek to acquire the flavor of a scholar. They often prove troublesome customers to deal with.[122]

Late in 1923, George settles on an organization dedicated to supporting *Isis* under his editorial eye. His notion is to enlist 250 people in the United States for the organization, offering them the journal in return for an annual fee of $4. D. E. Smith thinks that the scheme might work, especially if it is connected with Section L (history of science) of the American Association for the Advancement of Science, whose secretary is George's former student Frederick E. Brasch.[123] Early in December Smith, citing the agreement of his colleague Frederick Barry, writes to a small circle of colleagues, asking if they will sign a circular letter to create a history of science society. [124]

On the morning of 11 January 1924, George telegraphs D. E. Smith in care of Ginn & Co., George Arthur Plimpton's publishing house. George notes privately: "Smith: Kindness itself." George sees Smith as the agent of success: "John C. Merriam asked him to go to Washington & he did not hesitate to do so, fearing all the way that I had got into trouble. A true friend."[125] The Boston members of the History of Science Society, plus George and Smith, meet at the American Academy of Arts and Sciences, Harvard's pocket parnassus, on 12 January. "This may be considered the beginning of the new society."[126]

The society organizes a governing council, admits members, and disburses funds. It establishes a pattern of annual meetings alternately with the American Association for the Advancement of Science and the American Historical Association, a pattern continuing into the last third of the 20th century.[127] By late winter, the society counts 268 members.[128]

---

122. HGS, Arnold C. Klebs to George Sarton, 27 August 1916.

123. HGS, David Eugene Smith to George Sarton, 12 November 1923. Smith is clear that George is the architect of the notion.

124. HGS, David Eugene Smith to George Sarton [circular letter], 4 December 1923. George Sarton, "Preface to Volume VI," *Isis, 6* (1924), 4–8, for the chronology.

125. Diary, 13 January 1924.

126. Agenda.

127. HGS, Circular letter to Members of the Council, 14 March 1924, on the stationery of David Eugene Smith. The Council: Isaiah Bowman, Ernest William Brown, Henry Crew, David Starr Jordan, George Ellery Hale, Berthold Laufer, Duncan Black Macdonald, J. Playfair McMurrich, John C. Merriam, George F. Moore, Edgar Fahs Smith, Michael Rostovtzeff, Lynn Thorndike, William H. Welch, F. J. E. Woodbridge, in addition to President Lawrence J. Henderson, Vice Presidents James H. Breasted and Florian Cajori, Secretary David Eugene Smith, Treasurer Frederick E. Brasch, and George Sarton as editor of *Isis*.

128. Agenda.

George makes sure that the society enters the constellation of American learned life. It is an early member of the American Council of Learned Societies, for example. But the society remains a cipher, existing largely to guarantee the publication of *Isis*, until the 1970s, when it acquires an endowment and becomes a prize to be contested by people with an administrative turn of mind.

Raymond Clare Archibald, a member of the governing council as well as editor of the *American Mathematical Monthly* at Brown University, is driven to distraction by George's administration of his journal. He writes in 1925: "We think so widely differently on budget matters." Archibald, a Canadian who 15 years earlier had published a very large book on a very small Canadian theme in Thomas Carlyle's life, urges George to limit the number of pages in each issue—the principal cause of his perennial deficit. If *Isis* is to be the property of the society, then each volume must be limited to 350 pages; if it is not the society's property, then legal documents must define the obligations of the society in the event of George's disappearance from the scene.[129] Archibald's perception of George's strong editorial hand appeared nearly a decade earlier to Arnold C. Klebs, who observes to George about a programmatic text he wrote: "The one drawback it seems to me to have is that it bears so very much the imprint of your own profound scholarship and with that the improbability of permanency."[130] *Isis* does achieve a permanent place in the annals of scholarship, but it remains George's child, and its first half century cannot be understood without understanding him.

With the sixth volume of *Isis*, in 1924, the History of Science Society publishes the journal. George reprints the first circular of the society. It begins by recounting George's labors: "The first five volumes [of Isis] in which some ninety scholars collaborated, contain 3,116 pages, 70 papers, 527 reviews, and 9,196 bibliographic notes." George adds that the first number of the new volume was composed "as a propaganda number" before the society was formed, and it contains as an article only George's appeal for the New Humanism.[131]

In his programmatic appeal, George begins with his principles: Human progress depends on advancing knowledge; the progress of each discipline relates to other disciplines; the progress of science is a social phenomenon (proven by numerous simultaneous scientific discoveries) expressing the unity of humanity; the purpose of humanity is the production of "immaterial values, such as beauty, justice, truth"; and the people who guide

---

129. HGS, Raymond C. Archibald to George Sarton, 15 January 1925. Raymond Clare Archibald, *Carlyle's First Love: Margaret Gordon Lady Bannerman, An Account of Her Life, Ancestry and Homes, Her Family and Friends* (London, 1910).

130. HGS, Arnold C. Klebs to George Sarton, 19 April 1916.

131. George Sarton, "Preface to Volume VI," *Isis, 6* (1924), 4–8.

human progress do so in obscurity, creating their works in secret. George sings about his own destiny:

> Who cares to know the great business men and the financiers of Greece or Rome or of the Renaissance? Their very names are forgotten. The very few of them who escaped oblivion, did so only because they patronized the disinterested activity of scholars, artists and scientists,—men whom they probably treated as pets or servants. Yet in spite of the high regard which mankind has for those who minister successfully to its material needs, as soon as they are dead and its judgment is no longer influenced by these needs, it throws them back into the background and calls their servants—artists and scientists—into the center of the stage.

Listen, capitalists, if you wish to get into heaven, be generous in supporting good works: "The few men who enrich its [humanity's] spiritual life are its true representatives in the light of eternity." George believes that the progress of learning depends on synthesizing scientific knowledge, and that all the world's peoples have contributed to the synthesis, as scientists recognize. The history of science, as the measure of progress, when taken along with the history of art and the history of religion would explain the growth of human intellect. Here are the things about the past that really matter.

George's history of science is the path to the New Humanism: "It undertakes to bring together for the first time, scientists, historians, philosophers, sociologists; to coördinate and harmonize their points of view; to make the accomplishment of their higher task easier in spite of the increasing wealth of knowledge." The New Humanism unites knowledge with idealism. It does not revere the past by seeking oblivion in it, as advocated by Ruskin, Tolstoy, and Rabindranath Tagore; it looks to a new and unique future: "Whatever salvation there be, is forwards not backwards."[132] George's sermon, encouraged by Robert Millikan's religiosity, is unmistakable: The first reference in his article is the Gospel According to St. John. Helen Bigelow Merriman sees the point immediately. She offers that until the birth of a race of unselfish people, humanity needs consolation and inspiration "that appeals to the many minds incapable of being thrilled by the thought of a distant perfected humanity—Do we not need Religion in short?"[133]

If *Isis* as a house-organ opens without a real article, it contains more of George's long reviews. George's earliest publications are in this genre, and his longer writings often seem to be a collection of shorter reviews. George uses reviews to promote his vision of the discipline. The practice is unseemly for an editor, who is better served by paladins. In his reviews,

132. George Sarton, "The New Humanism," *Isis, 6* (1924), 9–34, quotations on pp. 15, 32.

133. HGS, Helen Bigelow Merriman to George Sarton, 30 May 1924.

George, who sees himself as a pariah in the world of letters, borrows on occasion from Jeremiah, Cassandra, and Savonarola. Early in 1924 he uses a review to assert his authority.

The objects of George's scrutiny are the first volumes of Lynn Thorndike's history of magic and experimental sciences. Thorndike, a professor of history at Western Reserve University, worked on the treatise for 20 years. Thorndike's study, George emphasizes, treats experimentation only incidentally, without discussing experimental details, and especially without phrasing the experiments as part of the accumulated wisdom of humankind. Judgment is severe: "To call such a book a 'history of experimental science' is a joke, and a very bad one at that." Readers without complementary information will "be unable to realize how completely they are deceived. This imposing book will thus become an abundant source of misunderstanding, and will work considerable harm." Part of the problem concerns words. *Experimenta* in medieval times means remedies or recipes. "Thorndike, investigating medieval ideas on magic, found the word *experimenta* so often repeated that he was imperceptibly beguiled to believe that he was studying experimental science as well and that the developments of magic and science were closely interrelated." Science aims at truth, while magic aims at power irrespective of the truth, or at nothing at all. George finds no evidence to support the view that magic slowly transforms itself into science, as Thorndike claims; the contrary is more reasonable, where science and magic are in continual conflict. The history of science is the record of progress toward the truth, "permanent conquests over chaos." Very little of this record is contained in Thorndike's book. Cultivation of science requires dedication, in the manner of weeding a flower garden, but Thorndike focuses on the weeds: "It is their story that he tells us; he would have us believe that they helped the flowers to grow! Nonsense!" Magic influenced science in medieval times no more than magic influences science in the 20th century. "The ancestors of the scientists of to-day were not the conceited and foolish magicians, but the craftsmen trying honestly to solve concrete and definite problems."[134] George seeks to exorcize the demon of the occult, which attracted him during his first years at Harvard.

Early in 1924 George solves one of his problems. He becomes a United States citizen on 11 February. His witnesses are the Hendersons. May follows as a US citizen ipso facto, but Mabel is naturalized later, on 14 July.[135] George wanders from her. On 24 January, the Sartons dine at Louise Inches's house. Louise dines with the Sartons on 24 February. On

---

134. George Sarton, "Lynn Thorndike, *A History of Magic and Experimental Science during the First Thirteen Centuries of Our Era*," *Isis*, 6 (1924), 74–89, on pp. 79, 80, 83, 84, 87.

135. Diary, 11 February 1924. Mabel's naturalization noted in her passport issued 11 May 1929, in YMS.

the 28th, George attends a concert with her. On the 29th, he records sex. Mabel travels to Washington on 11 March, returning on the 22nd. In her absence, George takes May to a concert, has dinner with Lucy Stanton, and then has sex. He dines with Stanton on 6 and 8 April. While Mabel is indisposed on 20 April, he goes to the symphony with May and Agnes Hocking. Stanton calls on 27 April with a student from Georgia. On 3 May, George attends the symphony with Louise Inches; the next day all three Sartons lunch with Inches.

George closes his apartment at 24 Agassiz Street at the end of June 1924. He is moving into Divinity Hall while Mabel and May spend the coming year in Europe. The family goes to Ogunquit.[136] George returns to Cambridge alone on 2 July. On 4, 5, and 6 July he records sex. On 9–14 July Mabel stops by Cambridge on the way to Washington.[137] George takes May to Intervale. The family reunites in Ogunquit for the month of August.[138]

The division of the family for 9 months is ostensibly due to financial considerations. Divinity Hall will not cost George much, and Mabel and May can be house guests, of sorts, with the Limbosches. In this way, the Sartons will be able to pay off their debts. But George could have lived as cheaply in Belgium as at Harvard—it would not have been difficult to justify the absence to his employer in Washington. In fact the situation is more complex than either Mabel or George acknowledge. George is wary of his old friends the Limbosches, and he has not achieved any of the goals that he announced to his European colleagues 5 years previously. Mabel wants to resume her *métier* with Céline. The evidence suggests that the Sarton family is undergoing a trial separation. It will be the "hard winter in Belgium" that May carries with her for the rest of her life.[139]

136. Diary, 24 June 1924.

137. Agenda.

138. Agenda, 19 July, 2 August, and 31 August 1924

139. May Sarton, *A World of Light: Portraits and Celebrations* (New York, 1976), p. 19.

CHAPTER 15

Two Solitudes

The Sartons knew solitude from the time of their youth. George grew up without much parental guidance, and Mabel endured long separations from her mother and father, who lived overseas. During their courtship and the early years of their marriage in Belgium, they often went separate ways. The pattern continued into the war years, when George worked in London and then lectured across half of North America while Mabel lived with relatives in England and traveled on business between Boston and Washington. Notwithstanding the uncertainty of their times and the distinctive nature of their personalities, the Sartons in their separations herald a new pattern of life for 20th-century families where each spouse is an intellectual.

Traveling to Europe with May, Mabel writes to George from the ship on 13 September 1924. She is touched that Katharine Taylor, the director of Shady Hill School, sent $10 and that she and other teachers provided a box of candies. Mabel, who borrowed money for the trip, is in a sombre mood. "I believe *tipping* has something to do with it. I hate it so."[1] On 25 September she arrives in Ipswich. On the way she and May visit the British Empire Exhibition at Wembly, where Phoebe Stabler planned some of the gardens. The section on Burma is the most beautiful. Mabel enjoyed the Indian paintings and the African buildings. As a birthday present, she buys George a small leather-covered box for his desk.

George is ensconced in his room in Divinity Hall, which "Mabel took immense pains to have...put in order." He is reading Corra Harris's *My Book and Heart* (Boston, 1924) "with great pleasure."[2] The book of aphorisms by the wife of a Georgia Methodist circuit-rider nurtures George's religious sentiments. We may speculate if it was suggested to him by Lucy Stanton, his favorite Georgian in Boston. At the end of September and early in October George dines at Stanton's home. At the time of his socializing with Stanton, he records sex.[3]

1. An early meeting of Reiner Leven discussed "an end to tipping." Agenda, 1 December 1906.

2. Diary, 21 September 1924. Helen Bigelow Merriman disagrees, offering that some of the book "is marred by a poor style, and the use of words that may be common in the Southern vocabulary." HGS, Helen Bigelow Merriman to George Sarton, 27 October 1924.

3. Agenda, 23 and 28 September 1924.

On 4 October Mabel replies to a letter of George's which we do not have. She observes that they respond to shocks in distinct ways. George gets over them after a few hours, but Mabel takes a long time to react. Their separation of 9 months will be justified only if they can pay their debts, May learns French, and Mabel is able to ease the life of her mother. Céline Limbosch will hire her at F 400 per month to provide designs at Mabel's own convenience. Among other projects, Mabel wants to visit all the existing *béguinages* and photograph them, perhaps write an article on them. She is reading Somerset Maugham's *Of Human Bondage* (New York, 1915). On 12 October, "Both May & I miss you dreadfully. We seem *to have lost our anchor*." Again: "I think you would be a little astonished if you knew how she misses you." Mabel is homesick, but not in the way she had expected: "I *know now* that I have become an American in heart as well as name. I mean America is my home, where I *want to live*."

Raymond and Céline Limbosch have a new friend, Marcel Letendard, a character reader, who analyzes May and announces his findings.

> I am *very sorry* we had this examination because May has understood parts, chiefly the faults, & is dreadfully depressed and nervous. She heard him say she had a "little tiny brain" & that it worked too fast etc. etc. If I had had any suspicion he would say such things & that May would hear & understand I would not have allowed it as she is in a particularly impressionable mood & needs encouragement & faith in herself, peace of mind. I cannot say enough how bad I think it has been for her. Of course we are all working now to undo the evil, but it is not easy.

Letendard detects signs of overwork in May. Mabel also notices them and as a result stops May's music lessons. On 21 October, May misses "the happy, peaceful times of her school life & our own home-life." Her neighborhood Belgian school is "mechanical & *extremely dull*," so Mabel looks elsewhere. One possibility is the innovatev Ecole Decroly, run by a woman who visited America and knows Shady Hill. On 25 October, Mabel has settled on Institut belge de culture française in Uccle.[4] It is

> a charming house that breathes the very spirit of quiet happiness & work & simplicity. I had hardly been there ten minutes that I felt it was quite, quite different to the other schools I had seen. Mlle [Marie] Gaspar with whom I talked is not tall, is middle-aged, dressed in insignificant black like Mme Curie, with nice brown eyes, rather high cheek bones, something kind, frank, intelligent & earnest in her face. The

4. *Phoenix*, pp. 120–35, for May Sarton's reminiscences of the school. On the organizers of the school, Marie Closset and Marie Gaspar: Pierre van den Dungen, "Parcours singuliers de femmes en lettres: Marie Closset, Blanche Rousseau et Marie Gaspar," *Sextant*, nos 13/14 (2000) [*Liber amicorum Andrée Despy-Meyer*], 189–209.

> house is exquisite, some really beautiful things, & a sense of peace & order & simplicity.

George, on 19 October, is unhappy about Mabel's decision to work, and he condemns it with a biblical allusion to the usurper and *chef de cuisine* Jacob:

> Very disturbed to hear that you had accepted to work regularly for Céline. It does not seem very wise to sell your birthright for a mess of potage. I would rather send you myself $20 more each month (it amounts to that), though I would prefer not. I hoped this year would be for you a year of rest and recreation; of course I knew that part of your time would have been devoted to work—that is as it should be—but I hoped that you would have defended your freedom.

Instead of sending Mabel $100, he still wants to deposit $30 to her account in Cambridge and send her only $70. He is trying to complete his *Introduction* as fast as possible. The Carnegie Institution did not give him money for secretarial assistance—a discouraging decision, since as early as 1915 George recognized the need for a secretary who knows both French and English.[5] Beginning on First August 1924, his 2 secretaries, 3 counting the Belgian managing director for *Isis* Léon Guinet, who was a sometime *lycée* professor in Paris, have been working at his own expense—$300. The printer has raised prices, and *Isis* will not be covered completely by the History of Science Society. "Be sure that I will do my best & fight like a man."

On books, "I loved *Of Human Bondage* inspite of its cruel realism, because of the intense human sympathy which it contains. I love it for the same reason that I loved *The Growth of the Soil*." He is reading physicist Michael Idvorsky Pupin's autobiography—the rise of a Central European immigrant to the heights of Columbia University—with great interest.[6] He was moved by Beethoven's Fifth at the symphony. On First November, George is concerned about May's health and also Mabel's. He urges them to go to Switzerland or France if they think that it would be better for them, even if the expense were greater. He thanks Mabel for the chrysanthemums she sent him, which he recalls from his childhood as the flowers sold before the cemetery visited on All Soul's Day with his Aunt Hélène. He has been entertaining. Physicist Charles Fabry on one night, and a group on another night—historian of religion George Foote Moore,

---

5. Diary, 24 December 1915. "I will not be able to publish *Isis* alone. Even without *Isis*, I will have enough work for a secretary." George presumably means that the Carnegie Institution did not increase its commitment of $100, which he notes in a letter to Mabel of 21 December 1924.

6. Michael I. Pupin, *From Immigrant to Inventor* (New York, 1923).

Charles Haskins, David Eugene Smith, and Harry Walter Tyler— to discuss plans for an institute for the history of science. Henderson has animated Yale and Harvard to bid against each other for the institute. "Harvard would give me a valuable site near the campus & every facility. We need a million dollars, this including of course the necessary environment (we would not spend more than $150,000 on the building). My former plans called for two million dollars, thus I have shown a willingness to compromise." Henderson is against the million-dollar search. George does not mind the absence of an institute. "Much as I would wish the Institute to be established, I realize that it would mean hard work for me. I would have far less freedom than I have now, and many administrative worries of all kinds." His salary, however, would be higher. He proposed $7000. Henderson wanted a total expenditure of only $200,000. At the last meeting, Haskins said George should get $8,000. "Well, that would make a difference to us, wouldn't it, mate?"

On 2 November 1924, Mabel writes that May is feeling at ease in her new school. Seeing her old friends makes Mabel realize, in French, "how much I am taken with you, how much we have evolved together, you and I, in the same direction." She cannot draw money from the bank where she deposited his check unless he writes to the bank, and Mabel asks George to do so at once. She has seen Irénée van der Ghinst and his son, Marco; Marianne "is always away, playing with a quartette in Paris & elsewhere." Irénée and his wife are estranged; the son is without a mother and sees his father only at meals and on Sunday. Mabel has read *Life of a Peasant*, by Tolstoy's sister-in-law, which she likes very much.[7] *Of Human Bondage* "is a book of tragic beauty such as I have rarely if ever read. It is *stark* in its truth." She concludes, "*Life is cruel*, it is brutally cruel. The dawn is pure & lovely! Yes, for us who have time to look at it." About finances, she prefers that George send the $30, "for one real reason only: If you don't, you might disbelieve that I could save them here." She thinks that it will be 2 or 3 years before they receive war reparations.[8]

George reports on 9 November 1924 that he is trying to find "an intelligent collaborator" to help him with Chinese thought in the Middle Ages. He is lonely, but "I must remain strong and fit for the many years to come, for you and May, for my self-imposed task." He has purchased a phonograph to play classical music, for he attends, he emphasizes, only 2 concerts each week. The cost is $100, including an initial purchase of records. For the rest of his life, George enjoys listening to classical recordings at home. As he often does during this time, he will dine at the home of Charles Macfie Campbell, psychiatrist and later director of the

7. *My Life, as Told by the Peasant Anissia to T. A. Kouzminskaya*, revised and corrected by Leo Tolstoy, trans. Charles Salomon (New York, 1924), also in a French edition, 1923.

8. *LM, II, s.v.* Van Oye.

Boston Psychopathic Hospital—"there is no house where I go with more pleasure."[9] George is concerned about May's health and is relieved that her problems may be explained by the onset of menarche.

Mabel, on 16 November, has arrived again in Ipswich. She is less concerned with May's mental state than her physical state. "May is proud. She suffers from the continual feeling of inferiority, of helplessness because she does not understand at school. Of course the remedy is in her hands, but her hands are a little weak just now." Mabel hopes that the local doctor "will find out what is wrong with her stomach or intestine. The question of her spine & thigh is serious & must be remedied as soon as possible." Clear-sighted Mabel doubts that George's institute will materialize for a long time. "I would *hate* 'more money & more worry' for you. Please try not to accept such conditions." Mabel wants time for them to live: "Nothing will satisfy me that does not give you a *fair share* of freedom to do your work & *live*, & me too." She wonders if George would resign from the Carnegie Institution upon the institute's creation. She longs for the "sympathy and counsel" of Louise Inches and Katharine Taylor. "Céline is terribly, almost cruelly exacting on those who work for her." Mabel is buying Céline an empty barrel to use as a kennel for her dog, who "needs one so pitifully."

George on 30 November 1924 reports that he had Thanksgiving dinner with the Campbells. He asks if Mabel saw that Paul Albert Besnard has become the first painter to be elected to the Académie française; the Besnards lunched with him at the home of Fogg Art Museum director Edward Waldo Forbes. As a postscript, George sends along an accounting of his expenses in the near future. In French:

> Monthly: Mabel \$100, Self \$100; Secretaries \$100.
> Trimestrial: Mother \$80, Guinet \$50, Storage \$36, Rent \$57.
> Extra: Clothing (1 suit, 1 overcoat) \$100, Insurance \$250, George's passage to Europe \$150; Their passage to America, \$450.

On 4 December 1924, George writes that he is off to the symphony ("I must still doll up") featuring Mozart, Debussy, Prokofieff, and Rimsky-Korsakov. Three generations later, just such a concert could be staged. On 7 December, George returns

> Letendart's worthless "analyse psychologique." He is an intelligent man, as most charlatans are (they could not prey upon their contemporaries if they were not more clever than most of them), & much of what he says is very plausible. Quacks do not know the difference between plausibility

9. Diary, 19 December 1920, 8 January 1921, and 7 August 1943. Campbell was a close colleague of George P. Amsden's, George's physician friend from 1915. George S. Amsden Papers, New York Weill Cornell Medical Center Archives, New York, Finding Aid.

> and truth, or if they do, they ignore it. And as most people want to know the truth, even when it is unobtainable, and prefer any definite answer to a doubtful one, quacks are very popular.

Beginning in 1923, George uses *Isis* to debunk "unreason which is rampant around us, even in the most enlightened communities."[10] But at the same time, George respects William McDougall's open-minded approach to parapsychology. He dined on Sunday, "as usual" with the Campbells. He also dined with the Boutons and Effie Merriman. Late in November and early in December, George records sex 4 times in his agenda.

Mabel, on 4 December, has not written because she is disoriented. In the evening, she was awakened by the sound of someone calling out her name. May is doing better, but "schoolwork tires her beyond measure, and they use a system of instruction that I do not like: *Learn everything by heart*." Furthermore, "May has a much larger sensitivity than the children here, not only because she is older, but her nature is very, very different." Mabel writes about the gifts she has assembled for the children's St Nicholas Day. And on 11 December, May writes to George about what she received. She encloses one of her last French dictations, much better than the first, in which she had 24½ mistakes. She thanks George for the papers and clippings he sends every week. In describing her gifts, she shifts into French, "quite without thinking," she quickly notes.

George, on 14 December, reports on the books he has given as Christmas gifts. Then on a shopping expedition:

> I went to Bawston the other day to buy a new suite of clothes, which I badly needed, & a pair of shoes. The buying of the suite was a record. I dropped into the shop, saw a suite, tried the coat and as it fitted me all right, bought it without further ado. The whole transaction did not last more than 2 minutes & the salesman was highly pleased. He told me that I was a perfect 39, and this transparent lie pleased me too! Which shows how foolish I am, for to be a perfect 39, I ought to be at least a foot taller to begin with, and this qualification is so damaging that it is not necessary to speak of others.

George returns to May's letter:

> I was quite amused to read in May's dear little letter that one of her teachers was Mme Closset. I take this to be Marie Closset (alias Jean Dominique) who wrote some very interesting verse. Life is strange! When I read her verse, years ago, or the kind little letter which she once wrote me, I did not dream that the same person would soon be teaching French to my American daughter!

10. George Sarton, "The Pillory," *Isis*, 5 (1923), 146; 7 (1925), 116–18, quotation from the latter.

Between St Nicholas and Christmas Mabel writes to George. May suffers from the way Belgian children view the world. They are harder, less sensitive than she is. May's teacher and Mabel have decided to require, as homework, only French reading and short writing assignments. Céline and Raymond Limbosch have become hardened to the world, their hearts are closed to everyone except their family, of which Mabel and May are part. She fears that Raymond would kill himself if he did not have a family, although she cannot sympathize with him. Raymond condemns her as "a blind and inveterate optimist." If she should die, she does not want May to spend more than an occasional vacation with the Limbosches. "There is an atmosphere which, later, *when she understands more*, will hurt her, I think." May finds nothing sinister or menacing about the Limbosch household in her memoirs.

George, on 21 December, has begun a recently published account by the French ambassador to Russia. It is recommended by Mabel's friend, Helena Baranovitch *principessa* Droutskoy-Sokolinsky, who lives in Italy and whom the Sartons support financially.[11] George asks Mabel how much money she wants to send to Baranovitch and her daughter each year. His financial picture is improving. The Carnegie Institution has given him $600 for secretarial help, up $100 from 1924 (a sum that George uncharacteristically omitted in a previous letter to Mabel about financing *Isis*). On Christmas morning: "This is the first Christmas that we spend so far away from one another & I hope to God it will be the last until death does part us."

George frequently asks Mabel and May to write, often frantically. He complains on 29 December 1924: "I had set my mind on a word from you for Christmas—a word that would be Christmas to me—& as it did not come I was terribly depressed. Nothing else counted, because I had not had that which I wanted, the only thing which I wanted." Mabel, on 29 December, writes that she and May love the books that George sent. "Oh beloved husband, I do miss you so. You can never, never guess how 'exiled' I feel because you are so far away. I begin to feel more & more that your heart is indeed my home & though some invisible part of me, knows it is safe kept in it, still I need your actual presence terribly." Mabel agrees that Letendart is a charlatan, but "Ray. feels that the University world," she switches to French, "the 'insignificant' people of science are unjust to this man and do not want to allow him to present his empiricism." The head of the Asile d'aliénés is, however, allowing Letendart to examine normal and crazy people, and also children. Mabel thinks that Letendart is a kind of intuitive reader of characters, who believes that there is truth in ancient

11. The name and title from accounts of George's, detailing people and causes he supported in the early 1950s, in YMS. Maurice Paléologue, *La Russie des tsars pendant la grande guerre* (Paris, 1921).

'sciences,' from which he takes his inspiration, even though he refuses to divulge his theories. A lawsuit is looming over Letendart, and Raymond is on his side. Mabel will "refuse to be called as a witness" at the expense of offending Raymond. "I am not sure that it is not best that such a man *should* be 'broken' sorry tho' I would be for him as an individual."

On 7 January 1925 Mabel clarifies her commitment to help support Baranovitch. She did agree to collect and send a definite sum of money for 3 or 4 months, but she clearly stated that she could not really continue to contribute in this way. "I do not feel a *moral* obligation to help them in an emergency." She is sending F 200 as a New Year's gift, and also some clothes. She does not see how they can help Elsie Masson or Claire van Halmé. "We still have a steep hill to climb financially, dearest, this year." Mabel has been able to save 30% of what George sends her. Regretfully Céline perceives May as a "spoiled child," perhaps because Mabel neither reprimands nor praises May publicly. Mabel thinks that "May has suffered beyond measure from the ignorance concerning & mockery of (not nasty but irritating) all things Anglo-Saxon." She responds by boasting about her school friends in Cambridge. The result is an inferiority complex. Mabel has tried to impress on May the slogan of George and George's father: "Do be hard on herself but very tolerant of others." Mabel "was deeply touched because she said one night, 'Our home was always such a *pleasant* place.' I wonder will you feel as I do, that that is fine praise for you & for me?" Mabel includes a reckoning of her earnings in 1924, except from the Gillespies: Winsor and Milton schools paid her $500, and she earned $20 from Céline.[12] On 12 January 1925 Mabel writes that she will send F 20 monthly to Elsie Masson (whose husband Emile died in 1923), if George can increase his monthly payment to $105 ($5 coming from the amount that Margaret Gillespie sends George in her name).

George, on 4 January 1925, returned from Washington "rather tired & dull. The [history-of-science] meeting was as successful as it could be, & my personal success was greater than I deserved, but those things do not give me much pleasure. What I enjoyed most was seeing quietly a few friends, though they are all getting as old as the hills." He visited Colonel Fielding Garrison, the bibliographer of medical history, who "suffered much from his two years stay in the Philippines." He enjoyed a visit to the Freer Gallery. The director John E. Lodge showed him around. "I suppose you heard of the death of his father, Senator [Henry Cabot] Lodge? He ought to have died many years ago for the good of the world." But George likes the son.[13] He dines with the Gillespies. Then to what George hates, shopping:

---

12. In fact Mabel began teaching drawing and design at the Winsor School in 1922, according to Jane Hoeffel Otte, Archivist at the School.

13. George thinks that Lodge's work is carried out "with admirable patience and modesty." George Sarton, "Art as an Approach to Asia," *Yale Review, 15* (1926), 540–52, on p. 549.

> I had first planned to buy a new overcoat & hat, but after more reflexion I compromised, as I am always ready to do with regard to trifles, and decided to buy only a hat. My idea was to buy a hat so fine that it would almost hypnotize the onlookers & they would not notice the coat. But when I went to the Coop to purchase it (Oh boy! Ain't a grand and glorious feeling!) the shop was closed for the sake of inventory. My old hat jumped upon my head out of sheer joy, & I was very much relieved, too.

In sum, George writes a "sad" letter, and when George is sad, he seeks refuge in the arms of a woman.[14] On 11 January, he writes that he dined at the home of historian of architecture Arthur Kingsley Porter. Among the people there are George Henry Chase, the professor of ancient art, and Miss Ada Louise Comstock, president of Smith College, whom he knew when she was dean. He is reading one after another of Anton Chekhov's works, and he begins "to like him immensely." George urges Mabel to send him titles of interesting books in French, "but there are so preciously few which I really like."

On 18 January 1925, George apologizes for his vexing letter of Christmas, where he laments having been forgotten. It was unreasonable for him to have complained. "But I cannot repress my feelings, at least not beyond a certain point; and when I feel very happy or unhappy I must say so." He is glad Mabel likes the books he sent. On the 19th, 20th, and 24th he notes sex in his agenda.[15] On 25 January, George dines with Lucy Stanton, and after dinner they see a movie. Edwin Herbert Hall invites him to lunch with physicist Peter Debye. Then to Kingsley Porter, where he meets again the Polish-Italian-English lady who knows Baranovitch from Florence. He sees the Whiteheads (Alfred North Whitehead arrived at Harvard in 1924 after teaching applied mathematics for 10 years at Imperial College in London) at the McDougalls.[16] He has received a new copy of Henri Pirenne's history of Belgium (the German soldiers having stolen his copy). The first volume considers *béguinages,* which would be useful for Mabel. He wonders if they should visit Italy in the summer. It might not be good for May, who still does not know French well, and it could be better for him to work on Arabic than on Italian. "Italian is so easy that I can pick it up later in life. Arabic is so difficult that I must try to master it as promptly as possible." He wonders if Mabel would consider spending the summer in the foothills of the Atlas Mountains. For advice, he will write to geographer Emile Gautier, "the funny old professor at the University of Algiers, who dined once or twice at our home." The great attraction of Algeria is the Sahara. He recalls Eugène Fromentin's

14. Agenda, 7 January 1925.

15. Agenda, 19 and 20 January 1925.

16. Agenda, 25 January 1925.

"enthusiastic & magnificent description of it," whether in his paintings or in his travel accounts.

Mabel, on 16 January 1925, asks George to follow what a nurse of hers told her as a child: "Take heart of grace, dearie." To her, the phrase means

> a lovely promise of renewal. I had the feeling as a child of literally taking my heart & turning it upside down & all the troubles fell out of it! And then a sweet feeling like a spring breeze blowing in, in early morning filled it up, a feeling of *promises,* things to come, things I did not even deserve, a real, divine "grace" poured into me, sometimes slowly. But it *came.*

Mabel arranged for flowers to be sent to George through Mary Bouton, and also for his Christmas gift, a pewter ash-tray. She sent a package of Belgian delicacies to the teachers at Shady Hill School. She asks George to show them May's picture of St Nicholas and his donkey, for it illustrates the legend Mabel described to them. Mabel is no longer close with the Limbosches. She would like May, in French, "to be in her own 'home,' which she has never had in her life." She wonders if George's loss of interest in the Carnegie people is due to the new History of Science Society. On 17 January, Mabel does not know how to approach Katharine Taylor about teaching at Shady Hill. "From what you say we could not afford to *pay* May's tuition & yet I would give *a great deal* for May to have one more year there." She could also teach at the Winsor School, 2 days weekly, netting her $1100, enough for May's tuition and a servant. She does not want George to be cross if she asks about money. "I do want to be able to ask you 'square' questions & get frank answers, without there entering into it, doubts of each other, nor reproaches, nor bitterness." In a postscript: "Is there *no* news of the Hist. of S. Soc. Are they getting any money? Are they in hopes of *any*?"

On 21 January 1925, Mabel writes from the home of the Willems in Ghent, where she and May are staying. She and Victor Willem have spoken about blacks in the United States. He thinks that it is, in French, an "insoluble problem." Mabel spoke about laws against interracial marriage. "I also spoke about the fear of people like Miss Shadd who, if she married, feared having a black child, or in any case a 'colored' one." Willem thinks that American scientists have shown that there is no reversion to "the pure race," and so this fear is unfounded. "Miss Shadd had a father clearly 'colored,' a mother white in appearance but perhaps also really colored, i.e., of mixed race where white predominated." She asks George to send her literature on the subject, perhaps meaning George's article and reviews on eugenics and heredity appearing in *Isis* before the war. In French, "Do you believe that it is proved that in the union of blacks and whites, their descendants inherit the weaknesses of the two races, rather than the strengths?" In French, *race* is perhaps translated best by strain or breed when it does not refer to nationality, but here we see how the 19th century has made race a thing in itself, to be feared when not to be admired—even among thoughtful writers.[17]

Mabel is sending George a copy of Ernest Pérochon's *Nêne*,[18] which she likes as much as *Growth of the Soil* and *Of Human Bondage*. On 30 January 1925, back with the Limbosches, Mabel describes the death of one of Céline's young employees. It was her first Europen funeral. Mabel recalls staying with her mother when her father died. "I think it is a terribly dreary, hopeless ceremony which must be heart-breaking to the remaining members of the family." There was no religious service for her father in the hospital chapel. "I don't know why I write you all that. You know what it is like. It is like many ceremonies. It has a trivial side & yet it is the expression of something fine, sometime of human solidarity, a pitiful longing to give & to receive comfort." Mabel still thinks that May's "déviation," in French, meaning her curved spine, must be cured. Perhaps exercises will do it. Mabel reaffirms that she will teach at the Winsor School, but not at Shady Hill, because of the higher salary, but *"I will not teach more than 2 days a week for love or money!!"*

George writes on First February: "I agree with you that the fact that we met was a miracle." Then he provides a succinct statement of his view that a woman should quietly accept the impulses and aspirations of her husband. A successful marriage "is becoming more & more difficult as woman is more emancipated. I do not object at all to her emancipation. This is a real progress, but progress is never easy and painless. We have to pay the price of it in full. And in this case the price of woman's emancipation is but too often failure to find a mate or, if one be found, to be happy with him." He can send her an extra $5 monthly, but not more; he will deposit $30 monthly to her account in Cambridge, apparently not trusting Mabel to save the amount in Belgium. He has not repaid the Baekelands. He obtained a copy of *Peasant Art of Austria* for Mabel, who had requested it, but the price was steep.[19] "This will be your next Christmas gift. Thus try to forget all about it." George missed dining

17. George writes from this time: "They preserve, deep within them, that instinct of race, that pride of family, which stiffens men and helps them to walk straight and go through life without hesitancy. Such traditions, mind you, represent a considerable economy of moral energy." George Sarton, "Basques [Eskual-herria]," in *Studies in the History of Culture*, ed. Waldo Gifford Leland (Menasha, Wisc., 1942), pp. 63–8. Manuscript of the article in NYPL, dated S. S. Roussillon, September 1925.

18. The novel won the Prix Goncourt in 1920. It appeared in New York in an English translation by the author in 1922.

19. Probably, Charles Holme, ed., *Peasant Art in Austria and Hungary* (London, 1911), published by the periodical *Studio*.

with Jean Capart, Director of the Musée du Cinquantenaire in Brussels. Designer Denman Waldo Ross commended a lecture of George's at the Fogg Museum to mathematician George David Birkhoff, "who is very naive." On 8 February, George reaffirms that under no circumstances will he cease to send an allowance to Mabel's mother. He will indeed increase his contribution by £4, if as Mabel says her uncles cannot pay it. Margaret Gillespie has reported that she paid Mabel $359.51, plus an unknown amount in clothes. George hates to discuss financial matters, but he sees much such discussion ahead. "I have no reason to believe that my income will ever materially increase. Any benefit from the H.S.S. must be used for the improvement of *Isis*. On the other hand, expenses increase all the time; more obligations and liabilities are being heaped on us. This is not pleasant." Then George writes, on 10 February, "a love letter." It is short.

On 6 February 1925 Mabel is glad to learn about the opening of the Isabella Stewart Gardner Museum in Boston, and glad for George's extra $25. She will now explain why she needs it. At Christmas she noticed a lump in her breast. A specialist cut it out. It is not cancerous, but it could have been a precursor of cancer. The operation was done with a local anesthetic. For the summer, Mabel wants May to be in a French-language place. Mabel has decided to withdraw May from school, given her spinal exercises. Irénée van der Ghinst has decided to divorce his wife. "It is a miserable story."[20] She recalls that the Hotsons bought a copy of *Peasant Art in Austria* for the Shady Hill School, George no doubt forgot. Mabel's designs draw on peasant themes. "I do want it, *badly*, that is the only comfort." She plans to join the Society of Bibliophiles and Iconophiles in Belgium. Mabel reflects on George's socializing in Cambridge: "R. & Cé do not know any, or very few interesting people (I don't mean 'well known'). They are so wrapped up in their children, Céline in her business & Raymond in his gloomy philosophising, that they ask nothing else of life." Mabel sees clearly that she and George experience a rich life in Cambridge. They choose to live *la vie bohème*, Mabel declining to teach full-time while nevertheless employing a servant. It is a life that would appeal to many scholars today. "Good-night, Dearest Husband, comrade, lover."

On 15 February 1925, Mabel thanks George for forwarding magazines, but she asks him not to send *Vogue* or *Vanity Fair*. "They are not much good to me here." Raymond Limbosch attended a meeting in Paris and "was shocked at the crudity of the terms used apparently by every-one, even talking to women, also more deeply shocked to hear every-one talking of the next war (with Germany)." Mabel has been asked to design an altarpiece for the exhibition held that year in Paris devoted to the deco-

20. But Paul Pelseneer reports late in 1918 that the Van der Ghinsts are divorced. HGS, Paul Pelseneer to George Sarton, 17 December 1918.

rative arts; she will in fact undertake the project, which receives a gold medal.[21] On 21 February, Mabel replies to George's financial strictures. "I feel discouraged & sad. But we must cheer up. We have been in worse case than this, dear husband, both as to money & as to understanding each other. And with a little patience & love on both sides, we shall come through, I know." George writes on 15 February that the Campbells do not appreciate his Christmas gift, *Of Human Bondage*. "They did not take to it! Campbell told me that it was all wrong, that for ex. the woman (with whom Pinlip has been all the time in love) is not correctly drawn." George is glad that May likes Ghent. "I was so intensely loyal to Ghent when I was a little boy, & I was so entirely of my own accord. Nobody around me shared that loyalty!" George pays his income tax on 18 February, no doubt a sad event, and follows it with sex.[22]

On 22 February George replies emotionally to the news of Mabel's operation: "As far as it is in my power, never again shall we be separated by such distances in time and space." George sends along plans for the summer. He wants to meet Mabel in Middelburg, in the Netherlands, and then go south to the Pyrénées. George earned money by lecturing at Wellesley. Afterwards they dined in the student dining hall. "It is embarrassing enough to be the only man among so many women, but the noise they made was prodigious. As soon as the benedicite was said they all began to chatter at the tops of their voices!" He attends a dinner and lecture by Leslie Hotson on the death of Christopher Marlowe.[23] He is off to hear Percival Chubb at the Ethical Society, after which he will dine him at the Four Tynes on Beacon Hill. The other guests: Lucy Stanton and Dr Moss the bacteriologist.

Mabel writes a long letter to George on 26 February 1925. Mabel acknowledges that she cannot contribute to the money they send to her mother because she is using the income from Margaret Gillespie to pay off debts. They still have great difficulty discussing financial matters: "Why then can we not meet and plan, and review, & even 'discuss' these things without continually hurting each other?" George worries that their "obligations & liabilities" increase while his income remains constant; Mabel wishes that she could earn enough so that George would not feel bound to sacrifice for her and May. "I feel strongly that you are an exceptional man whose work is so absorbing & *so un-ending* that it *ought not* to be

21. May Sarton, *A World of Light: Portraits and Celebrations* (New York, 1976), p. 60. NYPL, Madeleine van Thorenburg to Mabel Sarton, 18 December 1925, congratulating Mabel for the gold medal. Exposition internationale des arts décoratifs et industriels modernes de Paris, 1925, *Catalogue officiel de la Section belge* (Brussels, [1925]), p. 91, the altarpiece made by the Ecole de métiers d'art of the Benedictines at Maredsous and produced by Céline Dangotte, Mabel Sarton listed as "collaborateur." A note in the private archives of Susan Sherman, Riverdale, Bronx, New York, indicates a prototype of machine embroidery for the altarpiece.

hindered by family ties. Yet I am more & more convinced you need your family—could not be happy without it & without a real home." Mabel has used her income "to make ends meet, in many ways that you often were not aware of." In the future, she wants to be a part of George's financial projections. "I have thought of the Insurance question. The last time I spoke of it, you reproached me for selfishly forgetting May's future if you disappeared (those are the sort of poisoned darts that leave a wound hard to heal!)."

On 8 March 1925 George sends a spring letter. His bouquet of roses will have reached Mabel, along with Selma Lagerlöf's autobiographical *Mårbacka*.[24] George suggests reading the book aloud to May. He recalls reading d'Alembert's introduction to the *Encyclopédie* to his old grandmother. "Neither of us understood a word of it, but we liked it all the same." He gave a lecture at Smith College. He studied the college carefully with a view to whether it would be appropriate for May. Then to the matter of a research assistant. Previously George recruited his student Yuen Ren Chao to negotiate around Chinese texts. Chao assigned Chinese words a number and provided a definition, so that George could refer to them later. George has now found a Chinese student to complete the task.[25] George needs, and cannot afford, additional help—a *famulus*, or student who could intervene in all aspects of a professor's life, as the position was known in Germany at the time. Indeed, Philip Jourdain's assistant, James Marc Child, offered his services to George in 1920. Child elaborated:

> I shall esteem it a great honour to cooperate with you: but I suggest that it be done in the following way, the way in which I worked for Jourdain. As an historian, I am still in swaddling-clothes; Jourdain's method was to send me a definite job, a book or an idea that he wanted working out, like this:—"Enclose...; gut it, criticize —'s opinion; trace its influence on — and — ; and place it historically for growth of concept, and for importance." This was making me of higher class than a mere literary hack, but much of the same nature. In fact he looked upon me, and called me, his "mathematical detective."[26]

With an assistant like Child, George might have been able to produce his book on Leonardo.

22. Agenda, 18 February 1925.
23. J. Leslie Hotson, *The Death of Christopher Marlowe* (Cambridge, Mass., 1925).
24. Selma Lagerlöf, *Mårbacka*, trans. Velma Swanston Howard (Garden City, N.Y., 1924).
25. F. K. Huang, a philosophy student at Harvard, is his choice to replace Chao. George Sarton, *Introduction to the History of Science, volume 1: From Homer to Omar Khayyam* (Baltimore, 1927), p. 48.
26. HGS, James Mark Child to George Sarton, 1 January 1920.

On 11 March Mabel as the Wild Bird, apologizes for a recent discouraging letter:

> This is just to send you a loving smile right across the world, to tell you, that though I am unhappy & beat my wings against my own limitations sometimes, oftenest, I am happy, & that happiness makes me feel, instead of life being a cage & I a wild bird in it, that you have built me a nest of your love, where I *choose* to be & could never be happy elsewhere.

Mabel has comforted her mother in Ipswich, helping her complete paperwork for her pension, and returns to the Limbosch house in Belgium. Mabel thinks that Raymond is close to embracing Catholicism. Céline finds it hard to run her business, which is not producing enough to support the family. In French:

> It is sad to be among friends who have changed so much that I have more or less lost them. I love them above all for what they *were*. And so one can't stop loving those whom one loved for such a long time, when they were good for you, and they were very good for me. They are struggling, and they argue. They need so much affection and love to protect them against the bitterness that threatens them.

George writes on 15 March 1925 that he dined with Denman Ross, whose cousin (who is his companion) has gone insane. "Happily for him his work has a very strong grip on him, & he works hard, this being now his one source of happiness." Ross paints with his Japanese butler, also a painter. George has received *Nêne* and likes it very much, although he finds it a bit melodramatic. He hopes that it signals more good writing in French. Professor Gautier in Algiers reports that the summer is comfortable there; alternatively, they could spend the summer in Rabat, Morocco. To live in one of the medieval Moroccan cities, for George, would be "better than anywhere else in the world." George is close to finishing the first volume of his *Introduction*. He must revise the Greek part, "if it be only for the sake of one scholar in a thousand."

George reports on 22 March that the weather is lovely, and he adds a note that is uncharacteristically pessimistic: "But can it last? We have had such a mild winter that a hard spring would not be unfair." He dined on Sunday with Mr Sand and Miss Wilson, an architecture student. After dinner they danced to a Victrola. "I did not do it too badly considering that I had not danced for more than 25 years!" Here we read, and Mabel reads also, the thoughtlessness of a husband: Mabel loved dancing, she wrote 25 years earlier to Céline.[27] George dines with philosopher Horace Meyer Kallen, who reports that his sister Deborah is working at the Cooperative

27. NYPL, Mabel Elwes to Céline Dangotte, 14 June 1900.

School in Jerusalem. George asks Mabel to write to Deborah, in anticipation of their trip to the Mediterranean. He calls on the Richardses; the Conants are in Munich, where he has a half-sabbatical to do organic chemistry. The Richards family seem glad to see George, "but their welcome is not like the one I receive at the Campbells. They are both of a more retiring disposition." George's dour view of the weather presages sex 2 days later.[28]

Mabel, on 21 March from Rhode, thanks George for the bouquet of 15 red roses. May is recovering from the grippe, caught when Mabel was in England. Mabel is reading *Mårbacka,* which she and May enjoy. Mabel has no present for George, but she offers him "my very heart—all I have.... Raymond is really sad & vexed because you will be in Belgium so short a time, but I very much fear you will not understand each other. He is so violent, so passionate & yet often (to me) so confused in his arguments. C'est de la métaphysique." Mabel is glad that George has found a replacement for Chao. Louise Inches reports that the Chaos have "blossomed out into great prosperity in Paris." Then the observation of a professional *couturier:*

> Dearest Husband. You really make me sad when you go about with a summer coat in winter for economy. That is bad economy. If you get pneumonia or something, where would we be? Also it hurts me in some mysterious way (as it would hurt me to see May un-necessarily shabby or unsuitably dressed). I feel a real pang in my inside when I think of you shabby & poorly dressed. That is wrong. It is bad management. You earn enough to be properly dressed. Well, I don't want to scold you, but "it must not happen again."

From clothes to money. Mabel urges George not to send £20 to her mother, but instead send the amount to the Baekelands to work off their debt. George, on 29 March, called on the Porters and met the wife of the Harvard Egyptologist George Andrew Reisner. He learns that the climate of Egypt in the summer is tolerable. Generally Egypt and Palestine would be fine for the summer. The appearance of a recent essay of George's is giving him "a lot of trouble." People ask him about the person figuring in it, Jean, and "it is rather hard to have to confess that this Jean never existed except in my own imagination, & that this defense of sincerity is introduced by an enormous lie." In the article, appearing in *Scribner's,* George's Jean enlightens him in Brittany about the true way to view people and the world. It is Ruskin's sense of being genuine and honest, qualities that, when taken to the extreme, constitute genius. George has mastered the art of the potboiler.[29]

---

28. Agenda, 24 March 1925.

29. George Sarton, "Transparency: A Way of Looking at People," *Scribner's Magazine,* 77 (1925), 308–13.

Mabel writes to George on 28 March 1925 that she finds May "a treasure of sweet good sense, and delicate intelligence. She is passionate & over-sensitive, but capable of tempering her judgments, of balancing her reason against an excess of sentiment in a way that often astonishes me." Mabel worries that May has been injured "by the character & upbringing" of the Limbosch children, who seem hard and indifferent and materialistic, and who are spoiled by Céline's effusive praise and attention. May is the better for it all. She "adores *you*, & appreciates your qualities as perhaps she would never have done without this absence." Switching into French, Mabel remarks that sometimes George really does not understand all that Mabel wants for May:

> No, you really do not understand your little daughter well if you think she needs "lots of praise." And when you say, "Well, kid, it might be worse" about something she made, you are mistaken if you think that she does not know perfectly well your judgment. You are more transparent than you think, ô philosopher, and the eyes of your little daughter know the depths of your soul.

May's teacher Marie Gaspar highly recommends Romain Rolland's *Mahatma Gandhi* (Paris, 1923), and she passed Mabel a copy of Marie Closset's little book, *Les Livres et les enfants* (Brussels, 1911). Gaspar finds May exceptional, in that her sensitivity and intelligence are matched by her emotional warmth, and that her writing is strong. Mabel wants to invite her and Closset to dinner when George is in Brussels. She wonders if George can find a French version of Lagerlöf's book, which she would like to present to May's school; she is sending the author a drawing of one of Lagerlöf's characters made by May. Mabel is glad that George gave Chubb "a little party & I am sure Lucy managed it."

On 3 April 1925, Mabel reports that Céline is transporting the entire household to the new Limbosch house in Knokke for 2 or 3 weeks, renting out the Rhode residence. Mabel will find a room so that she can be near her work, notably the chapel embroideries. Mabel will go anywhere in North Africa with George if the climate is agreeable, and she asks if they can travel there at an easy pace through France.

> I cannot answer your letter further, dearest. I have so much still to do. I got up at 6 this morning. It is spring weather. The birds sing quite tremendously at about 5 before anyone else is moving. By 7 they have accomplished many important things. The miracle of flowers has begun. In *one night* the forsythia shook its bells open! Wild plum (tiny starry white flowers with [illegible] leaves and stalks) is a delicate delight in many places. May has seen *primroses* along a brook (I have never, never seen them in Belgium).[30]

30. Early in the 20th century, geneticist Hugo de Vries devoted a great deal of study to mutations of evening primrose–an invader from Texas—in the Low Countries.

George, on 5 April, is disappointed that the Rockefeller Foundation has decided against funding his history-of-science institute.[31] "Now, to loose a million dollars is rather tough, and though I had already dismissed the matter from my mind, it kept creeping in insidiously and disturbed my aequanimity." Religious services have become his solace:

> I love to go to the university chapel on Sunday. I love the singing of the boys & the organ playing and I love to unite with other men & women in worship. The university chapel is almost the only place where I can do it, because they do not expect me to subscribe to any dogmas. It suffices to be a Christian. But am I a Christian? I often wonder. However I am not sure that they would insist even on that. I believe that even an half-and-half Christian-Buddhist, as I am, would still be allowed to come in. And I do so with pleasure & look forward to it.

Historian of chemistry Hélène Metzger in Paris, who sent George F 1000 for *Isis* several years ago, has now offered F 1000 more. George prefers that she buy 16 or 17 subscriptions for French university libraries. She is the only person besides Solvay who offered material help for *Isis*. George writes on 12 April, Easter Sunday, with a vision about May: "It has often occurred to me that we would perhaps escape from oblivion only because of her, as her parents." George asks Mabel to buy a lace collar as a gift in honor of May's birthday (a book for May is coming), which he will pay for. Then to May's destiny, and also his own:

> True greatness is such a rare combination of gifts! It is not enough to have the gifts, they must balance and complete each other in such exquisite way, that the right combination is very seldom attained. It may not be entirely correct to say that "no man was ever great who believed himself to be so," but it is clear that this belief, this conceit, would spoil the combination and make him distinctly smaller. In the same way, I do not think that we help those whom we love by—especially the undeveloped ones—by thinking *too* highly of them. It is enough to make them feel that we *trust them* to make the best of all that is in them.

George wants to spend time with Mabel and May in the Netherlands, showing May the Mauritshuis art museum. He imagines May having a career in a museum. May will have to earn her living, and so for her college is "not a luxury, but almost a necessity." His Easter dinner consisted of sandwiches and a slice of pie. It is a sad time, and George notes sex the following day.[32]

Mabel writes from Knokke on Easter Sunday. The children spent a lovely day at the beach. After much difficulty, she has almost completed the chapel embroideries. She recalls the day, many years previously, when

31. HGS, David Eugene Smith to George Sarton, 2 April 1925, for the refusal.

32. Agenda, 13 April 1925.

she and George climbed a dune to see the marshes there, "mysterious & misty. I think that was one of the first moments that I was conscious how dear you were to me & though I believed then that you loved someone else, I, as it were, *stole a bit of happiness* & lived those few minutes in a wealth of joy!" She has heard from Elsie Masson and sent her the first 4 installments of their promised support. "She was touched by your article."[33] Mabel's valediction in French: "Vieux Rat whom I love, wise man, dear 'transparent' heart."

"The great event for me this week," Mabel writes on 27 April 1925, is a letter reappointing her in the fall as an instructor at the Winsor School for 2 days per week, at an annual salary of $1100. Mabel "will hand it over to the common fund & we will make a budget together." Mabel is especially pleased because she had declined another offer of employment, and George "never said one word about my refusing it." She will make plans to travel to Rabat for the summer, so "no more changes *please*." She and Céline returned from Knokke to find the garden "a dream of loveliness—starry with pale yellow narcissus and fleecy with pear cherry blossoms. I have never seen it so lovely." Mabel's embroideries are still not finished, but neither is the pavilion where they are to be installed. There is great excitement in Belgium about Paul von Hindenburg's election in Germany and his desire to lead a war of revenge against France.

George moves out of Divinity Hall early in June 1925. He looks forward to a few weeks in Belgium, and then on the Pyrénées with his family. There he will write "On Intellectual Toleration" as a preface for volume 8 of *Isis*,[34] a new proposal for an institute for history of science and civilization, and "Ugly Ducklings," a meditation on educating children which in the event appears only 14 years later.[35] Mabel fears that the happiness of meeting George will be denied her. George the rationalist is unsympathetic: "I do not believe in forebodings except as indication of real conditions." Mabel's letter frightens George. Never again does he want to be separated from her for so long. "We have thrown away a year of common happiness."[36]

33. Presumably, Sarton, "Transparency," where the late protagonist Jean is Breton. The Sartons remain friendly with Elsie Masson, sending her Thornton Wilder's *Woman of Andros* (New York, 1930) for Christmas 1930. Information courtesy of Marielle and Didier Giraud, Pontivy, France. YMS, Mabel Sarton to May Sarton, 24 March [1936], reporting Elsie Masson's praise for May's poems; Mabel Sarton to George Sarton, 30 August 1937, where Mabel received a "lovely letter from Elsie Masson." Mabel writes to Elsie Masson after the Second World War, and May visits her in Pontivy.

34. George Sarton, "Sur la tolérance intellectuelle," *Isis, 8* (1926), 241–53.

35. George Sarton, "Ugly Ducklings," *Isis, 30* (1939), 6–16, in which George conveys his belief that talented children should be allowed to follow their karma. He concludes: Ugly ducklings–misfit children–can anticipate a hard fate "of uninterrupted labor and devotion.... To be the mother of a child of genius is not in any sense a worldly triumph, but a glorious sacrifice."

36. Diary, 3 June 1925.

Conclusion

Sarton Resartus

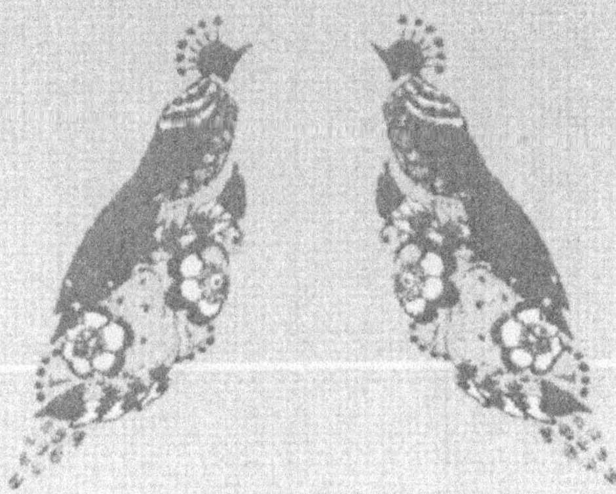

CONCLUSION

Sarton Resartus

Sometime shortly after 3 August 1903, George Sarton notes in his diary how he wants to be remembered: "He loved beauty and life, but he did not fear death. He wrote little but thought a great deal. He acted well in life. He was a friend of order, in a general sense, a friend of progress, above all a friend of logic and precision." In 1922 he emphasizes: "When people will ask 50 or 100 years hence what I did, I hope they will be told: 'He founded the history of science, he established it as an independent and organized discipline.'"[1] On the eve of rejoining his family in Europe in 1925, his life had indeed been played out according to his desires.

Mabel Sarton, in 1925, had also satisfied her desires. On 9 June 1908 Mabel writes to George about the reservations she brings to an eventual union between them: Her most fervent hope, more than having children, is "to *work*, to produce, to give form to my ideas and my feelings through design and painting." Maintaining her independence and creativity is paramount:

> When we spoke the other day, you compared what a man sacrifices of liberty to what a wife sacrifices. I really didn't understand your point. A wife who wants also to work makes an infinitely greater sacrifice than a man *if she has children* and if does her duty. You and I, for example, by getting married, I will love you enough I *know* never to slow down your work, but inevitably I would have to give up my own for at least 10 years—and then? *Would I be able* to pick it up again, would I have the strength, even the *desire*, given that I am beginning to acquire only the *basics* necessary for it? Would I be able then to study my craft, and not merely turn out products?

Mabel gave birth to three children; her designs received prizes and critical acclaim, and they netted her a small but important income.

For the next quarter century, the Sartons lived following the rhythm that had been established by 1925. In January 1933, for example, Mabel writes to George, who is lecturing in California, about spending time in New York with daughter May. She meets Abraham Flexner for dinner, for Flexner's daughter Eleanor is acting in a troupe under May's direction, the Apprentice Theater. Mabel's letter repeats a number of themes that

1. HGS, George Sarton to John Charles Merriam, 1 April 1922.

figure in the Sartons' life together over the preceding decades: Mabel's focus on physical form and harmony, her ecumenical spirit and independent thoughts, her wide knowledge of European culture, and especially her rôle as George's emissary and social facilitator. She comments with an even temper on Anton Chekhov's plays, Eva LeGallienne's accidental scarring and her lover Jo Hutchinson, and George's morose assistant Alexander Pogo, a native of Russia:

> May & I were invited to dine with Mr Flexner & his daughter Eleanor & then all go to the theatre together to see "Three Sisters." We dined at the Longchamps restaurant just around the corner from here—a very nice, rather quiet place & not terribly expensive (a delicious dinner for $1.00). May & I were there first (you have brought us up so well, we always *are* first) May was a bit depressed & had been pouring out her troubles to me here in my room before we left so the time passed in talking & they were soon there. He is a very little man, clean shaven & very bald, with a kind & shrewd face. His daughter teases him continually—too much so for my taste—she is a bit too cock-sure in a very young & gently scornfull way. He told amusing stories & it was rather a scrappy conversation, but I liked him & I was quite proud because at the theatre he talked to me a little of his scheme: he would like to draw together in a friendly & simple atmosphere a group of scholars who could meet & discuss & get to know each other; he has wonderful remembrances of such meetings at Johns Hopkins with Dr [William Henry] Welch, Osler & others, but I think that was a spontaneous, almost accidental gathering of kindred spirits, & that one cannot *deliberately* bring about such a thing. I did not pour cold water on his scheme because I know too little about it—& no doubt this bit he spoke of is a side-issue. His wife [Anne *née* Crawford] comes of a *Baptist* family & her marriage was evidently a terrific blow to them. She is very interested in plays as such & I believe writes (criticisms??). She has been ill with grippe & has just been sent off to Florida.
>
> He spoke to me with warm appreciation of May & says his wife is greatly impressed with her clear-cut ideas & power of expressing them. He seems very glad his daughter has May to teach her. I was a little sorry to be with them for this particular play. I hate discussing one which [illegible] me, in between acts & immediately after. For one thing I am slow to sort out my own impressions & crystallise them into words. The play is a very beautiful one—as you know, more beautiful than "Cherry Orchard"—taking in more varied temperaments. It was played with a "group-perfection" that moved ones heart & mind to joy & tears—I think Jo Hutchinson quite equals Eva Le G[allienne] in this. For the first time I noticed a painful change in Miss Le G's *face*—her cheeks look tight-drawn—almost artificial—& May says that is from the burns but she played so beautifully I quite forgot all about it after the first few minutes. It is a distressing play—but I don't agree that it is 'only Russians' who suffer from this helplessness before life, they suffer more & oftener from it perhaps, but I am sure many people have to struggle against it. (I confess

> I kept thinking of [Alexander] Pogo as the person I know who has it most strongly.) I went back-stage afterwards but could not see Miss Le G—I talked awhile with "Robbie" who welcomed me warmly—& with Jo. & met *Mrs* le Gallienne [Danish journalist Julie *née* Norregaard] whose face I did not much care for, but we scarcely spoke together.

Mabel expresses reservations about the Institute for Advanced Study in Princeton; if she had fonder thoughts about the institute or its location, she might have applied her social skills toward persuading Flexner to create a chair for George.

After 1925, and to the end of her life, Mabel travelled periodically to England and Belgium; George, accompanied by Mabel, took sabbaticals from Harvard. In 1933 they considered moving to California—by then Mabel's partnership with the Gillespies in Washington had essentially failed, and the Carnegie Institution would have continued George's salary in the event of a move—but the culture of Cambridge exerted too great an attraction. Mabel funneled her aesthetic sensibility into conoisseurship, establishing the origin of ceramics in a wide range of styles which she purchased in antiques shops. George edited, wrote, and lectured. Their lives were not without consequence, but it is their work accomplished by the middle 1920s that draws our attention. Mabel's early designs find an honored place in the Smithsonian Institution.[2] The first volume of George's *Introduction to the History of Science*, a work of character and originality that receives little attention today among narrowly focused specialists, is cited by scholars offering general observations about history.[3]

Making allowances for time and place, the Sartons in 1925 resemble their beloved Carlyles 90 years earlier, when Thomas Carlyle, having rewritten *The French Revolution* from memory at age 40, was turning down offers of employment as he struggled to make ends meet. Since his twenties, Thomas Carlyle hoped for an academic appointment at the universities of Edinburgh, London, St Andrew's, and Glasgow in everything from mathematics, physics, and astronomy to moral philosophy, rhetoric, and metaphysics; George knocked on doors across Europe and America, asking for a permanent professorship. (Their ambitions were gratified only late in life.) Thomas Carlyle's early autobiographical novel, *Sartor*

2. E. H. Gustafson, "Mabel Sarton Writing Desk," *Magazine Antiques, 151* (1997), 36. Virtually no trace remains of Mabel's pedagogical activity. She hoped to have some of her children's art appear in a publication by Geoffrey Holme at the London journal, *Studio.* Mabel Sarton to George Sarton, 5 July 1929. None of Mabel's pupils appears in R. R. Tomlinson's *Picture Making by Children* (London, 1934), a book published by *Studio* and edited by C. G. Holme.

3. J. L. Cisne, "How Science Survived: Medieval Manuscripts' 'Demography' and Classic Texts' Extinction," *Science, 307* (2005), 1305–7, citing George's *Introduction*; Nicholas D. Pyenson and Lewis Pyenson, "Treating Medieval Manuscripts as Fossils," *Science, 309* (2005), 698–9.

Resartus, enjoyed a mixed reception, as had George's early autobiographical novellas. In 1835, Thomas Carlyle vacationed in Scotland away from his wife, Jane; in 1925, Mabel spent much of a year in Europe away from George. Thomas Carlyle's remarkable work in the present tense appeared in 1837, the first volume of George's landmark *Introduction to the History of Science* in 1927.[4] Jane Carlyle was known and liked as the intellectual equal and firm supporter of her husband. In the same way, and notwithstanding her desire for an independent identity, Mabel Elwes Sarton provided support and understanding without which, it is safe to conclude, George Sarton would have lost his way.

Mabel Sarton was acutely aware of her sustaining rôle. On 23 November 1924, she affirms:

> In my thoughts I saw you so clearly trudging along & I saw myself beside you, sometimes having to run to keep up, & sometimes walking hands in yours a step in front of you, helping you along. I followed "us" for miles & miles, with me playing back & forth like a little dog, & I felt so deeply proud & happy, because you *really needed* me, to help you along towards your goal. If I took a bye-path, you kept looking for me & you walked near slowly. It sounds quite commonplace written down, but it is something I cannot forget, & your letter gave it to me. You have to be strong for your work, & for us, but *you need us* too, even for your work.

She knows that George loves them. "I feel quietly sure you need us, not just to round out your life, a sort of ornament to it, beautiful but outside it, no, as a necessary part. That is like a quiet, sure hand that comforts the frightened part of me."

The success of a marriage may lie in mutual support without subordination. May Sarton is severe in criticizing her father's neglect of her mother, but the preceding pages have shown that the Sartons enjoyed a deeper and more complex relationship than has been recognized. George and Mabel raised their daughter thoughtfully and generally went along with her choices, both professional and personal. Support does not necessarily imply an absence of dissent. Mabel fiercely defended her artistic sentiment, and she possessed the rare facility of being able to dispute George's opinions gracefully. She writes to George in January 1933 from New York, where she sought (among other things) to identify several ceramic pots that she had acquired in Syria, where the Sartons lived on a sabbatical of George's:

> I spent a good many hours (from 10.15 to 3.30) at the Metropolitan—chiefly in the rooms devoted to Near Eastern art—then to the Ottoman

4. Rosemary Ashton, *Thomas and Jane Carlyle: Portrait of a Marriage* (London, 2003), pp. 85–8, 92, 173–4.

> collection (where there are only a few N. E. Things)—then downstairs to see the exquisite little Greek terra-cotta figure—& the archaïc Apollo. Frankly I am disappointed at the latter—it appears to me heavy & sensual & "maladroite"—heavy in spite of the flattened thighs & simplified modelling—awkward in spite of deliberate "design" & stylisation on muscles—(Sensual is not the word I want: it seems so physically "literal" in spite of stylisation, perhaps that is more it?)
>
> Maybe it will grow upon me. I expected some of the subtlety & astounding elegance & *spiritual* vividness of the smiling women at the Acropolis & was accordingly disappointed & perhaps this mood shut my eyes to other beauty. I did not find your little horse but I shall certainly call upon him.

Here is the sharing of independent opinions that we saw when the Sartons took in museums in the Netherlands on their honeymoon.

Mabel supported George without embracing his ecumenical optimism, which was grounded in what he took to be a rejection of the present. For example, George writes to Mabel on 17 July 1914: "How I envy people without imagination. They are the only ones who live in reality. But I live the illusions and clouds. I don't live in the present, the only reality, but in the past or in the future."[5] In a letter of 23 November 1924, Mabel addresses the point directly. She writes: "I *cannot* see the unity. I *cannot* accept the horrors & unspeakable anguish which puzzled, racked humanity (& animals too) goes through. I *don't* understand & cannot say I am [illegible] to believe there is an explanation." Suffering deprives her of happiness.

> Nature? Yes, outwardly so alluring, so beautiful, & always beneath the beauty, pitiless cruelty, pain, decay, death. The beauty appears to me, & the happiness, an inexplicable accident, a surface thing, a deception which makes the reality worse & more hideous, like the flower a butcher places on the bleeding carcass. There is only one thing I adore & know to be beautiful & true—love—unselfish love—the instinct to protect & *to give*.

Religious explanations are only creations of the human mind, not the projection of a deeper truth, as George suggests. "The whole difference is that you accept & I cannot. You believe there is a pattern, an answer, a unity. I see darkness, with one simple shining frail thread." George

> will say: after death & decay, absorption & resurrection. And I shall say, the pain & death & decay will *go on* repeating themselves. What is the good of beginning over again if for most creatures the sum of pain & fear is so much greater than the happiness? And it is not *our own* that is

5. George Sarton to Mabel Sarton, 17 July 1914.

> the worst. The moment we love, we suffer infinitely more for the beloved than ever for ourselves, & most often we are quite helpless.

Throughout her life, Mabel meditated on the meaning of love. Her affection for George grew deeper at the end. On 26 August 1946 she wrote about it in her characteristic way of connecting small particulars—crossed meaning in a conversation with husband and daughter—with large themes:

> I am writing this because I shall not have a chance before you leave in the morning & it is during the night that I decided I must tell you something of what is in my heart. How are you to know otherwise?—I know so little still of what is in *your* heart because it is not in your nature to open it wide even to me.
>
> I have hesitated more than once to speak of some little thing I had done that I suspected hurt you because I felt it was best for *you* to let it be forgotten quickly—but twice over the last days I have seen a hurt look in your clear face that has haunted me.
>
> You did not understand how it was we laughed when you began telling the story of Mrs Mussolini;—you see you used the expression "only the big tops" were to go to the ceremony. Now I think you meant the 'big bosses' but this expression "big tops" was extremely expressive & *funny* & quite new—a Sartonesque invention!—& it set a *mood* for the story which was entirely wrong. Now that was not our fault—nor of course *yours*. But it hurt you—& the more so that we were *two* against *one*—You must not feel this. I am for you *both*. I love you both with yearning for your happiness,—but you, my husband are more necessary to *my* happiness than May, dearly as I love her. I can let May go to live far away if necessary—I cannot imagine what life would be if you went away.

On Christmas morning 1948, after undergoing an operation for the recurrence of breast cancer,[6] Mabel writes to George:

> I have been thinking of you, and of this last year, and remembering how it troubled me that last night in hospital before the operation that I had never told you *how happy* I have been;—I have sometimes tried to tell you how I loved you, & for this there are other ways than the poor way of words: (just the *tone* of your voice when you call me "Mother dear,"—the tragic, *bereft* look in your eyes when you left me in that hospital room,—these told me more than even your dearest words & so I hope it has been with you)—but to convey to you, the rare not to be measured, happiness, that you have given me, that does need to be said, or you might never guess how deep it was—(& if I had died, one of my last bitter thoughts would have been "he will never know")—but as I lie thinking of this, I

6. Mabel Sarton to George Sarton, 31 July 1948, where Mabel reports from Belgium on the discovery of a lump in her breast.

> realize that happiness is something that refused to be caught in any net of words—that as May says in her poem "Growth of happiness"—it is an inward, invisible thing as mysterious as life itself, nourished from without but also from roots that go wide & deep, far out of sight. In a thousand ways you have nourished those roots, but all I can find to say is "how happy I have been because of you."
>
> One more thing I'd like to tell you: I am humbled, by what people write about my courage: I believe that there is a sort of miracle still granted to us, by the faith we place in those we love: When people believe us to be better than we are, we become better, in a quite extraordinary degree. Because you & May counted on me to be of good courage, it helped me, & continued to help me, to have at least *some*,—so go on believing me to be much better than I am!—(I am often impatient, sometimes discouraged,—but never mind! Often the mood passes, with the help of your confidence in me.)

Mabel associated love with creative work. She comments in a letter of 14 January 1928:

> Yes, it is very true that to lose oneself utterly in love or work or play is to be purely happy, but the remembrance of love, (not its deliberate analysis) can have a poignant, trembling loneliness of its own, like the reflection of a star in a deep pool,—& there is a pure & almost impersonal joy in the realization that some piece of one's work is beautiful and true, a joy which nothing can surpass—it is literally a divine gift, undeserved possible?—one does not feel any personal aggrandisement, rather a kind of illumined humility. One works & works with intensity (or laboriously) with a sort of blindness that gropes & struggles & suffers & doubts, & then one day, out of the rough incompleteness, emerges a blossom, a crystal of truth, that you have to recognize as such, with a throb of joyful recognition.—It does not so much matter that it be a large & important thing, or a small & to other men's eyes, insignificant thing, so long as, within its scope, it has this quality of "completeness." It does not come often, but seems to me the very reason of living. I get it once in a while, with my work of designing. I would starve if I never had it.

During the first 20 years of her marriage, material circumstances notwithstanding, Mabel continually thought about designing. She applied her talents to illustrating books, constructing furniture, and producing stained-glass windows, but embroidery increasingly drew her attention. A traditional prerogative of the leisured classes, embroidery extended to the bourgeoisie of the 19th century. Early in the century the rage was for crudely-worked worsted-woolen yarns from Germany, called Berlin-work in England. By mid-century, thousands of people in London were working in the hand-embroidery trade, part of William Morris's artistic crusade intended to refine, revive, and extend this kind of handwork. Stimulated by Morris and others, the Royal School of Art Needlework opened in the 1872 at South Kensington, London, promoting crewel stitching (which became known in America as the Kensington stitch). Popular interest turned to traditional peasant themes, whether from Russia (marketed by the firm Broderie Russe) or England (in the form of smocking, an English craft focusing on yoke, collar, and sleeves).[7]

Notwithstanding the luddite injunctions of Morris and his aristocratic devotees, invention mechanized embroidery by the last third of the century. More precisely, the inventors of St Gall, Switzerland, dominated the embroidery market with an efficient mechanical loom, whose distinctive feature was a shuttle in the form of a ship's hull (from which the looms became known as Schiffli machines); by 1884 the St Gall embroiderers had more than 20,000 mechanical looms in service.[8] St Gall's command of the industry extended to the middle of the 20th century, when Swiss machines came to America in large numbers. To say that embroidery became mechanized is not to suggest that human hand and eye were superannuated. Into the early 20th century, machines required an operator to follow a pattern and insert design threads, as well as workers to replace bobbins and thread needles with silk—a perpetual, tedious task; by the eve of the First World War, the pattern was controlled by a system of cards. Both machine and hand embroidery, inspired by traditional themes, were in demand through the 1920s.[9]

The American bourgeoisie appreciated traditional motifs from northern Europe, and in the years leading into the Great Depression, they sought to wear handwork that they had neither the time nor the imagination to craft. In her designing, Mabel Sarton benefitted from this fashion of the times. Although she produced machine embroidery cards for the Dangotte firm in Ghent, in America she initially focused on hand designs—executed in the late teens and 1920s by highly trained and motivated European émigrées. In collaboration with her Washington partners, the Gillespies, and following her experience in Ghent, Mabel tried to expand into machine manufacture. The results were disappointing. She writes to George on 14 January 1928:

---

7. Barbara Morris, *Victorian Embroidery* (London, 1962), pp. 7–44; Amelia Peck, "Candace Wheeler: A Life in Art and Business," in *Candace Wheeler: The Art and Enterprise of American Design, 1875–1900*, ed. Amelia Peck and Carol Irish (New York, 2001), pp. 2–86, on p. 23.

8. Clara Spieser and Milo E. Legnazzi, *La Broderie dans la mode; Die Stickerei im Wandel der Mode* (Zurich, [1948]), pp. 136–8

9. Constance Howard, *Inspiration for Embroidery* (Newton Centre, Mass., 1966), p. 12; Anne Wanner, "The Sample Collections of Machine Embroidery of Eastern Switzerland in the St Gallen Textile Museum," *Textile History, 22* (1992), 165–76.

> I have *more* than a week's work before me, to leave the minimum of what Margaret does really need. We have decided that it is *impossible* to get out machine embroidered models in time to shew them in New York, with the hand-done ones. The agent is arriving (from California) earlier than expected & we cannot so far find the chemical gauze which is necessary to "back" the machine embroidery. A law has recently been passed prohibiting its sale as poisonous. Manufacturers are in some cases making their own, but it is a crime to pass on the formula. Also we cannot find an operator & shall have to eventually get one from New York. Thus nearly all my work of this last week is lost, for the present, & I must do a *new set* to make up for the lost machine ones. I have not worked fast—indeed very slowly until now—but I feel so much better, that I believe I can do more this week & more easily than in the last 10 days. I planned 6 new models yesterday—the actual "make" of the dress & placing of the embroidery, which often Margaret prefers to do, but she had no ideas for once, whereas I had many. I enjoy very much planning the whole dress & can usually, I believe, then do a better design—one that "belongs" to the dress more inevitably.

The Sarton-Gillespie partnership continued into the 1930s before being swallowed up by the Great Depression.

George admired Mabel's embroidery, relating on 11 April 1915 that he is "very proud of it," but Mabel maintained a critical, aesthetic distance from her husband. She writes on 3 January 1933:

> I am sad you want to oust my Parsi dress even for this, for I love it very much & it has no hint of "boudoir" prettiness to me. Our dining-room is by no means a place of heavy-laden "boards," dark & rich panelling & so on, where this remark might apply. However I am open to persuasion & I suffer too much from your dislike or scorn of a thing to enjoy it happily. Your personality is too strong for me to waste my strength fighting over small things.
>
> —I do keep a kind of inner territory where even you cannot encroach & which is absolutely necessary to me. I know there are all sorts of things in it that you would dislike or be indifferent to. You'd be very kind & conciliating about them I know, but I prefer just not to submit them to you. There are gentle & delicate opinions & tastes that need silence & absolute protection during their period of growth & blossoming—just as there are others which need opposition to thrive upon.

Today embroidery is most commonly featured as marks of commerce, seals of affiliation, or badges of rank. These standard and dull emblems cannot begin to convey the possibilities of the embroiderer's art. Embroidery is a 3-dimensional construction. The choice and size of threads—silk, wool, linen, metal—is analogous to a painter's brushstrokes and the texture of his paint. To say that an embroidered design is a white flower on a blue field is to report almost nothing of significance. What are the stitches? To what extent is the thread twisted? How does the design play against the background as the fabric drapes a wearer? Mabel Sarton, who from young adulthood designed and executed much of her own clothes, was a master of these practical scruples. It is not stretching too long a bow to contend that her mastery of the discipline of stitching a pattern by hand complements George Sarton's stitching together of *Isis* and his *Introduction to the History of Science*. Needle stitches are his sentences, patterns are his themes, cloths are the sections of his publications. George Bernard Saw referred to "that consciousness of the horrible tediousness of the process which is the main drawback to the rational enjoyment of embroidery."[10] Related criticism dogs George Sarton's *Introduction.*

Although the Sartons shared an affinity for meticulous labor, Mabel's aesthetic sense developed independently of George's. Her career as an artist-designer divides into two periods. From the closing years of the 19th century until 1908 she is a conventional miniaturist. Her academic training in London and in Ghent taught her to draw and paint from life. Surviving sketches are in the nature of student exercises, and her miniatures offer straightforward likenesses. She sold miniature portraits, but the income was insufficient to provide a livelihood.

Mabel Sarton's studies in Zurich at the school directed by Julius de Praetere, between 1908 and 1910, mark a distinct break in her style. Through De Praetere she absorbed the tradition and social thought of William Morris as well as the Jugendstil from Germany, and she came into closer contact with the Flemish Symbolists, who were De Praetere's friends. At Zurich Mabel learned principles of design in a wide range of media—on paper, cloth, leather, wood, glass, and ceramics. Her surviving furniture designs—reflecting De Praetere's interest in working-class homes—have simple, square lines, lacking the stylized pinions and struts characterizing the arts-and-crafts tradition.

Beginning in 1910 she found her artistic signature in two complementary motifs: flowers from the gardens at Assels near Ghent and the Swiss meadows, where she felt most relaxed; and wild birds—her alter ego and preferred term of endearment. In the years leading up to the First World War, birds and flowers appear in her designs for the poetry volumes published by Raymond Limbosch, embroideries, stained glass, and furniture; it is likely that she painted them on the woodwork in her house at Wondelgem. The designs are harmonious and balanced, but their principal feature concerns the use of color. When, living in Washington in 1916,

10. Constance Howard, *Design for Embroidery from Traditional English Sources* (London, 1956), p. 111, Shaw commenting on May Morris's work in the Arts & Crafts exhibition of 1888.

Mabel recalls her garden at Wondelgem, she focuses on a canvas of color. She writes to George:

> The rose-leaves from your bunch, that I put in the yellow bowl fill the room with a sweet faint perfume. Somehow that perfume recalls to me vividly this morning our garden & house in Wondelgem. (I used to collect the faded rose petals in bowls there too).
>
> I want to thank you oh! so much, Dearest, for giving me that wonderful garden—and the dear, quiet house. Even if we never saw it again, it has added a precious store of such happy memories. I loved the garden more than you know—much more—not only its flowers—but its trees—their shapes—the wide spaces of lawn where May made a tiny spot of white.
>
> Do you remember how wonderful the dark copper beeches are when their new young leaves first come out & you stand under them & look up at the light through their leaves? —& do you remember the chestnut—a Spanish one with smallish leaves, to the left of the lawn as you look from the house (just by the little turning of the path to the stable)—that turned a deep golden, shimmering yellow—*all* its leaves at once—& then shed them one by one, with a little silvery, rustling noise that one heard distinctly on quiet autumn days?
>
> And the wild chestnut—young trees that grew in a group between the two lawns & that had the loveliest spikes of delicate greenie-white blossoms in spring-time?
>
> And the kitchen garden—my walk bordered with trembling nodding "ancolies"—& the little stone path with the nasturtiums spreading all over it, & the tangled borders that were a riot of flowers & colour?
>
> Oh yes. You remember— & so do I—& this morning—I can only think of that & be deeply, intensely grateful to you and Fate who granted me that garden and home with you and May—and a few short years of peace and love.[11]

Mabel turned her eye away from color as an abstraction; for her, color is connected with natural form. On 4 December 1924 Mabel reports to George about touring a large factory in Belgium where Congolese designs are reproduced by hand. In French: "There are many things I hardly like: very stark colors in large lines and squares. But there are others that are really very, very beautiful, and I think that they would sell well in America."

By mid-century, those stark colors and bold patterns had moved to the center of fashionable design, leaving little room for Mabel's peasant-themed embroidery. Mabel brought her avant-garde experience to America, but (as may be seen from the illustrations in the Appendix) during the years of uncertainty and poverty, she essentially elaborated pre-1914 themes. Notwithstanding trips to Europe, Mabel failed to perceive

11. Mabel Sarton to George Sarton, 29 June 1916.

that style in the 20th century was changing quickly; the public had developed a taste for innovation—leading some critics to postulate the end of the traditional notion of style itself.[12] In its continuity from 1909, Mabel's artistic enterprise matches her intellectual sensibility. Her thoughts and feelings from the 1920s often seem close to her outlook decades earlier. Mabel Sarton's is a life of great integrity. This admirable quality she shared with her husband.

* * *

Dryasdust, the fictional pedant writing about the past, appears as an object of derision in Thomas Carlyle's writings. In Carlyle's time and afterwards, history was an industry for publishing and editing texts. The historical discipline in Germany found a reference point in Georg Heinrich Pertz's *Monumenta Germaniae Historica* just as the discipline in the United States, in George Sarton's own time, drew strength from J. Franklin Jameson's series on the documents of American diplomacy. The past as dry chronicles pales alongside tales of heroic striving, Carlyle contended. In his large publications, which as much as his shorter and polemical works gave direction to British tastes, Carlyle focused on significant individuals like Johann Wolfgang von Goethe, Oliver Cromwell, and Frederick the Great.

George Sarton is Carlyle's heir in the sense that he directed history of science to focus on individual scientists. The *living* past, as George often expressed it, is what matters. He and Carlyle held that people have not changed since antiquity. George emphasizes in 1947: "The passions of men were very much the same in the fourteenth century as today. The men themselves were very much the same under their skins. A few saints, a few rascals, and a great mass of people who are good or bad according to circumstances; so it was, so it is, so it will ever be."[13] Neither have adversity and opportunity changed over time. Accomplishment in politics, art, or science can be measured by one standard—Carlyle listed heroes of various kinds, and George recovered important scientists in many disciplines. George's conclusion is expressed simply: Just as it is possible to identify significant religious,

---

12. John Willats has called into question the entire notion of historical period. He detects no progress in a culture toward a single, coherent style, but rather, over the past 200 years, increasing diversity of style. In his view, historians should be sensitive to a wide range of iconic representation at any given moment: "Different pictorial styles are not like different languages: some styles are better suited than others to perform particular functions." John Willats, *Art and Representation: New Principles in the Analysis of Pictures* (Princeton, 1997), pp. 352–3.

13. George Sarton, *Introduction to the History of Science, volume 3: Science and Learning in the Fourteenth Century* (Baltimore, 1947), p. 19.

literary, and political leaders through the ages, so one can identify significant scientists.

On the eve of the First World War, history of science was an established feature of European scholarship, with learned journals and university courses (we have seen that the courses also ran at many American institutions). For the most part the courses and the monographs associated with them were a chronicle of ideas disembodied from thinkers as living people. Monographs focused on apparently timeless fields of inquiry such as planetary models, theories of elasticity, or notions of the luminiferous ether; works from this period that are still read today by authors like Ernst Mach and John Theodore Merz are histories of ideas. People associated with the ideas received more attention in the éloges memorializing them than in works of critical scholarship. The éloges, for example of the Paris Academy of Sciences, were not infrequently polemical pieces, and authorized life-and-letters, like the biography of Isaac Newton by Thomas Carlyle's patron David Brewster, could suppress relevant documents (in Brewster's case, Newton's alchemy).[14]

George Sarton believed that people, things, and institutions constituted the essential part of history of science. The focus on people is evident from his early writings to his last ones. The chronological sections of his *Introduction to the History of Science*, for example, bear the title of "the most representative man of the period considered."[15] Heroes were with George from his youth, and as late as September 1914 he still planned to write a "'heroic' biography," but Wagner's heroes progressively receded in favor of lay saints.[16] In 1905 George was moved by the death of Louise Michel, the writer and anarchist member of the Paris Commune: "This saint! This martyr for our cause. This sublime woman."[17] In an early letter to Mabel, George asks for a print of Donatello's St George to help him study.[18] Indecisive about his future, he "doesn't know which saint to worship."[19] Saints are often martyrs for the cause, and George saw his life as one of martyrdom for a scholarly ideal. Saints are represented in icons, and in his plans for a future institute, George made room for graven images of

14. David Brewster, *The Life of Sir Isaac Newton* (London, 1831); Betty Jo Teeter Dobbs, *The Foundations of Newton's Alchemy, or, 'The Hunting of the Greene Lyon'* (Cambridge, 1975).

15. George Sarton, *Introduction to the History of Science, volume 1: From Homer to Omar Khayyam* (Baltimore, 1927), p. 39.

16. Diary, 5 September 1914.

17. Diary, 18 January 1905.

18. George Sarton to Mabel Sarton, 10 June 1907.

19. George Sarton to Mabel Sarton, 28 November 1908.

scientists.[20] When the family was virtually penniless, George assembled a collection of medals.[21] He writes at the end of his life: "I am deeply interested in saints because they are rebels against material comfort and money, defenders of spiritual life. They are the heroes of the human conscience."[22] George regretted the theft of only two items from his house in Wondelgem during the First World War: Moritz Cantor's *Vorlesungen*, with corrections added by hand, and a print of Jean-Jacques Rousseau with the dedication, "Aux Ames sensibles."[23]

George's interest in people went beyond the greats. As his students Robert K. Merton and I. Bernard Cohen have emphasized, part of his contribution was to recover smaller figures, weaving them into the fabric of science past.[24] The impulse is present in his early novels and his literary

---

20. In a typescript outlining an institute devoted to the history of science, George indicates the importance of "iconographical collections" and photographical facilities. NYPL undated [1916–1917]. The inspiration may come from Karl Lamprecht's historical institute in Leipzig, which George cites as a model. Lamprecht collected many thousands of drawings by children in an attempt to formulate a law comparing the evolution of civilization with child development. Lewis Pyenson, "Uses of Cultural History: Karl Lamprecht in Argentina," *Proceedings of the American Philosophical Society, 146* (2002), 235–55; Pyenson, "Prerogatives of European Intellect: Historians of Science and the Promotion of Western Civilization," *History of Science, 31* (1993), 289–315 on pp. 293–4.

21. Late in the summer of 1916, Mabel surprised him with the gift of a medal of Whistler from the Ogunquit sculptor, Victor D. Brenner, who designed Lincoln's face on the penny. It is his third medal (one of the others is by Mabel's friend Harold Stabler). Diary, 1 September 1916. The medal collection survived at YMS. Possibly the Stabler medal is by Phoebe Stabler, inscribed "Peace & Hope"; the other two medals George acquired by 1916 are perhaps of J. H. Fabre by F. Sicard (1910) and Oscar Montelius by Erik Lindberg (1913). F. O. Bailey Antiquarians, *A Fine Auction of the Estate of May Sarton, Sunday, May 19, 1996* (Portland, Me., 1996), pp. 7, 9.

22. HGS, Diary, 28 March 1953. He observes that standards for canonization vary over time. As an illustration, he chooses Isidore of Seville, "a great name in the history of the Spanish Church and in the history of mediaeval science," commenting that his 2 brothers and his sister were, incredibly, also saints.

23. Diary, 7 February 1954.

24. Merton's self-indulgent book *On the Shoulders of Giants: A Shandean Postscript* (1965; San Diego, 1985) devotes much space to George Sarton's pantheon "erected for the giants of science and learning as well as for the indispensable dwarfs who climb upon their shoulders" (p. 39). Cohen emphasized George's concern for small figures in a conversation at the Residencia de estudiantes, Madrid, in 1991. It is tempting to relate George's biographical bent to the *Dictionary of National Biography*, which Sir Leslie Stephen edited until 1891 and for which Stephen "wrote biographies of hundreds of lesser-known figures." Jeffrey Paul von Arx, *Progress and Pessimism: Religion, Politics, and History in Late Nineteenth Century Britain* (Cambridge, Mass., 1985), p. 59. For an assessment by a distinguished historian of chemistry of the successor to Sir Leslie's work: William H. Brock, "Chemists and the *New Oxford Dictionary of National Biography*," *Chemical Heritage, 23*, no. 1 (2005), 28–31.

taste, in his popular writings, and in his catholic vision of scholars and scholarship. George was a private person—a personal trait wrongly interpreted by those who knew him later in life as aloofness. He was no snob. He enjoyed meeting and talking with "civilians, Philistines, Pharisees and publicans," as he wrote to Mabel on 11 January 1925, an inclination that may not have endeared him to the Boston Brahmins upon whose charity he depended. His rehabilitation of small figures in history of science found support in Mabel's adherence to the arts-and-crafts ideal of William Morris—the satisfaction of an unsung and constructively occupied life. In a letter to George from December 1924, Mabel affirms that she has a career that entirely suits her. In French: "While knowing very well that it is humble work that will remain anonymous, I am proud of it, and I am in fine company with all the artisans of handicraft of the past."[25] George's interest in smaller figures is behind his support of Robert K. Merton's study of 17th-century science, which features a prosopography of Royal Society members.[26] People accordingly dominate George's lectures on the Renaissance, delivered at the end of his life, where he devotes "more space for lesser men, whose humbler contributions were nevertheless of great importance."[27]

From people to things. The years in Ghent taught George Sarton the value of laboratory equipment. He saw experiments as essential for introducing students "to the handicraft part, the most living part, of science."[28] One of his earliest professional publications emphasizes: "A history of science obtains its full importance only if it is continually confronted with the history of civilization or, at least, with the history of technological progress."[29] In his initial manifesto for *Isis*, George gives a place of honor to the history of technology.[30] His first lecture course at Harvard treated tools and machines.[31] The Wondelgem house ran on pumps and motors, the maintenance of which caused continual worry. On the eve of the First World War, George cautioned Mabel that the house had no

25. Mabel Sarton to George Sarton, December 1924, undated, between St Nicholas and Christmas.

26. Lewis Pyenson, "'Who the Guys Were': Prosopography in the History of Science," *History of Science, 15* (1977), 155–88, p. 166 for Merton.

27. George Sarton, *Six Wings: Men of Science in the Renaissance* (Bloomington, 1957), p. viii.

28. George Sarton, "The Teaching of the History of Science," *Scientific Monthly,* 7 (1918), 193–211.

29. George Sarton, "Albert Ladenburg," *Revue générale des sciences pures et appliquées, 22* (1911), 937–8.

30. George Sarton, "L'Histoire de la science," *Isis, 1* (1913), 3–46.

31. The course was recorded on note cards by Yuen Ren Chao and is now kept in the Niels Bohr Library of the American Institute of Physics in College Park, Maryland.

running water due to the failure of the dynamo that supplied electricity to a motor for raising water to a tank in the attic.[32] George was not himself adept at mechanical manipulation—he resisted mastering the arts of typing and driving an automobile—but he admired technological ingenuity even if he was not clever at inventing things.[33] Arriving in Washington in 1915, for example, he connected a past literary interest with a future scholarly focus on Arabic in the context of a technical invention. "While I was reading the *Arabian Nights* after dinner, and while I was dreaming like one of Francis Jammes' young creoles, I got at once the idea of a new kind of electric energy motor." He drew up a prospectus, only to find that the motor had already been patented.[34] No photographer himself, for Christmas 1920 he gave Mabel the present of an enlarging camera.[35] George was, after all, the son of a railway engineer whom he loved and grew to maturity in an environment suffused with heavy industry. Later in life he grew more distant from technology. He sums up the matter in a letter to his daughter, explaining why he has given up driving:

> I don't like machines of any kind—I don't like to think of them or about them. As a historian of science I find myself in a paradoxical situation for I hate this machine age—which is a result of science, and yet I love science. Scientists are not by any means innocent of the present chaos, though it is clear that they have no control over the applications of science, whether these be merciful or criminal.[36]

Perhaps we read here a desire for stability from someone in middle age.

People and things are brought together in institutions. From his student years George brought people together for a common goal. His organizing drive went from student societies to socialist groups. An intense desire to organize the literature of history of science is evident from the beginning in *Isis*, in his plans for a union catalogue of British libraries, and in his efforts during the First World War toward building institutes for the history of science and for propagating American culture in Europe. The History of Science Society, which prospers today, was founded for George Sarton in 1924. He is also among the proponents of élitist gatherings, such as the International Academy of the History of Science, and notwithstanding his "pariah" status, for nearly 40 years he often ate at the Harvard Faculty

32. George Sarton to Mabel Sarton, 17 July 1914.

33. NYPL, George Sarton to Raymond Limbosch, 6 July 1921. George doesn't have an automobile because he prefers "by far to walk on the legs that God has given me." Later he acquires a vehicle.

34. George Sarton to Mabel Sarton, 26 and 27 April 1915. George drew up a prospectus on 25 April 1915, corrected it on 27 April. YMS, for the correction on 27 April 1915.

35. Agenda, 11 December 1920.

36. YMS, George Sarton to May Sarton, 25 June 1939.

Club. One of his earliest publications in history of science is an institutional history of the Royal Society of London.[37] In George's mind, there was an enormous spectrum of institution, extending from royal societies to folkloric credence. In 1925, for example, he warmly greets the appearance of Marc Bloch's *Rois thaumaturges*, a work about the ability of kings in medieval Europe to cure scrofula by touching a victim.[38]

Words, for George Sarton, do not make the world; they describe it. In the first volume of his *Introduction,* he contrasts medieval conflation of words and things with the modern scientific outlook:

> It is only after centuries of apparently sterile but necessary quarrels and after the final establishment of the experimental method and attitude, that we have slowly learned to consider words as symbols, which, as far as scientific purposes are concerned, would be usefully replaced by arbitrary signs having no signification but the one explicitly defined....The emancipation of science from verbalism was as slow and painful a process as its emancipation from theology; it is not yet completed in either case.[39]

He makes the point another way in an essay on Leonardo:

> It may seem strange that [the] technical basis of the Renaissance has been constantly overlooked, but that is simply due to the fact that our historians are literary people, having no interest whatever in craftsmanship. Even in art it is the idea and the ultimate result, not the process and the technique which engross their attention. Many of them look upon any kind of handicraft as something menial.[40]

In his view, nothing deep would result from the abstract, philosophical approach to studying history of science, an approach dominated by literary writers.

The philosophical approach, under the influence of talented scholars like Gaston Bachelard and Alexandre Koyré, continued from the first half of the 20th century into the second half; Michel Foucault's reception in North America is emblematic of its triumph. The technical side of history of science, into which George Sarton never went deeply, continued into the 20th century in a direct line from Jean-Etienne Montucla through Walter William Rouse Ball to Bourbaki and to Otto

37. George Sarton, "Le 250e Anniversaire de la Société royale de Londres," *Revue générale des sciences pures et appliquées, 23* (1912), 306–12.

38. George Sarton, in *Isis,* 7 (1925), 520–1.

39. George Sarton, *Introduction to the History of Science, volume 1*, p. 7.

40. George Sarton, "Leonardo and the Birth of Modern Science," *The Life of Science: Essays in the History of Civilization* (New York, 1948), pp. 65–82, on p. 79.

Neugebauer and his students.[41] Neugebauer almost completely avoided discussion of personalities and biographies, an inclination shared with other scientifically trained colleagues like Clifford Truesdell.[42] Neither philosophy nor technical glossing, however, have kept pace with fascination for lives of scientists. Sartonian emphasis on lives found an extension in the *Dictionary of Scientific Biography*, an enormous project ably edited by Charles Coulston Gillispie which engaged a large part of the community of historians of science in the 1960s and 1970s. Only a minority of historians of science today fathom the depths of philosophy or restrict their attention to concatenation of scientific ideas, but many historians of science have published a book or an article focusing at least in part on the narrative of a life.[43]

By the time that the first volume of his *Introduction to the History of Science* appeared in 1927, George Sarton had achieved his style as a scholar. Much has been written about style in scholarship over the past generation, producing a large literature addressing whether style may be considered *sub specie aeternitatis* or whether, indeed, it exists at all.[44] However a reader evaluates the matter, there is no mistaking the tension in a text of George's. Human expressions of self-justification and moral imperative contest for attention with the dispassion and interdiscursivity required by science. The narrative seems to oscillate between grand themes and punctual circumstances, reinforcing George's belief that accomplished scientists insulate themselves from the gritty context of

41. Noel M. Swerdlow, "Montucla's Legacy: The History of the Exact Sciences," *Journal of the History of Ideas, 54* (1993), 299–328 and the broader treatment in José María López Piñero, "Las Etapas iniciales de la historiografía de la ciencia: Invitación a recuperar su internacionalidad y su integración," *Arbor, 142*, nos 448/559/560 (1992), 21–67.

42. Lewis Pyenson, "Inventory as a Route to Understanding: Sarton, Neugebauer, and Sources," *History of Science, 33* (1995), 253–82.

43. For a different view, Michael Shortland and Richard Yeo contend that while "in the early twentieth century history of science was still largely informed by a biographical approach....since the 1940s a concern with the 'merely personal' has been seen as peripheral to the aim of understanding the scientific enterprise as a collection of institutionally based cognitive disciplines." Shortland and Yeo, "Introduction," in *Telling Lives in Science: Essays on Scientific Biography*, ed. Shortland and Yeo (Cambridge, 1996), pp. 1–44, on pp. 4–5. According to the 2002 *Guide to the History of Science* published by the History of Science Society, however, biographical interest remains strong among laureates of the Society's many prizes over the past half century; the exception is the Pfizer book award, which since 1985 went only once to a biography.

44. For a discussion of the former view, elaborated by Hayden White: Lewis Pyenson, "Three Graces," in *The Strength of History at the Doors of the New Millennium*, ed. Ignacio Olábarri and Francisco J. Caspistegui (Pamplona, 2005), pp. 261–335. Arguing in favor of a dissolution of the notion of style in art during the modern period, Willats, *Art and Representation*, pp. 352–3. More generally, Lewis Pyenson, "National Culture and Styles," in *Oxford Companion to the History of Modern Science*, ed. John L. Heilbron (Oxford, 2003), pp. 557–9.

daily life. In George's view, science and scholarship are free creations of the human spirit. They depend on the imagination of individual people. The point is shared with music and the fine arts. People are not strangers to their own time, and much creation would be impossible without material support, which may impose conditions on the final product. But the ideas are not extrusions of material culture. Nor are they a code for particular forms of human interaction.

George Sarton continually emphasized that science and scholarship assimilate past understanding. In this sense, they are historical processes. The first duty of a scientist or scholar is to master the literature. Written words, as a result, provide the measure of a professional life. Yet putting thoughts on paper, which George saw as a noble obsession, undervalues the tacit understanding and spontaneous insight that George enjoyed in the give and take of oral exchange. If he had been more in tune with this tension in his personal life, George might have been led to examine the extent to which publication represents the justification of a new idea, rather than merely its revelation. George took pains to distinguish science from sorcery (about which he had little good to say), but he was certainly attracted to the revelations of religion, and he appreciated monuments to faith. Spiritism also attracted him. A man of George's constitution who listened to himself might have arrived at Alexandre Koyré's insights about the irrational and mystical origins of scientific writing.[45] Furthermore, had George faced up to his sexuality rather than merely noting it, he might have been able to provide an additional dimension to his passion for history of science.

If science and scholarship are moral undertakings, it followed for George Sarton that there is right and wrong in both enterprises. Hence the possibility of scientific sainthood. Hence also his uncompromising judgments, rendered in book reviews in *Isis*. To observe wrongheadedness, in his view, is a constructive gesture, serving to focus the attention of authors and readers on the proper task at hand. In George's mind, the work is distinguished from the person. Bad works merit dispatch to Coventry, but people are frail, and they deserve forgiveness rather than ostracism. When George himself felt wronged by someone, he raised a cup of cheer, in the refrain of "Auld Lang Syne," and turned the page.

Isolation, the state that *Isis* intended to ameliorate, is nevertheless the natural condition of a scientist. George followed the spirit of his times in believing that scientists and scholars should be left to their karma, and that funding for their work should be a matter of dharma.[46] Science is a

45. For example, Alexandre Koyré emphasized in the 1930s that among 16th and 17th-century writers like Paracelsus and Valentin Weigel, the astral body is the immaterial seat of life, an energetic fluid that gives unity to a body and provides life to the seed. Koyré, *Mystiques, spirituels, alchimistes du XVIe siècle allemand* (Paris, 1971).

solitary undertaking, often undertaken in secret. As it is with the master, so it should be with the acolyte. It followed for George that elaborate course requirements are an impediment to learning. In Belgium, he reflected on educating infant May, "that is, helping her go in her natural direction. Does not all the value of education reduce to that?"[47] Ten years later, he brought the method into practice as he taught his daughter:

> Our most pleasant afternoon was spent at the Boston Museum of Fine Arts. I had planned to show her the Western paintings but did not try in the least to impose anything upon her. In fact she guided me more than I guided her. I gave her all the information she asked for & I gave my own opinion too if she seemed to want it. We had arranged that she would make a note of the pictures she liked best and would buy postcards of them: This proved an excellent stimulant. It obliged her to look at things and compare them more carefully. I believe that our afternoons were not simply pleasant to her, but profitable as well. I am more and more convinced that education must be suggestive rather than dogmatic. I am anxious to show her all the great things that have been done & can be done; to open doors, & also to protect her, *to a certain extent*, against evil influences. That is all that we can do, but it is not inconsiderable.[48]

And on 9 November 1924 he writes to Mabel: "I have sometimes thought that the duty of education, or at least with regard to intelligent children, is simply to act as a Guardian Angel, who protects and gives hints—the slightest hints only. For example he would leave good books on the child's table & disappear; then during the night, bring other books & disappear again." Dogmatic about the nature of science, he was ecumenical in supplying sources and summaries pertaining to its history. His goal, imperfectly achieved, was to provide a published foundation for the use of succeeding generations rather than giving rise to a school of followers.

At the beginning of his permanent installation at Harvard, George Sarton published a piece on style. George extends an earlier, pre-war thought connecting style to soul by emphasizing: Style "does not simply characterize the individual who uses it, but expresses the thought and feeling of a great many more, maybe of a whole nation, of a whole period, or, better still, of the whole race. The larger the group, or the longer the time the soul of which it succeeds in evoking, the higher the style." This is how one should look at Greek or Chinese style. The

46. Brigitte Schroeder-Gudehus, "The Argument for the Self-Government and Public Support of Science in Weimar Germany," *Minerva, 10* (1972), 537–70, for one expression of the belief.

47. Diary, 31 August 1913.

48. George Sarton to Mabel Sarton, 8 July 1923.

eternal qualities of style are opposed to fashion. "It is the very essence of fashion to flee away as fast as snobbism and business may demand it." The villains are "frivolous women and greedy merchants," whose concern for fashion makes it impossible for them to achieve style. Style derives from both "tradition and originality," and it is a tricky business. "The greatest revolutionaries of art and letters only succeeded because they kept to a large extent within the traditions of the past; sometimes their revolutionary effort consisted simply in reconciling divergent traditions." Scientists penetrating the secrets of nature offer writings that rival the greatest paintings, and the power of their vision is a function of their style. He cites Alfred North Whitehead: Style is "the ultimate morality of the mind." The path ahead, George writes, is for writers and artists to be inspired by science. The inspiration will create "the efflorescence of a new literature more pregnant and possibly more beautiful than anything of the past." And for this to happen, young people must be taught by scientists as much as by humanists. Scientific training, by encouraging "a deeper respect for the truth," would encourage good style. George looks forward to new humanists, who can understand nature as well as humankind, people trained in clear writing. He wants clarity of expression. He wants to discourage misty and obscure writing, in non-fiction and in fiction. But most of all he wants heart. "Is not any writing which has been deeply meditated, and has at last burst straight from one's heart, a message of love?" He concludes, paraphrasing his early novella, *Une vie de poète*: "If one does not write with one's own blood, what is the use of writing at all?"[49]

George Sarton certainly put much of himself into his writing. As a young man he was interested in everything new, but he tried to avoid fashion, as he saw it. His meaning and his judgments are seldom obscure. The early years of struggle, in his marriage and in his career, forged the outlook that he carried forward from the middle 1920s, when he was in his forties. Emotional and material deprivation furnished him with a new need. Added to an affirmation of truth-seeking, a belief in social justice, and a veneration of the past was his cry for love, the intimacy that he labored so hard to achieve. In one of his last writings, he concludes a long evaluation of Leonardo da Vinci:

> Leonardo was a defender of reason, an enemy of superstition, but an idealist. There was in him an original and deep conviction that the only things that really matter are spiritual. The supreme discipline is that of

49. George Sarton, "Science and Style," *Scribner's Magazine, 69* (1921), 755–9. He intended the article for the *Atlantic Monthly*. Diary, 26 December 1920. George Sarton, "L'Histoire de la science et l'organisation internationale," *La Vie internationale, 4* (1913), 27–40, on p. 37: "Style is not a literary quality, it is a quality of the soul."

> love. As he put it, "The love of anything is the fruit of our knowledge of it, and grows as our knowledge becomes more certain."
>
> One might add, and no doubt he did so, that without love there can be no real knowledge.[50]

His notion of love involved intense spirituality—it incorporated his religiosity and his idealistic social activism. Through Mabel, he was able to express love in carnality, although he never mastered the intricate dance of life and life's work.

George Sarton's scholarly failures reveal much about his *modus operandi*. First among his failures is a concrete project for his patron, the Carnegie Institution of Washington. It was to be a comprehensive edition of Leonardo da Vinci, illustrating the contention that science and art are complementary creative domains. A 19th-century impulse moves the project. It derives on the one hand from the research ethic of German universities, and especially the separation of pure or fundamental science from applied science and technology, and on the other hand from the invention of art for art's sake, and especially the stereotype of the artist as an avant-garde outsider—a picture known and loved in Giacomo Puccini's *La Bohème,* itself a *fin-de-siècle* production.

George's contention received elaboration early in the 20th century by Albert Einstein. In an address of 1918, titled "Motives for Research," delivered on the occasion of Max Planck's sixtieth birthday, Einstein observes that scientists and artists both seek to "escape from everyday life with its painful crudity and hopeless dreariness, from the fetters of one's own ever shifting desires." In addition, the scientist or the artist each

> tries to make for himself in the fashion that suits him best a simplified and intelligible picture of the world; he then tries to some extent to substitute this cosmos of his for the world of experience, and thus to overcome it....Each makes this cosmos and its construction the pivot of his emotional life, in order to find in this way the peace and security which he cannot find in the narrow whirlpool of personal experience.[51]

Three years later Einstein elaborates in a short statement:

> What do artistic and scientific experience have in common? Where the world ceases to be the scene of personal hopes, wishes, wants, where we face it as free creatures, admiring, questioning, beholding, there we enter the realm of art and science. We do science when we reconstruct in

---

50. George Sarton, *Six Wings*, p. 233.

51. Albert Einstein, "Motives for Research," in *The Collected Papers of Albert Einstein, volume 7, English Translation*, trans. Alfred Engel and Engelbert Schucking (Princeton, 2002), pp. 41–5, on p. 43. Einstein does not mention George Sarton, but George cites Einstein: George Sarton, *Introduction to the History of Science, volume 3*, p. v.

> the language of logic what we have seen and experienced; but when we communicate through forms whose connections are not accessible to the conscious mind, yet we intuitively recognize them as something meaningful—then we are doing art. Common to both is the loving devotion, the being above the personal, removed from our will.[52]

Einstein, in common with George Sarton, knew the world of industry, but he emphasized the anteriority, or priority, that science and art enjoy in society. An extension of this notion is the principle that society owes a living to scientists and artists, a persistent trope in George's writings. Einstein's formulation of the complementarity of art and science, like George's, can assimilate quantum physics and surrealist art, but neither Einstein nor George Sarton felt comfortable with uncertainty, disharmony, or paradox.

The uncertainty and disharmony of the early 1920s was as inauspicious for undertaking a large project on Leonardo as it was for launching George Sarton's institute for the history of science. George mentions that his focus on Leonardo dissolved as he devoted himself to learning Arabic and compiling notes for a universal history of science.[53] A deeper incompatibility, nevertheless, underpins George's disaffection with the archetypical Renaissance man. At the end of his life, George understood that Leonardo, the apparent unifier of art and science, was in fact no scientist. Leonardo, a genius, was not a scholar. His knowledge of sources was thin, his commitment to observation was spotty (as in his fanciful illustrations of a human fetus and a human heart), and his sense of truth came from the world of tradesmen and courtiers rather than from philosophers. He employed new techniques but performed no experiments. He kept his ideas secret, rather than publishing them—for years he held back even his canvases. He was, in short, outside the *universitas* of science.[54] George admired outsiders who stood behind their ideas, not ornaments of courtly society like Leonardo who were diffident about contributing to the improvement of humanity.

An understanding of Leonardo's science derives from a painstaking study of his manuscripts. That requirement in itself would not have commended the project to George Sarton. He seldom sat with a primary source long enough to extract new insight from it. Like Alice's White Rabbit, wherever he was, he perpetually looked elsewhere. In the end he achieved general understanding of a sort granted to few scholars, but it was always

52. Albert Einstein, "The Common Element in Artistic and Scientific Experience," *ibid*, p. 207.

53. As he describes in Sarton, *Six Wings*, p. 306.

54. Sarton, *Six Wings*, p. 228. A recent review by Adam Gopnik is helpful. Gopnik, "Renaissance Man: The Life of Leonardo," *New Yorker*, 17 January 2005, pp. 82 ff.

at the expense of appearing as a dilettante to his specialist colleagues. (Dilettante, ironically, is the condemnation George often directed toward other people.) To begin a career under the banner of synthesis, as he did, is to invite disaster, and especially so at the beginning of the 20th century, when the path to truth was paved in specialist endeavor. George wrote a doctoral dissertation on Newton's celestial mechanics, and then plunged into studying Newton's predecessors and successors; no copy of the dissertation survives. He spoke for a discipline, but he was diffident about taking the time to train doctoral students himself. Humanity looms large in his writings, but it is a quality that often seems passed by in his private life. The astonishing fact is that, notwithstanding these failures, George Sarton succeeded in bringing Paul Tannery's understanding of history of science into a form that appealed to early 20th-century tastes.

Tannery, a *polytechnicien* who over the last third of the 19th century worked for the French tobacco monopoly, became dean of French historians of science by dint of unusual talent, personal drive, and a supportive wife. In his writings, restricted for the most part to the exact sciences, he ranged from classical antiquity to his own times. His work touched every part of the modern scholarly enterprise: critical editions of published and unpublished sources, translation, commentary, encyclopedia entries, response to queries, and international organizing. At the beginning of the 20th century, he forged a European scholarly discipline from scholars working in isolation across the continent. To Tannery, and to him alone, belonged the chair of history of science at the Collège de France when the chairholder, Auguste Comte's disciple Pierre Laffitte, died in 1904.[55]

Many administrators miss an opportunity to hire the right person for a position, but even in a context where personalities and professions of faith are valued more than brilliant publications, naming the positivist Grégoire Wyrouboff as Pierre Laffitte's replacement instead of the ardent Catholic scholar Tannery carries an air of malice. As George Sarton saw the situation, Wyrouboff was "a diligent physico-chemist and the leading positivist of his time," and the French minister of education wanted a lecturer and a Comtean who could speak about modern scientific theories.[56] Paul Tannery appealed to George because of Tannery's idealistic view that the vitality of science, indeed of culture, depended on clarity of spirit. In the éloge of Tannery written by Wyrouboff's successor Pierre Boutroux, which George published posthumously, Boutroux emphasizes Tannery's belief that past declines of science derive from "a cessation of the active thinking that animates science with spirit," as well as Tannery's

55. Pierre Speziali, *s.v.* Tannery in *Dictionary of Scientific Biography*, ed. Charles C. Gillispie (New York, 1970–80); Pyenson, "Prerogatives of European Intellect."

56. George Sarton, "Paul, Jules, and Marie Tannery (with a note on Grégoire Wyrouboff)," *Isis, 38* (1947/48), 33–51, on p. 40.

desire to find through historical research "the way of encouraging this creative thought which alone allows science to move forward." Science, in a concluding thought broken in Thucydian mid-sentence, is always threatened with decadence, "and can live only if those who watch over it remain constantly alert...." [57] In the text, Boutroux shares George's view that times of decadence are due to a failure of collective genius, and that historians should be activists in the present.[58]

In his éloge of Tannery, Pierre Boutroux alludes to Grégoire Wyrouboff's insistence that science experiences "periods of crisis which, in certain epochs, break the continuity of the development of science and which have, as an effect, the creation of new disciplines and methods."[59] Boutroux shared Wyrouboff's view of the irregular progress of science. His endorsement of Comtean stages reflects his appointment to the chair founded in homage to Comte the Positivist, but in practice Boutroux was a mathematician of distinction and an expert of mathematics past who knew pragmatic America first-hand (he taught at Princeton University for a number of years).

Positivism declined at the end of the 19th century, but its reach has extended (at least among scientists and the educated public) into our own time.[60] Given the persistence, what shall be made of George Sarton's appeal to "positive" knowledge and to the cumulative nature of science? George certainly venerated Tannery, but he was ambiguous about Auguste Comte. Reviewing 3 biographies of Comte in 1937 (one of them by George's friend, the English positivist Francis Sydney Marvin), George is clear about Comte's unappealing dogmatism, which he labels as obstinacy and mulishness. Comte shared this dogmatism with René Descartes and Francis Bacon: "The effort needed to create their theories exhausted their minds and ended by making them impervious to other thoughts." George emphasizes Comte's mental breakdown in youth, his abstention from reading as a mature man, and the scholasticism of his schematic diagrams.[61]

George Sarton's *Introduction to the History of Science* bears little resemblance to a Comtean vision of the past. Absent in George's writings

57. George Sarton and Pierre Boutroux, "L'Oeuvre de Paul Tannery," *Osiris, 4* (1938), 690–705, on pp. 700, 702.

58. George Sarton, "East and West," in *The History of Science and the New Humanism* (1937; New York, 1956), pp. 59–110, the failure of genius on p. 76.

59. Sarton and Boutroux, "Tannery," p. 701.

60. Lewis Pyenson and Susan Sheets-Pyenson, *Servants of Nature: A History of Scientific Institutions, Enterprises and Sensibilities* (London, 1999), pp. 419–23, for positivism's eclipse. Paul R. Gross and Norman Levitt's successful and entertaining book *Higher Superstition: The Academic Left and Its Quarrels with Science* (Baltimore, 1994) qualifies as a positivist commentary, from the point of view of two scientists, on postmodernist literary thought.

61. George Sarton, in *Isis, 26* (1937), 470–7.

are Comte's stages of human history; rather, the advancement of learning proceeds in fits and starts. "The intellectual progress of mankind would not be correctly represented by a constantly increasing function, but rather by a sort of sinusoidal curve moving steadily upward," he emphasizes as he discusses the middle of the 10th century. The history of humankind mirrors the life of an individual, who has times of invention and accomplishment as well as times of quiescence. Indeed, it is hard to maintain the momentum of a time of great discovery, which inevitably generates uncertainty. In George's view, all revolutionary times have a similar character, because accomplished scientists have always shared similar goals and possibly even methods:

> Science always was revolutionary and heterodox; it is its very essence to be so; it ceases to be so only when it is asleep. The mental anxieties of some of our contemporaries, who are afraid of knowledge, have been shared in all times by some other men of that kidney....Mankind does not go forward as a united body; on the contrary, each advance has to be paid for by a protracted struggle between those who long for more light and those who are afraid of it. The latter are far more numerous than the former, but less intelligent, and thus bound to be beaten in the end; this accounts at once for the sinusoidal advance and for its upward tendency, or in other words, for the slowness, but also for the continuity of human progress.[62]

People have always sought creative leisure and financial support for their projects, and they have always tried to remain intellectually at arm's length from their patrons. Some scientists thrive on money from bloodthirsty rulers; others perish. The nature of the support determines little about the form of significant discoveries. History is like theatre. Actors enter and depart. The drama—worthy of Carlyle—"occurs on a gigantic, superhuman scale; the actors are not puppets, but real men; the time of action is not counted in days, but in centuries; the scene is the whole world; the playwright is inscrutable Destiny."[63]

The religious side of Comte exerted an attraction on George, whose Catholic apostasy was a perpetual source of concern for his Aunt Hélène.[64] George makes the point in the last volume of his *Introduction*

62. George Sarton, *Introduction to the History of Science, volume 1*, pp. 646–7.

63. Sarton, *Introduction, volume 1*, p. 165.

64. On 25 October 1924 Mabel, in Belgium, writes to George that his Aunt Hélène wants to convert her: "I told her one day that you had 'un esprit beaucoup plus religieux que moi' & she replied that she was *sure you* would be a good catholic before you died!" George replies on 8 November: "I am not surprised that my aunt believes I will finally return into the bosom of the Roman church; it is from her point of view a logical and not implausible conclusion, but she is wrong all the same."

where he emphasizes that the history of science "is a history of the gradual emancipation of men from superstition and ambiguities, a history of the growth of light in dark or darkened corners, a history of our salvation not only from lies but also from other evils, from servitude and intolerance."[65] While the German army was advancing on Ghent during the First World War, George planned to begin his history of "philosophie positive," but only in retirement did he discharge his debt to the founder of positivism.[66]

In a Comtean reassessment of 1952, he focuses on Comte's Calendar of Great Men, the 560 "gods, heroes, and saints" of humankind, whose names are given to days, weeks, and months. Through the 14th century, roughly half of Comte's notables receive a paragraph in George's *Introduction*. George, who since his youth constructed calendars to mark his moral and intellectual progress, freely disputes Comte's choices: Reformers Martin Luther and Jean Calvin are out while Quakers George Fox and William Penn are in, Napoleon Bonaparte is out while Julius Caesar and José Gaspar Rodríguez Francia are in.[67] Most shocking is Comte's exclusion of Claude-Henri de Rouvroy, *comte* de Saint-Simon, his direct predecessor, allowing the conclusion that Comte was "ungrateful and graceless." Comte's penchant for schematic representations promoted the rigidity of his doctrine, a rigidity that befitted a prophet but that "was deadly for a man of science." Truly "crazy" was Comte's scheme for positivism and the Catholic Church (through the Jesuits) to rule the world together, suppressing Protestantism, deism, and scepticism (the 3 evils of modernity). Comte, who experienced clinical insanity at the beginning of his mission, ended quite daft, elevating his deceased lover Clotilde de Vaux into a goddess. George extends his criticism of the calendar to its 20th-century English revision, conceived by positivists like Francis Sydney Marvin as a manual for the history of civilization.[68]

A discussion of humankind's saints is not the only exculpatory theme in the late assessment of Comte. There is also the idolization of a woman, the result of Comte's having been "starved for feminine affection." The sentence that is most striking to our eye, however, is the one where George exonerates Comte, the "ignorant and crude" pioneer whose petitions lay at the origin of the chair of the history of science at the Collège

65. George Sarton, *Introduction to the History of Science, volume 3*, p. 18.

66. Diary, 5 September 1914.

67. Francia, receiving favorable notice by Thomas Carlyle, was the tormenter of Aimé Bonpland, who collaborated with Alexander von Humboldt; Humboldt does not figure in the Positivist calendar, notwithstanding Humboldt's attendance at Comte's courses in Paris.

68. George Sarton, "Auguste Comte, Historian of Science, with a Short Digression on Clotilde de Vaux and Harriet Taylor," *Osiris, 10* (1952), 328–57.

de France. He observes that pioneers by their nature are neither erudite nor refined. Assessment over the past generation places George closer to Comte than George would have liked to admit, whether in terms of allegiance to positivism as a doctrine or to Comte as a dogmatist. He was no unqualified follower of the Paris prophet, however, and though he unswervingly followed his own star he was more generous than most of his critics have allowed.

Reviewing Francis Sydney Marvin's biography of Comte, George praises the author as a defender of Comtean ideals: "Unification of science and of society, peace, Humanity." In 1919 George spoke with Marvin at London's Positivist church, and Marvin's obituary appeared in *Isis,* although significantly George did not write it.[69] George Sarton's affinity with Auguste Comte lies in the moral domain rather than in matters of interrogating the world, as John Theodore Merz observed the two sides of the founder of positivism: Merz's Comte sought the "reconstruction of society upon a new foundation...prepared and sustained by the study of history as well as of science," whereas positivism, in the hands of Comte's successors, focused on "simply the methods of the exact and natural sciences."[70] Comte as a Jacobin animator of social change appealed to George the Fabian socialist; Comte as the originator of the discipline of sociology attracted George the friend of sociologist Emile Waxweiler and George the affiliate of the Solvay institutes. In George's understanding, the sciences, by their nature undogmatic, constituted a moral undertaking.

John Theodore Merz observed that in some ways Comte is Hegel without Christ: "Not only the dogmatic, the architectonic, and the profound historical spirit are common to both, but even the abstract notions and the vocabulary of the idealistic [Hegelian] system may be easily translated into those of the positivist."[71] The assessment of Merz, in whose shadow George worked during the early years at the Carnegie Institution of Washington, facilitates understanding why George, by his own admission, was an idealist. While enjoying a precarious affiliation with the department at Harvard University which saw the advent of American pragmatism, George went out of his way to criticize the doctrine. In his early correspondence in America, George often referred to the importance of idealism. Clarifying his moral principles, indeed, became a practical necessity, for George labored as a Belgian propagandist, a promoter of world government, and an anti-militarist who kept a loaded revolver in his house at Wondelgem and urged American intervention on the side of

69. George Sarton, *Isis, 26* (1937), 477; Cecil H. Desch, "Francis Sydney Marvin, 1863–1943," *Isis, 36* (1945), 7–9.

70. John Theodore Merz, *A History of European Thought in the Nineteenth Century* (Edinburgh, 1896–1912), *4*, 679.

71. Merz, *European Thought, 4,* 186.

Allies during the First World War. The war won, he rejected the American boycott of German science, welcoming contributions in *Isis* by German scholars. George remained on cordial terms with scholars in Germany through the middle 1930s, when as a member of the Leopoldina academy in Halle he responded to a questionnaire about his racial origins, although to his credit the academy struck him from its records in 1942 for "anti-German propaganda."[72]

Among the literary techniques favored by George are antithesis (juxtaposing contrasting notions) and procatalepsis (anticipating and answering an objection). These rhetorical devices have the effect of advancing the argument by dissolving an apparent opposition—a dialectical inspiration. George's mind favored dialectics. His marriage was an attraction of opposites. George the epicurean scientist; Mabel, the abstemious artist. He the devotee of churches and creator of associations; she the freethinker and maker of fashion. As a student he found himself torn between the humanities and natural knowledge, a polarity that he resolved into art and science. George strove mightily to resolve the contradictions that he saw. His struggle to live in harmony with Mabel is projected in his focus on Leonardo, the virtuoso painter and visionary observer of nature. His program for harmonizing opposites took the form of advancing a minor academic calling, the history of science, as the foundation of a New Humanism, where science with a human face would be the cornerstone of an educated citizenry. It was an Enlightenment vision, grounded in the notion of human progress, the transparency of evidence, and the primacy of reason. If his prose seems uncompromising, it is because he was a truth-seeker disinclined to put up with sophistry.

The past half millennium is animated by the quarrel of Ancients and Moderns. It is present in rhetoric, prose, and poetry; in painting, sculpture, and architecture; in mathematics, astronomy, and medicine. It is reflected in the claims for professional enfranchisement by apothecaries, surgeons, nurses, engineers, and librarians. George Sarton received education by the first generation for whom accomplishment in research was the path to a university chair. His modern age was filled with conflict attributable to industrial capitalism. If science is concerned with truth, and if truth is goodness, then in his view science should be able to pave the way for a kinder, gentler society. But society remained in the grip of men trained in the literary tradition who were ignorant of and hostile to

72. George Sarton, "Autobiographical sketch prepared at the request of Herr Prof. Dr. Abderhalden, President of the Kaiserl. Leop. Carol. Deutsche Akademie der Naturforscher on August 25, 1935," in which George takes pains to describe his Aryan heritage (YMS). Sybille Gerstengarbe, "Die Leopoldina und ihre jüdischen Mitglieder im Dritten Reich," *Jahrbuch 1993 Leopoldina, 39* (1994), 363–410, on p. 370; I am grateful to Theodore L. Sourkes for the reference. George entered the Leopoldina on 20 July 1925. *Isis, 9* (1927), 117.

science. History of science became the way to resolve the opposition of science and letters.

The call to humanize science is also part of a long tradition, although even major statements toward this end, for example by Goethe and Lenin, tend to be grotesque and insubstantial. George Sarton's innovation was to wed the call to an academic discipline. His proposal to bridge the culture of science and the culture of letters is expansive and imperializing in the way of other newly constituted disciplines in the 19th and early 20th centuries, notably physics, sociology, and biology. The proposal, which George defended to the end of his life, became imbedded in the cultural fabric of the second part of the 20th century. It lies behind Charles Percy Snow's famous essay *The Two Cultures* and George Steiner's call for an enlightened "post-culture."[73] Even an uncompromising critic of George's *oeuvre* like Noel M. Swerdlow recognizes the relevance of promoting the history of science as a way of uniting the humanities, with their concern for personal morality, and the sciences, where the personal is intentionally excluded.[74]

Reasons that may be given for the mixed reception of George Sarton's New Humanism are embedded in the intellectual climate of late 20th-century America. Among these are the failure of secular ethics, the evaporation of socialism, and the transformation of science into applied technology.[75] In 1917 George's supporter Arnold C. Klebs saw the matter clearly:

> I have asked myself why has a scheme of Comte made so little headway from 1822 to our day so that we have to agitate it in its identical form but under a different name? I answer: there must be something deeply & radically antagonistic to it in human nature. It is illuminating enough to point out we have had the religious, the political phase & now must come the scientific phase. It appeals forcibly to our logic. We may also say that the present emotionality is only temporary & that reason must come into its own again & with it the domination of intellectual methods. While I see all this, as I said I have become sceptical & fear that we are going to see

73. C. P. Snow, *The Two Cultures and the Scientific Revolution* (New York, 1959); George Steiner, *In Bluebeard's Castle: Some Notes Towards the Redefinition of Culture* (New Haven, 1971), p. 133: "The absence of the history of science and technology from the school syllabus is a scandal." Neither Snow nor Steiner refers to George Sarton.

74. Swerdlow, "Montucla's Legacy," p. 326: "The history of science, once looked to for the meeting of the two cultures, is more and more falling into the realm of no culture (which, incidentally, makes Lord Snow's essay more timely than ever)." Swerdlow does not use the moral distinction, which I take from Max Weber. Lewis Pyenson, "What Is the Good of History of Science?" *History of Science*, 27 (1989), 353–89, p. 364.

75. For the last notion: Svante Lindqvist, ed., *Center on the Periphery: Historical Aspects of 20th-Century Swedish Physics* (Sagamore Beach, Mass., 1993), generally; Pyenson and Sheets-Pyenson, *Servants of Nature*, p. 438.

> a very long epoch of anti-intellectual procedures, a reappearance again of the religious phase with some new fetish.[76]

It took more than 60 years before one of George's rich devotees, Bern Dibner, endowed his vision at the Massachusetts Institute of Technology. By then, in the recent formulation of one scholar, the discipline of history of science had, as Klebs foresaw, left behind its positivistic stage, regressed through a philosophical stage, and was about the enter into a Comtean theological stage.[77]

George's interest in the human side of science qualifies as compensation for an inability to maintain close relations with his friends, first among them Céline and Raymond Limbosch, and a failure to achieve intimacy with his wife Mabel. George's correspondents pointed out his insensitivity to Mabel's needs, which also looms large in May Sarton's memoirs. The insensitivity is present at the very end. After the Second World War, on the day of Mabel's death, George writes to his old friend Horace Kallen about a commitment to lecture at the New School in New York. Mabel will die very soon, George reports, but her death "will not prevent me from doing my duty in N.Y. next Sunday." He continues: "My life has been a torture during the whole summer & autumn. We have been married almost 40 years, without ever a cloud except her poor health."[78] A dispassionate observer of George and Mabel Sarton might gloss: If never a cloud, then surely lightning struck out of the blue. Absent is Thomas Carlyle's remorse about his insensitivity to Jane Carlyle's needs. But different from Thomas Carlyle, George continually thanked his wife for acting as his conscience and his compass.

In his chosen discipline, George Sarton was a buccinator, an advocate, who (as the admired buccinator Francis Bacon claimed of himself) "rang the bell which called the wits together."[79] Bacon's wits were the Fellows of the Royal Society of London; Sarton's wits are found in the officers, laureates, and members of the History of Science Society. At the end of his life, he was respected by the general public for his scholarly advocacy.

---

76. HGS, Arnold C. Klebs to George Sarton, 19 November 1917.

77. The Sartonian Maecenas was also a bibliophile. Bern Dibner, *Heralds of Science* (Norwalk, Conn., 1955), p. 6: "To live in this age of science without an awareness of its fascinating origins is to miss much of the spirit of its attainments. To miss knowing specifically the work of its heroes, saints and martyrs is to lose the opportunity of gauging the greatness and accomplishments of this age as compared to other ages in human history." For the inversion of Comte, John Heilbron, "Science as a Subject of History," talk presented at the VII Trobada d'història de la ciència i de la tècnica, Barcelona, November 2002.

78. YIVO Institute for Jewish Research, New York, Papers of Horace Meyer Kallen, 562/1004, George Sarton to Horace Kallen, 19 November 1950.

79. Charles Singer, *A Short History of Scientific Ideas to 1900* (Oxford, 1959), p. 267, quoting Bacon.

There are not many people who sit as scholars, pursuing their calling in isolation, and who also speak for their enterprise in the forum. Those who mattered in the calling knew George's measure. In 1939 historian of science Max Meyerhof spoke for a dozen of his Arabist colleagues when he related that George was "in nearly everything 'our spiritual leader.'"[80] The forum dignified his undertaking: *Newsweek*, in 1947, called him "the world's leading historian of science"; shortly after his death, an advertisement for women's fashions in *Vogue* featured a twiggy model carrying a Dover edition of two of his short books.[81] The first volume of George's *Introduction* sealed his fate in America as a scholarly generalist. His great circle of colleagues provided an invigorating education for his daughter May and a continuing stream of dazzling personal contacts for his wife, a recognized commercial designer in America and in Europe whose talents were sought by élite private schools. Science progressed, of that George was sure. But it was above all in his marriage where George Sarton measured his personal progress by experiences shared and truths won.

80. Thomas F. Glick, "From the Sarton Papers: Paul Kraus and Arabic Alchemy," *Cronos*, 2 (1999), 221–44, on p. 240; Meyerhof refers to George's views about ecumenism.

81. "Lessons from History," *Newsweek*, 28 April 1947; the *Newsweek* and *Vogue* extracts in YMS. The Dover edition of *The Study of the History of Mathematics* and *The Study of the History of Science* appeared in 1957.

Appendix

The Art of Mabel Sarton

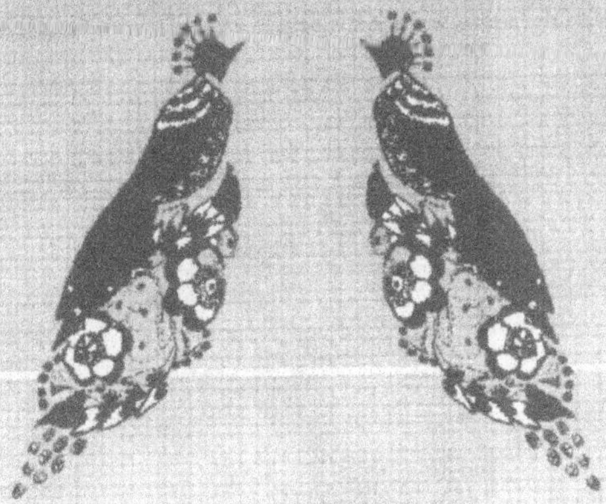

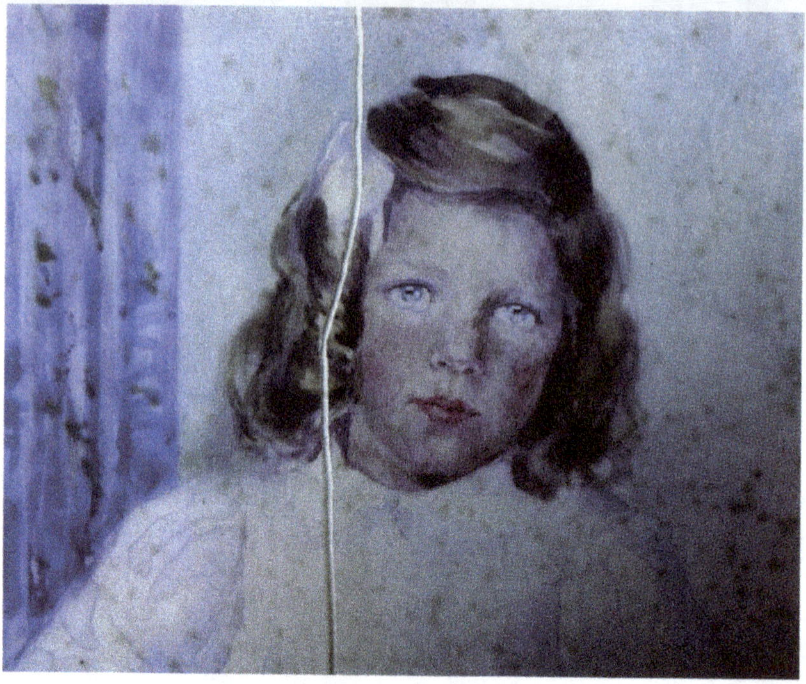

Fig. 19. Miniature on ivory, circa 1907, by Mabel Elwes. Thin sheets of ivory, easily receiving oil paint, were fragile. This portrait is approximately 4 inches broad.

Fig. 20. Ivory miniature, by Mabel Elwes.

Fig. 21. Framed miniature portrait by Mabel Elwes.

Figs. 22 abcd. Stained-glass flowers by Mabel Sarton.

Fig. 23. Sarton residence, 5 Channing Place, Cambridge, Massachusetts. Mabel comments: "In the dining room, shewing built in corner cupboard–the little chest on stand I had made & decorated when we lived in Washington 16 years ago."

Fig. 24. Chest designed by Mabel Sarton for her residence in Washington, mentioned in Fig. 23, showing bird-and-flower motif.

Fig. 25. Céline Dangotte, 19 January 1910, sketched by Mabel Elwes.

Fig. 26. Belfry (left) and church of St Bavon, by Mabel Elwes. The 14th-century Belfry, constructed by the guilds of Ghent, is home to one of Belgium's most renowned carillons . The tower of St Baafs or St Bavon was completed in the 16th century. St Bavon contains the famous 15th-century triptych, "The Adoration of the Mystic Lamb," by Jan and Hubert van Eyck. Mabel Elwes painted this water-color from her lodging in the Dangotte residence atop the Adolphe Dangotte store in the rue Courte du Marais, or Korte Meerstraat.

Fig. 27. Watercolor, by Mabel Elwes.

Fig. 28. Bird-and-flower motif, sketched on cloth. Much of Mabel Sarton's embroidery was executed on patches of cloth, later to ornament clothes.

Fig. 29. Bird-and-flower lace, by Mabel Sarton.

Fig. 30. Silken embroidered birds, by Mabel Sarton.

Fig. 31. Silken embroidered birds, by Mabel Sarton.

Fig. 32. Hand-embroidered blouse, avian pattern, by Mabel Sarton.

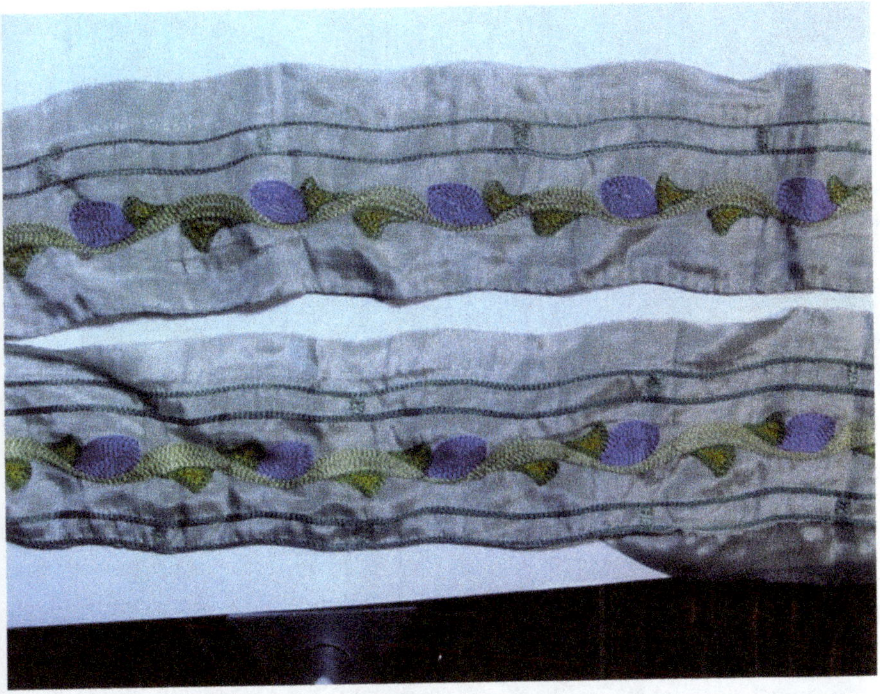

Fig. 33. Machine-embroidered appliqué, designed by Mabel Sarton.

Fig. 34. Flower-motif hand embroidery, designed by Mabel Sarton.

Fig. 35. Hand embroidery designed by Mabel Sarton, showing stencilled pattern.

Fig. 36. Peasant-style blouse, keyhole neck, designed by Mabel Sarton.

Fig. 37. Appliqué collar, designed by Mabel Sarton.

Fig. 38. Detail of partially-completed hand embroidery designed by Mabel Sarton.

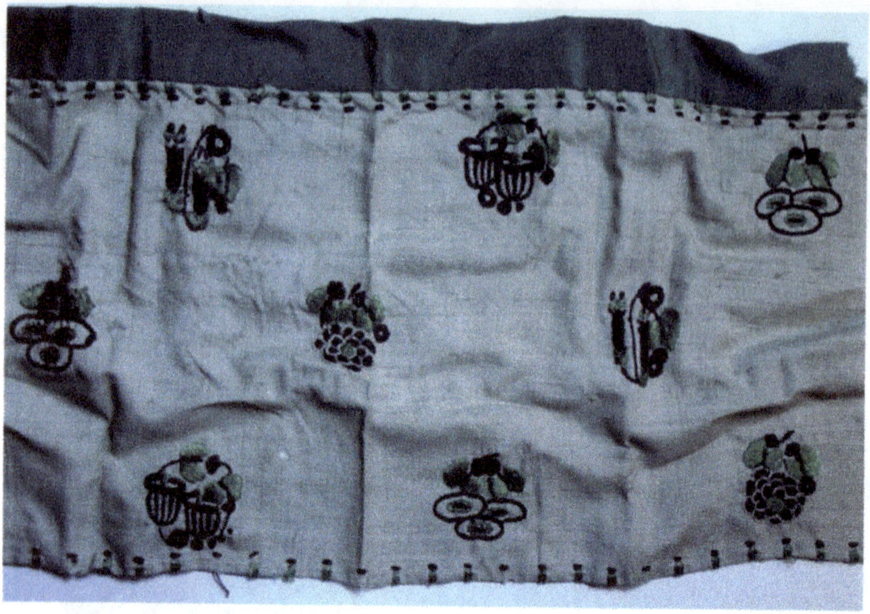

Fig. 39. Machine embroidery designed by Mabel Sarton. This pattern is evidently guided by hand.

Fig. 40. Hand-embroidered blouse designed by Mabel Sarton. The flower motif is highly abstract.

Fig. 41. Flower design, painted on gauze by Mabel Sarton. The design was intended for a painted dress.

Fig. 42. Machine-embroidered butterflies, designed by Mabel Sarton.

Fig. 43. Hand-embroidered bird-and-flower motif peasant blouse, designed by Mabel Sarton.

Fig. 44. Art-deco machine embroidery by Mabel Sarton.

Fig. 45. Art-deco machine embroidery by Mabel Sarton.

Fig. 46. Woolen embroidery, child's dress, designed by Mabel Sarton.

Fig. 47. Peacocks sketched by Mabel Sarton, possibly for her exhibition embroidery of 1910.

Fig. 48. Prototype of machine embroidery for an altarpiece designed by Mabel Elwes for L'Art Décoratif Céline Dangotte. The altarpiece, which received a gold medal at the 1925 International Exhibition of Decorative Arts in Paris, was formerly in the possession of the Abbaye d'Orval in Belgium.

Fig. 49. Cover of *L'Enclos*, by Raymond Limbosch. Mabel Sarton's design for the book, published in 1910, shows a bird-and-flower motif. This copy presented by Céline and Raymond Limbosch to Mabel Elwes and George Sarton on 3 December 1910.

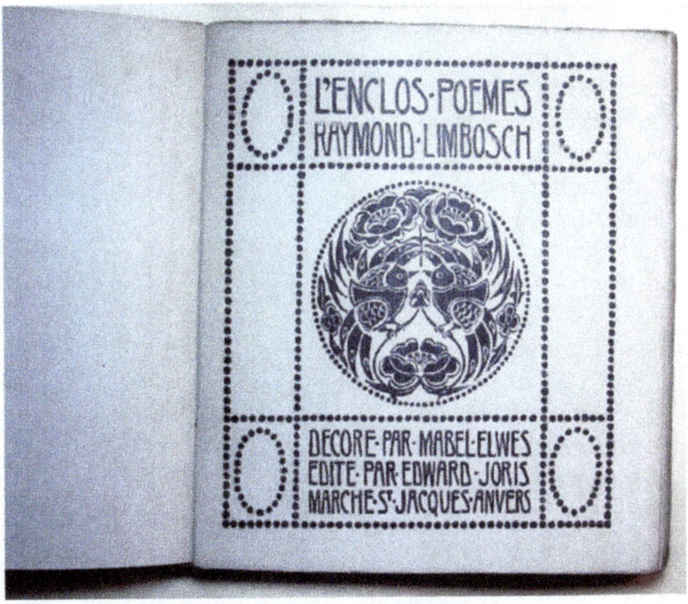

Fig. 50. Title page of *L'Enclos*, by Raymond Limbosch. A woodblock print by Mabel Sarton.

Fig. 51. Title page of *Le Bois d'oliviers,* by Raymond Limbosch. Mabel Sarton's motif of vine and flowers is in the tradition of Art Nouveau. This book is dedicated to Mabel Elwes and George Sarton. The beading and framing indicate that the book is a continuation of *L'Enclos*.

Fig. 52. Ex-libris, designed by Mabel Sarton for George Sarton, as recorded in George Sarton's diary, 31 August 1920. We may read the prominent nipple profile as an assertion of the artist's independence. By the 1970s, the History of Science Society preferred a sexless hieroglyph. Rosemary Regan and Richard H. Schallenberg, ed., *History of Science Society Directory of Members and Guide to Graduate Study* (New York, 1977), the cover illustration.

Index

Index

www.ingramcontent.com/pod-product-compliance
Lightning Source LLC
Chambersburg PA
CBHW071247180926
51613CB00002B/157
9780871692603